THE LATERALIZED BRAIN

SECOND EDITION

THE LATERALIZED BRAIN

The Neuroscience and Evolution of Hemispheric Asymmetries

SECOND EDITION

SEBASTIAN OCKLENBURG
MSH Medical School Hamburg, Hamburg, Germany

ONUR GÜNTÜRKÜN
Ruhr University Bochum, Bochum, Germany

ELSEVIER

ACADEMIC PRESS
An imprint of Elsevier

Academic Press is an imprint of Elsevier
125 London Wall, London EC2Y 5AS, United Kingdom
525 B Street, Suite 1650, San Diego, CA 92101, United States
50 Hampshire Street, 5th Floor, Cambridge, MA 02139, United States

Notices
Knowledge and best practice in this field are constantly changing. As new research and experience broaden our understanding, changes in research methods, professional practices, or medical treatment may become necessary.

Practitioners and researchers must always rely on their own experience and knowledge in evaluating and using any information, methods, compounds, or experiments described herein. In using such information or methods they should be mindful of their own safety and the safety of others, including parties for whom they have a professional responsibility.

To the fullest extent of the law, neither the Publisher nor the authors, contributors, or editors, assume any liability for any injury and/or damage to persons or property as a matter of products liability, negligence or otherwise, or from any use or operation of any methods, products, instructions, or ideas contained in the material herein.

ISBN: 978-0-323-99737-9

For Information on all Academic Press publications
visit our website at https://www.elsevier.com/books-and-journals

Publisher: Nikki Levy
Acquisitions Editor: Megan McManus
Editorial Project Manager: Kristi Anderson
Production Project Manager: Sruthi Satheesh
Cover Designer: Mark Rogers

Typeset by MPS Limited, Chennai, India

Dedication

In memory of Ariane Schwarz, who left us too soon.

Onur Güntürkün and Sebastian Ocklenburg

Contents

8. Recognizing yourself and others—the role of the right hemisphere for face and self-perception 241

9. Hemispheric asymmetries in emotion processing and social behaviors 271

10. Ontogenesis of hemispheric asymmetries 307

11. Hemispheric asymmetries over the lifespan: development and aging 337

About the authors

Author Picture: Onur Güntürkün is on the left, Sebastian Ocklenburg is on the right. © RUB, Marquard.

Sebastian Ocklenburg is a professor of research methods in psychology at the Medical School Hamburg, Germany. Born in Duisburg, Germany, he knew since primary school that he wanted to become a scientist. After obtaining his bachelor's and master's degrees in psychology, Sebastian conducted his PhD thesis on the genetics and neurophysiology of hemispheric asymmetries. For this work, he was awarded the prestigious Heinz Heckhausen award by the German Society for Psychology. Subsequently, he worked as a postdoc in the Bergen fMRI group in Norway, before returning to Germany. Since the very beginning of his career, Sebastian has held a deep fascination with understanding how hemispheric asymmetries develop. His work focuses on answering this question using an integrative multimethod approach, including neuroimaging, electrophysiology, and molecular genetics techniques. Sebastian is a member of the International Society for Behavioral Neuroscience (ISBN). He has authored more than 145 papers on hemispheric asymmetries and other topics in international peer-reviewed journals, including top journals like *Psychological Bulletin*, *Physiological Reviews*, *Science Advances*, *Neuron*, and *eLife*. Besides "The Lateralized Brain," Sebastian has written books on hugging and the climate crisis.

Affiliation: MSH Medical School Hamburg, Hamburg, Germany

Onur Güntürkün is a Turkish-born professor of biopsychology at Ruhr University Bochum in Germany. He once came to Germany with the aim to study psychology and to learn about the science of the mind. Onur is kept awake by questions like: "Why do humans and other animals have asymmetrically organized brains?," "Can different kinds of brains produce the same cognition?," or "What is the neural substrate of thought?." He spent many years of his life in different universities in Germany, the United States, Australia, Turkey, Belgium, and South Africa. He mostly uses humans and pigeons as his experimental subjects but also loves to do science with dolphins and magpies. Onur Güntürkün works with all kinds of research approaches that reach from simple fieldwork, via behavioral experiments, neuroanatomical approaches, and single-unit recordings up to brain imaging at ultrahigh magnetic fields. He is a member of the German National Academy of Sciences, holds two honorary doctorates, and received numerous major national and international scientific awards, among them the highest German and Turkish science awards. Onur authored more than 390 papers with more than 27,000 citations in top journals like *Nature, Science, Nature Reviews Neuroscience, TICS, Neuron, PNAS, Current Biology, PLoS Biology*, and *eLife*. Onur serves as a reviewer or as a member of editorial boards in a large number of journals and also reviews applications for 23 different national and international science organizations.

Affiliation: Ruhr University Bochum, Bochum, Germany

Foreword
by Lesley J. Rogers

I was very pleased to be asked to write a foreword to this second edition of the book by Sebastian Ocklenburg and Onur Güntürkün. Both authors have extensive knowledge of asymmetry in humans, as well as in nonhuman species, both have made major contributions to this expanding field of research, and both are leaders in research contributing to our understanding of interhemispheric integration of information and metacontrol of responding. Indeed, this second edition of their book is both engaging and comprehensive. It informs readers reliably of exciting recent discoveries, presented with clarity in both text and illustrations.

This book begins with an account of the discovery of brain asymmetry in humans and its functional expression in handedness and language, including the innovative studies of split-brain patients, through the 1960s–80s, so instrumental in enhancing knowledge of brain asymmetry in control of behavior in humans. Chapters on asymmetry in humans cover laterality of language, both typical and atypical, face recognition and self-perception, structural asymmetries, and evidence of neurological and psychiatric conditions associated with differences in asymmetry.

Beginning in the 1970s, when asymmetry of brain function was discovered in nonhuman species, attention shifted to include lateralization in nonhuman species. My own research had revealed asymmetry of brain function in domestic chicks and the unforeseen effects of light exposure of the embryo on development of visual asymmetry.[1,2] Then, Onur Güntürkün discovered asymmetry in the pigeon brain and made links between brain structure and behavior in that species.[3] Güntürkün's incisive research further revealed both similarities and crucial differences in the development and structural dependence of asymmetry in pigeons versus chicks (i.e., likely between an altricial and precocial species).[4,5] This comparative research on development of lateralization is one of the reasons for my long-time and fruitful scientific exchange with his research, as well as my more recent and valuable communication with Sebastian Ocklenburg, who has conducted influential research and published key papers about laterality in both humans and nonhuman species.[6]

Subsequent studies on laterality in a range of nonhuman species have shed light on the neural basis of functional asymmetry, its evolution, development, and function.[7,8] A unique feature of the book is its integration of our current knowledge of asymmetry in nonhuman brains with that of asymmetry in the human brain. Even research focused on hemispheric neglect in humans is edified by consideration of side biases in predator detection shown in several nonhuman species. Factors influencing the development of brain asymmetry in model species (zebrafish, chicks, and pigeons), a recent finding, are discussed in the book as a basis for understanding the roles of genetic expression and experience in the development of asymmetry.

Coverage of handedness is considered in a range of vertebrate species. Understanding the role of the mammalian corpus callosum in transferring information between the hemispheres is enhanced by comparison of hemispheric integration in species without a corpus callosum; namely, with fewer connections between the left and right sides of the brain.

Throughout the book, each topic is discussed in a clear and lively manner. The reader learns about new techniques implemented with increasing frequency and providing dramatic advances in understanding asymmetry.

Of crucial relevance to scientists, the evidence discussed is supported by comprehensive citation of the published research and, hence, the book is an indispensable reference volume for researchers. Students of the field will also find the detailed glossary of terms most useful. This second edition of the book is indeed an invaluable source of information and ideas. More general readers will find the book a lucid entry into an exciting area of neuroscience that integrates neuroanatomy, neurochemistry, neuroimaging, genetic association studies, and behavioral studies to understand brain asymmetry.

Lesley J. Rogers
University of New England, Armidale, NSW, Australia

References

1. Rogers LJ. Light experience and asymmetry of brain function in chickens. *Nature*. 1982;297:223–225.
2. Rogers LJ. Unfolding a sequence of sensory influences and interactions in the development of functional brain laterality. *Front Behav Neurosci*. 2023;16:1103192.
3. Güntürkün O. Lateralization of visually controlled behavior in pigeons. *Physiol Behav*. 1985;34:575–577.
4. Güntürkün O, Ströckens F, Ocklenburg S. Brain lateralization: a comparative perspective. *Physiol Rev*. 2020;100:1019–1063.
5. Manns M, Ströckens F. Functional and structural comparison of visual lateralization in birds – similar but still different. *Front Psychol*. 2014;5:206.
6. Ocklenburg S, Güntürkün O. Hemispheric asymmetries: the comparative view. *Front Psychol*. 2012;3. Available from: https://doi.org/10.3389/fpsyg.2012.00005.
7. Rogers LJ, Vallortigara G, Andrew RJ. *Divided Brains: The Biology and Behaviour of Brain Asymmetries*. Cambridge: Cambridge University Press; 2013. Available from: https://doi.org/10.1017/CBO9780511793899.
8. Güntürkün O, Ocklenburg S. Ontogenesis of lateralization. *Neuron*. 2017;94:P249–263.

Foreword to the first edition *by Giorgio Vallortigara*

The study of brain and behavioral asymmetries is a central topic in modern cognitive neuroscience, but it is also deeply rooted in the classical disciplines like anatomy and experimental psychology. This amazing and agile book by Sebastian Ocklenburg and Onur Güntürkün—two leading authorities in the field—conveys to the reader both a taste of the historically rich enterprise and the excitement of current research achievements.

Let me mention two among the lesser known antecedents of lateralization research, which I believe are instructive in different ways. The first is associated with the name of Gustav Theodor Fechner, the founder of psychophysics, who gave the name together with Ernst Heinrich Weber to the law that links sensation with the physical characteristics of the stimulus. Fechner was deeply interested in the issue of consciousness, and in 1860, he first wondered what would happen if the cerebral hemispheres were split completely. He thought that if the brain could be divided into two parts without compromising the life of its owner, the neural activities of the two separate parts would be accompanied by their own stream of consciousness. On the contrary, the psychologist William McDougall, among others, was convinced that splitting the brain could not result in the splitting of the mind because of the fundamental unity of consciousness. McDougall was dualist and so convinced that his mind would remain intact following division of his brain that he tried to persuade the legendary neurophysiologist Charles S. Sherrington to cut his brain in case he would be affected by some incurable disease—a request that, luckily for McDougall, was never fulfilled.

This theoretically conceivable, but seemingly practically undoable experiment, was eventually, as we know, performed in the 1960s. The preliminary work conducted on animals by Roger Sperry at Caltech revealed surprisingly little effects of the corpus callosum section. Then, when Sperry had the opportunity to examine some people whose brains had been split for medical reasons, it was apparent that, in Sperry's own words "(...) the surgery has left these people with two separate minds; that is, two separate spheres of consciousness."[1] To a certain extent, Fechner was correct. However, there was something more, for the two sides actually perform different functions, some of which are not available to the partner—for example, language. As discussed in the book, the two hemispheres in split-brain patients are not completely separated, for interhemispheric transfer over subcortical connections may occur.

These were fascinating observations, but as it happens, too frequently in the cognitive neurosciences, certain ideas or concepts become hostages of their own success. Some of the findings were misinterpreted in their relevance to the functions of the normal human brain and have nurtured a pop-neuroscience view, according to which we would have virtually independent sides to our brains and therefore two intellects. These misconceptions about the left and right brain are properly targeted and criticized in the book, as it has been also done by other prominent laterality researchers.[2]

There have been ups and downs in the interest of research in brain asymmetries and associated phenomena in the past century. The seminal work of Sperry on split-brain patients increased interest in asymmetries, as revealed in normal subjects by tachistoscopic and dichotic listening techniques. After this phase, a "research industry" of left-right differences in human cognitive functions emerged and then slowly declined. Similarly, after a long time during which the dogma that laterality was a uniquely human feature dominated, it progressively became clear that lateralization is widespread in the animal kingdom, leading to a new wave of research on animals looking for any kind of subtle left-right difference in brain and/or behavior. However, there is now a feeling of a real progress in laterality research, and there is hope that this new, more biologicallyoriented wave will take us close to an answer to the fundamental "why" and "how" of brain asymmetry. The authors of this book effectively went through the delicate balance of conveying both the excitement of novel ideas and the need for scientific rigor in assessing the new evidence.

The second example of antecedents of modern laterality research underlies the fact that science is full of rediscoveries or, in the words of Andre Gide, "Toutes choses sont dites déjà; mais comme personne ńécoute, il faut toujours recommencer." Consider in this regard the "discovery" of structural (and nowadays also genetic) asymmetries in the diencephalon of several species of vertebrates. Interestingly, some of these asymmetries were indeed very well-known by 19th-century scientists, but appeared and disappeared in the literature. The most striking case is the asymmetry of the habenular nuclei, lying dorsally in the diencephalon, which receives input from light-sensitive organs in the roof of the epithalamus. It was first described by Goronowitsch in 1833 in the sturgeon in this fish species that the right habenula is larger than the left one. Then Gaupp, in 1896, reported habenular asymmetries in frogs, Gierse in a cyclophone fish in 1904, and Dendi in petromyzons in 1906. Braitenberg and Kemali,[3] who summarized this story, noted that between 1900 and 1930, the instances of asymmetries in diencephalic regions were so common that lack of asymmetry, rather than its presence, was noteworthy. However, the standard compendium of comparative neuroanatomy of the time explicitly denied the presence of habenular asymmetries in amphibia, though recognizing its presence in fish[4]. Subsequent textbooks started to report that the habenular complex is symmetric from the amphibians up to the mammals. Rediscovery of diencephalic asymmetry occurred much later, in 1970, with the paper by Braitenberg and Kemali.[5]

These accounts are not only relevant for historians of comparative neuroanatomy, as diencephalic asymmetry is today under the focus of attention using molecular genetics tools. It has been argued that vertebrate cerebral lateralization originated in animals with extreme asymmetries of body, that in turn imposed asymmetries on the central nervous system, of which those in the diencephalon may represent an ancestral condition.[6]

The study of brain and behavioral asymmetry thus offers an extraordinary opportunity for a stroll through all the realms of modern neuroscience, from the *mirabilia* of modem neuroimaging techniques unveiling human connectome to the study of behavioral asymmetries in human kissing; from the intricacies of neuroanatomy of species like pigeons to the gesturing of nonhuman primates in their natural environment. A taste of all this, and much more, is provided with a lively and engaging style in the book of Ocklenburg and Güntürkün. A main outcome and satisfaction for the authors will be, I am sure, promoting a number of new recruits among young neuroscientists that will be attracted by this fascinating area of research.

Giorgio Vallortigara
University of Trento, Trento, Italy

References

1. Sperry RW. Brain bisection and mechanisms of consciousness. In: Eccles JC, ed. *BrRin and Conscious Experience: Study Week, September 28 to October 4, 1964, of the Pontificia Academia Scientiarum*. Berlin: Springer-Verlag; 1965:298–313.
2. Corballis MC. Left brain, right brain: facts and fantasies. *PLoS Biol*. 2014;12(1):e1001767. Available from: http:/Idx.doi.org/10.1371/journal.pbio.1001767.
3. Braitenberg V, Kemali M. Exceptions to bilateral symmetry in the epithalamus of lower vertebrates. *J Camp Neural*. 1970;138:137–146.
4. Kappers CUA, Huber GC, Crosby EC. *The Comparative Anatomy of the Nervous System of Vertebrates. Including Man*. New York: Macmillan; 1936.
5. Harris JA, Guglielmotti V, Bentivoglio M. Diencephalic asymmetries. *Neurosci Biobehav Rev*. 1996;20:637–643.
6. Rogers LJ, Vallortigara G, Andrew RJ. *Divided Brains. The Biology and Behaviour of Brain Asymmetries*. New York: Cambridge University Press; 2013.

Preface to the second edition

More than 6 years have passed since the initial release of *The Lateralized Brain*, and a remarkable amount of exciting researches on laterality have been published since then. These new developments and insights warranted a second edition of *The Lateralized Brain*. We have updated all chapters with new sections and sometimes with dozens of new papers published after 2017. Many chapters also contain new figures, and a completely new chapter on "Hemispheric Asymmetries over the Lifespan: Development and Aging" (Chapter 11) is included in the second edition that was not included in the first edition. The second edition heavily features meta-analyses and large-scale consortium studies by ENIGMA laterality and other consortia to highlight especially robust research. We hope that these updates and changes help students and scientists who read *The Lateralized Brain* to get an up-to-date view of laterality research in the early 2020s.

Bochum and Hamburg, October 2023

Sebastian Ocklenburg
Onur Güntürkün

Preface to the first edition

In 1982 Onur Güntürkün was working on his PhD and was in a really bad mood. His goal was to analyze the mechanisms of visual attentional shifts in pigeons. He assumed that descending telencephalic pathways could modulate via the ipsilateral optic tectum (superior colliculus) attentional resources of ascending visual systems. To prove this hypothesis, he had severed with tiny lesions under stereotaxic control specific descending fiber systems of one or both hemispheres in pigeons. Subsequently, he tested the animals for many months to quantify their deficits. The results were disappointing. In fact, they were a mess. Some birds had extreme problems reacting to relevant stimuli. Others showed no deficit at all. In the beginning, he had hoped that the picture would clarify if he had finished the histology. He hoped to discover that he had missed the target fibers in those birds that showed no deficit. He had hoped in vain. More than a year of hard work was about to go down the drain.

At that time, Pierre Deviche, a brilliant behavioral neuroendocrinologist (now at Arizona State University) was working in the same lab and just had come back from the 18th International Ornithological Congress in Moscow. Pierre listened to Onur's laments and then asked if he had placed the lesions on the left or the right side of the brain. Before waiting for the answer, he went on to report that an Australian scientist named Lesley Rogers had given a talk in Moscow, in which she claimed that chicks show pronounced asymmetries of visual control. This information was a sudden spark of insight. Of course, in half of the animals, the lesion had been placed on the left and in the other half on the right. Onur rushed back to the data sheets and reorganized them according to hemisphere. Now it all made sense: only the left-lesioned animals displayed severe deficits while the right-lesioned ones behaved like they were unharmed. Without this moment and without this insight, this book would not exist.

Many years later, Onur Güntürkün had his own lab and one day Sebastian Ocklenburg joined the team. Both shared the enthusiasm to understand cerebral asymmetries from a mechanistic point of view. Both wanted to combine behavioral, genetic, and neurobiological data in a comparative framework to achieve a comprehensive understanding of why the left brain differs from the right.

What motivated us to write this book was largely the absence of an up-to-date teaching resource on hemispheric asymmetries. Like many of our colleagues, we used "Left Brain, Right Brain," the classic textbook by Sally P. Springer and Georg Deutsch, as our main textbook in classes on hemispheric asymmetries for a long time. However, the last edition of this excellent book was published in 1998, and neuroscience since then has made dramatic advances. Among them was the widespread use of neuroimaging techniques like fMRI and DTI, the sequencing of the human genome, the application of molecular genetic techniques to answer neuroscientific questions, and a substantial increase in advanced research in nonhuman model species. These and other developments have substantially changed how we understand the development and neurophysiology of hemispheric asymmetries today, and we felt that a textbook on hemispheric asymmetries needs to include them.

This book is intended for a wide range of interested individuals. In the first place, we hope to provide educators with an up-to-date, accessible, but still in-depth teaching resource to introduce undergrad students of all levels to the fascinating topic of hemispheric asymmetries. Additionally, PhD students, postdocs, and senior researchers who seek an introduction to hemispheric asymmetries get a comprehensive resource that could help them in designing experiments and writing papers on hemispheric asymmetries. Moreover, interested individuals outside the scientific community who seek up-to-date knowledge on left-handedness, language lateralization, and other topics within the general field of hemispheric asymmetries could find it in this book.

Bochum, June 2017

Sebastian Ocklenburg
Onur Güntürkün

Acknowledgments

Writing both the first edition and the second edition of this book would not have been possible without the support of various people. We are most grateful to Jutta Peterburs, Yasmin El Basbasse, Maike Schwalvenberg, and Lisa Deneke, who spent countless hours correcting our errors, and Sandra Linn for obtaining copyright permissions and submitting chapters. Moreover, we thank Sarah Kritzler for editing our references and Maria Höfer for researching papers for us. A special thanks to Oliver Wrobel, the graphic artist who did an outstanding job of creating many of the figures used in the book, and to Delia Sophie Wulfmeyer who provided a beautiful *Anomalocaris* figure for Chapter 2. A special thanks also to Erhan Genç, who not only provided the amazing cover picture for the first edition of the book but also custom-made the new cover picture for the second edition. Moreover, we are indebted to Laura Blevins and Lena Sophie Pfeifer, who agreed to model for some of the pictures shown in the book, and John E. Desmond who allowed Sebastian to temporarily work in his lab at Johns Hopkins School of Medicine in Baltimore to finish writing this book. Chapters 1 and 3 were written in the Hanse Wissenschaftskolleg in Delmenhorst, where a monastery-like academic seclusion gave Onur and Sebastian the possibility to start a competition for being the first to finish a chapter. Onur was, for 3 months, a fellow of the Stellenbosch Institute for Advanced Study (STIAS, South Africa), where he enjoyed a similar situation to finalize a few more parts of this book.

The laterality research community has always been a group of friendly and supportive individuals, and we are thankful to many of its members for providing permission to reuse figures to illustrate their research in the book, reminding us of long-forgotten papers, and helpful discussions of the literature. Among them are (in no particular order): Marietta Papadatou-Pastou, Gina Grimshaw, Lesley Rogers, Giorgio Vallortigara, Chris McManus, Mike Nicholls, Kenneth Hugdahl, Michel Thiebaut de Schotten, Lorin Elias, Bill Hopkins, Silvia Paracchini, Martina Manns, Yegor Malashichev, Marc Brysbaert, Lise van der Hagen, Tulio Guadalupe, Clyde Francks, Tao Sun, Lisette Leliveld, Dominic Norris, René Westerhausen, Josef J. Bless, Marco Hirnstein, Turid Helland, Hans-Otto Karnath, Sabine Hofer, Alice Cronin-Golomb, Christian Grefkes, Eileen Luders, Florian Kurth, Lukas Heydrich, Julian Keenan, Roberto Caldara, Urs Maurer, Annuka Lindell, Lara Schlaffke, Marie Banich, Alan Pegna, Ruud Hortensius, Christian Bellebaum, Gjurgjica Badzakova-Trajkov, Marc Schönwiesner, Bruce S. Lieberman, David W. Frayer, Richard Robison, Jay Sayre, Loren Babcock, Charlotte Faurie, Wulf Schiefenhövel, Lâle Uluç, Lizzy Griffiths, James Danckert, Sarah Medland, Gareth Richards, Paul Thompson, Judith Schmitz, Julian Packheiser, Guy Vingerhoets, Evelina Fedorenko, Saima Malik-Moraleda, Robin Gerrits, Emma Karlsson, and David Carey.

Finally, we would like to thank the staff at Elsevier who guided us expertly through the complex undertaking of writing this book, especially Kristi Anderson.

CHAPTER

1

Brain asymmetries—two millennia of speculation, research, and discoveries

OUTLINE

Introductory short story

"I have been betrayed. I have been humiliated. They stole my ideas; they neglected my insights and never allowed me to stand on the pedestal of fame that I deserve. I am from the south of France, deep from the province. They never said it publicly, but their eyes gave it away: With my short and pot-bellied stature, my thick southern accent, and my lack of this special way of talking that they call 'distinguished intellect', I, Gustave Dax, will always be inferior to the academic elite of Paris. My father Marc Dax had discovered that language was governed by the left hemisphere. He was a surgeon during the Napoleonic wars and had seen repeatedly that wounds to the left but not to the right hemisphere affected language abilities. Yes, I know: Father's methods of analysis were crude.

The Lateralized Brain
DOI: https://doi.org/10.1016/B978-0-323-99737-9.00003-3

But still, they were accurate enough to make the point. Father had presented his findings at a congress in Montpellier in 1836. Now the scientists in Paris say that no trace of such a presentation can be found. In 1863 I submitted the collection of findings from my deceased father and more than a hundred further cases to the Academy in Paris. A huge database; absolutely solid and with only a single possible way of interpretation: Language is controlled by the left hemisphere. Thus, the human brain is asymmetrically organized. What a discovery! But what did the Academy do? They sat on my manuscript for 20 months to then reject it in the most dismissive way!

Then came this young fellow, Pierre Paul Broca. Elegant, brilliant, good-looking, and with a quality of his scientific work that was vastly superior to anything that I have ever seen. Broca gradually collected more and more cases and finally concluded in 1865 that language was controlled by the third convolution of the frontal lobe on the left side. Language had a cortical location, and this location was in the left hemisphere. This is what Broca said. And he simply neglected my manuscript from 1863. He also neglected my paper from 1865 which I had published 2 months before Broca's final publication. I was simply brushed aside. The Parisian elite can't stand that a provincial fellow won the race. Broca turned into a scientific celebrity, while none of these people ever heard my name. But mine is the priority and I will fight for it until my last day."

On June 12, 1966, a plaque was unveiled in front of the house in Sommières in which Gustave Dax was born. At the same time, the little place in front of this house was renamed "Le Place des deux Docteurs Dax." The ceremony to which the President of the World Federation of Neurology had come was to commemorate the discovery by Dr. Marc Dax (1770–1837) and his son Dr. Gustave Dax (1815–1893) on the role of the left hemisphere for language. This discovery had happened about a century before, and it was now high time to recognize the contribution of these two physicians. For some decades, when people waited for an empty seat in one of the nearby cafes, they had time to read that it was father and son Dax who had discovered the location of language. And that it had happened here, deep in the province. But sometime in the early 2000s, the plaque was removed to make space for an advertisement for a new pizzeria.

Early reports on brain asymmetries

The true scientific history of brain asymmetries starts in the second half of the 19th century. This was a time of extraordinary discoveries that completely changed the view on brain functions and set the scene for a modern account of neuroscience that persists today. The epicenter of most of these early breakthroughs on hemispheric asymmetries was France—the leading science nation of its time. However, long before this golden era, several scholars foreshadowed what was to follow.

The oldest scientific account of brain asymmetries stems from late antiquity and was written as a medical treatise by an anonymous author who repeatedly refers in his writing to some even older Greek medical texts. His treatise survived about 1000 years possibly by diverse transcriptions and was finally discovered in a codex that dates from the end of the 11th or the beginning of the 12th century. This codex bears the title

De semine (On sperm) and was possibly in possession of the monastery of Saint Pantaleon in Cologne (Germany). In the part on brain asymmetries, the author states:

> "There are accordingly two brains in the head. The one gives us our intellect, the other provides the faculty of perception. That is to say: the brain on the right side is the one that perceives, whereas the left brain is the one which understands" (translation by Lokhorst,[1] p. 302).

How did the author come to this conclusion? According to Lokhorst,[1] the ancient view was that the soul was constituted by pneuma—a subtle entity that spreads from the heart to the whole body. Since ancient people often observed that in dead people the left cardiac ventricle and the aorta are relatively bloodless after death, it was assumed that the left part of the heart is filled with pneuma and thus should be the seat of the soul. Then, it was further assumed that the left and right brain hemispheres are connected to the left and right heart ventricles, respectively. Since the intellect was seen as superior to the faculty of perception, the first scholarly text on brain asymmetries concluded that the superior faculty of intellect has its place on the left side of the brain, leaving the right hemisphere for the inferior faculty of perception.

A much more empirically driven early account of hemispheric asymmetries was the classic book by Arthur Wigan, *The Duality of the Mind*.[2] The discovery that sparked Wigan's conviction was a chance finding that he had made as a young physician during an autopsy[3]: The brain of a dead man of about 50 years of age had only one hemisphere. However, this person had conversed rationally and even had written verses just days before his death. Thus, a single hemisphere was sufficient to process all aspects of the mind. Arthur Wigan concluded that if... "one brain be a perfect instrument of thought—if it be capable of all the emotions, sentiments, and faculties, which we call in the aggregate, mind—then it necessarily follows that man must have two minds with two brains. And however intimate and perfect their unison in their natural state, they must occasionally be discrepant when influenced by disease, either direct, sympathetic, or reflex" (Wigan,[2] pp. 201–202).

Indeed, it is possible that the loss of one entire hemisphere during very early ontogeny can be (nearly) fully compensated.[4] However, from this fact, it does not follow that the two hemispheres harbor independent minds in the healthy brain. Wigan had published a very radical proposal that was not accepted and was therefore neglected by the scientific community of his time. More than one century later, however, his thoughts on two parallel minds started to reverberate in a new and fresh way when a young Ph.D. student in Pasadena, United States, tested a split-brain patient and discovered that this person indeed seemed to have two minds in one skull. However, this is a different story and will be discussed in Chapter 3.

At the beginning of the 19th century, the main scientific debate was on localization of function. Were the cortical fields doing the same everywhere in a holistic manner or was it possible to locate diverse functions in different areas? This debate also ignited the first discussions about possible asymmetries of cortical functions. Treviranus[5] was the first to discover that: "Der Mensch hat sehr zahlreiche, sehr tiefe und in beyden Hälften unsymmetrische Windungen. Am großen Gehirn der Affen findet das Gegentheil statt" (translation: "Humans have a multitude and deep convolutions that are asymmetrically organized in the two hemispheres. This is different from the forebrain of monkeys").[5] Soon, the observations of Treviranus were verified in the best possible way. In 1839, Louis Pierre Gratiolet and François Leuret published a groundbreaking book on the cortical organization of mammals, including humans.[6] Their goal was to understand the evolution of cognition based on the comparative analysis of cortical folding patterns. Thereby, they observed that the arrangement of cortical gyri became more irregular and more asymmetric with phylogenetic proximity to humans. Was it conceivable that asymmetries of cortical functions exist, and that they represent an advanced organizational pattern of the human brain? The first functional evidence for this idea was, meanwhile, collected in a very unlikely place.

Marc Dax (1770–1837) was a local physician in Sommières, a picturesque little town in the far south of France. Dax worked at a home for the elderly and was a careful observer with a keen interest in studying language. In 1836, 1 year before his death, he wrote a paper on the association between aphasic disorders and lesions of the left hemisphere for presentation at the "Congrès Méridional," a celebration of achievements from the south of France, which was to be held in nearby Montpellier. The paper contained more than 40 clinical observations collected since the year 1800. Dax reported on the loss of word memory in a cavalry captain after a saber blow to the left side of his head, the association of right-sided paralysis and the loss of speech after left a hemisphere stroke, and about many more similar cases.[7] Not all cases represented his own observations as some only had been told to him. In no case did he have verification through autopsy, and in several cases, he could not even be sure that the brain had been damaged at all. He knew that he was about to make a claim of exceptional importance. However, he also knew that his database was not fully convincing. As a result, he shied away from presenting his paper at the conference. Instead, his manuscript collected dust at his home. He also had, however, given several copies to colleagues. As we will see, these copies later played a significant role in a bitter fight on the priority of the discovery of language asymmetries.

The discovery of language asymmetry

According to the official version, brain asymmetries of language were discovered by Pierre Paul Broca (1824–1880) between 1861 and 1865. Broca's findings must be understood within the context of three important scientific disputes of his time.[8,9] The first dispute was ignited by the second volume of the book by Leuret and Gratiolet.[10] This book came with an extended analysis of the comparative description of the cortex and also contained a review of cortical ontogeny. The authors made the case that the left hemisphere was developing faster than the right one (an idea that is disputed today[11]). As a result, it became possible that behavioral asymmetries like handedness were a mere result of the differential developmental speed of the cortex.[12] Thus, suddenly two highly respected scientists alluded to the possibility that a well-known behavioral asymmetry (handedness) may have grounding in neuroanatomy. The second dispute was about the spectacular growth of the frontal lobes in humans, compared to other primates (another idea that is disputed today[13]). Could it be that functions like language that are specific to humans can be found here?[14] The third dispute was laid out before Broca's discoveries and can be called *the* most important clash of neuroscience of the first two-thirds of the 19th century.

The central question was: "Do different functions have different cortical localizations?" The person who brought this discussion to the table was seen by some as a genius but by others as a charlatan.

Are functions localized at cortical level?

Back in 1800, Franz Joseph Gall (1758–1828), a German neuroanatomist and physiologist, published his conception that came to be known as phrenology.[15] Based on casual observations of people in his immediate surroundings, Gall developed the idea that 27 psychological faculties—like courage, vanity, memory for words, or the sense of speech—had separate cortical localizations. Individuals with a preponderance of some of these psychological qualities had a higher cortical volume at the respective localization and thus developed a cortical bump at this spot. Gall was convinced that the elevations on the human skull were caused by pressure exerted from the cortex underneath. Consequently, a personality inventory could be achieved by reconstructing the uneven landscape of the human skull—a method that Gall named Cranioscopy and that would later be known as phrenology (Fig. 1.1).[16] Armed with this idea, Gall started to analyze the heads of cruel murderers, great poets, spiritual leaders, and many more to come up with the cortical localization of 26 mental faculties like vanity, mercy, courage, and many more. Accompanied by an assistant, a servant, two monkeys, colored plaster casts, and skulls, he then started on a hugely successful tour through the public lecture halls of Europe that earned him a fortune and forever branded the term "phrenology."[17] The scientific community was outraged by his poor conclusions and his sloppy way of mapping cortical functions. Even today, more than

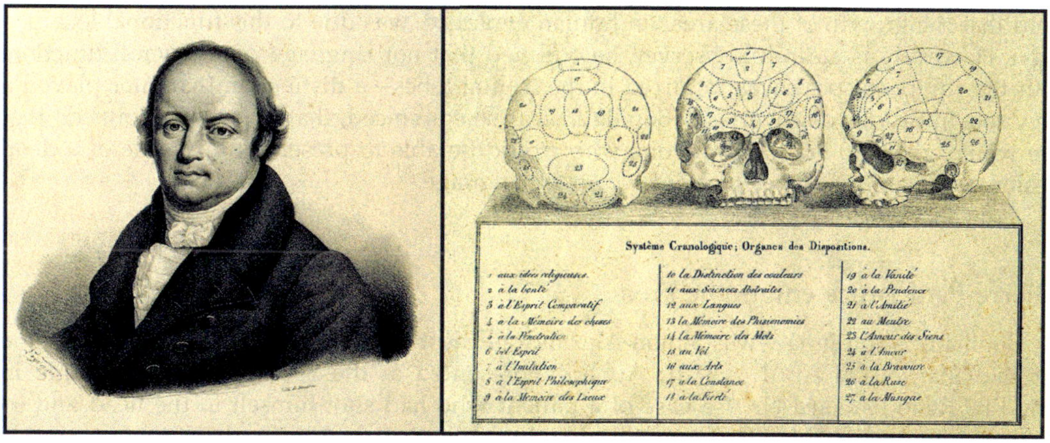

FIGURE 1.1 Franz Joseph Gall and his concept of phrenology. According to Gall, small elevations on the skull hint at pronounced mental faculties that are located in the underlying area of the cortex. On the right side, some of these faculties are shown alongside with their location on the skull. Some examples are (1) religious thoughts; (2) benevolence; (4) memory for items; (9) memory for places; (18) pride; (19) vanity; (24) love; and (25) bravery. Language (location 12) is in the lateral upper part of the eye sockets, beneath the location of the memory of physiognomies (location 13) and that of words (location 14). *Source: Reproduced from the Wellcome Library, London, licensed under CC-BY 4.0.*

200 years after the appearance of Gall's book, critiques of modern brain imaging studies use the term "phrenology" as a catchword to express their rejection of a kind of science that too easily associates superficially assumed psychological entities with certain brain regions.

The strongest argument against Gall was put forward by Marie-Jean-Pierre Flourens (1794–1867), a young anatomist, who conducted careful lesion experiments with pigeons and many more animal species. After ablating diverse forebrain regions of these animals, he observed no loss of a specific function, irrespective of where a small lesion was placed. If the lesions grew larger, the animals increasingly lost their abilities to sense or to initiate movements by their own will. However, when nudged, they still were able to walk a few steps or fly a short path without any problem. Flourens concluded that the cerebral hemisphere must be the area where all senses are integrated and from where higher mental functions emerge. These hemispheres were, however, not arranged as a map of separate functions but as holistic chambers where all processes intertwine.[18]

After these events and discoveries, only a person with great courage and an immaculate scientific record could openly defend the idea that the cortex had functional localizations. Jean-Baptiste Bouillaud (1796–1881) was such a person. Bouillaud was the president of the French Académie de Médicine, dean of the Faculty of Medicine in Paris, and commander of the Legion d'honneur. Bouillaud was convinced that brain diseases were scientific windows into the mechanisms of the mind but had to be analyzed with the utmost care and verified by subsequent autopsies. He also was convinced that not so much single cases but large numbers of carefully conducted studies would reveal the truth. Toward the end of his career, Bouillaud had accumulated about 500 perfectly analyzed cases with diverse brain injuries and could derive detailed insights from this wealth of data. Bouillaud strongly defended the claim that speech was governed by the frontal lobes and that the growth of these areas in human evolution was due to this functional localization. However, as a careful observer, he outlined that not language as an overall function, but the ability to speak was localized in the frontal lobes—a distinction that later played a key role in Broca's discoveries.[19] Bouillaud was so convinced, that he even announced that he would give 500 Francs to anyone who would be able to present an example of a deep lesion of the anterior lobes that did *not* affect language.[20]

Pierre Paul Broca enters the scene

Bouillaud's son-in-law was Simon Alexandre Ernest Aubertin, who was chef de clinique at the Charité. On April 4, 1861, Aubertin presented at the Société d'Anthropologie in the Rue René Panhard Nr. 1 a case of a patient who had shot himself in the head and so exposed his brain. This man was still alive but dying. Aubertin decided to run a simple experiment with this patient by gently pressing a spatula on the anterior cortex while the man was talking. Immediately the man stopped, but continued talking as soon as the spatula was taken away.[21] This presentation was followed by a heated debate between Aubertin and Gratiolet. Aubertin strongly believed in functional localizations. Gratiolet accepted developmental asymmetries of the cortex but denied functional localizations. In the audience was Pierre Paul Broca, the young secretary of the Société. Broca leaned toward the localizationist view of Bouillaud and Aubertin since his own microscopic studies had shown how heterogeneous the microarchitecture of the cortex is. In addition, his clinical cases seemed to only make sense when assuming differences in function across the cortex. What was missing was a perfect clinical case in which a close relationship between structure and function could be shown. This case came 8 days after the heated debate that followed Aubertin's presentation.

On April 12, 1861, an epileptic patient named Leborgne was transferred to Broca's surgical ward at the hospital Bicètre. The man had lost his speech long before and could only utter the word Tan-Tan, which had become his nickname. Due to gangrene in his paralyzed right leg, Leborgne was bound to die soon. Broca invited Aubertin to jointly analyze the patient. Aubertin accepted this invitation and was impressed by the clarity of the case. He predicted that Leborgne had a lesion in the frontal cortex. A few days later, Leborgne died. An autopsy was done, and Broca could reveal that the starting point of the brain disease was the third convolution of the frontal lobe on the left side. In two successive presentations in front of the Société d'Anthropologie, Broca could demonstrate, after meticulous analysis, a close relationship between the functional language deficit and the anatomical reconstruction of the lesion.[22,23] Shortly thereafter, another patient was brought to Broca. His name was Lelong, and he could only utter the words "oui," "non," "tois," "toujours," and "Lelo." His language comprehension was intact, and he had no paralysis of the mouth or the tongue. Lelong died a few months later and an autopsy revealed a lesion at exactly the same location (Fig. 1.2)[25] as in Leborgne.

This time it was Broca's turn to come under heavy fire from Louis Pierre Gratiolet. The line of the battlefield was like before: Broca defended localization and dismissed asymmetry; Gratiolet was willing to accept asymmetry but regarded the localizationists' claims as nonsense and mere phrenology: "If intelligence is one entity, then this must equally be true of the brain. Therefore language, which is inherently and inseparably a characteristic of intelligence, must be a diffuse function of the brain" (p. 60).[26]

Broca started to collect more cases. He developed his trademark approach that made him one of the towering scientists of his time: Broca was very careful in his observations and his thinking. He always thought of all possible counterarguments and tested them. He was cautious not to call a deficit an "aphemie" (what later became today's aphasia) without checking all options that could produce a similar deficit without really reflecting a deficit in the production of language. In addition, he meticulously conducted autopsies to verify all possible lesions in the brain. In 1863, Broca presented eight cases with lesions in the third convolution of the frontal lobe with severe aphasia.[27] He commented with amazement that they were all in the left hemisphere, but that he would refrain from drawing too strong conclusions from this before having collected more cases (Fig. 1.3).

Finally, on June 15, 1865, Pierre Paul Broca presented a large number of cases before the members of the Société Anthropologie. All the collected cases had left hemisphere lesions in a region that was later called the Broca area of the brain. In all the analyzed cases, the patients had suffered from a deficit in producing language without having major difficulties understanding language. Broca concluded his insights with the sentence: "Nous parlons avec l'hémisphère gauche" ("We are talking with the left hemisphere") (Broca,[28] p. 384). This sentence is the official starting point of brain asymmetry research (Fig. 1.4).[29]

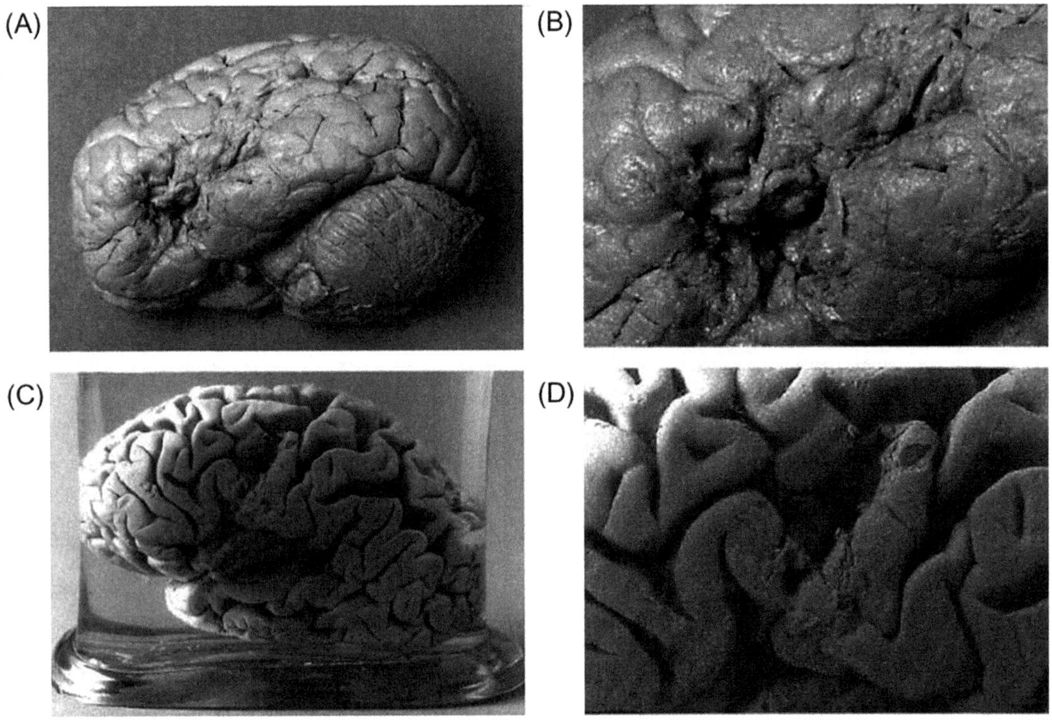

FIGURE 1.2 The brains of Leborgne (A and B) and Lelong (C and D) as a whole and with a higher magnification of the location of the lesion in the third convolution of the frontal lobe.[24] *Source: Reproduced with permission of the publisher from Dronkers NF, Plaisant O, Iba-Zizen MT, Cabanis EA. Paul Broca's historic cases: high resolution MR imaging of the brains of Leborgne and Lelong.* Brain. *2007;130:1432–1441.*

The fight for priority

While Broca had reached his apex of scientific esteem, his glory was dangerously dimmed by a severe accusation: On March 24, 1863, Gustave Dax, the son of Marc Dax, had sent a manuscript to the Académie de Médecine entitled "Observations Tendant à Prouver la Coincidence Constante des Dérangements de la Parole avec une Lésion de l'Hémisphère Gauche du Cerveau" ("Observations seeming to prove the constant coincidence of speech disorders with a lesion of the left hemisphere of the brain"). The manuscript contained his father's original collection from 1836 as well as new material that Gustave Dax had gathered. Together, they made a substantial case that language had to be left hemisphere-based.[30] As per convention, the title of the manuscript was publicly announced, but the manuscript itself was kept confidential and given to three reviewers (Jean-Baptiste Bouillaud, Jules-Augustin Béclard, and Françisque Lélut). These three needed more than 20 months to completely dismiss the work. As the head of the committee, Françisque Lélut presented his decidedly negative conclusion by pointing out that the manuscript was a grotesque return to the pseudoscience of Gall by proposing that functions were differently localized in the two hemispheres. In addition, it contained a large

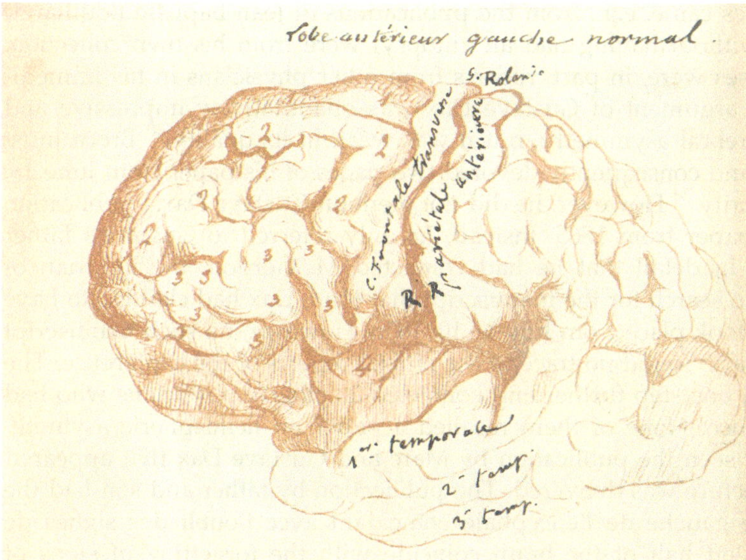

FIGURE 1.3 Drawing of the lateral view of the brain from Pierre Paul Broca. The lesions of his patients with "aphemie" were always located in the third convolution (labeled "3") of the frontal lobe on the left side. *Source: Portrait by P. Petit, reproduced from the Wellcome Library, London, licensed under CC-BY 4.0.*

FIGURE 1.4 Pierre Paul Broca (1824–80) in his later years. *Source: Reproduced from the Wellcome Library, London, licensed under CC-BY 4.0.*

number of flaws like wrong references to the work of Bouillard and so forth.[30] All in all, the review of the academy was more than a simple rejection; it was a humiliation for Gustave Dax.

However, Gustave Dax did not give up. On April 28, 1865, he published his father's memoir and his own collection in the Gazette Hebdomadaire de Médecine et de Chirurgie, a popular French medical journal.[31,32] Now everybody could officially see their work. It became apparent that the majority of cases reported by Gustave Dax were from published literature (64 cases came, e.g., from the publications of Jean-Baptiste Bouillaud) and only 16 patients (all without having had an autopsy) were from his own collection. However, even these 16 cases were, in part, reports from other physicians in his immediate vicinity. Therefore, the argument of Gustave Dax was quantitatively impressive and made a strong point for cerebral asymmetry, but it was weak in its quality.[30] Broca must have seen this publication and consequently devoted two pages of his paper from June 18, 1865, to the problem of priority.[28] However, he did not mention Gustave Dax's publication attempt from 1836 or his paper from 1865. Instead, he only referred to Gustave's father Marc Dax. Broca outlined in detail that he had contacted M. Gordon, the librarian of the faculty at Montpellier, to search for the manuscript that Marc Dax had claimed to have read at the conference that took place from July 1–10, 1836. However, no such manuscript could be revealed. Gordon also found no trace in the official reports of the conference. The diligent librarian then went one step further and contacted 20 elderly physicians who had participated in the conference. None of them recalled a paper on hemispheric asymmetries. Broca very likely had seen the publication by Marc and Gustave Dax that appeared 2 months before his 1865 lecture was delivered. The publication by father and son had the title: "Lésions de la moitié gauche de l'encéphale coïncidant avec l'oubli des signes de la pensée" (lesions of the left half of the brain coincide with the forgetting of signs of thought).[31] It is not very likely that Broca could have overlooked a publication with this title. It is more likely that he was frustrated by the poor quality of the work of Marc and Gustave Dax and wanted to prioritize the discovery of brain asymmetries.

Until his last day, Gustave Dax insisted that his father had given the lecture at the 1836 conference in Montpellier and thus deserved priority. Up to now, no trace of such a presentation has been discovered and it seems likely that Marc Dax indeed did not travel to Montpellier but instead distributed his manuscript to some like-minded scientists in his vicinity. Indeed, Raymond Caizergues, the grandson of Fulcrand Cesar Caizergues, the previous dean of the faculty of Montpellier, announced that he had found the 43-year-old manuscript by Marc Dax under piles of papers from his grandfather that had been left untouched.[33]

So, who deserves priority, Broca or Dax? We possibly should dismiss the 1836 conference presentation by Marc Dax of 1836 as it that was never published, and so could not have been known to other scientists. However, from 1863 to 1865, Gustave Dax did everything a scientist could do to announce his father's and his own findings. He also always announced and published his views faster than Broca. However, Pierre Paul Broca introduced a completely new quality to a field that would eventually be called neuropsychology. His behavioral observations were exhaustive, cautious, and highly self-critical. His neuroanatomical lesion reconstructions were detailed and accurate and revealed the possible sequences of events during the spread of cerebral damage. Obviously, fellow scientists were impressed by his diligence and readily followed his conclusions. Still, Broca came second.

The many mysteries of handedness

Right-handedness dates back to pre-Neolithic and Neolithic times,[34,35] where it can be recognized by the combination of traces on stone tools[36] and left its cultural traces in all

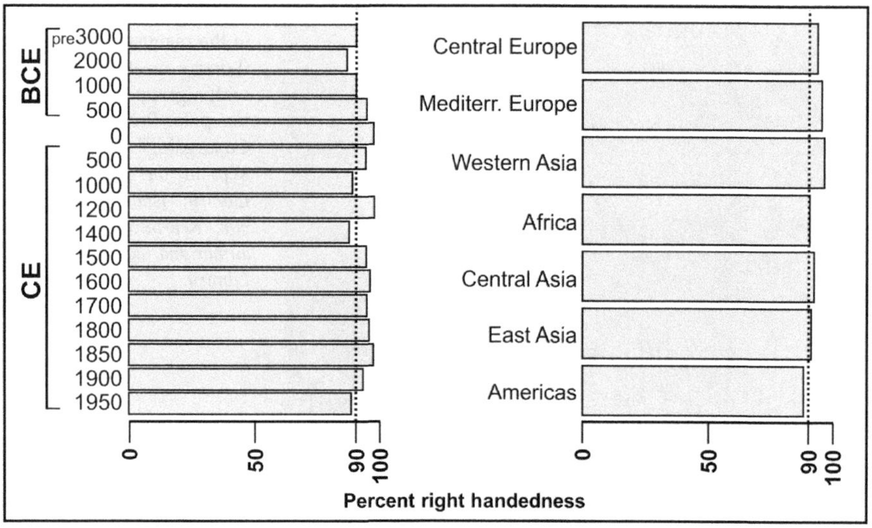

FIGURE 1.5 Historical distribution of handedness as manifested by works of art of 50 centuries (left). Geographical distribution of handedness as manifested by works of art (right). BCE = before common era, CE = common era (drawn according to Coren & Porac,[38]). *Source: Reproduced with permission from AAAS and the author from Coren S, Porac C. Fifty centuries of right-handedness: the historical record.* Science. *1977;198(4317):631–632. https://doi.org/10.1126/science.335510.*

classic texts like, for example, the Bible.[37] Right-handedness is also visible in all kinds of visual arts. Coren and Porac[38] analyzed close to 1200 drawings, paintings, and sculptures from over 50 centuries. Each item was examined for figures displaying a clear hand preference. By this procedure, they discovered that 92.6% of the depicted humans used their right hand (Fig. 1.5, left).

In the same study, the depictions were also analyzed for geographical location. Again, no systematic change in right-handedness was discovered across cultural territories (Fig. 1.5, right).

Fig. 1.6 shows a clear example of tracking the incidence of handedness in historic paintings. It depicts an Iranian Safavid court milieu with men on horseback playing polo in front of a pavilion.

Six of the men hold the polo stick in their right hand, while only one is using his left hand. Polo was invented in Iran, and this game is depicted in many Persian and Ottomanian miniatures throughout the centuries.

The predominance of right hand use is so glaringly visible in all cultures that there is practically no society that has not developed its own explanations for this left–right difference. Some of these explanations may sound strange to us but would have made sense to people in those times. Only two will be sketched here to show how many diverse ideas to explain handedness were put forward by scientists of the 19th century.

Buchanan[40] was convinced that the walking patterns of humans were affected by the asymmetry of their inner organs. Since both the liver and the right lung are heavy organs, he concluded that our body's center of gravity was shifted to the right.

FIGURE 1.6 Copy of a painting in the manuscript Majalis al-'Ushshaq (Lover's meeting): Fakhr al-Din Iraqi with a group of youths who are playing polo. Probably by Kamal al Din Gazurgahi,[39] c.1580, Shiraz. *Source: Kept at the Topkapi Palace Museum Library, Istanbul, Hazine 829 Folio 90r. Reprinted by permission of the author and the Topkapi Palace Museum Library.*

Consequently, we have to lean slightly to the left to restore balance. However, this puts the right arm in an upper position, providing a higher degree of freedom of movement on the right side. Since less weight has to be held by the right leg, it can also move more freely. The resulting pattern, according to Buchanan,[40] is right-handedness and right-footedness.

Philip H. Pye-Smith proposed a hypothesis based on natural selection. He assumed that symmetry played a role in the ancient human form. However, once warfare included using weapons like swords and shields, the tables were turned due to the left-side position of the human heart: "If a hundred of our ambidextrous ancestors made the step-in civilization of inventing a shield, we may suppose that half would carry it on the right arm and fight with the left, the other half on the left and fight with the right. The latter would certainly, in the long run, escape mortal wounds better than the former, and thus a race of men who fought with the right hand would gradually be developed by a process of natural selection. Such a race would naturally use the right hand also when they discovered how to draw and to write" (Ref. [41], p. 145). As strange as this hypothesis may sound today, it reoccurred more than a century later in modern science but in an inverted

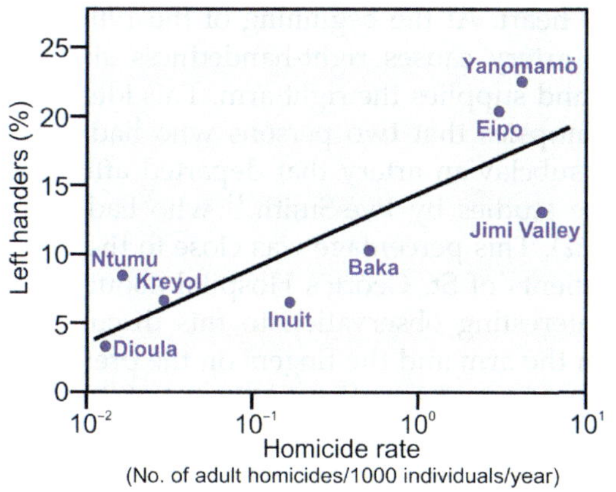

FIGURE 1.7 Percentage of left-handers and homicide rates in traditional cultures. Homicide rate was computed as the number of homicides per 1000 individuals per year.[42] *Source: Reproduced with permission of the publisher from Faurie C, Raymond M. Handedness, homicide and negative frequency-dependent selection.* Proc Biol Sci. *2005; 272(1558):25–28.*

manner: Faurie and Raymond[42] would show that left-handers occur more often in traditional and violent societies (Fig. 1.7).

The higher the homicide rate, the higher the frequency of left-handedness. It ranges from 3% in the most pacifistic societies to 27% in the most violent war-like tribes. The authors argued that this was a pattern well known from sports: Left-handers have an advantage in sports that involve a dual confrontation such as fencing and tennis but not in noninteractive sports such as gymnastics.[43] The advantage of left-handers is frequency-dependent. If they are rare, their competitors have a hard time becoming accustomed to their inverted movement patterns. The higher their proportion grows in the population, the smaller is their advantage. Possibly, the mechanism for traditional warrior societies is the same. The more hostile and murderous their interactions get, the more left-handers have an edge over their enemies—until their number grows too high.

This advantage of left-handers in pitched battles was already recognized in the 14th century by the Ottomans. Sultan Murad I. ordered the formation of a left-handed and highly trained subgroup of the Janissaries—an elite infantry that constituted the first modern standing army in Europe. The carefully selected left-handed Janissaries were called "Solak" (left-handed in Turkish) and their number was limited to just 400. During campaigns, these formed a half circle around the sultan to defend him against attacks. To signal the enemy their presence as an elite troop with an exceptionally high success rate in man-to-man combat, they wore a colorful uniform with a feather on their turban.[44]

As easy as it might appear to recognize handedness, it was as difficult to understand its causes. Chapter 6 outlines the myriads of dead ends into which scientists were trapped until they started to see a clearer picture. Two of these dead ends will be discussed here first, before embarking into the first studies on the cerebral causes of handedness.

Is right-handedness caused by an uneven blood supply to the arms?

The first surmises on handedness started about 2500 years ago. Aristotle wrote a full chapter on this issue and developed his theory that the right arm was supplied with especially warm and fresh blood since the right cavity of the heart was the largest and highest.[9]

Attention later shifted to the major arteries of the heart. At the beginning of the 19th century, it was supposed that the right subclavian artery causes right-handedness since it departs closer to the heart, has a larger diameter, and supplies the right arm. This idea was revitalized by Hyrtl,[45] who discovered during autopsies that two persons who had been left-handed during life evinced an atypical right subclavian artery that departed after the left one from the heart. Hyrtl also mentioned the studies by Pye-Smith,[41] who had seen such an unusual pattern in 4 of 296 autopsies (1.3%). This percentage was close to the 4.4% left-handers that Ogle[46] found by asking 2000 patients of St. Georg's Hospital about their handedness. Hecht and Langstein[47] added an interesting observation to this discussion: They discovered that blood pressure was higher in the arm and the fingers on the preferred side. Even more interestingly, original left-handers who had been forced during childhood to switch their preferred hand still had higher blood pressure values on the left side!

At the turn of the 19th to the 20th century, scientists were increasingly convinced that not blood supply but rather something in the brain causes handedness. Therefore, interest in the asymmetrical position of arteries vanished. In an attempt to put an end to this discussion, Gaupp[48] reviewed several cases of right-handed persons in whom the autopsies revealed that their right subclavia departed from the aorta after the left one. However, the interest in a possible correlation between blood supply and handedness never completely ended, and Rosolanka and Rosolankova[49] proposed this correlation as a "new idea" to the scientific audience.

Are genes making the right arm strong?

Although an understanding of genetic mechanisms at the molecular level only started to emerge in the first half of the 20th century, basic knowledge about heritability was established throughout the 19th century. Not surprisingly, most authors of the late 19th century were sure that the mystery of handedness was decisively simple: Neither the brain nor the blood supply does the job, it was simply the right (or left) arm itself that had stronger muscles in humans at birth. So, handedness should be a very simple, peripheral effect. Hasse and Dehner[50] painstakingly analyzed 5141 German soldiers and discovered that only 7% had a longer left arm, while 75% had a longer right arm. When concentrating only on left-handed individuals, 57 out of 58 had a longer left arm. Meanwhile, Theile[51] discovered that arm bones were generally heavier on the right side. Stier[52] described that the circumference and the measured force were larger on the right side of his subjects. By reviewing a larger number of previous studies, Ludwig[53] concluded that the length of the right arm was on average 1 cm longer in 75% of his subjects. However, is the strength and length of the right arm the reason or the consequence of right-handedness?

Pye-Smith,[41] who came up with the sword-and-shield theory, had assumed that a selection for right-handedness resulted in morphological asymmetries of the extremities, resulting in stronger right arms. German authors tended to see it the other way around. For example, Stier[52] argued that the finding of morphological asymmetries has no bearing on causality discussions. Criticizing the conclusions of a major study by Hasse and Dehner,[50] he pointed out that only 1% of the soldiers in his study had defined themselves as left-handed, but still, 7% had a longer left arm. Gaupp[48] added a very interesting observation to this discussion: He measured the arm lengths of right-handed people who had worked for decades in professions in which they mainly used their left hands (musicians, porcelain painters) and saw that most of them had longer left arms. Thus, morphological asymmetries of the extremities are a fact but possibly result from asymmetrical hand use without inducing it. 73 years after Gaupp,[48] Means and Walters[54] published a study on hand- and foot-length asymmetries and demonstrated a correlation with handedness. Without knowing that this discussion had already taken place and had ended about three-quarters of a century earlier, they presented their data as novel insight into handedness research.

Handedness in the brain

The groundbreaking discoveries of Broca[28] on language asymmetry put an end to almost all attempts to explain handedness without reference to the brain. In a transition period, Ogle[46] tried to reconcile the arterial and the cerebral arguments by assuming that the larger left-sided carotid artery provided more blood supply to the left hemisphere, thereby promoting faster growth and a more complex gyral pattern. Many authors did not follow this argument and were highly skeptical about possible left–right differences in blood supply to the brain.[55] Instead, diverse aspects of the brain were measured to identify any correlation to handedness. Among them, for example, brain weight,[56,57] cortical surface,[58] and skull length[59] were proposed. The differences discovered were small and highly variable. In addition, the diverse publications contradicted each other. The scientists were also deeply frustrated since they knew that their methods to study gross aspects of the whole brain were too coarse for their aims. The great success story of language asymmetry could not be repeated for handedness. For many decades, the search for the cortical fundaments came to a halt.

However, what if scientists did not analyze cortical areas but the connectivity between brain structures? The possibility to properly analyze fiber tracts only became available during the late 19th century. It was in 1884 when Carl Weigert established, after many long attempts, a useful and reliable method to stain myelinated axons.[60] Before this breakthrough, only far less consistent methods had been available. One of them was introduced and employed by Flechsig[61] to identify the course of the corticospinal tract. This fiber bundle descends from the primary motor cortex and some surrounding cortical areas toward the spinal cord. On its way, this huge pathway assumes a pyramid-shaped appearance in cross sections of the lower brainstem. This is why it is called the pyramidal tract. At the lower end of the brainstem, at least 80% of pyramidal fibers cross to the other side, while others remain ipsilateral. The pyramidal tract then travels down the spinal cord and via interneurons activates the spinal motor neurons that then stretch muscles to move the body. Those pyramidal tract axons that activate motor neurons of the hand muscles bypass the spinal interneurons and directly synapse on spinal motor neurons. Thus, our cortex has absolute control over our hand movements, and this is special to humans with their unmatched dexterity. This makes the pyramidal tract a perfect candidate for handedness research. Flechsig was well aware of this and therefore had high hopes to become the *Broca of handedness*. Indeed, he discovered that in 40% of all his cases, the crossover of the pyramidal tract fibers was asymmetric such that fewer right-sided axons crossed to the other side but instead remained ipsilateral. So, more axons should be present on the right side to activate spinal motor neurons. This is a great finding, no doubt. However, it was not possible for Flechsig to correlate his postmortem findings with the handedness of his subjects during their lifetime. In addition, the number of 40% was odd. If the asymmetry of crossing was the biological fundament of right-handedness, this number should be much higher.

Brodal[62] could replicate Flechsig's findings with more advanced staining methods. Then, Yakovlev and Rakic[63] attempted the next logical step: They looked at a large number of brains ($n = 105$) and managed to obtain information on the handedness of their subjects before their death. To their dismay, no convincing correlation between crossing patterns and handedness was found. Kertesz and Geschwind[64] increased their sample to 158 and tried to verify handedness with a detailed questionnaire that they gave to the relatives of the person whose brain had been histologically analyzed. According to their results, preferred crossing from left to right was visible in 73% of their subjects. The reverse could only be seen in 17%, while 10% had no preference. These numbers roughly matched the behavioral handedness data. However, significant disappointment occurred during their attempt to correlate anatomy with behavior: There was no difference between right- and left-handers in their patterns of crossing. That should have been the end of a century-long search to discover a link between handedness and the crossing pattern of the corticospinal tract. Yakovlev[65] still did not give up. He now tried it the other way around. He obtained the brains of four left-handed persons and hypothesized that if the crossing pattern determined handedness, he would expect that in these brains more pyramidal tract fibers should cross to the left side. What did he find? Well, reality can be mean. In all four left-handed cases, the pyramidal fibers preferentially crossed to the right. That was the end of that story. Scientists gave up looking for a relationship between pyramidal crossing patterns and handedness.

Shortly after these frustrating experiences, magnetic resonance imaging (MRI) became available, which, for the first time, provided the possibility to measure neuroanatomical details in living subjects so that behavioral asymmetries could also be tested. Many MRI studies that tried to correlate all kinds of neuroanatomical variables with handedness were then published. Several studies observed a larger extent of the right frontal lobe in right-handers that was less pronounced in left-handers. For the occipital lobe, the pattern was reversed.[66–68] None of these early findings could be properly replicated in further studies with larger samples.[69,70] To some extent, the imaging data of the 1980s resembled the morphological studies on overall brain measures conducted exactly one century before: 100 years apart, scientists tried to measure practically everything and link it to handedness. No hypothesis-driven thinking was involved and almost any results obtained were automatically linked to handedness, irrespective of whether it made sense or not. It is high time for a fresh approach. Were scientists able to come up with one? The answer is given in Chapter 6.

Face asymmetries

On April 8, 1820, the Greek farmer Yorgos discovered a large female statue while working in his field on the island of Milo. He informed his neighbor M. Louis Brest, an agent of the French Council, about his discovery. Brest immediately recognized the value of the Hellenistic statue and arranged its transportation to Istanbul where the French ambassador Marquis de Rivière bought it for a mere 550 Francs from the Turkish Ottomanian Emperor Mahmud II. On March 11, 1821, Marquis de Rivière gave the Venus of Milo as a present to King Louis XVIII in Paris, who donated it to the Louvre Museum, where it is still exhibited today. At the beginning of the 19th century, Napoleon had occupied major parts of Europe and started to plunder tremendous amounts of historical artifacts and pieces of art to bring them to the Louvre. As a result, this museum turned into the cultural center of the Western World. After Napoleon was defeated in Waterloo in 1815, the European powers took back everything and more, leaving the Louvre mostly blank. The Venus of Milo, therefore, came just in time to again attract visitors to the Louvre and it was hyped as the most precious piece of art from antiquity. This early example of clever advertisement worked tremendously well, and people started again to visit the Louvre with its mostly empty rooms, just to see this statue. One of them was Friedrich Goeler von Ravensburg who subsequently published a monograph on the Venus of Milo.[71] From a detailed analysis of all aspects of the statue, he concluded that the Venus of Milo represented an outstanding piece of naturalistic art and was therefore probably made from a living model. Despite his general admiration, the author had to note that the central line of the statue's face was slightly displaced. He therefore advised all visitors to the Louvre Museum that it was better to observe her in profile from the right-hand side, where she seemed to express more warmth and tenderness (Fig. 1.8).

Goeler von Ravensburg's book was published during a heated European debate on bodily asymmetries. Max Nordau (1849–1923), a famous physician and social critic of his time, wrote in his two-volume book on "Degeneration" that:

> "Die Entartung verräth sich beim Menschen durch gewisse körperliche Merkmale… Solche Merkmale sind…in erster Reihe die Asymmetrien, d.h. die ungleiche Entwicklung der beiden Hälften des Gesichts und des Schädels… (translation: The degeneration of humans becomes visible through some bodily signatures… These are… foremost the asymmetries, i.e., the unequal development of the two halves of the face and the skull)."[72]

Along the same line of argumentation, Francis Galton (1822–1911) was also convinced that the genetic background defines both the face as well as character and intelligence. To prove the link between facial features like asymmetry and a criminal mind, he asked his friend Sir Edmund Du Cane, Surveyor-General of Prisons, to grant him thousands of photographs of three classes of criminals: those convicted of (1) murder, manslaughter, and burglary; (2) felony and forgery; and (3) sexual crimes. Browsing through the pictures, he was happy to see that these three groups looked differently to him. What was now needed was an objective and quantifiable method to analyze these photographs. His solution was to optically superimpose the various drawings (Fig. 1.9A) to obtain the average result which then should reveal the typical face of a criminal (Fig. 1.9B). To his dismay, he realized "that the features of the composites are much better looking than those of the components. The special villainous irregularities in the latter have disappeared, and the common humanity that underlies them has prevailed. They represent not the criminal, but the man who is liable to fall into crime. All composites are better looking than their components because the averaged portrait of many persons is free from the irregularities that

FIGURE 1.8 The Venus of Milo in the Louvre. *Source: Wikimedia Commons, Ricardo Tulio Gandelman licensed under CC-BY 2.0.*

variously blemish the looks of each of them." (p. 97)[73] (Fig. 1.9C). Francis Galton had discovered an effect that subsequently was forgotten for more than a century until Langlois and Roggman (1990) discovered it again, namely, that averaged faces are perceived as attractive.[74]

Given this background, the asymmetry of the Venus of Milo could not be seen in a positive light in the 19th century. Indeed, 7 years after Goeler von Ravensburg, anatomist Henke[75] published a harsh critique of all attempts to idealize the Venus of Milo and advised that an anatomist should be consulted first before indulging in reflections on works of art. Henke observed that the legs of the statue were of different lengths and that the pelvis was oblique. He suspected that the model had limped during her life. However, the worst, he stated, was the face of the Venus of Milo, where the two lines connecting both pupils and both lateral ends of the lips were neither parallel to each other nor perpendicular to the nose (Fig. 1.10).

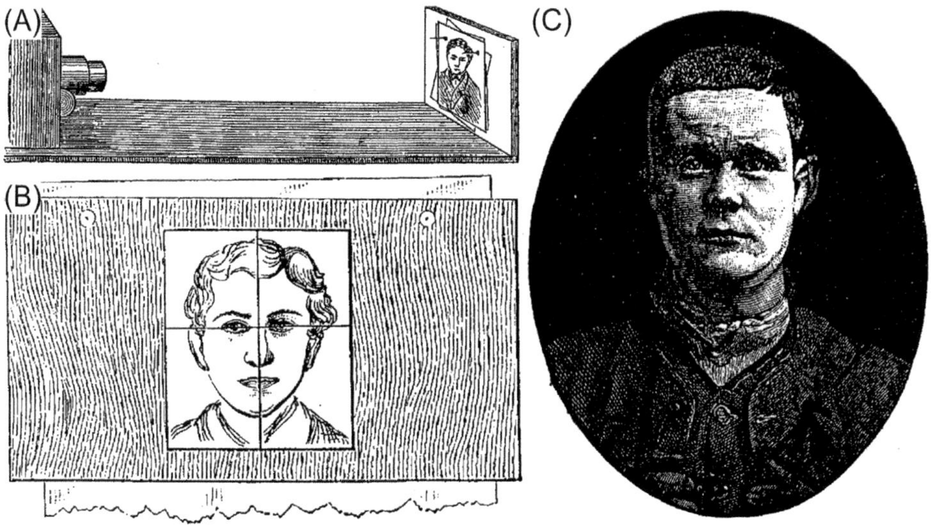

FIGURE 1.9 Face analysis system of Galton (1878).[73] (A) A pack of photographs is printed to the same distance between the eyes and mounted opposite to the camera such that the eyes are superimposed. If the exposure time of one portrait would be, say, 80 s, and if 8 pictures are to be superimposed, the cap of the camera objective is taken away for 10 s for each portrait, thereby creating a superimposed average face on the photo plate. (B) Schematic view of one portrait in the correct position. (C) A woodcut from only three portraits. *Source: Reproduced with permission of the publisher from Galton F. Composite portraits.* Nature. *1878;18:97–100.*

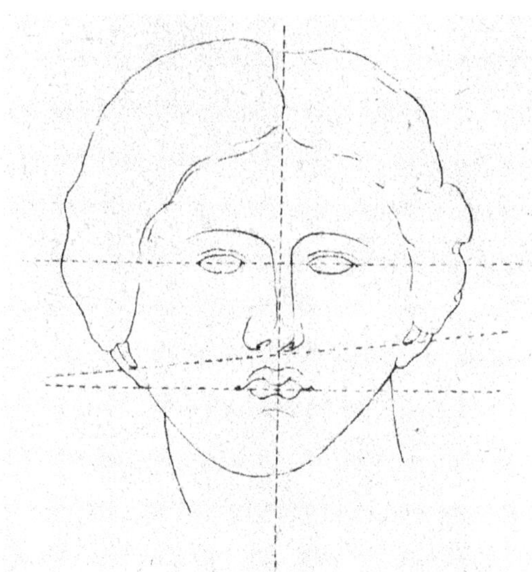

FIGURE 1.10 Analysis of the asymmetries in the face of the Venus of Milo by Henke[75]. Published under a Creative Commons Public Domain Mark (CC-PDM) license and used in accordance with this license.

Henke summarized that the Venus of Milo was far from perfect since asymmetries may occur in life but always as a matter of abnormality. Probably it was this last sentence that stimulated the first empirical study on the lateralization of the human face.

Christoph Hasse from the anatomical institute in Breslau, Poland, published a monograph on the Venus of Milo.[76] He admired the naturalistic style of the statue and felt that the conclusions his colleague Henke had come to were wrong. Hasse photographed the face of the Venus of Milo and those of several people in his laboratory standing behind a square grid and revealed a large number of similar asymmetries (Fig. 1.11).[77]

Three of these asymmetries seemed to be the most important to him: In all cases studied, the left eye was on average 3.5 mm nearer to the medial vertical than the right one. Additionally, the line connecting both pupils was generally not horizontal, and the left forehead was about 5 mm wider. These asymmetries were identical in the face of the Venus and in those of other people he studied. Hasse speculated that the pupil-line asymmetry would compensate for the asymmetry of the pelvis and the slight bow of the spinal cord which he had often observed and which tilts most people's heads to one side of the body vertically. He suspected that this compensation would just align both eyes on the horizon. In his opinion, the larger left side of the forehead was a result of the larger left brain hemisphere, which in turn seemed to be connected to the right-handedness of the majority of his subjects.[46] To place Ogle's findings on a firm data basis, Hasse and Zakrzewski[78] studied the pelvis of a large number of women in a gynecological hospital and observed the same pattern of asymmetry as displayed by the Venus of Milo.

Additionally, Hasse and Dehner[50] studied the asymmetries of arms, legs, and backbones in 5141 men of the German armed forces. This is one of the largest samples ever analyzed in lateralization research and antecedes by nearly a century today's studies on

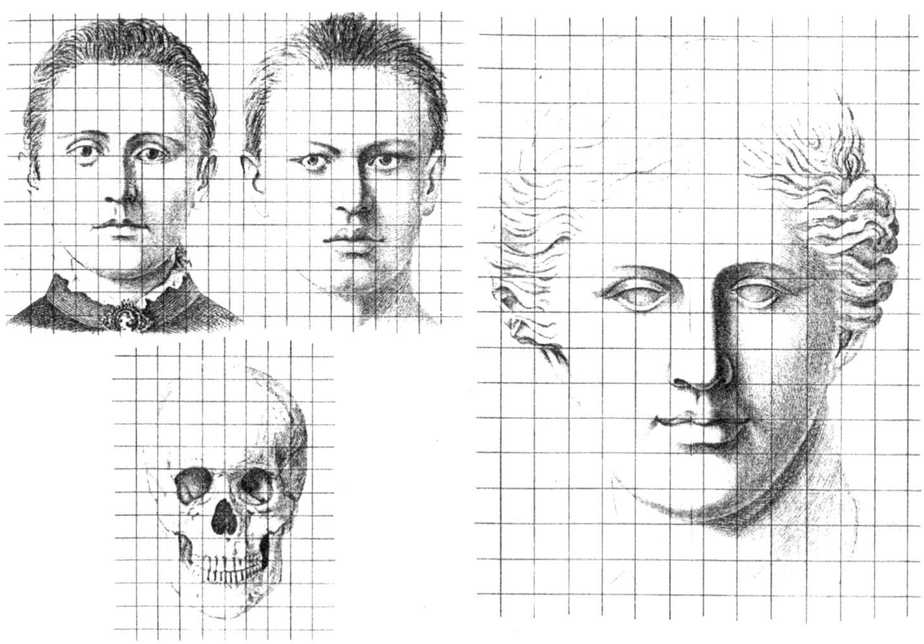

FIGURE 1.11 Depiction of the asymmetries of the face of the Venus of Milo along with highly similar asymmetries in the faces of two research assistants in the laboratory of Christoph Hasse and a skull from the anatomical collection of the university[77] (CC-PDM).

the morphological asymmetries of the extremities.[54] Hasse and Dehner[50] observed a right bow in 52% of their sample and a left bow of the spinal cord in 16%. The backbone was completely straight in only 32% of the individuals studied. This result supported the hypothesis that the deviation of the eye line from the horizontal could compensate for the tilt of the spinal cord. The authors concluded that they were able to contradict the long-held opinion of anatomists and artists that the human body was symmetrical.

Stimulated by the results obtained by Hasse, psychiatrist Hallervorden[79] became interested in a more dynamic analysis of facial asymmetries in expressing emotions and introduced a new method which is still widely in use. For each expression photographed, he printed an original and a mirror-reversed picture and then created composites by cutting both the original and the reversed along the midline of the face. The author described the left–left pictures as always being of a primitive and emotional character, while the right–right composites gave the impression of insightfulness. In left-handed subjects, this pattern was reversed. Therefore, Hallervorden erroneously concluded that the left hemisphere is not only dominant for language but also for the expression of personality. Later he modified this hypothesis and assumed that a general brain mood was expressed in both halves of the face.[80] The facial asymmetry would then arise due to the speech ability of the left hemisphere and its concomitant higher educational level. The more active and intellectual appearance of the right–right composites would then not be caused by a hemispheric asymmetry for facial expression but as a side effect of speech lateralization. Although Hallervorden was primarily interested in situation-dependent changes of mimics as a psychodiagnostic tool, he always studied faces at rest. Thus, he could not reveal the more intense emotional expressions in left–left pictures that will be described in Chapter 9.

All of these studies clearly demonstrated that asymmetries of the face are not a sign of aberration or degeneration, as previously thought. However, are they, despite being normal, a phylogenetically primitive trait? To find an answer, Liebreich[81] traveled to various countries and measured the faces of people of diverse cultures. He started his voyage with the firm assumption that "inferior races" (all peoples who were not Western European) had more pronounced facial asymmetries. To his surprise, it was Europeans who had the most pronounced facial asymmetries. This forced him to come up with a new explanation. In his book, he surmised that facial left–right differences could simply result from the uneven pressure on the face of an unborn baby by their mother's pelvis. This asymmetry of compression occurs due to the normal upside-down embryonal position, with the head facing down toward the birth canal and the left arm and face positioned close to the vertebral column of the mother. Liebreich[81] assumed that this pressure was increased by the tight costumes European women wore, resulting in even more pronounced asymmetries of the face.

Some decades later, Wolff[82] suspected that the two halves of the human face represent two divergent personality traits with the left half being the "wish" image and the right one the "official" image. He noted that strangers, after seeing the original picture of a person, mainly classify the right–right composites as being more like the whole face (Fig. 1.12).

However, the photographed persons identify themselves more with the left-hand half of their face. In the discussion of his results, the author follows a psychodynamic approach, assuming that the left hemisphere controls language so that the right hemisphere could, by

FIGURE 1.12 Test material from the study by Wolff[82] (CC-PDM). Original portraits are depicted on the left. Right–right composites (from the right side of the face) are shown in the middle and left–left composites are shown in the right column. Strangers see the right–right pictures as more similar to the whole face, while the photographed persons identify themselves more with left–left composites.

default, be the seat of the unconscious. This is an interesting proposition. Although wrong in a strict sense, it anticipates many studies on the asymmetry of the self that were conducted much later and that will be reviewed in Chapter 8.

The studies reviewed above were incited by discussions on the asymmetries of an ancient Greek statue. Indeed, already Winckelmann[83] had noted that statues of the Hellenistic era displayed subtle asymmetries, like the left testicle generally being larger than the right. About 200 years later, art historians cataloged a large number of facial asymmetries in statues from antiquity.[84] However, it was the discussion on facial asymmetries of the Venus of Milo that finally transgressed scientific inquiries from the field of art history into experimental asymmetry research in Germany. The last experimental study within this new area of science was published by Wolff.[82] This paper is written in English and contains no citations to previous work. Consequently, all subsequent English-speaking authors assumed that Wolff was the first to work on this topic. As a result, about half a century of research on facial asymmetries was forgotten until it was rediscovered toward the end of the 20th century.[85] Today, facial asymmetries are seen as reflections of the motor side of speech control by the left hemisphere (Chapter 6) and right-hemispheric processes related to self-recognition and emotion (Chapters 8 and 9).

Summary

The first truly scientific studies on left–right differences of the human brain started in the 19th century and sparked a whirlwind of ever-faster conducted experiments that still persists. Many details of brain asymmetries that were subsequently discovered are outlined in this book. In this chapter, the very beginnings of brain asymmetry research were reviewed for the three main areas of study namely language, handedness, and facial asymmetries.

The discovery of language asymmetry in many ways constitutes a special case. The soap opera-like circumstances of the relevant events and the deep conceptual changes in the understanding of brain functions that followed from these discoveries represent the central historical narrative of modern lateralization research. However, the events surrounding the discovery of language asymmetry also laid the foundation for many basic assumptions that paved the way for how research on left–right differences was conducted in the following century. Since humans are the only species that have language, and since language is lateralized, cerebral asymmetries were seen as an advanced phylogenetic trait that only exists in humans. This incorrect assumption blocked animal lateralization studies until the late 20th century and resulted in the fact that, still today, only a few animal models for brain asymmetries exist (Chapter 2). The extremely lateralized control of speech was also the decisive discovery of split-brain studies that fostered the view of the "two minds in one skull" (Chapter 3). Language asymmetry was also central in the development of our current understanding that, firstly, functional brain asymmetries are mostly about degree and not of kind, and, secondly, that both hemispheres make differential contributions to a multifaceted ability such as language. This will be outlined in Chapter 5. Last but not least, language asymmetry also helped to discover the major role of left–right differences of inter- and intrahemispheric pathways that constitute the blueprints of functional brain asymmetries (Chapter 4).

Handedness is the most easily visible aspect of the asymmetry of the human brain. However, as outlined in this chapter, the seemingly obvious link between lateralization of hand and brain was only drawn much later on. As simple as handedness may appear, even today, its neural fundamentals are just beginning to emerge. There is one strange aspect of handedness that will be outlined in greater detail in Chapter 6. Scientists thought that research concerning the reasons for handedness was "finished" and subsequently lost interest in pursuing deeper inquiries. At the turn of the 19th to the 20th century, most scientists were convinced that the asymmetrical crossing of corticospinal fibers was the neural foundation of handedness. In the second half of the 20th century, psychologists were sure that a single gene was responsible for the handedness of the individual. Both theories turned out to be wrong. Presently, we are, to some extent, back to square one and are again searching for the very fundamentals of lateralized hand use. In addition, there is also a dark and mostly misunderstood side of handedness: Handedness covaries with all kinds of pathologies of the human brain. Why? We still do not have a full explanation but will review what is known in Chapter 13.

Facial asymmetry research is possibly the ideal topic to outline the intricacies surrounding the historical fates of scientific discoveries. When the statue of the Venus of Milo became the publicly adored female figure from antiquity to the late 19th century, a dismissive report about the many bodily asymmetries of the woman who was possibly the living model for the ancient artist was published. This publication sparked not only a heated debate but also a large number of studies on facial asymmetries and their possible link to left–right differences in brain functions in humans. All of these studies except for one were published in German. When German lost its status as one of the leading science languages after World War II, all of these discoveries were forgotten and collected dust in the basements of university libraries. They were rediscovered about 70 years later and now show that many of the early descriptions foreshadowed today's studies on asymmetries of self-perception (Chapter 8) and on the facial signals of emotions (Chapter 9).

Sample questions

- Who should be regarded as the discoverer of language asymmetry—father and son Dax or Pierre Paul Broca?
- What is the twofold implication of the scientific work of Broca for neuroscience?
- What were the initial theories about handedness before scientists discovered links between brain function and handedness?
- What is the advantage of left-handers in fencing, tennis, and person-to-person combat?
- Which neuroanatomical features were seen as promising candidates for the neural basis of handedness?
- How did research on facial asymmetries start?
- Were asymmetries of face and body initially seen as a normal part of nature or as a sign of abnormality?

References

1. Lokhorst GJ. The first theory about hemispheric specialization: fresh light on an old codex. *J Hist Med Allied Sci.* 1996;51(3):293–312. Available from: https://doi.org/10.1093/jhmas/51.3.293.
2. Wigan AL. Duality of the mind. *Lancet.* 1844;43(1078):154. Available from: https://doi.org/10.1016/S0140-6736(02)71566-3.
3. Puccetti R. Two brains, two minds? Wigan's theory of mental duality. *Br J Philos Sci.* 1989;40(2):137–144. Available from: https://doi.org/10.1093/bjps/40.2.137.
4. Muckli L, Naumer MJ, Singer W. Bilateral visual field maps in a patient with only one hemisphere. *PNAS.* 2009;106(31):13034–13039. Available from: https://doi.org/10.1073/pnas.0809688106.
5. Treviranus GR. Über die Verschiedenheiten der Gestalt und Lage der Hirnorgane in den verschiedenen Classen des Thierreichs. In: Treviranus GR, ed. *Untersuchungen über den Bau und die Functionen des Gehirns, der Nerven und der Sinneswerkzeuge in den verschiedenen Classen und Familien des Thierreichs.* Johann Georg Heyse; 1820:4–5.
6. Leuret F, Gratiolet LP. *Anatomie Comparée Du Système Nerveux: Considéré Dans Ses Rapports Avec L'intelligence.* J.-B. Baillière et fils; 1839.
7. Joynt RJ, Benton AL. The memoir of Marc Dax on aphasia. *Neurology.* 1964;14(9):851–854. Available from: https://doi.org/10.1212/WNL.14.9.851.
8. Gainotti G. Some historical notes orienting towards brain mechanisms that could underlie hemispheric asymmetries. *Cortex.* 2023;163:26–41. Available from: https://doi.org/10.1016/j.cortex.2023.03.001.
9. Harris LJ. Side biases for holding and carrying infants: reports from the past and possible lessons for today. *Laterality.* 2010;15(1–2):56–135. Available from: https://doi.org/10.1080/13576500802584371.

10. Leuret F, Gratiolet P. *Gratiolet, Pierre i. e. Louis Pierre.* Anatomie comparée du système nerveux: Considéré dans ses rapports avec l'intelligence. J.-B. Baillière et fils; 1857.
11. Chiron C, Jambaque I, Nabbout R, Lounes R, Syrota A, Dulac O. The right brain hemisphere is dominant in human infants. *Brain.* 1997;120(Pt 6):1057–1065. Available from: https://doi.org/10.1093/brain/120.6.1057.
12. Marshall LH, Magoun HW. *Discoveries in the Human Brain: Neuroscience Prehistory, Brain Structure, and Function.* Humana Press; 1998.
13. Gabi M, Neves K, Masseron C, et al. No relative expansion of the number of prefrontal neurons in primate and human evolution. *PNAS.* 2016;113(34):9617–9622. Available from: https://doi.org/10.1073/pnas.1610178113.
14. Gere C. Curating aphasia: Pierre Paul Broca's Museological Science. *Interdiscip Sci Rev.* 2013;38(3):200–209. Available from: https://doi.org/10.1179/0308018813Z.00000000047.
15. Gall FJ. *Philosophisch-Medicinische Untersuchungen Über Natur Und Kunst Im Kranken Und Gesunden Zustande Des Menschen.* 2nd ed. Baumgärtnersche Buchhandlung; 1800.
16. Gibson WC. Pioneers in localization of function in the brain. *JAMA.* 1962;180:944–951. Available from: https://pubmed.ncbi.nlm.nih.gov/13898444/.
17. Güntürkün O, Ströckens F, Ocklenburg S. Brain lateralization: a comparative perspective. *Physiol Rev.* 2020;100(3):1019–1063. Available from: https://doi.org/10.1152/physrev.00006.2019.
18. Flourens P. *Recherches Expérimentales Sur Les Propriétés Et Les Fonctions Du Système Nerveux.* Crevot: *Dans Les Animaux Vertébrés*; 1824.
19. Bouillaud J. Recherches cliniques propres à demontrer que le sens du langage articulé et le principe coordinateur des mouvements de la parole résident dans les lobules antérieurs du cerveau. *Bulletin de l'Académie Royale de Medecine.* 1848;13:699–719.
20. Critchley M. *Aphasiology and Other Aspects of Language.* Edward Arnold; 1970.
21. Finger S. *Minds Behind the Brain: A History of the Pioneers and Their Discoveries.* Oxford University Press; 2000.
22. Broca P. Perte de la parole, ramollissement chronique et destruction partielle du lobe antérieur gauche du cerveau. *Bull Soc Anthropol.* 1861;2:235–238.
23. Broca P. Remarques sur le siège de la faculté du langage articulé, suivies d'une observation d'aphémie (perte de la parole). *Bulletin et Memoires de la Societe anatomique de Paris.* 1861;330–357.
24. Dronkers NF, Plaisant O, Iba-Zizen MT, Cabanis EA. Paul Broca's historic cases: high resolution MR imaging of the brains of Leborgne and Lelong. *Brain.* 2007;130(Pt 5):1432–1441. Available from: https://doi.org/10.1093/brain/awm042.
25. Broca P. Nouvelle observation d'aphémie produite par une lésion de la moitié postérieure des deuxième et troisième circonvolutions frontales. *Bulletins de la Société anatomique de Paris.* 1861;6:398–407.
26. Gimènez-Roldán S. A critical review of Broca's contribution on aphasia: from priority to Leborgne the hatter. *Neurosci Hist.* 2017;5:58–68.
27. Broca P. Localisation des fonctions cérébrales: siége de langage articulé. *Bulletins de la Société d'Anthropologie de Paris.* 1863;4:200–208.
28. Broca P. Sur le siège de la faculté du langage articulé (15 juin). *Bulletins de la Société Anthropologque de Paris.* 1865;6:377–393.
29. Harris LJ. Cerebral control for speech in right-handers and left-handers: an analysis of the views of Paul Borca, his contemporaries, and his successors. *Brain Lang.* 1991;40(1):1–50. Available from: https://doi.org/10.1016/0093-934x(91)90115-h.
30. Roe D, Finger S. Gustave Dax and his fight for recognition: an overlooked chapter in the early history of cerebral dominance. *J Hist Neurosci.* 1996;5(3):228–240. Available from: https://doi.org/10.1080/09647049609525672.
31. Dax G. Lesions de la moitie gauche de l'encephale coincident avec l'ouble des signes de la pensee (lu au Congrés Méridional tenu à Montpellier en 1836). *Gazette hebdomadaire de médecine et de chirurgie.* 1865;2(2): 259–260.
32. Dax G. Notes sur le même sujet. *Gazette Hebdomaire de Medecine et de Chirugie.* 1865;2:262.
33. Caizergues R. Note pour servir à l'histoire de l'aphasie. *Mont Méd.* 1879;42:178–180.
34. Kushner HI. *On the Other Hand: Left Hand, Right Brain, Mental Illness, and History.* Johns Hopkins University Press; 2017.
35. Spenneman DR. Handedness data on the European neolithic. *Neuropsychologia.* 1984;22(5):613–615. Available from: https://doi.org/10.1016/0028-3932(84)90025-3.
36. Bargalló A, Mosquera M. Can hand laterality be identified through lithic technology. *Laterality.* 2014;19(1): 37–63. Available from: https://doi.org/10.1080/1357650X.2013.769559.

37. Fabbro F. Left and right in the Bible from a neuropsychological perspective. *Brain Cogn.* 1994;24(2):161–183. Available from: https://doi.org/10.1006/brcg.1994.1009.
38. Coren S, Porac C. Fifty centuries of right-handedness: the historical record. *Science.* 1977;198(4317):631–632. Available from: https://doi.org/10.1126/science.335510.
39. Uluç L. The Majalis al-'Ushshaq: Written in Herat, Copied in Shiraz, Read in İstanbul. In: Schick I.C., ed. M. *Uğur Derman 65 Yaş Armağanı/65th Birthday Festschrift.* Sabanci Universitesi, 2000:569–602.
40. Buchanan A. Mechanical theory of the predominance of the right hand over the left, or, more generally, of the limbs of the right side over those of the left side of the body. *Proc Phil Soc Glasgow.* 1862;5:142–167.
41. Pye-Smith PH. On left-handedness. *Guys Hosp Rep.* 1871;16:141–146.
42. Faurie C, Raymond M. Handedness, homicide and negative frequency-dependent selection. *Proc Biol Sci.* 2005;272(1558):25–28. Available from: https://doi.org/10.1098/rspb.2004.2926.
43. Faurie C, Raymond M. The fighting hypothesis as an evolutionary explanation for the handedness polymorphism in humans: where are we? *Ann NY Acad Sci.* 2013;1288(1):110–113. Available from: https://doi.org/10.1111/nyas.12159.
44. Özcan A. *Kânunnâme-I Âl-I Osman.* Kitabevi; 2003.
45. Hyrtl J. *Lehrbuch Der Anatomie Des Menschen Mit Rücksicht Auf Physiologische Begründung Und Praktische Anwendung.* 16th ed. Braumüller; 1882.
46. Ogle W. On dextral pre-eminence. *Med Chir Trans.* 1871;54:279–301. Available from: https://doi.org/10.1177/095952877105400114.
47. Hecht A, Langstein L. Zur Kenntniss der Rechts- und Linkshändigkeit. *Dtsch Med Wochenschr.* 1900;26(32):513. Available from: https://doi.org/10.1055/s-0029-1203960.
48. Gaupp E. *Über Die Rechtshändigkeit Des Menschen.* Gustav Fischer Verlag; 1909.
49. Rosolanka M, Rosolankova K. Asymmetrical blood-supply of upper extremities in the human-embryo as the determining factor of brain hemisphere dominance. *Act Nerv Super (Praha).* 1984;26(3):227–229.
50. Hasse C, Dehner D. Unsere Truppen in körperlicher beziehung. *Anatomische Abteilung: Archiv für Anatomie und Physiologie.* 1893;249–256.
51. Theile FW. Gewichtsbestimmungen zur Entwicklung des Muskelsys-tems und des Skelettes beim Menschen. *Verhandl. d. Kaiserl. Leopoldinisch Carolinischen deutsch. Akad. d. Naturforscher.* 1884;46:98–244.
52. Stier E. Untersuchungen Über Linkshändigkeit Und Die Funktionellen Differenzen Der Hirnhälften: Nebst Einem Anhang:. "Über Linkshändigkeit in Der Deutschen Armee". Gustav Fischer Verlag; 1911.
53. Ludwig W. *Das Rechts-Links-Problem im Tierreich und beim Menschen: Mit Einem Anhang Rechts-Links-Merkmale der Pflanzen.* Springer; 1932.
54. Means LW, Walters RE. Sex, handedness and asymmetry of hand and foot length. *Neuropsychologia.* 1982;20(6):715–719. Available from: https://doi.org/10.1016/0028-3932(82)90073-2.
55. Cunningham DJ. Right-handedness and left-brainedness. *J Anthropol Inst G B Irel.* 1902;32:273. Available from: https://doi.org/10.2307/2842821.
56. Braune CW. Die Gewichtsverhältnisse der rechten zur linken Hirnhälfte beim Menschen. *Arch Anat Physiol.* 1891;15:253–270.
57. Broca P. Instructions craniologiques et craniométriques de la Société d'anthropologie de Paris. G. Masson; 1875.
58. Weil A. Measurements of cerebral and cerebellar surfaces. Comparative studies of the surfaces of endocranial casts of man, prehistoric men, and anthropoid apes. *Am J Phys Anthropol.* 1929;13(1):69–90. Available from: https://doi.org/10.1002/ajpa.1330130125.
59. Hoadley MF, Pearson K. On measurement of the internal diameters of the skull in relation to the "pre-eminence" of the left hemisphere. *Biometrika.* 1929;21:85–123. Available from: https://doi.org/10.1093/biomet/21.1-4.85.
60. Weigert C. Ausführliche Beschreibung der in No. 2 dieser Zeitschrift erwähnten neuen Färbungsmethode für das Centralnervensystem. *Fortschritte Med.* 1884;2:190–194.
61. Flechsig PE. *Die Leitungsbahnen im Gehirn und Rückenmark des Menschen, auf Grund entwickelungsgeschichtlicher Untersuchungen.* Engelmann; 1876.
62. Brodal A. *Neurological Anatomy in Relation to Oligodendrogliomas. A review of two hundred cases.* Clinical Medicine. *Oxford: Clarendon Press;* 1948.
63. Yakovlev PI, Rakic P. Patterns of decussation of bulbar pyramids and distribution of pyramidal tracts on two sides of the spinal cord. *Trans Am Neurol Assoc.* 1966;91:366–367.

64. Kertesz A, Geschwind N. Patterns of pyramidal decussation and their relationship to handedness. *Arch Neurol.* 1971;24(4):326–332. Available from: https://doi.org/10.1001/archneur.1971.00480340058006.
65. Yakovlev PI. A proposed definition of the limbic system. In: Hockmann CH, ed. *Limbic System Mechanisms and Autonomic Function.* Charles C. Thomas; 1972:241–283.
66. Bear D, Schiff D, Saver J, Greenberg M, Freeman R. Quantitative analysis of cerebral asymmetries. Fronto-occipital correlation, sexual dimorphism and association with handedness. *Arch Neurol.* 1986;43(6):598–603. Available from: https://doi.org/10.1001/archneur.1986.00520060060019.
67. Chui HC, Damasio AR. Human cerebral asymmetries evaluated by computed tomography. *J Neurol Neurosurg Psychiatry.* 1980;43(10):873–878. Available from: https://doi.org/10.1136/jnnp.43.10.873.
68. Le May M, Kido DK. Asymmetries of the cerebral hemispheres on computed tomograms. *J Comput Assist Tomogr.* 1978;2(4):471–476. Available from: https://doi.org/10.1097/00004728-197809000-00018.
69. Koff E, Naeser MA, Pieniadz JM, Foundas AL, Levine HL. Computed tomographic scan hemispheric asymmetries in right- and left-handed male and female subjects. *Arch Neurol.* 1986;43(5):487–491. Available from: https://doi.org/10.1001/archneur.1986.00520050059023.
70. Pieniadz JM, Naeser MA, Koff E, Levine HL. CT scan cerebral hemispheric asymmetry measurements in stroke cases with global aphasia: atypical asymmetries associated with improved recovery. *Cortex.* 1983;19(3):371–391. Available from: https://doi.org/10.1016/s0010-9452(83)80007-0.
71. Goeler von Ravensburg FF. *Die Venus von Milo: Eine kunstgeschichtliche Monographie.* C.F. Winter'sche buchdr; 1879.
72. Nordau M. *Entartung. Erster Band.* Carl Dunder; 1892.
73. Galton F. Composite portraits. *Nature.* 1878;18(447):97–100. Available from: https://doi.org/10.1038/018097a0.
74. Langlois JH, Roggman LA. Attractive faces are only average. *Psychol Sci.* 1990;1(2):115–121. Available from: https://doi.org/10.1111/j.1467-9280.1990.tb00079.x.
75. Henke W. Glossen zur Venus von Melos. *Zeitschrift für bildende Kunst.* 1886;21:194–199.
76. Hasse C. *Die Venus Von Milo: Eine Untersuchung Auf Dem Gebiete Der Plastik Und Ein Versuch Zur Wiederherstellung Der Statue; Mit 4 Lichtdruck-Und 4 Lithographischen Tafeln.* Gustav Fischer Verlag; 1882.
77. Hasse C. Über Gesichtsasymmetrien. *Archiv für Anatomie und Entwickelungsgeschichte.* Veit & Company; 1887: 119–125.
78. Hasse C, Zakrzewski V. Das menschliche Becken und die normalen Kindeslagen bei der Geburt. *Zeitschrift für Geburtshilfe und Gynäkologie.* 1888;19:105–119.
79. Hallervorden J. Eine neue Methode experimenteller Physiognomik. *Psychiatrisch-Neurologische Wochenschrift.* 1902;4:309–311.
80. Hallervorden J. Rechts-und Linkshändigkeit und Gesichtausdruck. *Zentralblatt fur die gesamte Neurologie und Psychiatrie.* 1929;53:560.
81. Liebreich R. *Die Asymmetrie Des Gesichtes Und Ihre Entstehung.* JF Bergmann; 1908.
82. Wolff W. The experimental study of forms of expression. *Character Personality.* 1933;2(2):168–176.
83. Winckelmann JJ. *Geschichte der Kunst des Alterthums.* In der Waltherischen Hof–Buchhandlung; 1764.
84. Schneider L. *Asymmetrie Griechischer Köpfe.* Steiner; 1973.
85. Güntürkün O. The Venus of Milo and the dawn of facial asymmetry research. *Brain Cogn.* 1991;16(2):147–150. Available from: https://doi.org/10.1016/0278-2626(91)90003-q.

CHAPTER

2

Evolution of asymmetries

OUTLINE

Introductory short story

The Cambrian Sea, about 500 million years ago: A few tiny animals slowly crawl above the seafloor. Beside a rock, a worm-like creature walks on long, stilt-like legs and displays long needles on its back. A few meters away, a slug-like creature feeds on tiny morsels of dead animals. It wears long and upward-pointing curved spikes on its side. Further away are about a dozen trilobites with their flattened body structures and broad, helmet-like heads. Trilobites wear spines as well, but they also protect themselves with a two-layered cuticle that covers and shields their entire body. Why are all these animals so heavily armored? The reason becomes visible after a while: *Anomalocaris*, the apex predator of its time, appears on the scene. With its 1-meter-long body, it is about 10–20 times larger than most of its prey and possibly represents the largest animal of its time. But *Anomalocaris* is

The Lateralized Brain
DOI: https://doi.org/10.1016/B978-0-323-99737-9.00014-8

not only big, but it is also exceptionally maneuverable by using its many pairs of overlapping flaps. They enable it to dart through the ocean, take sharp curves, and even easily swim backward. In addition, it has a pair of movable eyes with outstandingly detailed vision. However, the most terrifying are its two inward-curved grasping limbs that are composed of many movable segments with spines pointing inward. Small animals are grasped with these limbs from the seabed and pulled into the round mouth. This mouth displays a circular array of sharp edges that can be opened and closed like the iris of a camera. *Anomalocaris* is hungry and has smelled a hormone that is used by trilobites to induce molting. During this time, the cuticle mantle of trilobites splits open and is then thrown away such that the animal can grow into its new, larger cuticle that is slowly hardening. This is the most vulnerable period for trilobites, and *Anomalocaris* is keen to exploit this. The predator is now hovering above the trilobites, picks a victim, quickly pulls it into its mouth, and tries to bite into the cuticle. However, *Anomalocaris* has a problem: Its teeth are made of a material that resembles human fingernails and are thus too soft to cut through the double cuticle armor of trilobites. However, *Anomalocaris* has one more card up its sleeves. While biting the right posterior end of the trilobite, its right limb firmly holds it at the left frontal side and quickly moves the little animal upward and downward. These movements are repeated until the cuticle cracks such that *Anomalocaris* can suck out its victim. However, today the predator is not lucky at first. The little trilobite manages to escape and falls to the ground. Quickly *Anomalocaris* darts forward again but this time picks another individual from the herd. The heavily wounded, but lucky, individual escapes.

Half a billion years later, Loren Babcock and Richard Robison, two paleontologists from the University of Kansas, United States, stare through their binoculars at a little fossilized trilobite and conclude that this animal had survived near-death trauma during its lifetime. It wears a major scar on its right posterior side, but this area is clearly callused and so reveals that it has gone through a period of healing. They carefully place the fossil into a box full of further trilobites that show scars that can be attributed to predation. They collected 81 trilobites that survived attacks. From the collection, 69% have bite marks on their right sides. The two scientists realize they are on their way to make a major discovery. *Anomalocaris* must have been right-limbed. Babcock and Robison[1] have discovered that behavioral asymmetries already existed in Cambrian times, 500 million years before the dawn of humans.

The Cambrian explosion and the dawn of behavioral asymmetries

Life emerged on our planet more than 3 billion years ago. The organisms of the first billion years were single-celled. These life forms relied on anaerobic energy pathways since oxygen levels were mostly below 1%. Even today some bacteria live an anaerobic and therefore nonoxygen-dependent life. With the slow increase of oxygen levels to about 3%, true animals emerged and the Ediacaran Period (635–541 million years before the present) of our planet began. These organisms were often soft-bodied and slowly fed on the microbe mats of the seabed. One of them was *Dickinsonia*, an animal with an oval, flat body shape with rib-like segments emerging from a central groove. *Dickinsonia's* left and right halves often did not perfectly mirror each other.[2] Thus, bodily asymmetries existed already at the very beginning of animal life, although their function is presently unknown.

With the beginning of the Cambrian Period (541–485 million years ago), oxygen levels started to reach up to 10% and provided plenty of possibilities for animals to grow larger and to be more mobile.[3] As a consequence, the Ediacaran fauna made place for new animals that were armored with hard, mineralized exoskeletons.[4] Since these exoskeletons are costly to produce, they signal a dire need for defense.[5] Indeed, some animals had stopped feeding on microbes and had turned to a richer diet—other animals. The occurrence of true predators with complex body forms and lifestyles signals the onset of the Cambrian explosion that produced an unprecedented burst of diverse lifeforms, some of which constitute the ancestors of today's animals. However, the Cambrian explosion also signaled the first evidence for behavioral asymmetries.

As outlined in the introductory short story, Babcock and Robison[1] were the first to discover the asymmetry of trilobite scars that can be attributed to predation. By only concentrating on bite marks that showed signs of healing, they could exclude that their results were biased by fractures that could have happened after fossilization. In total, they found 56 individuals with bite marks on their right (69%), 22 (27%) on their left, and three (4%) with bites on both sides (see Fig. 2.1).

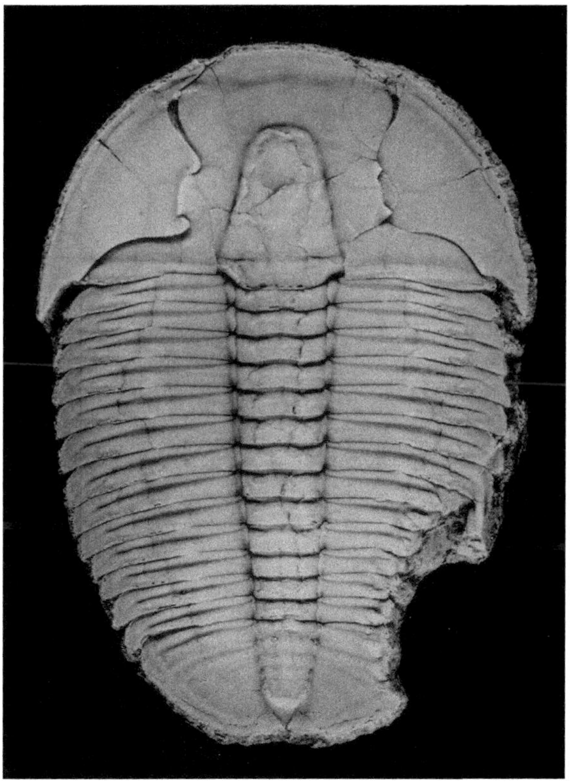

FIGURE 2.1 Middle Cambrian trilobite *Elrathia kingii*[6] from Utah showing a right-sided bite mark with a regenerated pleural tip (KUMIP 204773). Collection of the University of Kansas Museum of Invertebrate Paleontology. *Source: Reprinted with permission from the author and the publisher from Babcock L, Robison R. Preferences of Palaeozoic predators.* Nature. *1989;337:695–696.*

When they published their study, paleontologists had just discovered that a hitherto unknown giant predator called *Anomalocaris* had roamed the Cambrian oceans and had possibly also preyed on trilobites.[7] So, it was likely that the wounds on the flanks of the trilobites were inflicted, at least to some extent, by this animal. Although later studies make it likely that trilobites also cannibalized smaller individuals,[8] many scars recorded by Babcock and Robison[1] were possibly inflicted by a large predator like *Anomalocaris*.

Anomalocaris were soft-bodied animals with a length of about 1 m. They had a segmented body that was lined with a series of wide swim flaps which allowed high maneuverability. At its head were two movable compound eyes composed of more than 16,000 hexagonally packed lenses that strongly resembled the eyes of today's arthropods.[9] With this optical system, *Anomalocaris* could compete with current top visually capable insects. Prey was captured with a pair of spiny frontal limbs. These limbs and the circular plated mouthparts were about as hard as human fingernails.[10] Since the mouth could not be completely occluded[7] and the teeth might have been too soft and pliable to break the hard exoskeleton of a trilobite, *Anomalocaris* might have especially looked out for molting prey to suck it out.[11] In addition, it needed a behavioral strategy to crack the cuticle of trilobites. Nedin[12] observed a minor injury on the left frontal and a large biting wound on the right posterior part of the body in a trilobite specimen. A reconstruction of the events that might have produced these injuries revealed that *Anomalocaris* possibly lodged the right posterior part of the trilobite in its oral cone and used its right limb to hold the left frontal end of the prey to repeatedly flex the prey up and down until the cuticle cracked in the vicinity of *Anomalocaris'* mouth. The left limb of the predator could thereby simply keep the trilobite in position. If this model is correct, *Anomalocaris* would be the first animal on record that was "handed" and used a lateralized motor strategy (Fig. 2.2).

FIGURE 2.2 Depiction of a hunting scene in the Cambrian Sea. *Anomalocaris* just lodged a trilobite with the help of its frontal limbs in its oral cone. Since the teeth of the predator are too soft to crack open the cuticle of the trilobite, *Anomalocaris* uses its right limb to hold the left frontal end of the prey and repeatedly flex it up and down until the cuticle cracks open.

Reconstructing the evolution of handedness in vertebrates and beyond

The discovery of asymmetrical predation scars on trilobites and the reconstruction of the possible handedness of *Anomalocaris* is a rare and lucky incident. In most cases, the ancestors of today's animals are gone without leaving a trace of their behavioral asymmetries. A nice exception is constituted by the findings of Reisz et al. (2020) on the fossilized teeth of *Captorhinus aguti*,[13] a reptile from the Permian era that lived c. 289 million years ago and mostly fed on plants and little animals (Fig. 2.3). *Captorhinus* had multiple tooth rows on both upper and lower jaws. Novel teeth were constantly added on the inside (lingual) parts of the jaws, while old and worn teeth moved toward the outside (labial) parts. Such a pattern is obviously ideal to see which side has more worn teeth as a signature for an asymmetrical chewing pattern. Inspecting close to 100 jaws, the scientists discovered that *Captorhinus* preferred to feed on the right side of its jaw. Since in humans, chewing asymmetry correlates with hand, foot, ear, and eye side preferences,[14] the discovery of lateralized feeding behavior in a reptile that lived close to 300 million years ago provides evidence for a deep history of behavioral asymmetries in early vertebrate life.

An alternative to investigating fossil evidence is to observe handedness or limb preferences in diverse animal species to discover a pattern that resembles human behavior. Older studies that used such an approach are more anecdotal and can hardly be used for a contemporary analysis. For example, Ogle[15] observed monkeys in the London Zoo and found that 20 out of 23 monkeys consistently used their right hand to reach for food, while the remaining three preferred the left hand. Ogle concluded that monkeys have handedness that resembles that of humans. Unfortunately, he does not outline which species he observed and does not provide any quantitative information on the number of tests he conducted with each individual. A further and rather colorful report of this time is the observation of Thomas Dwight on footedness in parrots: "I had satisfied myself that every parrot had a favorite side."[16] Unfortunately, he went on saying: "... I have forgotten which side was used the most."[16]

Publications of the 20th and 21st centuries are usually much more detailed and can be used as a rich data source.[17] Up until recently, all reports were confined to vertebrates. However, the past 20 years have also seen a flurry of interesting publications on asymmetries in invertebrates.[18] We will outline some of these findings toward the end of this chapter. First, we will concentrate on publications reporting limb asymmetries in vertebrates only. We should ask ourselves how to best integrate the wealth of information that these many reports provide. One way is to collect these papers and reconstruct an evolutionary

FIGURE 2.3 Reconstruction of Captorhinus aguti, a Permian reptile from c. 289 million years ago that preferred to chew on its right-sided teeth. *Source: Picture produced by Nobu Tamura and published under a Creative Commons Attribution 3.0 unported license. Picture reproduced from the Wikimedia Commons (https://commons.wikimedia.org/wiki/File:Captorhinus_BW.jpg).*

pedigree from these observations. The most objective way to do this was scribbled in a simple notebook by the young German entomologist Willi Hennig during his time as a soldier and later as a prisoner of war during World War II. In 1950, he published his findings in German (*Grundzüge einer Theorie der phylogenetischen Systematik*), which became increasingly influential among entomologists. After its translation into English,[19] it started to be the most accepted procedure to systematize species and reconstruct evolutionary events in all areas of biology and beyond. Hennig's approach is called "Cladistics." A clade is usually defined as a group of organisms that have a common ancestor and that sit on a single branch of the tree of life. So, a cladistic analysis categorizes organisms based on derived traits that they share. Derived traits represent an innovation that can be traced back to the most recent common ancestor of the group. Importantly, the trait under analysis should not be present in more distant ancestors since it is then shared with many species of other clades. Thus, members that belong to one clade are believed to be closely related and to share a common innovation. Because the preponderance of one hand, paw, or foot over the other while reaching or scratching is possibly the most easily observed aspect of laterality, Ströckens et al.[20] ran a cladistic analysis on the many publications on limb asymmetries in vertebrates that were available. At this point, it is important to note that many studies on behavioral asymmetries in, for example, fish are available,[21] but that the number of publications on lateralized limb use is considerably smaller. Let us now talk about these few findings.

Fig. 2.4 shows some of the many systematic observations that have been published and could be used for the analysis. Clockwise, these examples show a pigeon that uses one of its feet to remove a piece of adhesive strip from its beak, a kea from New Zealand that handles food with its left foot, a red-necked wallaby that uses its left front paw to hold food, and a chimp that pulls food out from a narrow tube with its right hand using a stick.

After analyzing articles on 119 different vertebrate species, it became evident to Ströckens et al.[20] that 81 species (68%) had some form of handedness. A small number of these studies were conducted on fish, amphibia, and reptiles. For example, the blue gourami (*Trichogaster trichopterus*), a sweet-water fish from southeast Asia that is popular among aquarists, prefers to use its left ventral fin to explore novel artificial objects.[23] Robins and Rogers[24] discovered that green tree frogs (*Litoria caerulea*) showed a significant preference for using the right forelimb as the leading limb when climbing. Leatherback turtles (*Dermochelys coriacea*) mostly use their right hind paw to cover the egg chamber after laying eggs.[25]

Most studies on limb preferences come from observations of birds and mammals. For example, McGavin[26] observed 24 wild North Island kākās (*Nestor meridionalis septentrionalis*) eating food pellets from a feeder in a bird sanctuary. Like other *Psittaciformes*, kākās hold their food with one foot during feeding. Some individuals strongly preferred one foot over the other, but the numbers of left- and right-footed birds were about equal. Most reports came from observation of mammalian species like the one by Giljov et al.,[27] who quantified limb preferences in the red-necked wallaby (*Macropus rufogriseus*), a marsupial. These animals often take a bipedal position and hold branches or leaves in one of their front paws to feed on them. Of the 27 adult members of a group of wallabies, 74% showed a left-hand preference, resulting in significant population level left-handedness. After having briefly provided these examples from the literature, let us summarize a few key findings from the cladistic analysis by Ströckens et al.[20] First, 68% of all analyzed species showed

FIGURE 2.4 Examples of functional limb preferences in nonhuman animals. (A) A pigeon uses its left foot to remove a piece of adhesive strip from its beak. (B) A kea holds a piece of apple with its left claw. (C) A red-necked wallaby feeding from a bipedal position using its left paw. (D) A chimpanzee pulling out food from a tube using its right hand.[22] *Source: Part (C) is a reproduced with permission from the author and the publisher from Giljov A, Karenina K, Ingram J, Malashichev Y. Parallel emergence of true handedness in the evolution of marsupials and placentals.* Curr Biol. *2015;25(14):1878–1884; part (D) Courtesy of William Hopkins.*

some form of limb preference. Thus, humans are not the only species with biased hand preferences. It is conceivable that a publication bias exists (it is more likely to publish a report on limb preferences in a species when a significant asymmetry is detected). However, it remains a fact that many dozen species from all major vertebrate classes show some form of limb preference. Second, Ströckens et al.[20] could essentially distinguish three forms of patterns that are schematically depicted in Fig. 2.5:

1. *Absence of asymmetry* (all members of the population prefer to use both the left and the right limb with equal probability).
2. *Individual-level asymmetry* (some individuals only use the left limb, while others only prefer the right, but there is no asymmetry at the population level).
3. *Population-level asymmetry* (the majority of the population prefers one side).

Third, although the database is presently too sketchy, it is evident that there is nothing like a "common descent" for handedness. Limb preferences seem to have evolved in parallel for many different branches of vertebrates.

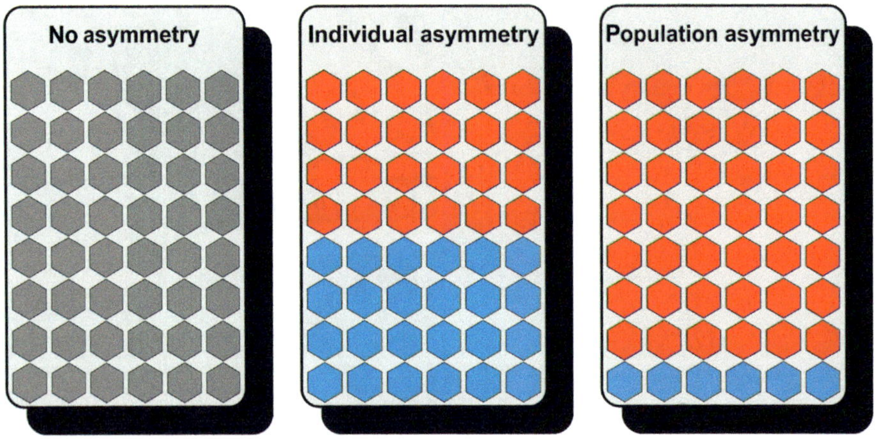

FIGURE 2.5 Three possible forms of limb preferences. Gray hexagons indicate individual animals without limb preference, red hexagons indicate left-limb preference, and blue hexagons right-limb preference. In the individual asymmetry pattern, half of the animals consistently prefer the left limb, while the other half consistently prefers the right limb. When population asymmetry exists, the majority of animals prefer the left limb.

Recently, this comparative approach was widened to encompass Bilateria—a very large group of animals that show at least as an embryo bilateral symmetry and share with humans the three canonical body axes (up-down, front-back, left-right). As outlined by Güntürkün et al. (2020), behavioral and bodily asymmetries are much more widespread than previously assumed.[28] Even tiny invertebrates like *Caenorhabditis elegans*[29] have it, but also slugs,[30] cuttlefish,[31] and various insect species.[18] Fig. 2.6 shows a cladogram-based depiction of such lateralized species of major taxa of the animal phylum. The implication of this finding is clear: Asymmetries of body, brain, and behavior are not a recent evolutionary feature, but about as old as the evolution of animals.

Handedness in primates

Compared to limb preferences in nonhuman animals,[20] motor asymmetries for hands and feet in humans are extremely skewed. They reach about 90% for using the right hand (see Chapter 6) and about 75%–88% for using the right foot.[32–34] Such an extreme population bias is unique among primates. Consequently, Ströckens et al. concluded that their "... findings ... support the position that population level asymmetries in limb preferences per se are a common feature among vertebrates, while the strong and consistent rightward population-level asymmetry observed for human handedness is not."[20] So, what is special about human handedness? Let us first look at primates in general before concentrating on the evolution of human handedness.

Humans are primates, and this order arose from ancestors that lived in the tropical forests of America, Africa, and Asia. Most primates are therefore arboreal, have excellent vision, and have large brains. In addition, they are very apt to use their hands for all kinds of tasks. In 1987, MacNeilage et al.[35] published a target article in which they assumed that

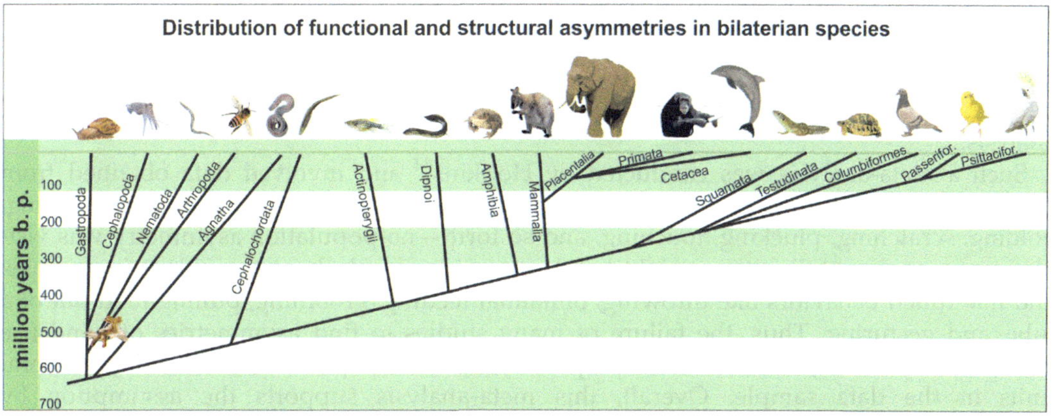

FIGURE 2.6 Cladistic depiction of asymmetries in bilateria. From left to right: Lateralized mating in pond snail (*Lymnaea stagnalis*), visual asymmetries in cuttlefish (*Sepia officinalis*), asymmetry in the nervous system and lateralized behavior in the nematode (*Caenorhabditis elegans*), asymmetry of olfaction in honey bees (*Apis mellifera*), handed curling in hagfish (*Eptatretus stoutii*), mouth asymmetry in lancelets (*Branchiostoma lanceolatum*), lateralized predator avoidance in Australian lungfish (*Neoceratodus forsteri*), asymmetrical organization of flight/fight-responses in zebrafish (*Danio rerio*), right-pawedness in toads (*Bufo bufo*), handedness in red-necked wallabies (*Notamacropus rufogriseus*), forefoot preference in Asian elephants (*Elephas maximus*), handedness in the food-tube task in chimpanzees (*Pan troglodytes*), visual asymmetry in dolphins (*Tursiops truncatus*), lateralized eye preference for monitoring and escape behavior in the common wall lizard (*Podarcis muralis*), asymmetrical visual responses to mirrors in tortoises (*Testudo hermanni*), lateralized visually guided foraging in pigeons (*Columba livia*), asymmetrical song control in canaries (*Serinus canaria*), and left-footedness in sulfur-crested cockatoos (*Cacatua galerita*). Anomalocaris as possibly the oldest evidence for behavioral asymmetry in animals is depicted on the line to modern arthropods, c. 500 million years before the present time. From Güntürkün et al., 2020, see this publication for further details.[28] *Source: Reproduced with permission from the authors from Güntürkün O, Ströckens F, Ocklenburg S. Brain lateralization: a comparative perspective.* Physiol Rev. *2020;100:1019–1063.*

the early condition in the primate lineage consisted of a left-hand specialization for simple visually guided movements like picking up an item and a right-hand dominance for fine manipulations of these items. They further assumed that this left-sided reaching preference disappeared in humans, while the right hand became dominant for all tasks for which only one hand was used. A few years later, Fagot and Vauclair[36] published a different interpretation of the distinction between simple manual behaviors (like picking something up) and more complex ones (like adjusting an object into a frame). They argued that simple behaviors displayed individual-level rather than population-level asymmetry, while complex manipulations go along with right-handedness at the population level. Now let us see if these theories reflect the result patterns on handedness in nonhuman primates.

In great apes (orangutans, gorillas, chimpanzees, and bonobos), only individual-level, but no population-level asymmetries were observed when tested in simple reaching tasks (chimps[37]; gorillas[38]). However, when primates (capuchin monkey[39]; chimpanzees[40]; olive baboons[41]) extract food from a narrow tube or gesture toward other monkeys (olive baboons[42]), a strong right-hand preference was recorded. This pattern overlaps with observations in humans, who are also more right-handed for fine-motor tasks and gesturing.[43] These reports indeed make a distinction between simple tasks (not a population-level asymmetry, but an individual-level asymmetry) and complex manipulations (right-hand population-level asymmetry) likely. Unfortunately, all of these studies are plagued by having small sample sizes. The solution to the last problem is a meta-analysis. This entails the integration of many published reports so that the samples can be combined in an overall analysis.

Such a meta-analysis was conducted by Hopkins[44] and involved data obtained from 1524 great apes. When only looking at simple behavioral patterns—for example, hitting, holding, scratching, plucking, touching, and so forth—no population asymmetry was visible in any species.[44] Significant right-hand asymmetries only became evident in complex and fine-tuned behaviors like throwing, bimanual feeding, grooming, pulling food out of a tube, and gesturing. Thus, the failure of many studies to find asymmetries of hand use was probably mostly due to an overrepresentation of simple nonlateralized behavioral units in the data sample. Overall, this meta-analysis supports the assumption by MacNeilage et al.[35] and Fagot and Vauclair[36] that right-handedness evolved in ancestral primates for precise movements only. It does not support the assumption by MacNeilage et al.[35] that the left hand was mostly used for simple reaching movements.

In a subsequent analysis, Regaiolli et al.[45] scrutinized the distinction between simple (e.g., reaching for food) and complex tasks (e.g., manipulating an object) further by researching whether monkeys and great apes show identical patterns. To this end, they observed 58 individuals of various monkey ($n = 32$) and ape species ($n = 26$) living in different groups. Regarding simple tasks, no population asymmetry was found in either of the two groups. For complex tasks, a group-level right-hand preference was found in great apes only. However, are we sure that population-level handedness results from the need for fine manipulation? In principle, it is also conceivable that this is only related to the fact that great apes are genetically closer to humans. Is it possible to decide between these alternatives? The answer is "yes." However, for this, we have to travel to Australia.

Giljov et al.[22] studied manual lateralization in kangaroos and wallabies. These animals are marsupials and are thus very distinct from placental mammals to which primates belong. The evolutionary trajectories of marsupials and placental mammals parted about 125 million years ago, long before the corpus callosum evolved. This massive fiber tract connects the hemispheres and plays a key role for brain asymmetries (see Chapter 3). Despite being so different from primates, several kangaroo and wallaby species display strong left-forelimb preferences at the population level across multiple kinds of behaviors during feeding.[22] So, like great apes and humans, kangaroos and wallabies have a strong population-level handedness, and they show this across different kinds of activities. A second, even more interesting conclusion becomes obvious when looking at the details of handedness in different marsupial species. Phylogenetic relatedness cannot explain their handedness since members of very different families show similar levels of handedness, whereas within the same family, different kinds of handedness coexist. Basically, what explained the presence or absence of strong handedness was what these animals do with their hands. Some marsupials are bipedal and can freely use their hands for manipulation. These species are capable of some sophisticated manipulations with their hands and display strong handedness with population asymmetries of about 3:1 (left/right). Contrary to this, species with weak limb preferences were mainly quadrupeds—that is, walk on four legs. Their left-right preference score at the population level was close to 1:1. Thus, strong population-level handedness seems to emerge every time an animal develops the need for fine manipulation and has the physical capability to do this. Not the degree of relatedness to humans but this simple functional selection pressure explains it.[46] This makes it clear why our population-level handedness reaches about 1:9 (left/right). The dexterity of human hands—*dexter* in Latin means "right"—is unique among all primates. We write, draw, and manipulate all kinds of tiny instruments with a precision that no other primate can achieve. This outstanding ability is consistent with a strong right-handedness at the population level.

However, remember that in great apes, simple actions revealed no sign of population asymmetry (1:1), while complex manipulations showed a 1:2 left/right population asymmetry.[44] What about human handedness during simple actions like picking something up or carrying something? The unbelievable answer to this question is that we do not really know! For example, the globally used Edinburgh Handedness Inventory that up to now has been used on possibly several hundred thousand subjects[47] has only 10 items, and all of them ask for some sort of tool use. This is mostly equally the case for other, less popular questionnaires. Only three studies did with humans what scientists do with nonhuman animals: They just observed hand use in the humans' natural environment. The most well-known of these studies is the one by Marchant et al.,[48] who quantified ethnographic video footage of the daily lives of people in three traditional cultures—Bushmen in Kalahari, Himba in Namibia, and Yanomamo from the Orinoco. This study revealed that hand actions that are not related to tool use produced a right-handedness of about 54% in the three populations—so essentially no population asymmetry. Hand use during tool use, however, was associated with an average of 84% right-handedness in the three societies. Two subsequent similar studies of other societies (Baka in Gabon, several ethnic groups in Burkina Faso, Eipo in Indonesia, Inuit around the Arctic Sea[49]; and Hadza in Tanzania[50]) produced highly similar results (Fig. 2.7).

These findings suggest that the pattern of population-level lateralization in humans is not that radically different from that of the great apes. Our conception of an extreme degree of population-level right-handedness is a result of our concentration on fine manipulations that go along with tool use. We thereby completely neglect the virtual absence of our right-handedness in simple tasks. After finally reaching a conclusion regarding this, we should now have a look at the evolution of human-specific right-handedness.

The evolution of human handedness

Are there any remains left with which we can reconstruct the handedness of our direct ancestors? While pondering this question, Toth[51] made a remarkable discovery. He realized that prehistoric stone artifacts from the Lower Pleistocene era (1.4–1.9 million years ago) revealed a preferential clockwise rotation of stone cores during flaking. When reconstructing how these sequences of actions were possibly performed, he recognized that this pattern could reveal the presence of right-handed toolmakers who held the stone core in the left hand while placing the powerful, but very precise stroke with the right (Fig. 2.8A). If correct, already before the dawn of modern Homo sapiens, right-handedness must have been part of our behavioral repertoire. Indeed, the Eipo, who live in the eastern highlands

FIGURE 2.7 Examples of right-handedness during tool use or detailed manipulation of objects as observed in the Eipo (Papua New Guinea) in 1974. (A) Cutting a tuber; (B) treating a toddler with magic pig fat. *Source: Courtesy of Wulf Schiefenhövel.*

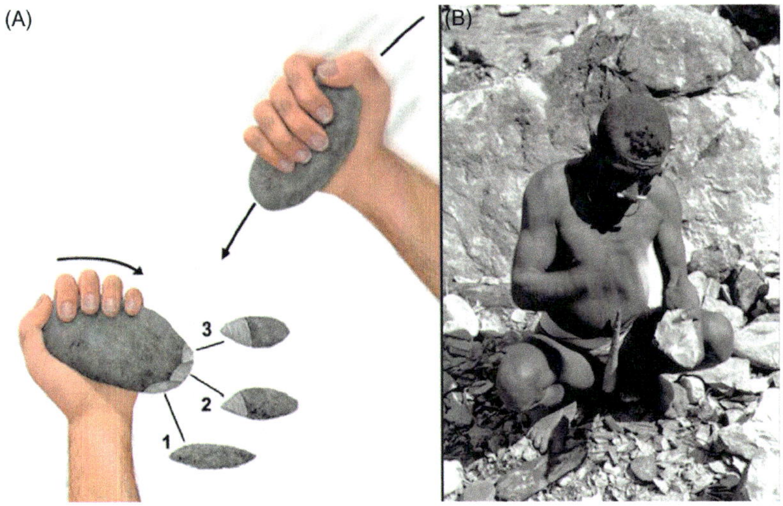

FIGURE 2.8 (A) Illustration of hand positions during flaking stones. In this example, the to-be-modified core stone is held in the left hand while the right hand strikes at precise positions with a hammer stone. A flake is thus detached. The core stone is rotated in a clockwise direction with the left hand. The flakes produced are thus right-oriented (according to Toth[51]). (B) An Eipo man flaking an adze with the identical technique in 1974. *Source: Courtesy of Wulf Schiefenhövel.*

of West Papua New Guinea, used to craft their adzes (a cutting tool) with this identical technique up to the 1990s (Fig. 2.8B).

Rugg and Mullane[52] put Toth's[51] assumption to the test and asked four left- and four right-handed knappers to produce stone tools as used in the Pleistocene era. A knapper is a person who is experienced in shaping stones through precise strokes that sequentially remove flakes until the final shape is achieved. Indeed, the handedness of the knappers could be inferred from the direction of the flakes that they produced.

Right-handedness did not only prevail during the Pleistocene era but also during the time of the much more recent Neanderthals. They produced sophisticated stone tools and reshaped them regularly when they were not sharp enough anymore. This reshaping produced characteristic flakes that can be used to reveal the handedness of the knapper. Indeed, Neanderthals were also mostly right-handed.[53] As beautiful as these findings about handedness during stone tool production are, they have one core problem: A single knapper produces many tools. So, if many cores and flakes from tool production are found at one site, it is impossible to know how many individuals produced them. Thus, we can safely conclude that our ancestors were mostly right-handed, but it is impossible to come up with a reliable handedness percentage.

This problem also exists for an equally beautiful source of information on ancient handedness: neolithic cave paintings of the upper Palaeolithic era—c.35,000–10,000 BCE. These sites often show "negative human hands" that are drawn by blowing pigments with a tube onto one hand that is pressed against the walls of a cave. Faurie and Raymond[49] demonstrated that 77% of these cave paintings display the left hand, possibly due to the right hand being engaged in the fine manipulative act of holding the tube (Fig. 2.9).

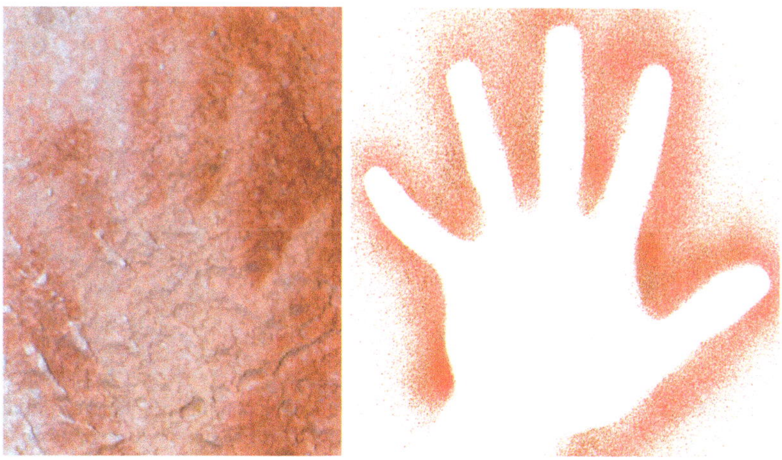

FIGURE 2.9 Left: Palaeolithic cave painting showing a left negative hand. Right: depiction of the left hand produced with a stencil used with the right hand from a right-handed modern person. *Source: Right photo reproduced with permission from the author and the publisher from Faurie C, Raymond M. Handedness frequency over more than ten thousand years.* Proc Biol Sci. *2004;271(suppl 3):S43–S45. Left photo by Heinrich Wendel (© The Wendel Collection, Neanderthal Museum, "Panneau des Femmes-Bisons" of the cave Peche-Merle).*

The authors gave French university students an identical task and obtained a percentage of 77.1% left-hand negative paintings. Thus, no difference was detected between the two proportions of left- and right-handers, separated by more than 10,000 years. This pattern is not only found in Europe but is present worldwide.[54] However, it is impossible to know whether each individual who visited the cave produced exactly one image of her/his hand—a very unlikely option—or whether a single individual produced all the images—which is equally unlikely. Thus, a predominance of right-handedness is clearly visible when looking at the worldwide pattern, but a percentage cannot be drawn.

This is different for a completely different kind of signature of prehistoric right-handedness: scratches on the front teeth of prehistoric skulls. Many scientists have described that the outer (labial) face of the front teeth shows strange and deep scratches that are not found on the side teeth and are also absent on the back (lingual) face of the front teeth. Most remarkably, many of these scratches show a distinct direction and run—from the perspective of the owner of the teeth—from the upper left to the lower right side[55] (Fig. 2.10A). It is likely that premodern humans held pieces of meat or other food between their teeth and pulled the outer part with one hand while cutting the meat with the other hand (Fig. 2.10B). If this was repeatedly done with a stone tool, they occasionally scratched the enamel of their teeth. When volunteers with mouthguards are asked to pull with the left hand and cut with the right, an identical pattern that runs from the upper left to the lower right is produced.[55] Thus, the owners of the prehistoric skulls were very likely right-handed during their lifetime.

The beauty of this signature of right-handedness is that it is clear that these premodern humans produced these scratches by themselves. We can therefore run an analysis on the percentage of right- and left-handed individuals based on the pattern of scratches on the front teeth of their skulls. According to Frayer et al.,[56] right-handedness in European premodern humans (*Homo antecessor*, *Homo heidelbergensis*, and *Homo neanderthalensis*) was 93.1%. This is virtually identical to the present proportion. A novel analysis could also

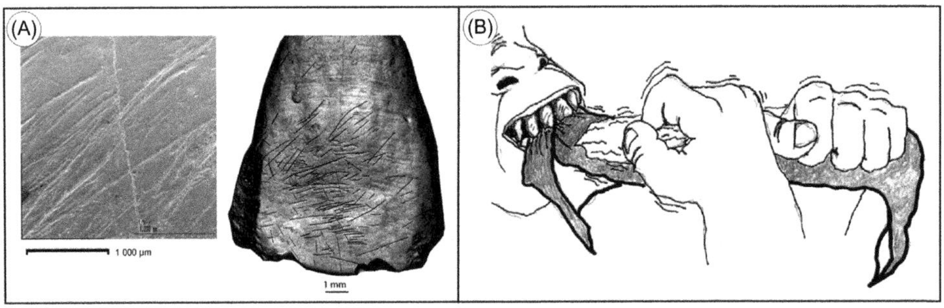

FIGURE 2.10 (A) Magnified image of labial scratches with a right oblique orientation from a hominid tooth along with a composite image of labial scratches on a different tooth. (B) Depiction of a right-hander pulling meat with the left hand and cutting with the right using a stone tool. Scratches in a characteristic oblique orientation on the front teeth can occur during this procedure. *Source: Part (A) is reprinted with permission of the author and the publisher from Frayer DW, Lozano M, Bermúdez de Castro JM, et al. More than 500,000 years of right-handedness in Europe.* Laterality. *2012;17(1):51–69. Part (B) is reprinted with permission of the author and the publisher from Frayer DW, Clarke RJ, Fiore I, et al. OH-65: the earliest evidence for right-handedness in the fossil record.* J Hum Evol. *2016;100:65–72. The drawing in part (B) was done by Jay Sayre.*

FIGURE 2.11 Woman from Cap Blanc who possibly was the left-handed extraordinary artist who carved and painted the neolithic frieze with the running horses and wisents. Created by Elisabeth Daynès who reconstructed this bust from the skull that was found in front of the frieze. Licensed via Science Photo Library.

demonstrate the identical scratch pattern on the front teeth of the skull of a 1.8 million-year-old representative of *Homo habilis*.[57]

A rare opportunity to come face to face with these early humans is given at Cap Blanc (France). This is a high rock shelter where 12,000–15,000 years ago, life-size animals like horses and wisents were beautifully carved from the soft limestone of the rock face and subsequently painted. In its time, this spectacular piece of glacial art was possibly visible from a far when standing in the valley. Close examination of the direction of tool blows makes it likely that the artist was left-handed.[58] A female human skeleton of 25–35 years was buried directly in front of the frieze. This position could imply that she was the artist. Indeed, an anthropological examination provided clues that she was left-handed. The reconstruction of her head by Elisabeth Daynés gives us the unprecedented chance to look into the face of this young woman and to realize that each of the skulls on which our analyses rest was once an individual human being (Fig. 2.11).

The evolution of limb preferences in vertebrates—a summary

Now, let's summarize our reconstruction of the evolution of limb preferences in four take-home messages.

- First, limb preferences are a widespread phenomenon in vertebrates. If a functional organization is so ubiquitous, it must have an evolutionary advantage. We will speculate about this point at the end of this chapter. However, presently, it is sufficient to keep in mind that limb preferences are widespread. The existence of hand and foot preferences in humans is therefore not a unique signature of humans.

- Second, limb preferences have no single common ancestor within vertebrates but appear in different, unrelated clades when the animals of these clades start to specialize in some form of action with their limbs that goes beyond basic locomotion. Thus, it is futile to attempt to identify the last common ancestor of limb preferences in vertebrates.
- Third, if limb preferences are present, they can be classified as individual or as population asymmetry. Population asymmetries can be skewed toward the left side as in marsupials and parrots or toward the right as in primates. It is possible that this is a chance event of evolution and that it does not matter whether a group of animals is biased toward one or the other side. An alternative view is that the side matters because other functionally lateralized neural systems need to connect to the hemisphere that dominates limb control. In humans, some scientists assume that right-handedness is causally linked to left-hemispheric language dominance (see Chapter 6). This debate is still open, but this link is increasingly questioned.
- Fourth, the strength of population asymmetry seems to be related to the manual complexity of the actions that the animals were able to produce. Parrots can produce extremely complex manipulations with their claws and consequently show population handedness scores of around 90%.[59] Monkeys are not very good at complex manipulations and show individual but not population asymmetry. Great apes have good manual dexterity and show a reasonable strength of population asymmetries for complex tasks. This is driven to an extreme in humans who have an outstanding ability for complex manual tasks. Consequently, we are highly skewed in our population asymmetry. Thus, the degree of population asymmetry depends on task complexity, while basic tasks produce no visible population asymmetries.

Reconstructing the evolution of vocalization asymmetries

In Chapter 1, we outlined the whirlwind of studies from the 19th century that resulted in the discovery of speech asymmetry in the human brain. Since only humans have language, a search for a similarly lateralized system in nonhuman animals seemed to be futile. At least that is what everyone thought. Well, nearly everyone. Otto Kalischer, a German neurologist in Berlin, hoped to find a similar phenomenon in parrots. He knew that parrots do not really speak in a human sense but rather repeat learned phrases. However, he was hoping that the motor side of their utterances could reveal a similar asymmetry as in humans. For 5 years, he painstakingly studied parrot brains to be able to identify the different substructures. Simultaneously, he taught many of his birds to learn phrases of different complexities. Finally, he placed lesions in different compartments of the left or the right hemisphere, hoping to find a place that could result in lateralized motor aphasia as known from patients with lesions of the Broca area. He was searching in vain. Left- and right-sided lesions had identical results. However, an area that almost resembled the position of the Broca area in the human brain seemed to be related to "speech" production in parrots. Thus, the general organization to produce learned vocalizations was like the human condition. However, a functional asymmetry was not discernable.[60]

For more than half a century, Kalischer's failed hopes were likely the final nail in the coffin of all attempts to reconstruct the evolution of language asymmetry by searching for similar findings in nonhuman animals. The situation suddenly changed with a publication by Nottebohm[61] on the song system in chaffinches. This was the first study that revealed functional asymmetries in the nervous system of a nonhuman animal.

Male songbirds use their song mainly for courtship and to mark their territory. Their song system consists of several forebrain areas that finally project to a brainstem nucleus that then projects via the hypoglossal nerve to the syrinx. The syrinx is the sound-producing organ in birds that has the same overall function as our larynx, although in many aspects, it is built quite differently. Nottebohm[61] cut the hypoglossal nerve that runs to the syrinx either on the left or on the right side. The difference was remarkable: Left sections resulted in some birds in the omission of around 90% of song elements. Other individuals were subsequently only able to produce a random burst of noise. In contrast, right-sided sections had very small or no effects. Thus, at least the final production pathway of the chaffinch song system was lateralized! The same surgery on very young birds that had not yet sung had no effects. Although the paper by Nottebohm[61] could only reveal an asymmetry in the final peripheral pathway of the song system, similarities to the lateralization of the human speech system were clearly visible. This resulted in a sudden resurgence of the search for nonhuman asymmetries of vocalization asymmetries that could meanwhile clarify many aspects of its evolutionary past.

Ocklenburg et al.[62] again used cladistics to systematize all the published data on vertebrate vocalization asymmetries to reconstruct their evolutionary pedigree. Let us now talk about their findings. In Chapter 5, we will outline in depth what we know about language asymmetry in humans. It will become clear that our language system is lateralized both at the level of perception as well as at the level of production. Although there are disputes about whether perception is less asymmetrically organized than production,[63] most authors agree that the whole human language system is fundamentally lateralized with more than 90% of the tested subjects showing a left-hemispheric language dominance.[64,65] We therefore outline below both asymmetries of perception and production of species-typical communicative vocalizations in nonhuman animals.

The results of the cladistic analysis by Ocklenburg et al.[62] are summarized in Fig. 2.12.

The numbers in the boxes show the number of species tested. First, no data exist on cartilaginous fishes and reptiles. The data on bony fish result from just two studies on one species each. In amphibians, there is only a single study that studied one frog species. These three studies discovered a left-hemispheric dominance to produce species-specific calls in these fish and amphibia species. Fourteen publications report vocalization asymmetry data in birds but unfortunately only in the order of songbirds. Five different species were studied, with most experiments conducted in zebra finches. These studies mostly observe a right-hemispheric dominance for processing heard and memorized songs in these finches. Some studies also discovered that each hemisphere has a differential contribution to song (see also Ref.[67] for a more recent study). There is no class-specific overall asymmetry pattern in birds but only species-specific asymmetries. Zebra finches are on average more prone to use the right hemisphere to process song, while, for example, canaries are more left-hemisphere-dominant.

For mammals, the data pattern is more abundant and widespread. Ocklenburg et al. (2013) list 12 different nonhuman mammalian species, among them six of the primate order. The analyzed species reach across five orders from mice to horses. Most data point

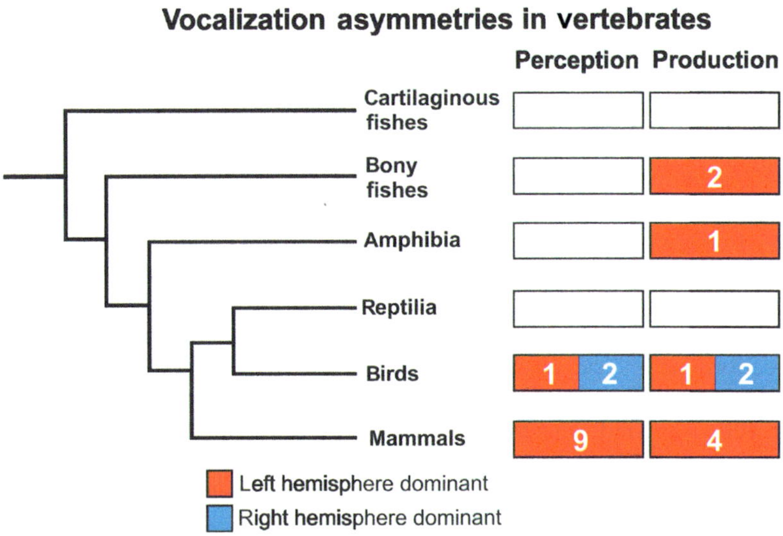

FIGURE 2.12 Cladographic overview of vocalization asymmetries in the vertebrate clade (based on Ref.[66]). The published studies are divided into those that analyzed the perception or the production side of vocalization asymmetries. The numbers depict the number of species studied. Red (left) and blue (right) colors depict the dominant hemisphere. All mammalian species were found to be left-hemispheric dominant for species-typical vocalization.

to a clear dominance of the left hemisphere for both the perception and the production of species-typical vocalizations. In some of these studies, a loudspeaker was used to broadcast a species-typical vocalization behind the back of the animal. The subsequent right- or left-turn of the animal toward the sound source is observed and interpreted as a left- or right-hemispheric perceptual dominance, respectively. For example, Böye et al. (2005) used this approach in California sea lions (*Zalophus californianus*) and found a right ear—left hemisphere—perceptual advantage for conspecific calls, while nonconspecific sounds did not elicit any lateralized reactions.[68] However, this procedure was later shown to be less reliable, even in detecting human language asymmetries,[69] while other approaches that used activation patterns of the brain or perceptual tasks were more consistent.

Since the cladistic analysis by Ocklenburg et al. (2013), many further experimental studies were published that consolidate the main conclusions. Only a few examples are given here to round up the picture. For example, dogs show hemispheric asymmetries for processing human speech and thus can analyze parameters of acoustic communication of a nonconspecific.[70,71] The mustached bat (*Pteronotus parnellii*) is an especially interesting case since it shows a differential specialization for temporal and spectral analyses in the left and the right auditory cortex, respectively.[72] At the same time, this perceptual asymmetry is more pronounced in males.[73] As discussed in Chapters 5 and 12, this pattern is also visible in humans, making it likely that deep and thus trans-species principles of acoustic communication could govern language asymmetries in humans. Such a deep principle could result from the fact that it is physically impossible to analyze the temporal (time) and the spectral (frequency) parameters of the acoustic input at the same time since analyzing each comes at the expense of the other.[72] As a result, all species that need to analyze each variable might have evolved asymmetrical specialization for time and frequency. Interestingly, brain asymmetries for auditory cues in both male and female mustached bats are organized according to this principle, but use different neural mechanisms.[74]

Unfortunately, these bat studies do not allow us to uncover the neural circuit level of asymmetries. However, this can be done in mice. When hearing their pups, female mice show a right ear advantage[75] and experience a higher expression of immediate early genes in the left auditory cortex.[76] Indeed, communication calls with first up and subsequent down sweeps of frequencies play a key role in mice and the left auditory cortex shows a selective preference for them.[77] However, how is this preference realized at the neural circuit level? Levy et al. (2019) could uncover this in a landmark study.[78] They demonstrated that left auditory cortical layer 2/3 neurons that code for high frequencies receive input from layer 6 neurons that are tuned to lower frequencies. Thus, such a hard-wired microcircuit in the left hemisphere would selectively detect a critical communication call with a high frequency up and a subsequent down sweep. In contrast, neurons in the right hemisphere are far less specialized.

Based on these and further studies,[79] some preliminary conclusions about the evolution of communication asymmetries can be drawn from the analysis by Ocklenburg et al.[62] and subsequent studies:

First, lateralization of acoustic communications is a widespread brain feature in vertebrates. So, our language system is not unique with respect to its asymmetry. It is unique due to other features that are inherent to the specificities of human language. These findings suggest that lateralization has been favored during evolution as a neural feature that increases neural efficacy and therefore overall fitness.[80] Concerning perceptual asymmetries, the dichotomy of temporal and spectral analyses could have been an evolutionary driving force.[81,82]

Second, there is no clear pattern for left-hemispheric vocalization dominance across all vertebrate classes. The avian data deviate from such a pattern, and the results in amphibians and fish should be considered with caution due to their scarceness and many methodological constraints. Thus, it is unlikely that vertebrates started their existence with left-hemispheric vocalization dominance and passed this trait down to their descendants. This shows that the directions of vocalization asymmetry can be more variable, possibly due to different genetic backgrounds and/or a stronger dependence on species-specific environmental or epigenetic factors.

Third, we can, however, safely say that there is convincing evidence for a left-hemispheric lateralization of production and perception of conspecific vocalizations in several mammalian species across five orders. The collected data are especially abundant within the order of primates. Concerning the phylogenetic origins of human language lateralization, the existence of vocalization-specific hemispheric specializations in several mammalian orders may imply that we inherited our left-hemispheric dominance for conspecific vocalization from our mammalian ancestors and then elaborated our unique language system on this common mammalian pattern. This finding has important consequences for many theories on the evolution of human language. Scholars like LeDoux et al.[83] or Corballis[84] have repeatedly stressed the link between our right-handedness and our left-hemispheric language dominance. They emphasized that human communication across cultures involves a rich repertoire of manual gestures, making it possible that language started with a gestural system and later increasingly incorporated spoken words. If this is correct, the left-hemispheric bias for language and handedness would not be a mere coincidence but a necessary component of a shared neural system. However, if language developed in humans from primate gestures via signed language to fully-fledged speech, we should expect to see it in primates only. The analysis by Ocklenburg et al.[62] as well as more recent studies demonstrate that mice, rats, gerbils, mustached bats, sea lions, dogs, and horses also display a left-hemispheric communication asymmetry. However, these species do not gesture and (as outlined before) do not necessarily show a right-limb preference.

The link between language asymmetry and handedness also dissolves when looking closer at human data. Packheiser et al. (2020) gathered data from 1554 individuals who had been tested in previous studies with dichotic listening to probe perceptual language lateralization and were also tested for handedness.[85] Only a very weak positive correlation between asymmetries of perceptual language asymmetry and handedness could be discovered that explained a mere 0.4% of the language lateralization variance. Subsequent Bayesian analyses found no evidence for a relation between language lateralization and handedness. Thus, as attractive a theory about manual gesturing that turned into spoken language may seem, it is possibly just one of the many evolutionary stories that we love to tell but that fall apart when scrutinized in greater detail.

The evolution of language asymmetry

We have seen that a left-hemispheric superiority for species-specific vocalization is widespread among mammals. It is, therefore, likely that we have inherited this trait from our primate ancestors. However, humans have a communication system that no other primate has: language. Furthermore, this language system shows multiple functional and structural asymmetries. Is there a way to reconstruct the emergence of these human-specific asymmetries during the process of the evolution of the human language system?

In 1987, South African anthropologist Tobias[86] discovered that the endocasts of *H. habilis* display a sulcal pattern in the location of the Broca area that resembled modern humans but was different from those of the older ape-like australopithecine brains. Tobias[86] concluded that the evolution of language asymmetry could be traced back via a careful analysis of the skulls of our ancestors. Consequently, many subsequent studies attempted to achieve this goal by scrutinizing fossil records. These attempts were unfortunately plagued by many problems. First, the details of sulcal or gyral imprints are not always identifiable in fossils, and even when they can be recognized, sulci do not always correlate precisely with the borders of functional cortical areas. Second, the number of structural asymmetries of the brain that can be reconstructed from endocranial casts is extremely limited and mostly not related to language.[87] Third, structural asymmetries become visible only when the complete surface measure of an individual gyrus or the sizes of axon tracts are reconstructed using modern imaging tools.[88] None of these are possible with fossils. Finally, language asymmetry does not produce a material culture like stone tools that can be analyzed.

There is another option: Using modern imaging techniques, we could compare the structural asymmetries of the human language system with those of corresponding regions in apes. However, comparing humans and chimps for language-related brain asymmetries can produce the logical fallacy that we think of chimps as representatives of ancient humans. This is obviously nonsense since both chimps and humans have evolved independently after descending from their last common ancestor. However, if we avoid thinking of chimps as ancient humans, a structural comparison of human and nonhuman primate brains in language-related areas can tell us a lot about the evolution of human language asymmetries.

Population-level asymmetries have been reported in great apes and monkeys for several anatomical measures of areas surrounding the Sylvian fissure.[89] The Sylvian fissure is positioned in the posterior language area and separates the frontal and the parietal lobe from the temporal lobe below. The area surrounding the Sylvian fissure is called the perisylvian region and hosts many language-related functions. The length of the Sylvian fissure is longer on the left side,[90] and the upper surface of the temporal lobe—the planum temporale—below the Sylvian fissure is larger on the left.[91] Thus, chimps already have some overall structural asymmetries in the left temporal region that resemble the human condition. This is also true for connectivity: Cheng et al. (2021) studied macaques, chimpanzees, and humans to investigate the connectivity patterns of the inferior parietal lobule (IPL) which plays a key role in tool use and other complex motor actions.[92] They discovered a leftward bias of IPL connections to vocalization/language areas along the Sylvian fissure in chimpanzees and humans, with this asymmetry being much more widespread in humans. Macaques showed no such pattern. Such a pattern makes it likely that left-right differences in neural connectivity have a phylogenetic past that anteceded the lateralized human language system. This arrangement possibly started with a hemispheric specialization in the last common ancestor of today's apes and humans and a subsequent elaboration during human brain evolution.[93] Thus, even before the evolution of human language asymmetry, different kinds of neural left-right differences existed that were possibly related to a species-typical communication asymmetry.

Humans seem to have dramatically expanded this asymmetry during the evolution of language. The overall extent of Broca's area reveals a prominent leftward asymmetry in humans[94] but not in chimpanzees.[95] In addition, cortical minicolumns in Broca's area[96] and the planum temporale of the temporal cortex[97] are wider in humans than in great apes (Fig. 2.13).

Cortical minicolumns are vertical stakes of about 100 strongly coupled and coherently working neurons. Wider minicolumns in the human brain could hint at denser synaptic interconnectivity in such a minicolumn. Last but not least, the whole temporal cortex has expanded in the human brain, pushing the visual associative areas into even more posterior positions.[99] Thereby, neural space was added in the temporal cortex that now hosts language functions (see Chapter 5). However, the temporal cortex must communicate with Broca's area to translate words into sentences and speech. For this transfer of information, a major fiber bundle comes into play: the arcuate fasciculus.

The arcuate fasciculus is the most widely investigated white matter pathway of the human language system. Initially, it was thought to only connect Broca's area to Wernicke's area but is now understood to be part of a larger tract, the superior

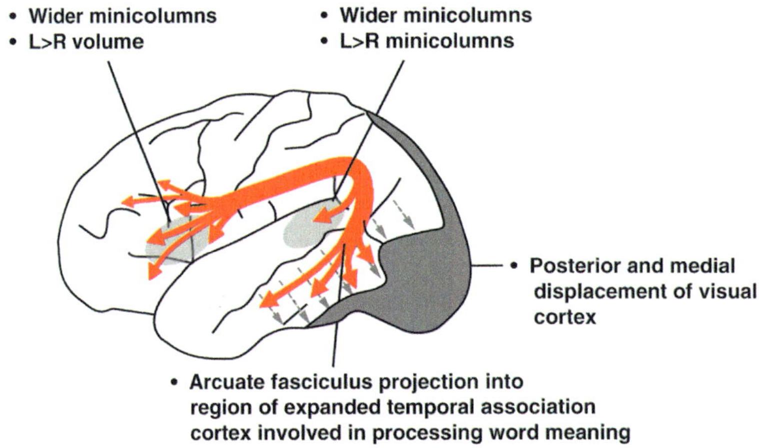

FIGURE 2.13 A summary depiction of left-sided asymmetries in the human language system that are not present in the brain of great apes. Broca's area (light gray region in the frontal cortex) has wider minicolumns and leftward higher volume in humans. Wernicke's area (light gray region in the temporal cortex) of humans also has wider minicolumns, and there is a leftward asymmetry in the column width in humans. In the left hemisphere, the arcuate fasciculus is more massive and has a much wider connectivity pattern between the frontal and temporal cortex. In humans, it expands to central and ventral expanses of the temporal cortex. As a consequence, major parts of the human temporal cortex underwent a massive expansion toward ventral and posterior directions during human evolution (dark gray expanse).[98] *Source: Reproduced from with permission from the author and the publisher from Rilling JK. Comparative primate neurobiology and the evolution of brain language systems.* Curr Opin Neurobiol. *2014;28:10–14.*

longitudinal fasciculus that connects the whole perisylvian region to the frontal lobe.[100] Functionally, the left arcuate fasciculus has been found to support complex syntactic processes of language.[101] Structural left-sided advantages of the arcuate fasciculus have been reported for several measures like its axonal integrity,[66,102] its greater relative fiber density,[103,104] tract volume,[105,106] and tract length.[107] Taken together, there is substantial evidence for a left lateralization of the arcuate fasciculus in humans. In addition, several studies investigating the relation of these leftward structural asymmetries to functional language asymmetries found positive correlations, although these were somewhat task-dependent and modulated by handedness.[88] Interestingly, the situation in nonhuman primates is very different. When the extent of the arcuate fasciculus is tracked with in vivo imaging techniques, it becomes apparent that it is much smaller and not asymmetrically organized in chimpanzees and macaques[108] (Fig. 2.13).

This pattern changes when individual neuroanatomical variations in the Broca complex homolog of chimpanzees are correlated with their ability to produce Attention-Getting (AG) sounds. These vocalizations serve to attract the attention of other individuals, are used in an intentional as well as referential manner, and require voluntary control of both orofacial musculature and larynx.[109] The frontal cortex shows an overall similar organization in old-world monkey and hominid brains.[110] Notable exceptions to these similarities are the frontopolar cortex and especially the ventrolateral prefrontal cortex. Here, in the region where Broca's area is located in humans, major differences become visible when comparing baboons, macaques, chimpanzees, and humans. The ventrolateral prefrontal cortex in humans has expanded to the extent that parts of it had to fold over the insula and are no longer visible from the outside. These buried parts are collectively called the prefrontal extent of the frontal operculum (PFOp). This area seems to be absent in nonhuman primates, although great apes display an organization with some resemblances. Interestingly, chimpanzees with orofacial motor abilities that enable an increased ability to produce AG sounds have an anatomical organization in (mostly) the left hemisphere that comes close to the human PFOp.[111] It is therefore conceivable that the last common ancestor of humans and chimps already had a lateralized ventrolateral prefrontal region for the organization of intentional vocalizations. We do not know if chimpanzees simply retained it or also elaborated on this heritage. However, certeinly this area dramatically expanded in the left hemisphere, adding multiple neural subcomponents that are required to produce the most complex motor task that any living animal can create: speech.

The evolution of vocalization asymmetry in vertebrates—a summary

Now, let us summarize what we said about the evolution of vocalization asymmetries in vertebrates:

- First, like for limb preferences, asymmetries of vocalization are widespread among vertebrates. So, human language asymmetry is not unique with respect to its lateralized pattern.
- Second, the overall picture of left and right dominances does not show a convincing bias toward the left hemisphere across vertebrate species. It is therefore likely that asymmetries of vocalization developed independently several times in the vertebrate lineage.
- Third, the situation is different in mammals. A left-hemispheric superiority for species-specific vocalizations can possibly be traced back to the beginnings of mammals and is consequently found in all mammalian orders studied so far. The primate order is one of them, and many primate species evince a leftward dominance in processing vocalizations. In apes, diverse structural asymmetries can be found that closely match the human language system condition. From all this, we can conclude that humans inherited the left-hemispheric asymmetry and a basic design of connectivity between cortical areas that now are part of our language system.
- Fourth, the evolution of our language system resulted in the emergence of a large number of additional human-specific neural entities within both the temporal and the frontal cortex. These have their human-specific morphological and connectional asymmetries, for example, the arcuate fasciculus and the Broca complex.

Asymmetry pays

Up to now, we have concentrated on limb preferences and vocalization asymmetries in vertebrates. However, hundreds of publications could meanwhile show diverse asymmetries of other systems in all kinds of vertebrate species. Most importantly, an increasing number of studies also report left-right differences in invertebrates. One of them is the nematode *C. elegans* with a tiny nervous system that consists of just 302 neurons. This little worm lives in moist soil and moves in a film of water, constantly smelling or tasting particles that signal food, danger, or mates. Some neuron pairs in *C. elegans* express their chemoreceptors with left-right differences.[29] This asymmetry provides the worm the ability to discriminate between distinct sensory cues.[112,113] The slug *Limax* memorizes food odors asymmetrically in its procerebral brain divisions,[30] and honey bees show left-right asymmetries in odor-driven sensory processing, behavior, and neural representation.[114] These are just a few examples of the dozens of different invertebrate species that demonstrate asymmetries of brain and behavior.[115]

If animals from nematodes to humans show asymmetries of brain and behavior, we should expect that left-right differences provide some important fitness benefits. The brain represents only about 2% of the body weight of an adult human but accounts for about 20% of the calories consumed by the body.[116] Thus, making efficient use of brain tissue should underlie strong evolutionary pressure. Vallortigara[117] suggested that the major fitness benefit of lateralization is to increase neural capacity because the specialization of one hemisphere for a cognitive function avoids useless duplication of cognitive functions between the two hemispheres.

One way to test these benefits is to exploit the individual variation in the degree of lateralization and correlate this with task performance. These studies show that chimpanzees with stronger individual handedness were more successful when fishing for termites using a tool.[118] Similarly, pigeons with stronger visual asymmetry were more successful when foraging for grain scattered among pebbles.[119] Also, individuals with pronounced language lateralization have a higher verbal IQ[120] and better reading skills.[121]

However, further analyses cast doubt on the simple assumption that increasing asymmetry levels increase perceptual and cognitive functions. Boles et al.[122] searched for possible benefits from high degrees of asymmetry in linguistic and spatial tasks and found positive effects for only some tasks. The most sobering news came from a meta-analysis with several hundreds of thousands of subjects on the possible effects of handedness on verbal and spatial tasks.[123] No positive effect of right-handedness on overall verbal ability and only a very moderate effect on spatial ability could be shown. Similar negative reports also came from studies in nonhuman animals.[124] For example, Whiteside et al. (2020) could not see a relation between footedness in pheasants and their performance in color discrimination and spatial tasks.[125] So, why do some studies reveal a benefit of being asymmetric, while others do not find this correlation?

There are possibly several reasons for these inconsistent findings. First, asymmetrically organized neural systems represent independent entities in which the degree of left-right difference in one system can be different from that of the other.[85,126–129] Therefore, asymmetries in one system should only be relevant to the specific function of this system. It does not make sense to correlate the extent of an asymmetry in one system with the behavioral performance level in a completely different system. The second problem concerns the question of how performance in a cognitive task is measured. In some studies, performance and lateralization are derived from the same task such that these values are not independent or at least confounded. Hirnstein et al.[130] took all these points into account and examined the relationship between language lateralization and performance in 1839 participants in a verbal dichotic listening task. The results revealed an U-shaped relationship between the degree of lateralization and overall accuracy: The stronger the left- or right-ear advantage, the better the overall accuracy. Thus, the strength of lateralization was a predictor for high performance, while direction—independent of right or left—was not.

However, these findings do not clarify the mechanism by which brain asymmetries increase performance. According to numerous studies, three such mechanisms are likely. First, asymmetry could selectively increase the perceptual or motor learning effect in one hemisphere. In birds, the eyes are laterally placed so that most of the visual input is derived from monocular vision. Thus, increased perceptual training of one eye/one hemisphere should ensure higher discrimination ability with this side. This is what was repeatedly found.[119,131,132] The situation is similar for the motor system in primates where some tasks are executed by only one hand or leg. This also holds for desert locusts (*Schistocerca gregaria*) that occasionally marauder the crops of farmers in West Asia and North Africa. When climbing a plant, locusts often must crawl across gaps. When doing so, the animals have a preferred forelimb to reach out to the next leaf. Half of the animals prefer their left, the other half their right leg (individual level lateralization). Bell and Niven (2016) showed that strongly lateralized locusts make fewer errors with their preferred forelimb.[133] Thus, even for the eighth biblical plague, the relation between the strength of asymmetry and performance holds, if the lateralized system is tested for its related performance. A small ontogenetic advantage of one extremity can result in further lateralized practice in a higher performance level with the preferred limb in conjunction with a higher likelihood to use this hand[118,134] or leg.[135]

The second mechanism for an advantage of asymmetry is directly related: Increased learning with one perceptual or motor system also decreases reaction times, resulting in a time advantage of the dominant side.[136–138] This is possibly the reason why in most tasks in which human subjects or other animals have to respond under high time pressure the dominant side has a significant time advantage both in computation[139] and in task execution.[140]

The third mechanism of advantage is parallel and complementary processing during task execution. For example, in poecilid fish that were artificially selected for a high degree of visual lateralization, asymmetric individuals were twice as fast as nonlateralized ones at catching live prey when the animals had to simultaneously monitor for predators.[141] Similarly, when lateralized and nonlateralized chicks are tested in a foraging task that requires them to find grains scattered among grit and, at the same time, monitor overhead for a flying model predator, the lateralized birds can conduct both tasks efficiently and in parallel.[142] By contrast, nonlateralized chicks perform poorly since they often miss seeing the predator, and when they see it they suddenly start mistaking grit for grains. Both lateralized birds and fish could attain this result by attending the feeding task with one eye while monitoring the predators with the other. Thus, hemispheric specialization seems to increase parallel processing by enabling separate processing of complementary information in the two hemispheres (Fig. 2.14).

However, there is still one mystery left: All of the abovementioned arguments would also work if half of humans were left- and the other half right-handed. So, why are 90% of us right-handed? To put it in more general words, why is there a population asymmetry for some functions in some species? Models based on mathematical game theory show

FIGURE 2.14 Asymmetry pays. Visual asymmetry in domestic chicks enables them to perform two tasks in parallel.[142,143] In this study, chicks had to find grains among pebbles and were simultaneously forced to be vigilant for birds of prey. Lateralized birds could do this well and found more grains, while non-lateralized ones failed. *Source: Reproduced with permission from the authors from Güntürkün O, Ströckens F, Ocklenburg S. Brain lateralization: a comparative perspective.* Physiol Rev. *2020;100:1019–1063.*

that an alignment of the direction of asymmetries in a population can be maintained as an evolutionarily stable strategy when individually asymmetric organisms benefit from aligning their behavior with other asymmetrical animals.[144,145] If these models are correct, we should see lateralization at the population level more often in social contexts, while lateralization at the individual level should occur with a higher likelihood in solitarily living species.[80,144,146] Indeed, population-level lateralization in turning responses in front of a predator is more frequent in shoaling fishes than in nonshoaling fishes.[147,148] However, it is important to keep in mind that even in social species, certain conditions might make individual asymmetries more beneficial. For example, when attacking schooling sardines, group-hunting sailfish show individual-level lateralization use of their saw-like bills to stab or slash their prey.[149] The reason is easy to understand: If sardines can predict the attack direction of sailfish, they have a good chance to escape. However, they lose this option when individual sailfish have individual but not population asymmetries. Thus, it is not that population lateralization is always related to social interactions, rather the nature of social interaction can generate evolutionarily stable strategies of lateralization at the individual or population level, depending on ecological contexts.[150]

This hypothesis can be beautifully tested in invertebrates. While colony-dwelling eusocial species like ants and honeybees exhibit population-level lateral biases, solitary species such as octopuses, terrestrial slugs, shrimps, and crayfish show individual-level asymmetries.[18] However, why then do animals without obvious social lifestyles such as spiders, mosquitoes, blowflies, and parasitic wasps display population-level asymmetries?[150–154] As suggested by Vallortigara and Rogers,[80] mathematical models,[144] new reviews,[115] and studies,[152] population-level asymmetries in these solitary species might be related to repeated associations with social or aggressive contexts like during courtship during evolutionary times. This, however, inoculates the theory against any possibility to be falsified. In the end, nearly all species, whether social or solitary, have to meet other individuals during mating or, at least, have an evolutionary past in which a social lifestyle was possible. Taken together, the theory by Vallortigara and Rogers[80] is a highly interesting step forward to explain the evolution of population asymmetries. It was also extraordinarily successful in stirring many studies on asymmetries in diverse species. However, possibly the last word on this theory has not been spoken yet.

Summary

The first clear evidence for behavioral asymmetries possibly dates back 500 million years to the hunting strategies of *Anomalocaris*—a fierce, top-apex predator of the Cambrian Sea. Thus, the evolution of asymmetries is by far not restricted to the slice of time since we humans exist. It is much older.

Both the analyses of the reports on limb preferences and those on vocalization asymmetries show that these left-right differences seem to have emerged and vanished again and again independently in the diverse ramifications of the vertebrate tree of life. It is therefore a futile attempt to try to identify the last common ancestor of vertebrate asymmetries. It is, however, not at all futile to reconstruct these sorts of genealogies in shorter branches of this tree, for example, within the order of primates. Both for handedness and for vocalization, an evolutionarily early emergence of left-hemispheric dominances for fine manipulations with the right hand and for controlling species-specific calls is likely. Thus, we have inherited the lateralization of handedness and left-hemispheric language system from our ancestors. However, due to our excellent dexterity, we have developed an extreme bias to use the right hand to such an extent that 90% of humans are right-handed. Such a strong bias is otherwise only known in some parrot species that show an equally strong population-based preference to use the left claw. Similarly, the evolution of our human-specific language system went along with the development of several unique left-hemispheric language areas and pathways that we, mostly, do not share with our direct phylogenetic relatives.

Last but not least, a comparative approach to left-right differences reveals three mechanisms why asymmetries pay. First, asymmetry can be shown to selectively increase the perceptual- or motor-learning effect in one hemisphere. Second, due to this unihemispheric learning effect, reaction times that are controlled by this hemisphere decrease. Third, brain asymmetries provide the great advantage of parallel and complementary processing during task execution. No wonder that brain asymmetries have such a long and diverse history throughout the evolution of life.

Sample questions

- What is the evidence for behavioral asymmetries in the Cambrian period and what kind of left-right difference did they reveal?
- Summarize the four main findings from the cladistic analysis of limb preferences.
- What is an individual- and what is a population-level asymmetry?
- What is the pattern of human handedness during the execution of simple tasks like picking up an object versus during using a tool like a screwdriver?

- What is the evidence for right-handedness in premodern humans? Describe three findings.
- Shortly outline the results of what can be said about the evolution of perception- and production-related species-typical communicative vocalization asymmetries in nonhuman animals and humans.
- Why does asymmetry pay? Give three main reasons.

References

1. Babcock LE, Robison RA. Preferences of palaeozoic predators. *Nature*. 1989;337(6209):695–696. Available from: https://doi.org/10.1038/337695c0.
2. Babcock LE. Asymmetry in the fossil record. *Eur Rev*. 2005;13(S2):135–143. Available from: https://doi.org/10.1017/s1062798705000712.
3. Sperling EA, Wolock CJ, Morgan AS, et al. Statistical analysis of iron geochemical data suggests limited late Proterozoic oxygenation. *Nature*. 2015;523(7561):451–454. Available from: https://doi.org/10.1038/nature14589.
4. Darroch SAF, Sperling EA, Boag TH, et al. Biotic replacement and mass extinction of the *Ediacara biota*. *Proc Biol Sci*. 2015;282(1814). Available from: https://doi.org/10.1098/rspb.2015.1003.
5. Fox D. What sparked the Cambrian explosion. *Nature*. 2016;530(7590):268–270. Available from: https://doi.org/10.1038/530268a.
6. Meek FB. Descriptions of fossils collected by the U. S. Geological Survey under the charge of Clerence King, Esq. *Proc Acad Nat Sci Phila*. 1870;22(1):56–64.
7. Whittington HB, Briggs DEG. The largest Cambrian animal, Anomalocaris, Burgess Shale, British-Columbia. *Phil Trans R Soc Lond B*. 1985;309(1141):569–609. Available from: https://doi.org/10.1098/rstb.1985.0096.
8. Selly T, Huntley JW, Shelton KL, Schiffbauer JD. Ichnofossil record of selective predation by Cambrian trilobites. *Palaeogeogr Palaeoclimatol Palaeoecol*. 2016;444:28–38. Available from: https://doi.org/10.1016/j.palaeo.2015.11.033.
9. Paterson JR, García-Bellido DC, Lee MSY, Brock GA, Jago JB, Edgecombe GD. Acute vision in the giant Cambrian predator Anomalocaris and the origin of compound eyes. *Nature*. 2011;480(7376):237–240. Available from: https://doi.org/10.1038/nature10689.
10. Daley AC. Anomalocaridids. *Curr Biol*. 2013;23(19):R860–R861. Available from: https://doi.org/10.1016/j.cub.2013.07.008.
11. Daley AC, Paterson JR, Edgecombe GD, García-Bellido DC, Jago JB. New anatomical information on Anomalocaris from the Cambrian Emu Bay Shale of South Australia and a reassessment of its inferred predatory habits. *Palaeontology*. 2013;56(5):971–990. Available from: https://doi.org/10.1111/pala.12029.
12. Nedin C. Anomalocaris predation on nonmineralized and mineralized trilobites. *Geology*. 1999;27(11):987. Available from: https://doi.org/10.1130/0091-7613(1999)027<0987:APONAM>2.3.CO;2.
13. Reisz RR, MacDougall MJ, LeBlanc ARH, Scott D, Nagesan RS. Lateralized feeding behavior in a Paleozoic reptile. *Curr Biol*. 2020;30(12):2374–2378, e4. Available from: https://doi.org/10.1016/j.cub.2020.04.026.
14. Serel Arslan S, İnal Ö, Demir N, Ölmez MS, Karaduman AA. Chewing side preference is associated with hemispheric laterality in healthy adults. *Somatosens Mot Res*. 2017;34(2):92–95. Available from: https://doi.org/10.1080/08990220.2017.1308923.
15. Ogle W. On dextral pre-eminence. *Med Chir Trans*. 1871;54:279–301. Available from: https://doi.org/10.1177/095952877105400114.
16. Dwight T. What is right-handedness? *Scribner's Mag*. 1891;9:465–472.
17. Boulinguez-Ambroise G, Aychet J, Pouydebat E. Limb preference in animals: new insights into the evolution of manual laterality in hominids. *Symmetry*. 2022;14(1):96. Available from: https://doi.org/10.3390/sym14010096.
18. Frasnelli E, Vallortigara G, Rogers LJ. Left-right asymmetries of behaviour and nervous system in invertebrates. *Neurosci Biobehav Rev*. 2012;36(4):1273–1291. Available from: https://doi.org/10.1016/j.neubiorev.2012.02.006.
19. Hennig W. *Phylogenetic Systematics*. University of Illinois Press; 1966.

20. Ströckens F, Güntürkün O, Ocklenburg S. Limb preferences in non-human vertebrates. *Laterality*. 2013;18 (5):536–575. Available from: https://doi.org/10.1080/1357650X.2012.723008.
21. Miletto Petrazzini ME, Sovrano VA, Vallortigara G, Messina A. Brain and behavioral asymmetry: a lesson from fish. *Front Neuroanat*. 2020;14:11. Available from: https://doi.org/10.3389/fnana.2020.00011.
22. Giljov A, Karenina K, Ingram J, Malashichev Y. Parallel emergence of true handedness in the evolution of marsupials and placentals. *Curr Biol*. 2015;25(14):1878–1884. Available from: https://doi.org/10.1016/j.cub.2015.05.043.
23. Bisazza A, Lippolis G, Vallortigara G. Lateralization of ventral fins use during object exploration in the blue gourami (*Trichogaster trichopterus*. *Physiol Behav*. 2001;72(4):575–578. Available from: https://doi.org/10.1016/s0031-9384(01)00417-6.
24. Robins A, Rogers LJ. Lateralized visual and motor responses in the green tree frog, *Litoria caerulea*. *Anim Behav*. 2006;72(4):843–852. Available from: https://doi.org/10.1016/j.anbehav.2006.01.022.
25. Sieg AE, Zandonà E, Izzo VM, Paladino FV, Spotila JR. Population level "flipperedness" in the eastern Pacific leatherback turtle. *Behav Brain Res*. 2010;206(1):135–138. Available from: https://doi.org/10.1016/j.bbr.2009.08.023.
26. McGavin S. Footedness in north island kākā (Nestor meridionalis septentrionalis). *Notornis*. 2009;56:139–143.
27. Giljov A, Karenina K, Malashichev Y. Limb preferences in a marsupial, *Macropus rufogriseus*: evidence for postural effect. *Anim Behav*. 2012;83(2):525–534. Available from: https://doi.org/10.1016/j.anbehav.2011.11.031.
28. Güntürkün O, Ströckens F, Ocklenburg S. Brain lateralization: a comparative perspective. *Physiol Rev*. 2020;100(3):1019–1063. Available from: https://doi.org/10.1152/physrev.00006.2019.
29. Hobert O, Johnston RJ, Chang S. Left-right asymmetry in the nervous system: the *Caenorhabditis elegans* model. *Nat Rev Neurosci*. 2002;3(8):629–640. Available from: https://doi.org/10.1038/nrn897.
30. Matsuo R, Kawaguchi E, Yamagishi M, Amano T, Ito E. Unilateral memory storage in the procerebrum of the terrestrial slug Limax. *Neurobiol Learn Mem*. 2010;93(3):337–342. Available from: https://doi.org/10.1016/j.nlm.2009.11.008.
31. Schnell AK, Bellanger C, Vallortigara G, Jozet-Alves C. Visual asymmetries in cuttlefish during brightness matching for camouflage. *Curr Biol*. 2018;28(17):R925–R926. Available from: https://doi.org/10.1016/j.cub.2018.07.019.
32. Carey DP, Smith DT, Martin D, et al. The bi-pedal ape: plasticity and asymmetry in footedness. *Cortex*. 2009;45(5):650–661. Available from: https://doi.org/10.1016/j.cortex.2008.05.011.
33. Packheiser J, Schmitz J, Berretz G, et al. Four meta-analyses across 164 studies on atypical footedness prevalence and its relation to handedness. *Sci Rep*. 2020;10(1):14501. Available from: https://doi.org/10.1038/s41598-020-71478-w.
34. Papadatou-Pastou M, Ntolka E, Schmitz J, et al. Human handedness: a meta-analysis. *Psychol Bull*. 2020;146 (6):481–524. Available from: https://doi.org/10.1037/bul0000229.
35. MacNeilage PF, Studdert-Kennedy MG, Lindblom B. Primate handedness reconsidered. *Behav Brain Sci*. 1987;10(2):247–263. Available from: https://doi.org/10.1017/s0140525x00047695.
36. Fagot J, Vauclair J. Manual laterality in nonhuman primates: a distinction between handedness and manual specialization. *Psychol Bull*. 1991;109(1):76–89. Available from: https://doi.org/10.1037/0033-2909.109.1.76.
37. Finch G. Chimpanzee handedness. *Science*. 1941;94(2431):117–118. Available from: https://doi.org/10.1126/science.94.2431.117.
38. Annett M, Annett J. Handedness for eating in gorillas. *Cortex*. 1991;27(2):269–275. Available from: https://doi.org/10.1016/s0010-9452(13)80131-1.
39. Spinozzi G, Castorina MG, Truppa V. Hand preferences in unimanual and coordinated-bimanual tasks by tufted capuchin monkeys (*Cebus apella*). *J Comp Psychol*. 1998;112(2):183–191. Available from: https://doi.org/10.1037/0735-7036.112.2.183.
40. Hopkins WD, Fernandez-Carriba S, Wesley MJ, Hostetter A, Pilcher D, Poss S. The use of bouts and frequencies in the evaluation of hand preferences for a coordinated bimanual task in chimpanzees (*Pan troglodytes*): an empirical study comparing two different indices of laterality. *J Comp Psychol*. 2001;115(3):294.
41. Vauclair J, Meguerditchian A, Hopkins WD. Hand preferences for unimanual and coordinated bimanual tasks in baboons (*Papio anubis*). *Cognit Brain Res*. 2005;25(1):210–216. Available from: https://doi.org/10.1016/j.cogbrainres.2005.05.012.
42. Meguerditchian A, Vauclair J. Baboons communicate with their right hand. *Behav Brain Res*. 2006;171 (1):170–174. Available from: https://doi.org/10.1016/j.bbr.2006.03.018.

43. Kimura D. Manual activity during speaking—I. Right-handers. *Neuropsychologia*. 1973;11(1):45–50. Available from: https://doi.org/10.1016/0028-3932(73)90063-8.
44. Hopkins WD. Comparative and familial analysis of handedness in great apes. *Psychol Bull*. 2006;132 (4):538–559. Available from: https://doi.org/10.1037/0033-2909.132.4.538.
45. Regaiolli B, Spiezio C, Hopkins WD. Three actions, two groups: looking for the origin of primate manual lateralization. *J Comp Psychol*. 2016;130(3):259–268. Available from: https://doi.org/10.1037/com0000031.
46. Vallortigara G. Handedness: what kangaroos tell us about our lopsided brains. *Curr Biol*. 2015;25(15): R672–R674. Available from: https://doi.org/10.1016/j.cub.2015.06.036.
47. Oldfield RC. The assessment and analysis of handedness: the Edinburgh inventory. *Neuropsychologia*. 1971;9 (1):97–113. Available from: https://doi.org/10.1016/0028-3932(71)90067-4.
48. Marchant LF, McGrew WC, Eibl-Eibesfeldt I. Is human handedness universal? Ethological analyses from three traditional cultures. *Ethology*. 1995;101(3):239–258. Available from: https://doi.org/10.1111/j.1439-0310.1995.tb00362.x.
49. Faurie C, Raymond M. Handedness frequency over more than ten thousand years. *Proc Biol Sci*. 2004;271 (suppl 3):S43–S45. Available from: https://doi.org/10.1098/rsbl.2003.0092.
50. Cavanagh T, Berbesque JC, Wood B, Marlowe F. Hadza handedness: lateralized behaviors in a contemporary hunter–gatherer population. *Evol. Hum Behav*. 2016;37(3):202–209. Available from: https://doi.org/10.1016/j.evolhumbehav.2015.11.002.
51. Toth N. Archaeological evidence for preferential right-handedness in the lower and middle Pleistocene, and its possible implications. *J Hum Evol*. 1985;14(6):607–614. Available from: https://doi.org/10.1016/s0047-2484(85)80087-7.
52. Rugg G, Mullane M. Inferring handedness from lithic evidence. *Laterality*. 2001;6(3):247–259. Available from: https://doi.org/10.1080/713754411.
53. Cornford JM. Specialized resharpening techniques and evidence of handedness. In: Callow P, Cornford JM, eds. *La Cotte de St. Brelade 1961–1978: Excavations by C.B.M. McBurney*. GEO Books; 1986:337–351.
54. Uomini NT. The prehistory of handedness: archaeological data and comparative ethology. *J Hum Evol*. 2009;57(4):411–419. Available from: https://doi.org/10.1016/j.jhevol.2009.02.012.
55. Bermúdez de Castro JM, Bromage TG, Jalvo YF. Buccal striations on fossil human anterior teeth: evidence of handedness in the middle and early Upper Pleistocene. *J Hum Evol*. 1988;17(4):403–412. Available from: https://doi.org/10.1016/0047-2484(88)90029-2.
56. Frayer DW, Lozano M, Bermúdez de Castro JM, et al. More than 500,000 years of right-handedness in Europe. *Laterality*. 2012;17(1):51–69. Available from: https://doi.org/10.1080/1357650X.2010.529451.
57. Frayer DW, Clarke RJ, Fiore I, et al. OH-65: the earliest evidence for right-handedness in the fossil record. *J Hum Evol*. 2016;100:65–72. Available from: https://doi.org/10.1016/j.jhevol.2016.07.002.
58. Clifford E, Bahn PG. *Everyday life in the ice age: a new study of our ancestors. Archaeopress Archaeology*. Monster Bookshop; 2022.
59. Brown C, Magat M. The evolution of lateralized foot use in parrots: a phylogenetic approach. *Behav Ecol*. 2011;22(6):1201–1208. Available from: https://doi.org/10.1093/beheco/arr114.
60. Kalischer O. Das Grosshirn der Papageien in anatomischer und physiologischer Beziehung. *Abhandlungen der königl. preuss. Akademie der Wissenschaften*. 1905;IV:1–105.
61. Nottebohm F. Ontogeny of bird song. *Science*. 1970;167(3920):950–956. Available from: https://doi.org/10.1126/science.167.3920.950.
62. Ocklenburg S, Hugdahl K, Westerhausen R. Structural white matter asymmetries in relation to functional asymmetries during speech perception and production. *NeuroImage*. 2013;83:1088–1097. Available from: https://doi.org/10.1016/j.neuroimage.2013.07.076.
63. Hickok G, Poeppel D. The cortical organization of speech processing. *Nat Rev Neurosci*. 2007;8(5):393–402. Available from: https://doi.org/10.1038/nrn2113.
64. Knecht S, Dräger B, Deppe M, et al. Handedness and hemispheric language dominance in healthy humans. *Brain*. 2000;12(12):2512–2518, 123 Pt. Available from: https://doi.org/10.1093/brain/123.12.2512.
65. Friederici AD, Alter K. Lateralization of auditory language functions: a dynamic dual pathway model. *Brain Lang*. 2004;89(2):267–276. Available from: https://doi.org/10.1016/s0093-934x(03)00351-1.
66. Ocklenburg S, Ströckens F, Güntürkün O. Lateralisation of conspecific vocalisation in non-human vertebrates. *Laterality*. 2013;18(1):1–31. Available from: https://doi.org/10.1080/1357650X.2011.626561.
67. Hausberger M, Cousillas H, Meter A, et al. A crucial role of attention in lateralisation of sound processing. *Symmetry*. 2019;11(1):48. Available from: https://doi.org/10.3390/sym11010048.

68. Böye M, Güntürkün O, Vauclair J. Right ear advantage for conspecific calls in adults and subadults, but not infants, California sea lions (*Zalophus californianus*): hemispheric specialization for communication. *Eur J Neurosci.* 2005;21(6):1727–1732. Available from: https://doi.org/10.1111/j.1460-9568.2005.04005.x.
69. Fischer J, Teufel C, Drolet M, et al. Orienting asymmetries and lateralized processing of sounds in humans. *BMC Neurosci.* 2009;10(1):14. Available from: https://doi.org/10.1186/1471-2202-10-14.
70. Andics A, Gábor A, Gácsi M, Faragó T, Szabó D, Miklósi Á. Neural mechanisms for lexical processing in dogs. *Science.* 2016;353(6303):1030–1032. Available from: https://doi.org/10.1126/science.aaf3777.
71. Andics A, Gácsi M, Faragó T, Kis A, Miklósi A. Voice-sensitive regions in the dog and human brain are revealed by comparative fMRI. *Curr Biol.* 2014;24(5):574–578. Available from: https://doi.org/10.1016/j.cub.2014.01.058.
72. Washington SD, Tillinghast JS. Conjugating time and frequency: hemispheric specialization, acoustic uncertainty, and the mustached bat. *Front Neurosci.* 2015;9:143. Available from: https://doi.org/10.3389/fnins.2015.00143.
73. Washington SD, Kanwal JS. Sex-dependent hemispheric asymmetries for processing frequency-modulated sounds in the primary auditory cortex of the mustached bat. *J Neurophysiol.* 2012;108(6):1548–1566. Available from: https://doi.org/10.1152/jn.00952.2011.
74. Washington SD, Pritchett DL, Keliris GA, Kanwal JS. Hemispheric and sex differences in mustached bat primary auditory cortex revealed by neural responses to slow frequency modulations. *Symmetry.* 2021;13(6). Available from: https://doi.org/10.3390/sym13061037.
75. Ehret G. Left hemisphere advantage in the mouse brain for recognizing ultrasonic communication calls. *Nature.* 1987;325(6101):249–251. Available from: https://doi.org/10.1038/325249a0.
76. Ehret G., Geissler D.B. Communication-call representation in the mouse auditory cortex: perception vs. recognition. In: Syka J, Merzenich MM, eds. Plasticity and Signal Representation in the Auditory System. Scholars Portal; 2005:85–96.
77. Geissler DB, Schmidt HS, Ehret G. Knowledge about sounds-context-specific meaning differently activates cortical hemispheres, auditory cortical fields, and layers in house mice. *Front Neurosci.* 2016;10:98. Available from: https://doi.org/10.3389/fnins.2016.00098.
78. Levy RB, Marquarding T, Reid AP, Pun CM, Renier N, Oviedo HV. Circuit asymmetries underlie functional lateralization in the mouse auditory cortex. *Nat Commun.* 2019;10(1):2783. Available from: https://doi.org/10.1038/s41467-019-10690-3.
79. Ruthig P, Schönwiesner M. Common principles in the lateralization of auditory cortex structure and function for vocal communication in primates and rodents. *Eur J Neurosci.* 2022;55(3):827–845. Available from: https://doi.org/10.1111/ejn.15590.
80. Vallortigara G, Rogers LJ. Survival with an asymmetrical brain: advantages and disadvantages of cerebral lateralization. *Behav Brain Sci.* 2005;28(4):575–589. Available from: https://doi.org/10.1017/S0140525X05000105.
81. Poeppel D. The analysis of speech in different temporal integration windows: cerebral lateralization as 'asymmetric sampling in time'. *Speech Commun.* 2003;41(1):245–255. Available from: https://doi.org/10.1016/S0167-6393(02)00107-3.
82. Zatorre RJ, Belin P, Penhune VB. Structure and function of auditory cortex: music and speech. *Trends Cogn Sci.* 2002;6(1):37–46. Available from: https://doi.org/10.1016/s1364-6613(00)01816-7.
83. LeDoux JE, Wilson DH, Gazzaniga MS. Manipulo-spatial aspects of cerebral lateralization: clues to the origin of lateralization. *Neuropsychologia.* 1977;15(6):743–750. Available from: https://doi.org/10.1016/0028-3932(77)90004-5.
84. Corballis MC. *From Hand to Mouth: The Origins of Language.* Princeton University Press; 2002.
85. Packheiser J, Schmitz J, Arning L, Beste C, Güntürkün O, Ocklenburg S. A large-scale estimate on the relationship between language and motor lateralization. *Sci Rep.* 2020;10(1):13027. Available from: https://doi.org/10.1038/s41598-020-70057-3.
86. Tobias PV. The brain of *Homo habilis*: a new level of organization in cerebral evolution. *J Hum Evol.* 1987;16(7-8):741–761. Available from: https://doi.org/10.1016/0047-2484(87)90022-4.
87. Balzeau A, Holloway RL, Grimaud-Hervé D. Variations and asymmetries in regional brain surface in the genus Homo. *J Hum Evol.* 2012;62(6):696–706. Available from: https://doi.org/10.1016/j.jhevol.2012.03.007.
88. Ocklenburg S, Friedrich P, Güntürkün O, Genç E. Intrahemispheric white matter asymmetries: the missing link between brain structure and functional lateralization? *Rev Neurosci.* 2016;27(5):465–480.
89. Marie D, Roth M, Lacoste R, et al. Left brain asymmetry of the planum temporale in a nonhominid primate: redefining the origin of brain specialization for language. *Cereb Cortex.* 2018;28(5):1808–1815. Available from: https://doi.org/10.1093/cercor/bhx096.

90. Hopkins WD, Pilcher DL, MacGregor L. Sylvian fissure asymmetries in nonhuman primates revisited: a comparative MRI study. *BBE*. 2000;56(6):293–299. Available from: https://doi.org/10.1159/000047213.
91. Keller SS, Roberts N, Hopkins WD. A comparative magnetic resonance imaging study of the anatomy, variability, and asymmetry of Broca's area in the human and chimpanzee brain. *J Neurosci*. 2009;29 (46):14607–14616. Available from: https://doi.org/10.1523/JNEUROSCI.2892-09.2009.
92. Cheng L, Zhang Y, Li G, et al. Connectional asymmetry of the inferior parietal lobule shapes hemispheric specialization in humans, chimpanzees, and rhesus macaques. *eLife*. 2021;10. Available from: https://doi.org/10.7554/eLife.67600.
93. Vickery S, Eickhoff SB, Friedrich P. Hemispheric specialization of the primate inferior parietal lobule. *Neurosci Bull*. 2022;38(3):334–336. Available from: https://doi.org/10.1007/s12264-021-00807-4.
94. Amunts K, Schleicher A, Ditterich A, Zilles K. Broca's region: cytoarchitectonic asymmetry and developmental changes. *J Comp Neurol*. 2003;465(1):72–89. Available from: https://doi.org/10.1002/cne.10829.
95. Schenker NM, Hopkins WD, Spocter MA, et al. Broca's area homologue in chimpanzees (*Pan troglodytes*): probabilistic mapping, asymmetry, and comparison to humans. *Cereb Cortex*. 2010;20(3):730–742. Available from: https://doi.org/10.1093/cercor/bhp138.
96. Schenker NM, Buxhoeveden DP, Blackmon WL, Amunts K, Zilles K, Semendeferi K. A comparative quantitative analysis of cytoarchitecture and minicolumnar organization in Broca's area in humans and great apes. *J Comp Neurol*. 2008;510(1):117–128. Available from: https://doi.org/10.1002/cne.21792.
97. Buxhoeveden DP, Switala AE, Litaker M, Roy E, Casanova MF. Lateralization of minicolumns in human planum temporale is absent in nonhuman primate cortex. *BBE*. 2001;57(6):349–358. Available from: https://doi.org/10.1159/000047253.
98. Rilling JK. Comparative primate neurobiology and the evolution of brain language systems. *Curr Opin Neurobiol*. 2014;28:10–14. Available from: https://doi.org/10.1016/j.conb.2014.04.002.
99. Orban GA, van Essen D, Vanduffel W. Comparative mapping of higher visual areas in monkeys and humans. *Trends Cognit Sci*. 2004;8(7):315–324. Available from: https://doi.org/10.1016/j.tics.2004.05.009.
100. Catani M, de Schotten MT. A diffusion tensor imaging tractography atlas for virtual in vivo dissections. *Cortex*. 2008;44(8):1105–1132. Available from: https://doi.org/10.1016/j.cortex.2008.05.004.
101. Friederici AD. Evolution of the neural language network. *Psychon Bull Rev*. 2017;24(1):41–47. Available from: https://doi.org/10.3758/s13423-016-1090-x.
102. Büchel C, Raedler T, Sommer M, Sach M, Weiller C, Koch MA. White matter asymmetry in the human brain: a diffusion tensor MRI study. *Cereb Cortex*. 2004;14(9):945–951. Available from: https://doi.org/10.1093/cercor/bhh055.
103. Nucifora PGP, Verma R, Melhem ER, Gur RE, Gur RC. Leftward asymmetry in relative fiber density of the arcuate fasciculus. *NeuroReport*. 2005;16(8):791.
104. Vernooij MW, Smits M, Wielopolski PA, Houston GC, Krestin GP, van der Lugt A. Fiber density asymmetry of the arcuate fasciculus in relation to functional hemispheric language lateralization in both right- and left-handed healthy subjects: a combined fMRI and DTI study. *NeuroImage*. 2007;35(3):1064–1076. Available from: https://doi.org/10.1016/j.neuroimage.2006.12.041.
105. Powell HWR, Parker GJM, Alexander DC, et al. Hemispheric asymmetries in language-related pathways: a combined functional MRI and tractography study. *NeuroImage*. 2006;32(1):388–399. Available from: https://doi.org/10.1016/j.neuroimage.2006.03.011.
106. Ocklenburg S, Schlaffke L, Hugdahl K, Westerhausen R. From structure to function in the lateralized brain: how structural properties of the arcuate and uncinate fasciculus are associated with dichotic listening performance. *Neurosci Lett*. 2014;580:32–36. Available from: https://doi.org/10.1016/j.neulet.2014.07.044.
107. Propper RE, O'Donnell LJ, Whalen S, et al. A combined fMRI and DTI examination of functional language lateralization and arcuate fasciculus structure: effects of degree versus direction of hand preference. *Brain Cogn*. 2010;73(2):85–92. Available from: https://doi.org/10.1016/j.bandc.2010.03.004.
108. Rilling JK, Glasser MF, Preuss TM, et al. The evolution of the arcuate fasciculus revealed with comparative DTI. *Nat Neurosci*. 2008;11(4):426–428. Available from: https://doi.org/10.1038/nn2072.
109. Hopkins WD, Taglialatela J, Leavens DA. Chimpanzees differentially produce novel vocalizations to capture the attention of a human. *Anim Behav*. 2007;73(2):281–286. Available from: https://doi.org/10.1016/j.anbehav.2006.08.004.
110. Amiez C, Sallet J, Giacometti C, et al. A revised perspective on the evolution of the lateral frontal cortex in primates. *Sci Adv*. 2023;9(20):eadf9445. Available from: https://doi.org/10.1126/sciadv.adf9445.

111. Amiez C, Verstraete C, Sallet J, et al. The relevance of the unique anatomy of the human prefrontal operculum to the emergence of speech. *Commun Biol.* 2023;6(1):693. Available from: https://doi.org/10.1038/s42003-023-05066-9.
112. Ortiz CO, Faumont S, Takayama J, et al. Lateralized gustatory behavior of *C. elegans* is controlled by specific receptor-type guanylyl cyclases. *Curr Biol.* 2009;19(12):996–1004. Available from: https://doi.org/10.1016/j.cub.2009.05.043.
113. Wes PD, Bargmann CI. *C. elegans* odour discrimination requires asymmetric diversity in olfactory neurons. *Nature.* 2001;410(6829):698–701. Available from: https://doi.org/10.1038/35070581.
114. Rigosi E, Haase A, Rath L, Anfora G, Vallortigara G, Szyszka P. Asymmetric neural coding revealed by in vivo calcium imaging in the honey bee brain. *Proc Biol Sci.* 2015;282(1803):20142571. Available from: https://doi.org/10.1098/rspb.2014.2571.
115. Frasnelli E. Brain and behavioral lateralization in invertebrates. *Front Psychol.* 2013;4:939. Available from: https://doi.org/10.3389/fpsyg.2013.00939.
116. Raichle ME, Gusnard DA. Appraising the brain's energy budget. *Proc Natl Acad Sci USA.* 2002;99(16):10237–10239. Available from: https://doi.org/10.1073/pnas.172399499.
117. Vallortigara G. The evolutionary psychology of left and right: costs and benefits of lateralization. *Dev Psychobiol.* 2006;48(6):418–427. Available from: https://doi.org/10.1002/dev.20166.
118. McGrew WC, Marchant LF. Laterality of hand use pays off in foraging success for wild chimpanzees. *Primates.* 1999;40(3):509–513. Available from: https://doi.org/10.1007/bf02557586.
119. Güntürkün O, Diekamp B, Manns M, et al. Asymmetry pays: visual lateralization improves discrimination success in pigeons. *Curr Biol.* 2000;10(17):1079–1081. Available from: https://doi.org/10.1016/s0960-9822(00)00671-0.
120. Everts R, Lidzba K, Wilke M, et al. Strengthening of laterality of verbal and visuospatial functions during childhood and adolescence. *Hum Brain Mapp.* 2009;30(2):473–483. Available from: https://doi.org/10.1002/hbm.20523.
121. Chiarello C, Welcome SE, Halderman LK, Leonard CM. Does degree of asymmetry relate to performance? An investigation of word recognition and reading in consistent and mixed handers. *Brain Cogn.* 2009;69(3):521–530. Available from: https://doi.org/10.1016/j.bandc.2008.11.002.
122. Boles DB, Barth JM, Merrill EC. Asymmetry and performance: toward a neurodevelopmental theory. *Brain Cogn.* 2008;66(2):124–139. Available from: https://doi.org/10.1016/j.bandc.2007.06.002.
123. Somers M, Ophoff RA, Aukes MF, et al. Linkage analysis in a Dutch population isolate shows no major gene for left-handedness or atypical language lateralization. *J Neurosci.* 2015;35(23):8730–8736. Available from: https://doi.org/10.1523/JNEUROSCI.3287-14.2015.
124. Agrillo C, Dadda M, Bisazza A. Escape behaviour elicited by a visual stimulus. A comparison between lateralised and non-lateralised female topminnows. *Laterality.* 2009;14(3):300–314. Available from: https://doi.org/10.1080/13576500802396693.
125. Whiteside MA, Bess MM, Frasnelli E, et al. No evidence that footedness in pheasants influences cognitive performance in tasks assessing colour discrimination and spatial ability. *Learn Behav.* 2020;48(1):84–95. Available from: https://doi.org/10.3758/s13420-019-00402-8.
126. Badzakova-Trajkov G, Corballis MC, Häberling IS. Complementarity or independence of hemispheric specializations? A brief review. *Neuropsychologia.* 2016;93(Pt B):386–393. Available from: https://doi.org/10.1016/j.neuropsychologia.2015.12.018.
127. Badzakova-Trajkov G, Häberling IS, Roberts RP, Corballis MC. Cerebral asymmetries: complementary and independent processes. *PLoS One.* 2010;5(3):e9682. Available from: https://doi.org/10.1371/journal.pone.0009682.
128. Häberling IS, Corballis PM, Corballis MC. Language, gesture, and handedness: evidence for independent lateralized networks. *Cortex.* 2016;82:72–85. Available from: https://doi.org/10.1016/j.cortex.2016.06.003.
129. Liu H, Stufflebeam SM, Sepulcre J, Hedden T, Buckner RL. Evidence from intrinsic activity that asymmetry of the human brain is controlled by multiple factors. *PNAS.* 2009;106(48):20499–20503. Available from: https://doi.org/10.1073/pnas.0908073106.
130. Hirnstein M, Hugdahl K, Hausmann M. How brain asymmetry relates to performance – a large-scale dichotic listening study. *Front Psychol.* 2014;4:997. Available from: https://doi.org/10.3389/fpsyg.2013.00997.
131. Ventolini N, Ferrero EA, Sponza S, Della Chiesa A, Zucca P, Vallortigara G. Laterality in the wild: preferential hemifield use during predatory and sexual behaviour in the black-winged stilt. *Anim Behav.* 2005;69(5):1077–1084. Available from: https://doi.org/10.1016/j.anbehav.2004.09.003.

132. Karenina K, Giljov A. Lateralization in feeding is food type specific and impacts feeding success in wild birds. *Ecol Evol.* 2022;12(2):e8598. Available from: https://doi.org/10.1002/ece3.8598.
133. Bell ATA, Niven JE. Strength of forelimb lateralization predicts motor errors in an insect. *Biol Lett.* 2016;12(9). Available from: https://doi.org/10.1098/rsbl.2016.0547.
134. Annett M. The growth of manual preference and speed. *Br J Psychol.* 1970;61(4):545–558. Available from: https://doi.org/10.1111/j.2044-8295.1970.tb01274.x.
135. Teixeira MCT, Teixeira LA. Leg preference and interlateral performance asymmetry in soccer player children. *Dev Psychobiol.* 2008;50(8):799–806. Available from: https://doi.org/10.1002/dev.20322.
136. Gülbetekin E, Güntürkün O, Dural S, Cetinkaya H. Visual asymmetries in Japanese quail (*Coturnix japonica*) retain a lifelong potential for plasticity. *Behav Neurosci.* 2009;123(4):815–821. Available from: https://doi.org/10.1037/a0016406.
137. Vallortigara G, Chiandetti C, Sovrano VA. Brain asymmetry (animal). *Wiley Interdiscip Rev Cogn Sci.* 2011;2 (2):146–157. Available from: https://doi.org/10.1002/wcs.100.
138. Xiao Q, Güntürkün O. Asymmetrical commissural control of the subdominant hemisphere in pigeons. *Cell Rep.* 2018;25(5):1171–1180, e3. Available from: https://doi.org/10.1016/j.celrep.2018.10.011.
139. Ocklenburg S, Güntürkün O, Beste C. Hemispheric asymmetries and cognitive flexibility: an ERP and sLORETA study. *Brain Cogn.* 2012;78(2):148–155. Available from: https://doi.org/10.1016/j.bandc.2011.11.001.
140. Verma A, van der Haegen L, Brysbaert M. Symmetry detection in typically and atypically speech lateralized individuals: a visual half-field study. *Neuropsychologia.* 2013;51(13):2611–2619. Available from: https://doi.org/10.1016/j.neuropsychologia.2013.09.005.
141. Dadda M, Bisazza A. Does brain asymmetry allow efficient performance of simultaneous tasks. *Anim Behav.* 2006;72(3):523–529. Available from: https://doi.org/10.1016/j.anbehav.2005.10.019.
142. Rogers LJ, Zucca P, Vallortigara G. Advantages of having a lateralized brain. *Proc Biol Sci.* 2004;271(suppl 6): S420–S422. Available from: https://doi.org/10.1098/rsbl.2004.0200.
143. Güntürkün O, Ocklenburg S. Ontogenesis of lateralization. *Neuron.* 2017;94(2):249–263. Available from: https://doi.org/10.1016/j.neuron.2017.02.045.
144. Ghirlanda S, Frasnelli E, Vallortigara G. Intraspecific competition and coordination in the evolution of lateralization. *Philos Trans R Soc Lond B Biol Sci.* 2009;364(1519):861–866. Available from: https://doi.org/10.1098/rstb.2008.0227.
145. Ghirlanda S, Vallortigara G. The evolution of brain lateralization: a game-theoretical analysis of population structure. *Proc Biol Sci.* 2004;271(1541):853–857. Available from: https://doi.org/10.1098/rspb.2003.2669.
146. Rogers LJ. Laterality in animals. *Int J Comp Psychol.* 1989;3(1):5–25.
147. Bisazza A, Cantalupo C, Capocchiano M, Vallortigara G. Population lateralisation and social behaviour: a study with 16 species of fish. *Laterality.* 2000;5(3):269–284. Available from: https://doi.org/10.1080/713754381.
148. Vallortigara G, Bisazza A. How ancient is brain lateralization. Comparative vertebrate lateralization. In: Rogers LJ, Andrew R, eds. *Comparative Vertebrate Lateralization.* Cambridge University Press; 2002:9–69.
149. Kurvers RHJM, Krause S, Viblanc PE, et al. The evolution of lateralization in group hunting sailfish. *Curr Biol.* 2017;27(4):521–526. Available from: https://doi.org/10.1016/j.cub.2016.12.044.
150. Rogers LJ, Frasnelli E, Versace E. Lateralized antennal control of aggression and sex differences in red mason bees, *Osmia bicornis. Sci Rep.* 2016;6(1):29411. Available from: https://doi.org/10.1038/srep29411.
151. Ades C, Ramires EN. Asymmetry of leg use during prey handling in the spider *Scytodes globula* (Scytodidae). *J Insect Behav.* 2002;15(4):563–570. Available from: https://doi.org/10.1023/A:1016337418472.
152. Benelli G, Romano D, Messing RH, Canale A. First report of behavioural lateralisation in mosquitoes: right-biased kicking behaviour against males in females of the Asian tiger mosquito, *Aedes albopictus. Parasitol Res.* 2015;114(4):1613–1617. Available from: https://doi.org/10.1007/s00436-015-4351-0.
153. Romano D, Canale A, Benelli G. Do right-biased boxers do it better? Population-level asymmetry of aggressive displays enhances fighting success in blowflies. *Behav Process.* 2015;113:159–162. Available from: https://doi.org/10.1016/j.beproc.2015.02.004.
154. Romano D, Donati E, Canale A, Messing RH, Benelli G, Stefanini C. Lateralized courtship in a parasitic wasp. *Laterality.* 2016;21(3):243–254. Available from: https://doi.org/10.1080/1357650X.2016.1150289.

CHAPTER

3

The connected hemispheres—the role of the corpus callosum for hemispheric asymmetries

OUTLINE

The Lateralized Brain
DOI: https://doi.org/10.1016/B978-0-323-99737-9.00007-0

Introductory short story

Martha was 13 when she was diagnosed with epilepsy. The seizures became worse over the years, and nothing her doctors tried could stop them. When Martha was an adult, her doctors told her that they wanted to apply a radical new treatment to stop the progression of her epilepsy. She was skeptical at first. The doctors explained that the surgery would involve cutting the corpus callosum, the neuronal tract connecting the two halves of her brain. This would be done to reduce the seizures and prevent her epilepsy from further damaging her brain. While she initially was not convinced, she consented to the procedure after her husband Rick urged her to trust the experts. After all, what did she have to lose? After surgery, she did not speak for some weeks. Apart from that, she recovered quickly, and the doctors were quite pleased with the outcome of the procedure. Months passed and Martha did not notice any residual impairment in her speech, senses, or movement, while the severity of the seizures seemed to be significantly reduced. When asked to participate in a series of scientific experiments aimed at understanding the functions of the two halves of the brain, she happily agreed. Over the next couple of months, Rick drove Martha to the university numerous times. There, scientists conducted experiments on her speech and movements, often testing the left and right sides of her brain separately. One chilly February morning she was told that they were going to try something new. The doctor asked her to sit down and look at a couple of photographs of people she knew well. The young doctor had arranged the pictures on her right side. He asked her to name the person she loved the most. Without hesitation, she said, "Rick, of course!" "And how would you rate your love for him on a scale from one to ten?" "Ten, of course!" Martha was mildly puzzled by the weird questions the young man in the lab coat was asking her. After she had rated her relations to several other people in the same manner, the doctor asked her to take a break while he rearranged some things. When she returned, he again asked her to have a seat, explaining to her that this time she would see pictures on her left side and that she was not allowed to speak to answer. Instead, he asked her to hand him 1 out of 10 little wooden numbers with her left hand to indicate how much she loved the person shown in the picture. The first photograph she saw was the one of Rick. This time, she needed much longer to give an answer. She picked up several different numbers with her left hand and put them back on the table again. When she finally handed the doctor one of the numbers, she saw his pupils widening with surprise. The wooden number she had put into his hand showed a one.

Neuroanatomy of the corpus callosum and other commissures

Viewing the human brain from the outside, one could get the impression that the left and the right hemisphere are two independent structures. While it looks as if they are completely separated by the medial longitudinal fissure—the large groove between the two halves of the brain (Fig. 3.1)—they are connected at several different locations. These connections are called "commissures" (from Latin "committere," which means "to join" or "to bring together"). They consist of tracts of nerve fibers (white matter, see Box 3.1) that connect structures in the left and the right hemispheres across the midline.

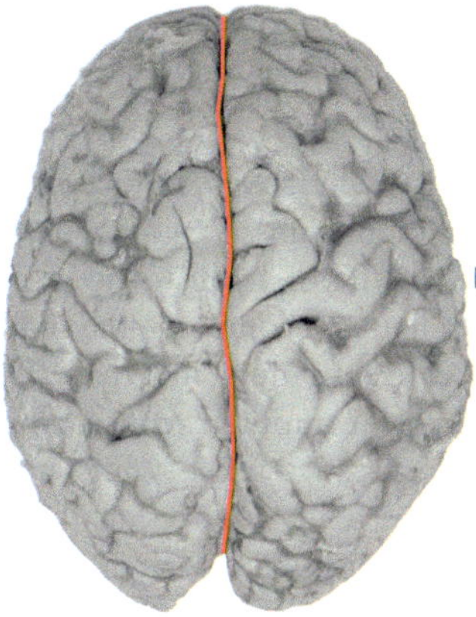

FIGURE 3.1 The medial longitudinal fissure separates the left from the right hemisphere.

The main commissures in the human brain (see Fig. 3.3) are:

- Corpus callosum: Connects the left and right frontal, temporal, parietal, and occipital lobes of the cerebral cortex.
- Anterior commissure: Connects the left and right temporal lobe of the cerebral cortex and the left and right amygdala.
- Hippocampal commissure (also called commissure of the fornix): Connects the left and right hippocampus.
- Adhesio interthalamica (previously also known as Massa intermedia): Connects the left and right thalamus.

The largest commissure in the brain of placental mammals is the corpus callosum (from Latin "corpus" meaning "body" and "callus" meaning "a toughened area of skin"—early neuroanatomists found the corpus callosum to be harder than the surrounding tissue, hence the name). The corpus callosum is by far the largest fiber tract in the brain. Its function was unclear for a long time, and early neuroanatomists suggested that it served as mechanical support to maintain the structural integrity of the ventricles.[3] Lashley jokingly suggested that its main purpose would be to keep the hemispheres from sagging.[4] Today, we know that the main purpose of the corpus callosum is to transfer information between the two hemispheres.[5,6] Therefore, it is crucial for bilateral integration of sensory, motor, or higher cognitive functions.[7]

In humans, the corpus callosum is composed of about 200 million fibers, of which about 70% are myelinated (see Box 3.1), while the remaining 30% are unmyelinated.[8] It mainly consists of homotopic interconnections (connecting the same areas in the left and right hemisphere) but also encompasses a smaller number of heterotopic connections (connecting different areas in the left and right hemisphere).[9] Callosal axons can lead to inhibitory and excitatory stimulation of postsynaptic targets, but also play modulatory roles when coupled to inputs from other networks.[10] The corpus callosum is topographically organized. Anterior cortical areas are connected through the anterior parts of the corpus callosum, and posterior cortical areas are connected through the posterior parts of the corpus

BOX 3.1

Gray and white matter.

The two major components of the central nervous system are gray matter and white matter (Fig. 3.2). While gray matter mainly contains neuronal cell bodies, white matter primarily consists of myelinated axons. Its main purpose is the transfer of information between different brain areas. The white color results from the high fat content of myelin, a substance that is built by oligodendrocytes (a type of glia cells). It insulates the axons like a rubber sheathing that is used to encase copper wire. This way it leads to much faster conduction of neuronal information than in unmyelinated axons—up to 100 times faster.[1] The so-called "network analogy" compares the brain to a network of computers in which gray matter represents computers conducting information processing, while white matter represents the cables connecting the computers. However, newer research indicates that white matter fiber tracts play more than a purely passive role in learning and cognition.[2]

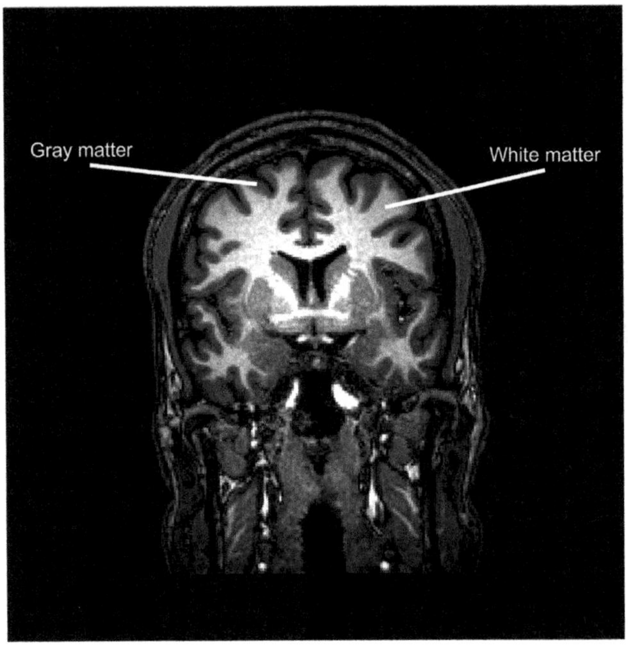

FIGURE 3.2 Gray and white matter in the human brain.

callosum. Moreover, fiber composition of callosal segments is different depending on the interconnected brain regions. Callosal segments connecting primary and secondary sensory brain areas have a higher frequency of larger and highly myelinated fibers than callosal segments connecting cortical areas relevant for higher order cognition.[11] Based on this

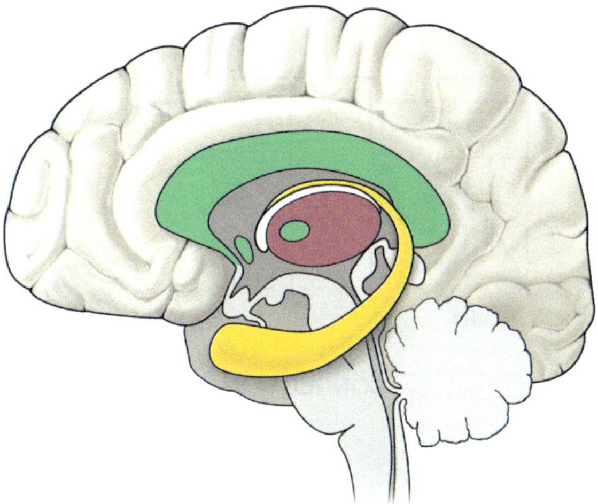

FIGURE 3.3 Commissures in the human brain. The corpus callosum (in green) is the largest commissure and connects the left and right frontal, temporal, parietal, and occipital lobes of the cerebral cortex. The anterior commissure (in green) connects the left and right temporal lobe of the cerebral cortex and the left and right amygdala. The adhesio interthalamica connects the left and right thalamus (in brown), and the hippocampal commissure connects the left and right hippocampus (in yellow).

topographical organization of the corpus callosum, several authors have suggested that the corpus callosum can be divided into segments. Probably the most widely used classification scheme was proposed by Witelson.[12] It was based on postmortem examinations of human and primate brains. The following seven segments of the corpus callosum are distinguished (Fig. 3.4C):

1. Rostrum ("beak"): This segment connects orbitofrontal areas.
2. Genu ("knee"): This segment connects prefrontal areas.
3. Rostral body: This segment connects premotor and supplementary motor areas.
4. Anterior midbody: This segment connects motor areas.
5. Posterior midbody: This segment connects somatosensory and posterior parietal areas.
6. Isthmus ("narrow place"): This segment connects the superior temporal and posterior parietal cortices.
7. Splenium ("bandage"): This segment connects the inferior temporal and occipital cortices.

In some publications, rostrum, genu, and rostral body are also combined to constitute the so-called anterior third.

In addition to this topographical model, several other ways to parcellate the corpus callosum have been proposed over the years.[14] In general, they can be differentiated into:

1. Anatomical division: A structural approach based on macrostructural features obtained from anatomical studies.
2. Histological division: Based on microstructural features obtained from histological studies, such as the range of axonal density.[11]
3. Geometrical division: Based on devising the corpus callosum into sections by size.[12]
4. Connectional division: A functional approach based on diffusion tractography data.[13]

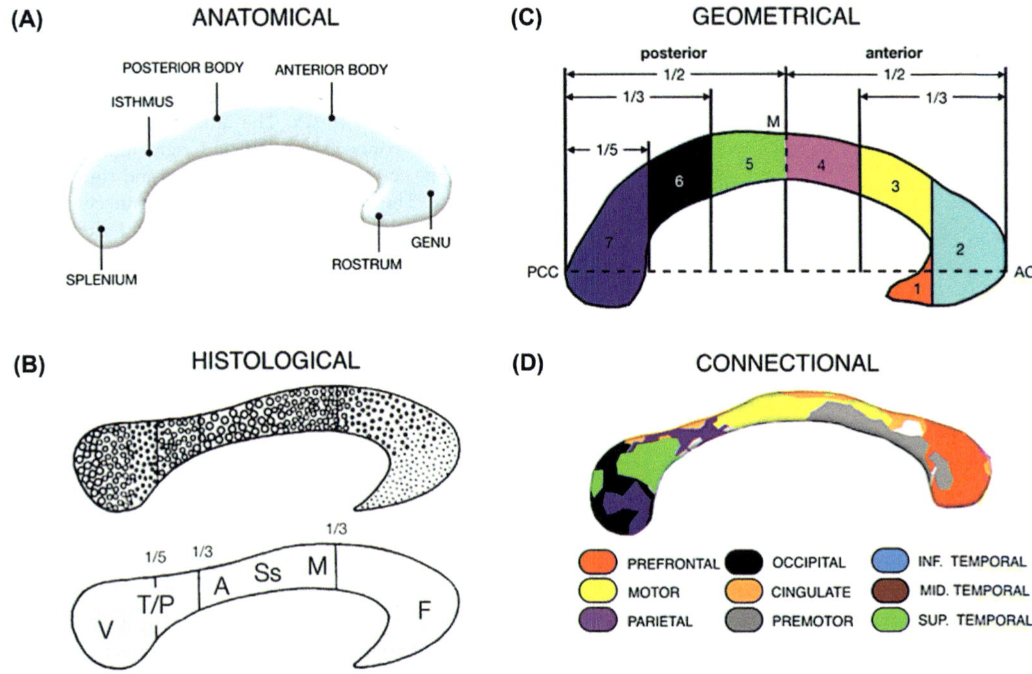

FIGURE 3.4 The four major types of parcellations of the corpus callosum. (A) Anatomical division based on gross anatomical features. (B) Histological division based on microstructural properties[11] (*F*, frontal; *M*, motor; *SS*, somatosensory; *A*, auditory; *T/P*, temporoparietal; and *V*, visual). (C) Geometrical division[12] (*ACC*, most anterior point of the corpus callosum; *PCC*, most posterior point of the corpus callosum; *M*, middle of the corpus callosum). (D) Connectional division based on diffusion tractography.[13] *Source: Reproduced from Friedrich P, Forkel SJ, Thiebaut de Schotten M. Mapping the principal gradient onto the corpus callosum. NeuroImage. 2020;223:117317. https://doi.org/10.1016/j.neuroimage.2020.117317, with permission of the authors and the publisher.*

Fig. 3.4 shows examples for each of the four parcellation schemes. In addition, it has recently been suggested to map callosal connections following the hierarchical organization of cognitive functions obtained from resting state data without enforcing boundaries between subsections like in the Witelson parcellation scheme.[14]

A commonly used connectional parcellation scheme for the corpus callosum was suggested by Hofer and Frahm [15] (see Fig. 3.5 for a comparison with the Witelson scheme). These authors used diffusion tensor imaging (DTI) data (Box 3.2) to conduct white matter tractography in order to create a classification scheme based on structural in vivo fiber connectivity (Fig. 3.6). They distinguished five vertical segments of the corpus callosum containing fibers projecting into functionally different cortical areas (Fig. 3.5):

1. Prefrontal connections
2. Premotor and supplementary motor connections
3. Motor connections
4. Sensory connections
5. Parietal, temporal, and occipital connections

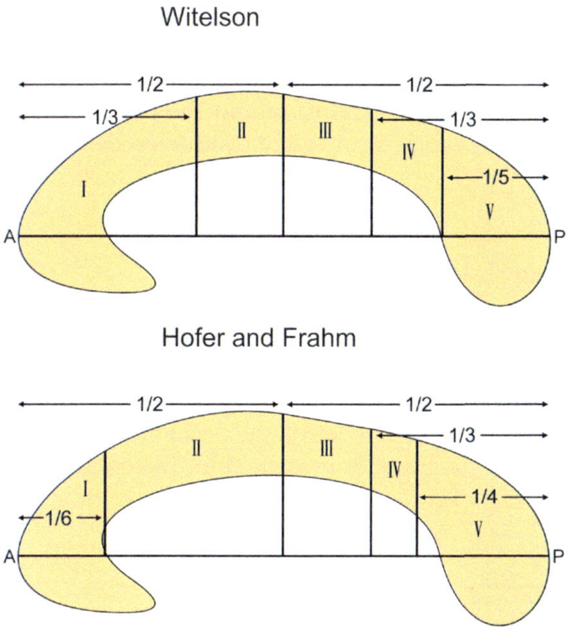

FIGURE 3.5 Comparison of the two most common models of the topography of the midsagittal corpus callosum. (Top) Witelson's classification. (I) Anterior third: prefrontal, premotor, and supplementary motor; (II) anterior midbody: motor; (III) posterior midbody: somesthetic, posterior parietal; (IV) isthmus: posterior parietal, superior temporal; (V) splenium: occipital, inferior temporal. (Bottom) Hofer and Frahm's classification. (I) Prefrontal; (II) premotor and supplementary motor; (III) motor; (IV) sensory; (V) parietal, temporal, and occipital. *A*, Anterior; *P*, posterior. *Source: From Hofer S, Frahm J. Topography of the human corpus callosum revisited—comprehensive fiber tractography using diffusion tensor magnetic resonance imaging. NeuroImage. 2006;32 (3):989–994. https://doi.org/10.1016/j.neuroimage.2006.05.044, used with permission from the authors and the publisher.*

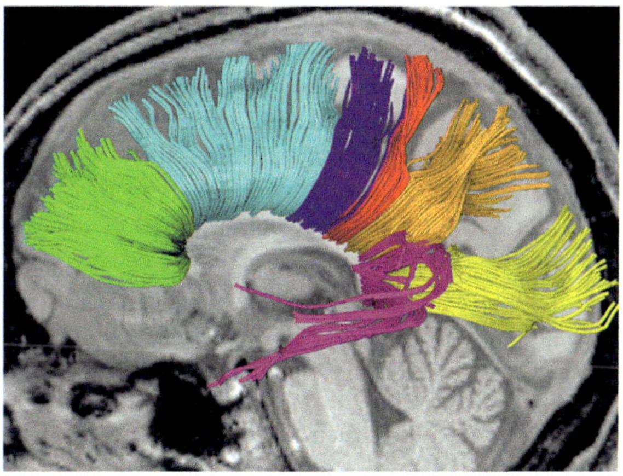

FIGURE 3.6 Sagittal view of a 3D reconstruction of callosal fibers comprising bundles projecting into the prefrontal lobe (coded in green), premotor and supplementary motor areas (light blue), primary motor cortex (dark blue), primary somatosensory cortex (red), parietal lobe (orange), occipital lobe (yellow), and temporal lobe (violet). *Source: Altered from Hofer S, Frahm J. Topography of the human corpus callosum revisited—comprehensive fiber tractography using diffusion tensor magnetic resonance imaging.* NeuroImage. *2006;32(3):989–994. https://doi.org/10.1016/j.neuroimage.2006.05.044, used with permission from the authors and the publisher.*

A brain, divided: the split-brain procedure

In the 1960s, pharmaceutical aid for epileptic patients was in its infancy, and for many patients, not much could be done. One major problem was callosal propagation of epileptic foci, which is the spreading of the initially unihemispheric source of the seizure over the corpus callosum to homotopic brains areas in the other hemisphere. This could lead to interhemispheric reciprocal excitation and therefore more severe seizures and possibly more extensive damage to affected brain regions as well as a higher risk of accidental physical injuries during seizures.[11] To prevent this expansion of epileptic activity, the idea came about that cutting the corpus callosum could represent a "last option" in patients with exceptionally severe epilepsy who did not respond to any other form of treatment.[17] As not much was known about the function of the corpus callosum and the possible consequences of such a procedure at that time, animal experiments were conducted before the procedure was applied to human patients.

Split-brain research in animals

Early split-brain research in animals was mainly conducted in the visual system of cats and monkeys. In the mammalian visual system, both eyes receive input from the left and the right visual fields. However, the optic fibers cross partially at the optic chiasm so that the nasal portions of the visual input (the parts on the sides towards the nose) are processed by the hemisphere contralateral to the visual field the input is presented in. In contrast, the

BOX 3.2

Key technique: diffusion tensor imaging (DTI).

DTI is the main neuroimaging technique used to visualize white matter pathways in the human brain in vivo.[16] DTI pictures are acquired by scanning participants' brains in a magnetic resonance imaging (MRI) scanner using a specific DTI sequence (Fig. 3.7A). The MRI scanner is a medical device that uses magnetic fields to generate images of the human brain (or other body parts). In a DTI scan, the direction of movement of water molecules in the brain is measured. The technique is based on the principle that this movement of water molecules is not equal in all directions. Instead, it is modulated by obstacles in the tissue. For example, the myelinated membranes of fiber tracts represent a natural barrier for the diffusion of water molecules. This leads to so-called anisotropic diffusion, that is, the water molecules have a higher probability of moving in the same direction as a fiber tract than against the direction of this tract. For the analysis of DTI data, the recorded brain images are subdivided into so-called voxels, for example, three-dimensional pixels of a given size (e.g., 2 by 2 by 2 mm). For each voxel, three values (the eigen values λ_1, λ_2, and λ_3) are calculated, indicating the diffusion of water molecules within this voxel along the x, y, and z axis. Based on these values, the fractional anisotropy (FA) of the voxel is determined. FA has a value between 0 and 1, with higher values indicating stronger diffusion along one of the three axes compared to the other two. By combining the FA values of all voxels in one picture and color-coding their main diffusion direction (red: left to right, green: posterior to anterior; and blue: foot to head), whole-brain FA maps can be created (Fig. 3.7B). These FA maps can be used to estimate the position and size of specific fiber tracts in the brain such as the corpus callosum. To this end, different tractography algorithms are used that generate visual representations of the investigated fiber tracts (see Fig. 3.7C).

BOX 3.2 (*cont'd*)

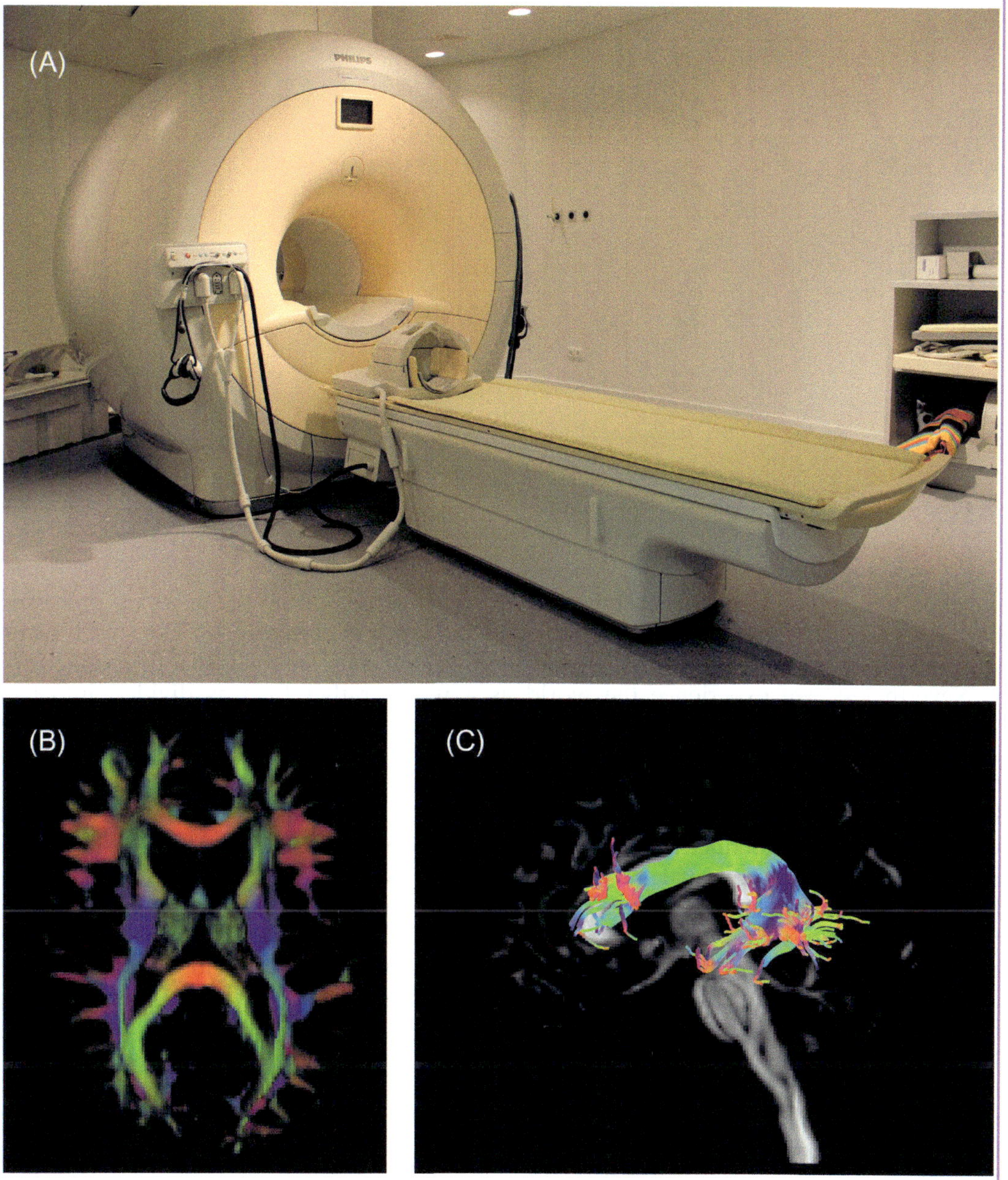

FIGURE 3.7 (A) To acquire DTI pictures, participants' brains are scanned in a magnetic resonance imaging (MRI) scanner using a DTI sequence. (B) Whole-brain FA-maps are created by combining the FA-values of all voxels in one picture. Color encodes the main diffusion direction (red: left to right, green: posterior to anterior; blue: foot to head). (C) Using these FA-maps, the position and size of specific fiber tracts in the brain can be estimated and visualized.

FIGURE 3.8 In the mammalian visual system, information in the left visual hemifield is processed in the right visual cortex and vice versa.

distal portions of the visual input (the parts on the sides away from the nose) are processed by the hemisphere ipsilateral to the visual field the input is presented in. Thus, each hemisphere initially only receives visual information from the contralateral visual field (Fig. 3.8).

Nevertheless, we experience the world as a whole rather than as divided into two visual fields. Could it be that the corpus callosum is critical for this integration? This groundbreaking idea was developed by Roger Sperry (1913–94), a neuroscientist at the California Institute of Technology ("Caltech") in Pasadena, USA. Sperry proposed to test this hypothesis with Ronald E. Myers from the Department of Anatomy at the University of Chicago. Myers then conducted a seminal study in which he tested the function of the corpus callosum in transmitting visual information between the hemispheres in cats.[18] In six cats, both the optic chiasm and the corpus callosum were transected so that contralateral information from the eyes was not transmitted and the hemispheres did not exchange information anymore. After recovering from surgery, the animals were trained to perform simple pattern discrimination tasks (e.g., distinguishing a circle from a square), while one of their eyes was blindfolded. As soon as the animals made the correct choice on more than 85% of the trials, the other eye was blindfolded. Instantly, performance dropped to chance level (see Fig. 3.9). Thus, callosal transection led to a failure to transfer simple visual information. These results were in contradiction to an earlier study in cats, in which only the optic chiasm but not the corpus callosum had been transected.[19] Myers concluded that his results clearly demonstrated the important role of the corpus callosum for visual learning and information transfer. This assumption was also supported by the finding that the cats could be trained on two conflicting problems with the two eyes with no indication of interference between the two learning processes. Myers findings were supported by further studies in cats[20,21] and monkeys.[22]

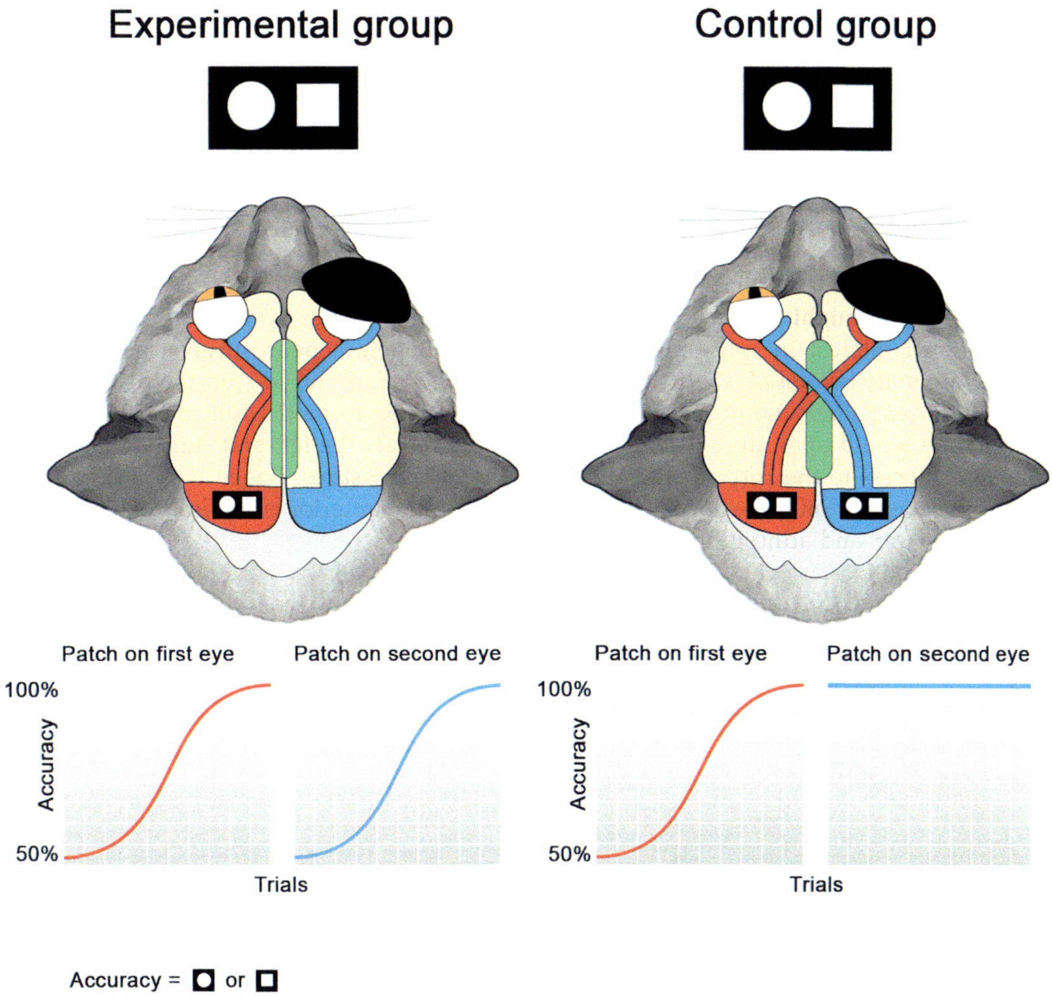

FIGURE 3.9 The Experiment by Myers.[18,19] Experimental groups: Cats in which both corpus callosum and optic chiasm were cut were able to learn a visual discrimination task with one eye blindfolded. When the eye patch was moved to the other eye, they showed no carryover effects and started learning the task anew. Control group: Cats in which only the corpus callosum or the optic chiasm or none of the two was cut retained their task performance when the eye patch was moved to the other eye.

Split-brain surgery in humans

The animal experiments showed that commissurotomy prevented transfer between the hemispheres. Aside from that, there seemed to be few (if any) side effects. The cats and monkeys ate, walked, and behaved normally just like animals that still had a corpus callosum. Commissurotomy in humans as a means to treat severe epilepsy had already been performed in the 1940s by William van Wagenen and Yorke Herren in a group of patients in Rochester. Apparently, the procedure neither affected the severity of seizures nor the neurobehavioral performance of the patients[23,24] so that Van Wagenen and Herren stopped performing it. After testing these patients in a series of experiments, Andrew Akelaitis of the University of Rochester School of Medicine, USA, wrote in 1944:

> Following complete section of the corpus callosum, no disturbance in visual, auditory and tactile gnosis was observed and praxis and language functions were unimpaired. *From Akelaitis*[25]

It was later speculated that these first commissurotomies were unsuccessful because the corpus callosum might have been only partially severed.[26] After Sperry's animal research had renewed interest in split-brain surgery, Joseph Bogen, a neurosurgeon at the Loma Linda University School of Medicine in Los Angeles, USA, re-introduced the procedure as a means to treat otherwise intractable epilepsy.[27] Bogen performed a full commissurotomy in which he would open the skull and mechanically retract the left and right hemispheres to provide access to the corpus callosum and other commissures. Bogen would then use a scalpel to sever not only the corpus callosum but also the anterior commissure, hippocampal commissure, and adhesio interthalamica[3] (Fig. 3.10).

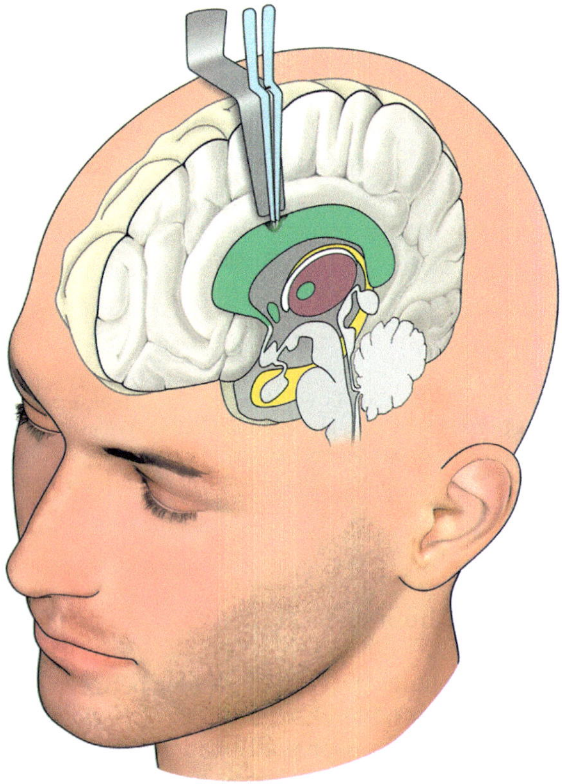

FIGURE 3.10 Split-brain surgery in humans.

Language in split-brain patients

The first patient to undergo split-brain surgery in Los Angeles was WJ, a 42-year-old World War II veteran who had started to experience blackouts with worsening convulsions after his parachute had failed to open completely during a bombing raid in the Netherlands in 1944. His fall led to several injuries and a prolonged period of unconsciousness. Since his condition had worsened over time and the pharmacological control of seizures was impossible, split-brain surgery was performed, and his forebrain commissures were dissected.[27] After surgery, WJ showed generalized weakness, akinesis (inability to initiate movements), and mutism (inability to speak), but these symptoms cleared after a few weeks. The surgery was deemed a success when the patient did not experience any major seizures for 6 months afterward and repeatedly remarked that he felt better than he had in years. Starting about 6 weeks after surgery, Michael Gazzaniga, a PhD student of Roger Sperry, started to investigate WJ's cognitive abilities[28,29] using the visual half-field technique (Box 3.3) and other neuropsychological tests.

Studies with the visual half-field technique revealed that split-brain patients had a division of thought and possibly even of consciousness. WJ responded normally to visual stimulation in the right visual half-field, which is processed by the left hemisphere, but showed agnosia for all stimuli presented in the left visual half-field, which is processed by the right hemisphere. He was also unable to name, draw a rough resemblance, or to otherwise describe objects shown in the left visual half-field. Written material was easily read and understood in the right visual half-field but elicited no response when presented in the left visual half-field. Thus, it became evident that only WJ's left hemisphere could talk. Essentially, all communication had been confined to an interaction with only half the person. While the tests on WJ were only a single-case study, a general predominance of the left hemisphere for language production and perception was found in nearly all split-brain patients tested in the decades following this initial discovery. However, it is important to note that this left-hemispheric language dominance is relative, not absolute. A few patients showed right-hemispheric language dominance, and there are certain aspects of language that the right hemisphere can process (and typically does)[31]; more on this fascinating phenomenon will be discussed in Chapter 5.

Interesting insight into the role of different parts of the corpus callosum comes from a study on JW, a young split-brain patient in whom the surgery had been performed in two consecutive steps.[32] In this patient, the posterior half of the corpus callosum including the splenium was sectioned in the first surgery. Since seizures were recurring after this intervention, the remaining anterior portion of the corpus callosum was sectioned in a second operation 10 weeks later. JW's perceptual and cognitive functions as well as sensory transfer over the corpus callosum were evaluated before and after the first and the second surgery. The patient was tested with a visual naming task in which a word (e.g., "knight") was presented in the left visual half-field. Before the corpus callosum was cut, the patient's naming ability was unimpaired, and he responded by saying the word that was shown. However, after the initial posterior section, the primary visual areas of the right hemisphere were disconnected from left-hemispheric language areas. At this stage, the patient was unable to name the object. However, some transfer of information clearly still occurred, as he seemed to have an understanding of the concept of the object that was shown.

BOX 3.3

Key technique: the visual half-field technique.

The visual half-field technique was the most important methodological tool used to investigate hemispheric asymmetries in split-brain patients. In a visual half-field paradigm,[30] the subject is seated in front of a screen and asked to fixate their gaze on a cross that is presented in the center of the screen. A word or picture is then presented very briefly (less than 200 ms) on either the left or the right side of the fixation cross, and the participant is asked to respond to it quickly, for example, by verbally expressing what they saw or by pressing a button to categorize the stimulus. This setup takes advantage of the way the visual system in humans and other mammals is organized (Fig. 3.11). In principle, both eyes receive input from both the left and the right visual field. The nasal portions of the visual input (the parts on the sides toward the nose) are processed by the hemisphere contralateral to the visual field the input was presented in. In contrast, the distal portions of the visual input (the parts on the sides away from the nose) are processed by the hemisphere ipsilateral to the visual field the input was presented in. Thus, everything on the left side of the screen is projected into the right hemisphere and vice versa. By presenting visual information in only one visual field, the information is at first only processed by the contralateral hemisphere, but this unilateral processing cannot be retained for very long as saccadic eye movements (quick eye movements toward one side) can occur. Since this type of eye movement typically has a latency of about 200 ms, the stimuli in visual half-field paradigms are usually presented for a duration shorter than 200 ms to ensure unilateral processing.

FIGURE 3.11 The visual half-field paradigm. Participants sit in front of a screen with their chin placed on a chin rest to prevent head movements. They are asked to fixate a central fixation cross, and stimuli are presented in the left or right visual half-field. Due to the organization of the visual system, stimuli presented in the left visual field are processed by the right hemisphere and vice versa.

For example, the patient said:

> I... have a picture in mind but, can't say it.... Two fighters in a ring.... Ancient wearing uniforms and helmets.... on horses trying to knock each other off... Knights? *From Sidtis et al.*[32]

After the anterior part of the corpus callosum had also been cut, the patient would typically respond that he did not see anything. These results revealed that the frontal part of the corpus callosum does not transfer sensory information but rather higher order cognitive information such as concepts and meanings.

Visuospatial abilities in split-brain patients

After establishing dominance of the left hemisphere for verbal functions, it was a natural next step to investigate whether the right hemisphere in split-brain patients showed any complementary dominance for other cognitive functions. Bogen and Gazzaniga investigated visuospatial functions in WJ and NG, two right-handed split-brain patients, before and after surgery.[33] Before surgery, both patients could copy geometric figures better with their right hand. However, the commissurotomy obliterated their postoperative ability to draw three dimensional objects with the right hand controlled by the left hemisphere. Thus, it can be assumed that before surgery, the left hemisphere had relied on the visuospatial processing capabilities of the right hemisphere to perform the task. Specialization of the right hemisphere for visuospatial processes was also confirmed using other tasks, e.g., arrangement of colored blocks into complex spatial figures.

Information transfer in the split-brain: the role of subcortical commissures

Our attentional resources are limited. Since the two hemispheres in split-brain patients work independently, one could assume that they also have independent attentional resources, for example, that the difficulty of cognitive operations in one hemisphere does not influence the performance of the other hemisphere at the same time. To test this hypothesis, Holtzman and Gazzaniga developed a paradigm in which dual task performance was assessed in split-brain patient VP and healthy controls.[34] On each trial of this experiment, two series of geometric shapes were shown simultaneously in the left and right visual field so that they were processed by the left and the right hemisphere of the patient independently. At the end of each stimulus series, a probe stimulus appeared in one visual field, and the participants had to indicate whether this stimulus matched any of the stimuli shown before in the visual field it appeared in. The patient was slower and made more errors when the series of geometric shapes shown in the relevant visual field was longer. Interestingly, he was also affected by the number of stimuli shown in the irrelevant visual field, as presentation of more stimuli in this visual field also decreased performance. Thus, split-brain patients show the same attentional limitations as individuals with an intact corpus callosum, suggesting that these attentional limitations are the product of a system that communicates to some extent via subcortical commissures.[35]

Motor timing in split-brain patients also seems to be controlled by a common subcortical module. Franz and his coworkers[36] tested split-brain patient JW and three healthy controls with a motor task, in which the participants had to draw stimuli (rectangles that were open on one side) that appeared on a computer monitor in front of them, one stimulus with the right hand and another stimulus with the left hand. In control subjects, the performance in this task was impaired when the two rectangles had the opening in different positions (e.g., to the top and to the right) or differed in size. In contrast, JW's performance was not affected by these variables. Thus, split-brain patients seem to have independent motor programs in their two hemispheres.

However, the patient did show temporal coupling of the left- and right-hand responses, as the two hands ended up drawing the three segments of the rectangles in the same rhythm.

Further studies investigating interhemispheric interaction in split-brain patients[37,38] found that:

1. Patients were able to say if two horizontal lines shown to both hemispheres were broken at the vertical meridian or not but were unable to say if the left line was higher or lower than the right line.
2. Patients were able to decide if two numbers shown separately to the hemispheres added up to 10 or more but were unable to say what the sum was.
3. Patients were able to say if two letters shown to each hemisphere made a word or not but were unable to say the word.
4. Patients could vaguely describe a person shown to the right hemisphere but could not say who this person was.

Thus, the two hemispheres in split-brain patients are not completely separated. Especially in split-brain patients operated decades ago, subtle interhemispheric transfer over subcortical connections seems to occur. However, this transfer is not fine-grained or declarative but rather appears to be category-based and with low resolution.

The Poffenberger task: findings in split brain patients

Support for the idea that some form of transfer of visual information occurs in split-brain patients also comes from research with the so-called Poffenberger task. The Poffenberger task is named after Albert Poffenberger (1885–1977), an American psychologist who published the task in his doctoral work in 1912.[39] The Poffenberger task is a simple visuomotor reaction time paradigm, in which the participant must react to visual stimuli presented in the left or right visual half-field by pressing a button using either the left or the right hand. If the response is to be made with the hand ipsilateral to the visual stimulus, the visual areas that perceive the visual information and the motor areas controlling the reacting hand are in the same hemisphere. On these so-called "uncrossed" trials, percept and response can be integrated within one hemisphere. If, however, the visual stimulus and the hand used to react to it are on different sides (so-called "crossed" trials), communication between the two hemispheres must occur to make a response. By comparing crossed and uncrossed trials, the crossed–uncrossed difference (CUD) can be calculated as an estimate of interhemispheric transfer time. Typically, healthy subjects show a small but statistically significant CUD (generally less than 10 ms). If no interhemispheric transfer occured in split-brain patients, they would be unable to solve the task on crossed trials. This is, however, not the case. Recently, René Westerhausen from the University of Oslo performed a meta-analysis of studies using the Poffenberger task in different groups of split-brain patients.[40] Overall, 22 studies with 38 unique cases were included in the meta-analysis. The mean CUD was 60.6 ms for patients with full commissurotomy, 43.5 ms for patients with complete callosotomy, and 8.8 ms for patients with partial anterior-medial callosotomy. Patients with full commissurotomy and patients with complete callosotomy had significantly longer CUDs than patients with partial callosotomy or healthy controls. This shows that cutting the corpus callosum leads to a very substantial increase in the time needed for interhemispheric integration of simple visual stimuli. Importantly, it does not completely prevent such transfer, further supporting the idea that subcortical connections between the hemispheres also play a role in interhemispheric information integration. The findings from the Poffenberger task meta-analysis were also in line with the results of a neuroimaging study that assessed functional connectivity before and after callosotomy in 22 epilepsy patients.[41] In this longitudinal study, it was found that callosotomy leads to markedly reduced interhemispheric connectivity. Interestingly, some interhemispheric functional connectivity remained even after complete callosotomy, specifically in the primary sensorimotor and visual areas relevant for performance in the Poffenberger task. This residual connectivity was mediated by extracallosal pathways.

Hemispheric asymmetries in information processing

Another fascinating insight from split-brain research is the finding that the two hemispheres seem to exhibit different styles of information processing. Based on somewhat anecdotal evidence, Gazzaniga suggested that the left hemisphere is an "interpreter," which constantly makes sense of its own actions, partly without knowing why these actions were done.[31] This idea was based on the results of a simultaneous concept test, in which the patient was shown two pictures, one to the left hemisphere and one to the right. They were then asked to choose from a number of pictures which were associated with the pictures they saw in the two visual fields. For example, patient PS was shown a picture of a chicken claw in the right visual half-field (left hemisphere) and a picture of a snow scene in the left visual field (right hemisphere). The patient then chose a picture of a chicken and a picture of a shovel from the pictures in front of him. When asked about his choice, the patient replied:

> Oh, that's simple. The chicken claw goes with the chicken, and you need a shovel to clean out the chicken shed. *From Gazzaniga*[31]

Thus, the left hemisphere which gives the verbal response interprets the picture of the shovel within the context of keeping chickens, consistent with its knowledge but without any information about the snow scene perceived by the right hemisphere.

Experimental evidence supporting the idea of the left hemisphere as "interpreter" comes from a probability guessing experiment in split-brain patients.[42] At the beginning of each trial, a row of arrows was presented, pointing either to the left or to the right side, thereby signaling in which visual field a stimulus would occur. Two stimuli could appear: either a green square toward the top of the screen or a red square toward the bottom of the screen. The patients had to press a button to indicate which stimulus would appear. After their guess, one of the two stimuli was presented. For the right visual half-field, the top stimulus had a probability of 80% and the bottom stimulus a probability of 20%.

For the left visual half-field, the top stimulus had a probability of 70% and the bottom stimulus had a probability of 30%. The sequences of stimuli in the left and right visual fields were randomly generated and independent of each other.

There are two possible strategies to react in this experiment:

- Frequency matching: The subject chooses the two alternatives with their respective probabilities, for example, pressing the button for the top stimulus in 80% of the trials and the button for the bottom stimulus in 20% of the trials. This is not an optimal strategy in this task, leading to 68% correct guesses for the right visual field ($0.8 \times 0.8 + 0.2 \times 0.2$).
- Maximizing: The subject always chooses the more frequent alternative, for example, pressing the button for the top stimulus in 100% of the trials, irrespective of the stimulus that was presented in the trial before. This is the optimal strategy in the paradigm, leading to 80% correct guesses for the right visual field ($0.8 \times 1.0 + 0.2 \times 0$).

In line with the idea of the left hemisphere being an "interpreter," the split-brain patients used a frequency matching strategy on right visual half-field trials but a maximizing strategy on left visual half-field trials. This indicates that the left hemisphere seems to try to generate rules in a random series of events, while the right hemisphere sticks to a less complex (and more efficient) outcome-oriented strategy. Thus, the left hemisphere can be inferior to the right one in guessing random events. Interestingly, pigeons also choose the maximizing strategy when tested in this paradigm,[43] thus performing better than the left hemisphere in humans.

Split-brain research today

> The great pleasure and feeling in my right brain is more than my left brain can find the words to tell you. *Roger Sperry*[44]

In 1981, the Nobel Prize in Physiology or Medicine was awarded to Roger Sperry for his seminal contributions to neuroscience. Split-brain research has been pivotal to our understanding of hemispheric asymmetries and to this day can still yield fascinating insights into the asymmetric brain.[26] In the 2020s, researchers continue to publish both empirical and theoretical articles on split-brain patients.[45–47] However, split-brain research is not as central to this area of research as it used to be. This is for several reasons. On the one hand, split-brain surgery has always been a rare procedure. Pharmacological treatment has now mostly been replaced with medication,[48] although it still is used in rare cases to treat drug-resistant epilepsy.[49] Patients from the initial cohorts are getting older and less available for testing, and while there are patient cohorts outside of the United States of America—for example, in New Zealand[50] and Italy[51]—most researchers simply do not have access to these populations. In addition, the typically small sample sizes of these cohorts, combined with heterogeneous comorbidities, complicate the use of more complex experimental designs, as they might lead to a lack of statistical power. Since the lack of statistical power due to small sample sizes has been identified as one of the main reasons for the ongoing replication crisis,[52] researchers today tend to be more critical of findings obtained in very small patient cohorts than they used to be in the last decades. Last but not least, the advance of neuroimaging techniques such as fMRI has made research on hemispheric asymmetries in healthy subjects much easier, as the dominant hemisphere can now be determined simply by comparing right- and left-sided brain activations during task performance. That said, using fMRI and other advanced neuroimaging techniques in split-brain patients has the additional potential to further advance our understanding of the role of the corpus callosum for hemispheric asymmetries,[53] and it can only be hoped that more researchers will be able to study these unique patients with modern neuroscientific methods.

Patients with agenesis of the corpus callosum: natural split-brains or not?

In addition to split-brain patients, another group of patients have been important in the quest to elucidate the role of the corpus callosum for functional hemispheric asymmetries. In patients with agenesis of the corpus callosum (AgCC), the corpus callosum is developmentally either completely (complete AgCC) or partially (partial AgCC) absent (see Fig. 3.12).

AgCC is a rare congenital condition that occurs in about 1 in 4000 individuals.[55,56] AgCC is genetically highly heterogenous and can result from a disruption of any of the multiple steps of prenatal development of the corpus callosum.[57] Accordingly, it has been associated with a large number of chromosomal, autosomal, and X-linked defects.[58,59] Moreover, environmental factors such as alcohol or toxin exposure in utero have also been suggested to cause AgCC.[60] Depending on the underlying developmental defect, patients with AgCC often show a wide range of other developmental abnormalities, such as seizure disorders, cysts, or lipomas.[58] Children with AgCC frequently show developmental delays, mental retardation, and learning disability, but it has been suggested that these problems are mainly caused by other cerebral anomalies associated with the underlying

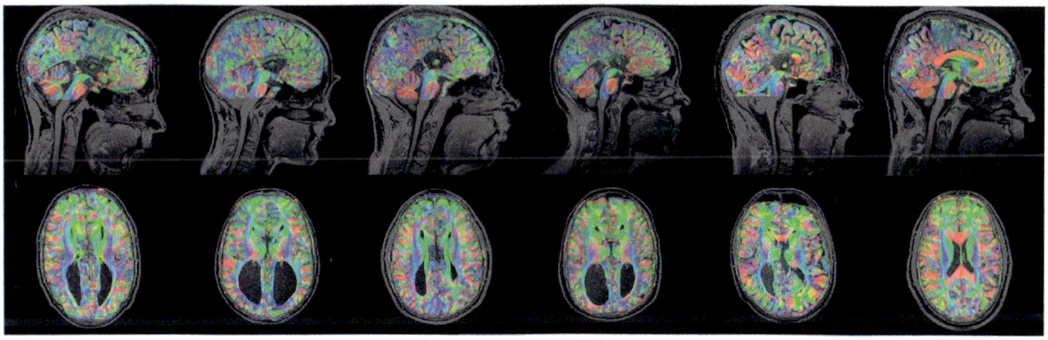

FIGURE 3.12 Sagittal (top) and axial (bottom) view of the DTI-based images of four patients with complete callosal agenesis, one patient with partial callosal agenesis (the anterior part of the corpus callosum is still intact), and one healthy participant with an intact corpus callosum. The colors represent fiber orientations in different directions: Right–left (red), anterior–posterior (green), and superior–inferior (blue). The corpus callosum of the healthy participant is clearly visible in red.[54] *Source: From Genç E, Ocklenburg S, Singer W, Güntürkün O. Abnormal interhemispheric motor interactions in patients with callosal agenesis. Behav Brain Res. 2015;293:1–9. https://doi.org/10.1016/j.bbr.2015.07.016, used with permission from the authors and the publisher.*

genetic defect, not the AgCC per se.[61] However, patients with isolated AgCC—for example, those who show only the absence of the corpus callosum but no other brain malformations or general defects of the central nervous system—frequently show only subtle cognitive impairments, if any, but no major neurodevelopmental defects.[62]

Research on hemispheric asymmetries in patients with AgCC has focused on patients with isolated AgCC. While it could be assumed that congenital absence of the corpus callosum would lead to neurobehavioral effects similar to those observed in split-brain patients, one has to keep in mind that these patients are different from split-brain patients in a number of ways. First, their brains have had time to adjust to the absence of interhemispheric transfer over the corpus callosum. In split-brain patients, the corpus callosum is typically severed in adulthood after the brains have depended on the corpus callosum for interhemispheric transfer for decades. In contrast, patients with AgCC are born without a corpus callosum, potentially allowing for more neuroplastic restructuring processes to compensate for the absence of the corpus callosum. Therefore, these patients typically show less severe neurobehavioral effects than split-brain patients.

The foremost evidence for neuroplastic restructuring processes in the brains of patients with AgCC is the existence of a white matter pathway that does not exist in healthy subjects or split-brain patients: the Probst bundle. This structure, named by Moritz Probst, an Austrian neuroanatomist who first described it after postmortem examinations of acallosal brains,[63] is constituted by callosal fibers that failed to pass to the other hemisphere and connect the cortex from anterior to posterior (Fig. 3.13). The existence of these fibers was later confirmed by DTI studies,[64,65] although their functional significance still remains unclear.

In addition to the existence of Probst bundles, Tovar-Moll and her coworkers[66] suggested more extensive rewiring in the brains of AgCC patients. Using DTI, these authors identified two anomalous white matter tracts in AgCC patients which connected the posterior parietal cortices via the posterior and anterior commissures, possibly allowing AgCC patients greater

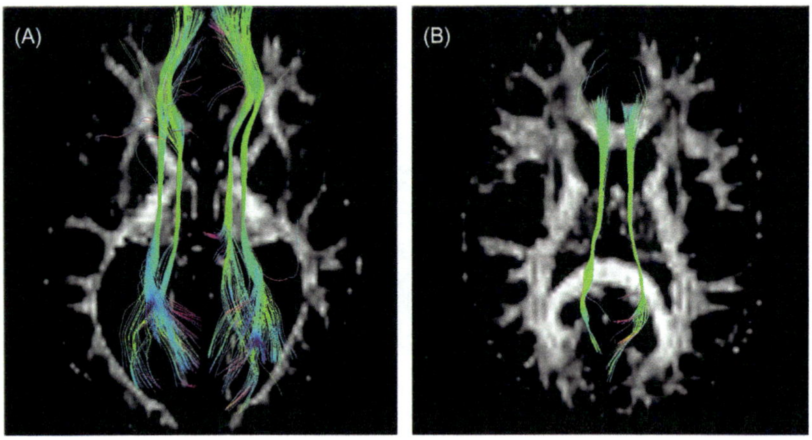

FIGURE 3.13 DTI-based pictures of the Probst bundles. (A) In a patient with complete agenesis of the corpus callosum, Probst bundles are visible in addition to the cingulum. (B) In healthy controls, only the cingulum is visible. The colors represent fiber orientations in different directions: Right–left (red), anterior–posterior (green), and superior–inferior (blue).

capacity for interhemispheric transfer than split-brain patients. Moreover, AgCC patients do not show the same neurobehavioral performance as healthy subjects.[62] While it was assumed for a long time that other commissures could completely compensate for the absence of the corpus callosum, careful experimental assessment of the cognitive capabilities of these patients showed that this assumption was incorrect. Importantly, acallosal individuals typically display longer interhemispheric transfer times than healthy participants.[67,68]

In the Poffenberger task, patients with callosal agenesis show a much larger CUD (up to 60 ms) than healthy individuals (less than 10 ms).[67] Thus, in healthy individuals, information transfer occurs rapidly over the corpus callosum. In contrast, AgCC patients have to use other, less efficient interhemispheric pathways, resulting in slower interhemispheric transmission times.[68]

Does this unusual way of transferring information between the hemispheres also affect hemispheric asymmetries? The answer is probably yes. A few studies have investigated functional hemispheric asymmetries in AgCC patients, and in general, the results suggest that the ontogenetic absence of the corpus callosum reduces—or sometimes even reverses—the development of brain asymmetries. For example, patients with callosal agenesis show a tendency to be more ambidextrous than healthy controls, that is, they present with reduced functional lateralization for fine motor processes.[69] Accordingly, mice without a corpus callosum show less consistent paw preferences than wild mice, indicating that paw preferences are strongly influenced by the environment.[70] Moreover, behavioral data suggests that left-hemispheric language dominance is affected by ontogenetic absence of the corpus callosum. The typical left-hemispheric language dominance found in the dichotic listening task (see Chapter 5 for a detailed description) seems to be reduced[69]—or even reversed[71]—in patients with callosal agenesis. More evidence suggesting a less lateralized, more bi-hemispheric development of usually distinctly lateralized linguistic functions was provided in a recent fMRI and DTI study on speech production and perception.[72] However, it has to be noted that there are also studies that failed to find an effect of AgCC on hemispheric asymmetries, and even others that show increased functional asymmetry. For example, an fMRI study on syntactic decision-making and subvocal verbal fluency in a sample of six high-functioning AgCC patients with complete absence of the corpus callosum reported mostly normal activation asymmetries.[73] In contrast, a single-case study reported strong structural and behavioral asymmetries in one male patient with callosal agenesis.[74] In conclusion, the extent to which callosal information transfer affects the development and maintenance of hemispheric asymmetries seems to be individually different and somewhat task-dependent.

Hemispheric asymmetries in species without a corpus callosum

Only placental mammals (e.g., mice, monkeys, and humans) have a corpus callosum. Marsupials (e.g., koalas, kangaroos, and wombats) and nonmammalian vertebrates lack this structure,[7] giving rise to the colloquial statement that these species represent "natural split-brains."[75] Interestingly, recent studies indicate that marsupials exhibit functional hemispheric asymmetries comparable to placental mammals. For example, eastern gray and red kangaroos show forelimb preferences at the population level in a variety of

FIGURE 3.14 Two red-necked wallabies feeding from a bipedal position. *Source: From Giljov A, Karenina K, Malashichev Y. Limb preferences in a marsupial, Macropus rufogriseus: evidence for postural effect. Anim Behav. 2012;83(2):525–534. https://doi.org/10.1016/j.anbehav.2011.11.031, used with permission from the author and the publisher.*

behaviors, for example, unimanual feeding.[76] Similar results have also been found for the red-necked wallaby[77] (see Fig. 3.14).

Moreover, a recent anatomical study suggested that despite not having a corpus callosum, *Marsupialia* and *Monotremata* share distinct features of interhemispheric connectivity with Mammalian species that have a corpus callosum.[78] Specifically, the study showed that in fat-tailed dunnarts (*Marsupialia*) and duck-billed platypus (*Monotremata*), neuronal connections through the anterior commissure were topographically organized, similar to the corpus callosum. The authors of the study therefore proposed that the interhemispheric connectome in mammals predates the evolution of the corpus callosum, which may explain why marsupials show similar functional hemispheric asymmetries compared to mammals with a corpus callosum.

Outside the *Mammalia* class, functional hemispheric asymmetries have also been observed in several different reptile, amphibian, and fish species,[79] a large number of bird species,[80,81] and even in invertebrates[82] (see Chapter 2 for a detailed discussion of hemispheric asymmetries in non-human animal species). These findings clearly show that functional hemispheric asymmetries can develop even in brains that do not have a corpus callosum and that in those species interhemispheric transfer of neuronal information is conveyed by other structures, for example, the commissura anterior. This idea was recently supported by Sara Letzner from Ruhr University Bochum, Germany, who investigated the connectivity and neurochemistry of the commissura anterior in pigeons.[75] She found that in this bird species, the commissura anterior interconnects a wide network of forebrain structures, thus proving that the idea of calling birds "natural split-brains" is wrong.

The intact corpus callosum and behavioral asymmetries

Naturally, the findings in split-brain patients sparked interest in the research community to investigate the role of the intact corpus callosum for functional hemispheric asymmetries. However, before the widespread availability of MRI, it was notoriously difficult for researchers to obtain information on both the structural properties of the corpus callosum and functional hemispheric asymmetries in the same subject, with postmortem examination often being the only way to gain any information on the brain structure. In a classic study in *Science*, Sandra Witelson circumvented this issue with an elegant experimental design, in which she tested handedness in seriously ill cancer patients who agreed to postmortem brain examination in the event of death.[83] After the patients' passing, their brains were removed from the skulls, fixated, and bisected sagittally. The size of the corpus callosum was measured by tracing the outlines of the corpus callosum in photographs of the midsagittal section. Witelson found that the corpus callosum was 11% larger in left-handed and ambidextrous individuals compared to right-handers, a difference that represents more than 20 million individual fiber tracts. She concluded that the fact that in most left-handers, language and motor functions of the dominant hand are represented in two different hemispheres increased the need for interhemispheric interaction, resulting in an enlarged corpus callosum. Several subsequent neuroimaging studies have generally found a larger size of the corpus callosum in left-handers or inconsistent right-handers. For example, it was found that left-handers with left-hemispheric language dominance had a larger corpus callosum than right-handers,[84] or that left-handers have larger areas of the anterior body, posterior body, and isthmus of the corpus callosum than right-handers.[85] However, some studies did not find differences in callosal size between left- and right-handers,[86] and one study even reported a larger size of the corpus callosum in right-handers compared to left-handers.[87] Moreover, a series of meta-analyses on 24 different studies and databases did not find any significant associations between handedness and corpus callosum morphology.[88] The authors therefore concluded that left-handers and mixed-handers do not show an increase in callosal connectivity.

It has also been suggested that it is not the left- and right-handedness per se that is linked to corpus callosum size, but rather the degree of handedness, that is, how strongly an individual prefers to use one hand over the other (see Chapter 6). This hypothesis was supported by a large-scale neuroimaging study[89] analyzing the correlations between callosal thickness and degree of handedness. The size of the anterior and posterior midbody was found to correlate negatively with degree of handedness, with weaker handedness being related to a larger size of the segments of the corpus callosum.

In addition to handedness, other forms of behavioral lateralization have been investigated in relation to the size of the corpus callosum. For example, it has been found that lateralization in shape and texture discrimination is positively correlated with callosal size,[90] and that the size of the posterior corpus callosum correlated negatively with language lateralization.[91] Several studies have investigated the role of the corpus callosum for performance in the dichotic listening task, the most important behavioral measure of language lateralization (see Chapter 5 for a detailed description of this task), consistently reporting that participants with a larger corpus callosum showed less left-hemispheric language dominance in this task.[92,93] Moreover, microstructural properties of the corpus callosum have been linked to the electrophysiological correlates of dichotic listening in a comparable manner. A study in 103 right-handers found that higher microstructural integrity of the posterior third of the corpus callosum was associated with decreased latency asymmetry of the N1, an event-related EEG component linked to attention toward language stimuli.[94]

Assessing the functional role of the corpus callosum for hemispheric asymmetries

One of the most important functions of neuroimaging studies relating structural parameters of the corpus callosum to behavioral lateralization is to assess the functional role of the corpus callosum for hemispheric asymmetries. Most axons in the corpus callosum rely on glutamate, an excitatory neurotransmitter. In view of this, it could be concluded that information transfer over the corpus callosum likely serves an excitatory function for the contralateral hemisphere. However, most of these excitatory axons target inhibitory interneurons, highlighting the possibility that callosal information transfer can also have an inhibitory function on the contralateral hemisphere.[95] Due to this finding, both excitatory and inhibitory models for the role of the corpus callosum for the emergence of hemispheric asymmetries in healthy brains have been proposed.

The best-known excitatory model was proposed by James L. Ringo from the University of Rochester, USA. He assumed that functional hemispheric asymmetries arose due to the conduction delay during interhemispheric information transfer.[96] The bigger the brain, the longer this delay, as conduction time is stable, but the distance over which the information must be transported gets larger. Thus, a bigger brain would lead to stronger functional hemispheric asymmetries since in that case, time-sensitive processes would be more likely to be performed by faster unihemispheric circuits. If both hemispheres were involved in generating behavior that is critical for survival in a big-brained organism (e.g., fleeing from a predator), the organism would be at an evolutionary disadvantage, as the conduction delay over the corpus callosum is large and the motor output could happen too late. This model predicts that mammals with large brains such as humans (but also elephants and whales) should have a higher degree of hemispheric specialization than animals with smaller brains. The model was supported by the finding that the area of the corpus callosum correlated negatively with the extent of ear asymmetry in a dichotic listening task, indicating that the bigger the corpus callosum, the smaller is the functional lateralization.[97] Moreover, a recent meta-analysis of MRI studies on all types of functional hemispheric asymmetries reported that cortical regions that showed enhanced asymmetries in task activity also showed reduced connections with the other hemisphere.[98] In line with the Ringo hypothesis, this finding suggests that functional hemispheric asymmetries may have evolved to avoid excessive conduction delay.

In contrast to Ringo's excitatory model, inhibitory models assume that the dominant hemisphere inhibits the subdominant one via the corpus callosum, leading to a reduction of functional lateralization with smaller callosal size.[99,100] This model was supported by an fMRI study that investigated the relation between the midsagittal surface area of the corpus callosum and language-related brain activation asymmetries, finding that participants with a larger corpus callosum showed stronger left-hemispheric language dominance in inferior frontal and posterior temporal brain regions.[101] Also, greater FA in the corpus callosum has been associated with increased functional language lateralization.[102,103]

Moreover, white matter fiber degradation in the corpus callosum due to alcoholism has been shown to lead to a reduction of functional visuomotor asymmetries,[104] further supporting the idea of an inhibitory function of the corpus callosum.

While both models are supported by empirical evidence, it is probably too simplistic to assume that a large fiber tract like the corpus callosum is either completely inhibitory or completely excitatory, and recent evidence indeed suggests that the corpus callosum likely exerts both excitatory and inhibitory influences on the contralateral hemisphere.[95,99] For example, specific callosal fiber types, subregions, or targeted interneurons could differ with regard to the extent of inhibitory or excitatory properties and thus differentially affect functional hemispheric asymmetries. This idea was supported by a study in healthy elderly participants which found significant negative correlations between functional language lateralization and the area of the most anterior and the most posterior of the five callosal segments but not with the area of the three intermediate subsegments.[105]

Interhemispheric integration and metacontrol

In typical behavioral experiments used to assess hemispheric asymmetries, each hemisphere is tested separately, and dominance is determined by comparing performances of the left and right hemisphere. However, the situation that only one hemisphere processes information—for example, in a tachistoscopic divided visual field task—is highly artificial and almost never occurs in real-life situations. Thus, it has been argued that a more realistic understanding of the role of the corpus callosum for hemispheric asymmetries and cognition in general can only be obtained by assessing how the different types of processing in the two hemispheres are combined to yield a single behavioral response.

In a classic study, Stuart Dimond and Graham Beaumont from the University College in Cardiff, UK, investigated whether subjects showed better performance when both hemispheres participated in a task compared to only one.[106] Subjects were asked to report pairs of random black digits with either both digits shown in the left visual field, both in the right visual field, or one in the left and the other one in the right visual field. Subjects reported about 61% of pairs correctly for left visual field presentation and about 69% of pairs correctly for right visual field presentation. Surprisingly, accuracy for stimulus presentation in both visual fields at the same time surpassed that for the better unilateral presentation, as subjects reported 75% of stimuli correctly. This shows that distribution of perceptual load between the hemispheres increases accuracy, implying that interhemispheric integration is beneficial for maximizing performance.

To investigate the role of task demands for this effect, Banich and Belger[107] developed a paradigm in which cognitive load was manipulated by varying the difficulty of the decision process. Subjects observed a stimulus display composed of three letters arranged in the shape of an inverted triangle (see Fig. 3.15). The two upper letters were located in the left and right visual field, while the lower letter was presented more medially than the others and either in the left or the right visual field.

In the less demanding physical identity condition, only upper case letters were presented, and subjects had to decide whether two of the three letters were physically

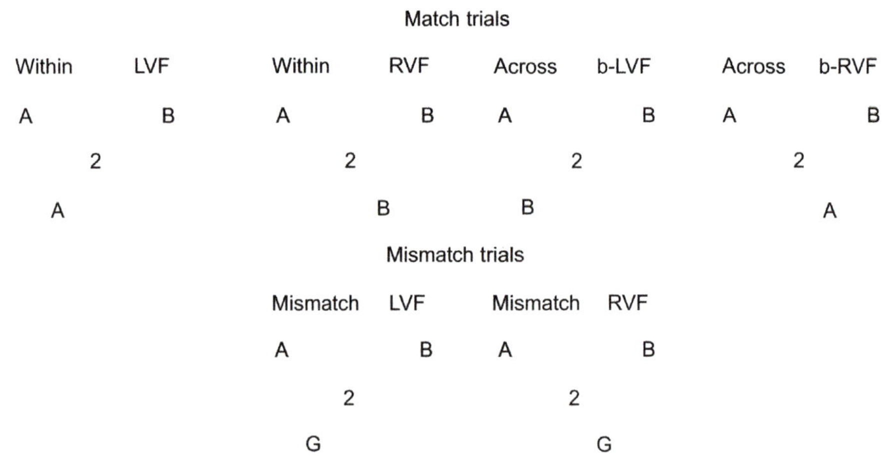

FIGURE 3.15 The four "match" and two "mismatch" trial types used in the study by Banich and Belger.[107] *Source: From Banich MT, Belger A. Interhemispheric interaction: how do the hemispheres divide and conquer a task? Cortex. 1990;26 (1):77–94. https://doi.org/10.1016/s0010-9452(13)80076-7, recreated with permission of the author and the publisher.*

identical (match trial, e.g., G G R) or not (mismatch trial, e.g., G R Q). For half of the match trials, the lower letter matched the upper letter presented in the same visual field ("within" trial), whereas in the other half of the match trials, the lower letter matched the upper letter presented in the other visual field ("across" trial). In the more demanding name identity condition, upper and lower case letters were used as stimuli, and subjects had to decide whether two of the three letters had the same name (e.g., G g). The results showed that dividing processing between the hemispheres on "across" trials led to better performance only for the more demanding name identity condition. In the easier physical identity task, subjects performed better on "within" trials. These findings indicate that performance is enhanced by interhemispheric integration for complex tasks, while simple perceptual tasks can be performed more efficiently by faster intrahemispheric circuits, that is, without information exchange over the corpus callosum.

Interhemispheric integration refers to the two hemispheres working together to solve a common problem, but what happens if they have different ideas about what to do? This particularly interesting aspect of interhemispheric integration over the corpus callosum is the focus of research on metacontrol. Metacontrol refers to the neural processes that determine which hemisphere takes control over cognitive operations when conflicting information on how to react are presented to the two sides of the brain. In human subjects, research on this phenomenon was pioneered by Jerre Levy of the University of Pennsylvania, USA, and Colwyn Trevarthen of the University of Edinburgh, Scotland. These researchers tested four split-brain patients who had to match tachistoscopically presented stimuli according to either structural appearance or functional category.[108] Stimuli were chimeric pictures based on black line drawings of common objects (e.g., a cake or hat). The two halves of each stimulus were made up of two different half-pictures (e.g., the left side of a cake and the right side of a hat). The middle of the stimulus was presented in the center of the screen so that the left half of the stimulus was processed only by the right hemisphere and the right half of the stimuli was processed only by the left hemisphere. Participants were asked to match the tachistoscopically presented chimeric pictures with whole pictures of the same objects which were laid out on a table in front of them. To determine which hemisphere takes control under which conditions, participants received three different types of instructions on a given run of trials:

1. Ambiguous instructions: "Select the one which is similar to, goes with, or is like in some way the one you see." This instruction favors no particular mode of processing and should introduce no hemispheric asymmetries.
2. Structural instructions: "Pick the thing that looks similar to what you see." This instruction favors a visuostructural mode of processing and should lead to dominance of the right hemisphere.
3. Functional instructions: "Pick the thing you would use with or that normally goes with what you see." This instruction favors a conceptual functional mode of processing and should lead to dominance of the left hemisphere.

The results showed that with ambiguous instructions, split-brain patients made both appearance and function matches. However, structural instructions led to matches based on appearance and right-hemisphere control, while functional instructions led to matches based on function and left-hemisphere control. These findings show that task instructions that favor a mode of processing preferred by one hemisphere can determine which hemisphere takes control of cognitive operations. Hemispheric metacontrol thus allows the brain to flexibly react to changing task demands by using the hemisphere that is best suited to solve the problem at hand. Subsequently, metacontrol has been shown for face processing in healthy human subjects,[109,110] visual recognition in split-brain monkeys,[111] and color discrimination in pigeons,[112–114] indicating that it is a widespread principle of neural organization in vertebrate brains.

Sample questions

- What is the corpus callosum and which four ways have been used to parcellate it?
- Name the seven segments of the corpus callosum according to the Witelson parcellation scheme and the five segments of the corpus callosum according to the Hofer and Frahm parcellation scheme. Why do these parcellation schemes differ?
- How did research on split-brain patients advance our knowledge about hemispheric asymmetries?
- What is the Poffenberger task and what are the main results obtained with this task in split-brain patients?
- What is callosal agenesis?
- What is the functional role of the corpus callosum for functional hemispheric asymmetries?
- How can interhemispheric integration be experimentally assessed?
- What is metacontrol? Please describe an experiment to investigate this phenomenon.

References

1. Fields RD. White matter in learning, cognition and psychiatric disorders. *Trends Neurosci.* 2008;31(7):361–370. Available from: https://doi.org/10.1016/j.tins.2008.04.001.
2. Scholz J, Klein MC, Behrens TEJ, Johansen-Berg H. Training induces changes in white-matter architecture. *Nat Neurosci.* 2009;12(11):1370–1371. Available from: https://doi.org/10.1038/nn.2412.
3. Bogen JE. The neurosurgeon's interest in the corpus callosum. In: Greenblatt SH, Forcht Dagi T, Epstein MH, eds. *A History of Neurosurgery: In Its Scientific and Professional Contexts.* Thieme; 1997:489–516.
4. Lashley KS. *Brain Mechanisms and Intelligence: A Quantitative Study of Injuries to the Brain.* University of Chicago Press; 1929.
5. Zaidel E, Iacoboni M, eds. *The Parallel Brain: The Cognitive Neuroscience of the Corpus Callosum.* MIT Press; 2003.
6. Friedrich P, Fraenz C, Schlüter C, et al. The relationship between axon density, myelination, and fractional anisotropy in the human corpus callosum. *Cereb Cortex.* 2020;30(4):2042–2056. Available from: https://doi.org/10.1093/cercor/bhz221.
7. Suárez R, Gobius I, Richards LJ. Evolution and development of interhemispheric connections in the vertebrate forebrain. *Front Hum Neurosci.* 2014;8:497. Available from: https://doi.org/10.3389/fnhum.2014.00497.
8. Fabri M, Pierpaoli C, Barbaresi P, Polonara G. Functional topography of the corpus callosum investigated by DTI and fMRI. *WJR.* 2014;6(12):895. Available from: https://doi.org/10.4329/wjr.v6.i12.895.
9. Clarke JM, Zaidel E. Anatomical-behavioral relationships: corpus callosum morphometry and hemispheric specialization. *Behav Brain Res.* 1994;64(1–2):185–202. Available from: https://doi.org/10.1016/0166-4328(94)90131-7.
10. Innocenti GM, Schmidt K, Milleret C, et al. The functional characterization of callosal connections. *Prog Neurobiol.* 2022;208:102186. Available from: https://doi.org/10.1016/j.pneurobio.2021.102186.
11. Aboitiz F, Montiel J. One hundred million years of interhemispheric communication: the history of the corpus callosum. *Braz J Med Biol Res.* 2003;36(4):409–420. Available from: https://doi.org/10.1590/S0100-879X2003000400002.
12. Witelson SF. Hand and sex differences in the isthmus and genu of the human corpus callosum. *Brain.* 1989;112(3):799–835. Available from: https://doi.org/10.1093/brain/112.3.799.
13. Catani ST. *Atlas of Human Brain Connections.* Oxford University Press; 2012.
14. Friedrich P, Forkel SJ, Thiebaut de Schotten M. Mapping the principal gradient onto the corpus callosum. *NeuroImage.* 2020;223:117317. Available from: https://doi.org/10.1016/j.neuroimage.2020.117317.
15. Hofer S, Frahm J. Topography of the human corpus callosum revisited—comprehensive fiber tractography using diffusion tensor magnetic resonance imaging. *NeuroImage.* 2006;32(3):989–994. Available from: https://doi.org/10.1016/j.neuroimage.2006.05.044.
16. Mori S, Tournier J-D. *Introduction to Diffusion Tensor Imaging.* 2nd ed. Academic Press; 2014.
17. Gazzaniga MS. Cerebral specialization and interhemispheric communication: does the corpus callosum enable the human condition? *Brain.* 2000;123(7):1293–1326. Available from: https://doi.org/10.1093/brain/123.7.1293.
18. Myers RE. Function of corpus callosum in interocular transfer. *Brain.* 1956;79(2):358–363. Available from: https://doi.org/10.1093/brain/79.2.358.
19. Myers RE. Interocular transfer of pattern discrimination in cats following section of crossed optic fibers. *J Comp Physiol Psychol.* 1955;48(6):470–473. Available from: https://doi.org/10.1037/h0044224.
20. Miner N, Sperry RW, Stamm JS. Relearning tests for interocular transfer following division of optic chiasma and corpus callosum in cats. *J Comp Physiol Psychol.* 1956;49(6):529–533. Available from: https://doi.org/10.1037/h0048804.
21. Myers RE, Sperry RW. Interhemispheric communication through the corpus callosum: mnemonic carry-over between the hemispheres. *AMA Arch NeurPsych.* 1958;80(3):298–303. Available from: https://doi.org/10.1001/archneurpsyc.1958.02340090034004.
22. Gavalas RJ, Sperry RW. Central integration of visual half-fields in split-brain monkeys. *Brain Res.* 1969;15(1):97–106. Available from: https://doi.org/10.1016/0006-8993(69)90312-6.
23. Van Wagenen WP, Herren RY. Surgical division of commissural pathways in the corpus callosum. *Arch NeurPsych.* 1940;44(4):740. Available from: https://doi.org/10.1001/archneurpsyc.1940.02280100042004.
24. Mathews MS, Linskey ME, Binder DK, William P. van Wagenen and the first corpus callosotomies for epilepsy. *J Neurosurg.* 2008;108(3):608–613. Available from: https://doi.org/10.3171/JNS/2008/108/3/0608.
25. Akelaitis AJ. A study of gnosis, praxis and language following section of the corpus callosum and anterior commissure. *J Neurosurg.* 1944;1(2):94–102. Available from: https://doi.org/10.3171/jns.1944.1.2.0094.

26. Gazzaniga MS. Forty-five years of split-brain research and still going strong. *Nat Rev Neurosci.* 2005;6 (8):653–659. Available from: https://doi.org/10.1038/nrn1723.
27. Bogen JE. Cerebral commissurotomy in man. *Bull Los Angels Neuro Soc.* 1962;27:169–172. Available from: https://ci.nii.ac.jp/naid/10018562360/.
28. Gazzaniga MS, Bogen JE, Sperry RW. Some functional effects of sectioning the cerebral commissures in man. *Proc Natl Acad Sci U S A.* 1962;48(10):1765–1769. Available from: https://doi.org/10.1073/pnas.48.10.1765.
29. Sperry RW. Some effects of disconnecting the cerebral hemispheres. *Biosci Rep.* 1982;2(5):265–276. Available from: https://doi.org/10.1007/BF01115112.
30. Bourne VJ. The divided visual field paradigm: methodological considerations. *Laterality.* 2006;11(4):373–393. Available from: https://doi.org/10.1080/13576500600633982.
31. Gazzaniga MS. Principles of human brain organization derived from split-brain studies. *Neuron.* 1995;14 (2):217–228. Available from: https://doi.org/10.1016/0896-6273(95)90280-5.
32. Sidtis JJ, Volpe BT, Holtzman JD, Wilson DH, Gazzaniga MS. Cognitive interaction after staged callosal section: evidence for transfer of semantic activation. *Science.* 1981;212(4492):344–346. Available from: https://doi.org/10.1126/science.6782673.
33. Bogen JE, Gazzaniga MS. Cerebral commissurotomy in man: minor hemisphere dominance for certain visuospatial functions. *J Neurosurg.* 1965;23(4):394–399. Available from: https://doi.org/10.3171/jns.1965.23.4.0394.
34. Holtzman JD, Gazzaniga MS. Dual task interactions due exclusively to limits in processing resources. *Science.* 1982;218(4579):1325–1327. Available from: https://doi.org/10.1126/science.7146915.
35. Gazzaniga MS. Perceptual and attentional processes following callosal section in humans. *Neuropsychologia.* 1987;25(1):119–133. Available from: https://doi.org/10.1016/0028-3932(87)90048-0.
36. Franz EA, Eliassen JC, Ivry RB, Gazzaniga MS. Dissociation of spatial and temporal coupling in the bimanual movements of callosotomy patients. *Psychol Sci.* 1996;7(5):306–310. Available from: https://doi.org/10.1111/j.1467-9280.1996.tb00379.x.
37. Lambert AJ. Interhemispheric interaction in the split-brain. *Neuropsychologia.* 1991;29(10):941–948. Available from: https://doi.org/10.1016/0028-3932(91)90058-g.
38. Sergent J. A new look at the human split brain. *Brain.* 1987;110(Pt 5)(5):1375–1392. Available from: https://doi.org/10.1093/brain/110.5.1375.
39. Poffenberger AT. Reaction time to retinal stimulation, with special reference to the time lost in conduction through nerve centers. *Am J Psychol.* 1912;23(1):1–73. Available from: https://doi.org/10.2307/1413279.
40. Westerhausen R. Interhemispheric integration after callosotomy: a meta-analysis of Poffenberger and redundant-target paradigms. *Neuropsychol Rev.* 2022. Available from: https://doi.org/10.1007/s11065-022-09569-w.
41. Roland JL, Snyder AZ, Hacker CD, et al. On the role of the corpus callosum in interhemispheric functional connectivity in humans. *PNAS.* 2017;114(50):13278–13283. Available from: https://doi.org/10.1073/pnas.1707050114.
42. Wolford G, Miller MB, Gazzaniga MS. The left hemisphere's role in hypothesis formation. *J Neurosci.* 2000;20 (6):RC64. Available from: https://doi.org/10.1523/JNEUROSCI.20-06-j0003.2000.
43. Hinson JM, Staddon JE. Matching, maximizing, and hill-climbing. *J Exp Anal Behav.* 1983;40(3):321–331. Available from: https://doi.org/10.1901/jeab.1983.40-321.
44. www.nobelprize.org. The split brain experiments. https://educationalgames.nobelprize.org/educational/medicine/split-brain/background.html.
45. Corballis MC, Corballis PM. Can the mind be split? A historical introduction. *Neuropsychologia.* 2021;163:108041. Available from: https://doi.org/10.1016/j.neuropsychologia.2021.108041.
46. Haan EHF, de, Corballis PM, Hillyard SA, et al. Split-brain: what we know now and why this is important for understanding consciousness. *Neuropsychol Rev.* 2020;30(2):224–233. Available from: https://doi.org/10.1007/s11065-020-09439-3.
47. Hausmann M, Corballis MC, Fabri M. Revisiting the attentional bias in the split brain. *Neuropsychologia.* 2021;162:108042. Available from: https://doi.org/10.1016/j.neuropsychologia.2021.108042.
48. Wolman D. The split brain: a tale of two halves. *Nature.* 2012;483(7389):260–263. Available from: https://doi.org/10.1038/483260a.
49. Markosian C, Patel S, Kosach S, Goodman RR, Tomycz LD. Corpus callosotomy in the modern era: origins, efficacy, technical variations, complications, and indications. *World Neurosurg.* 2022;159:146–155. Available from: https://doi.org/10.1016/j.wneu.2022.01.037.

50. Hausmann M, Corballis MC, Farbi M. Line bisection in the split brain. *Neuropsychology*. 2003;17(4):602–609. Available from: https://doi.org/10.1037/0894-4105.17.4.602.
51. Prete G, Fabri M, Foschi N, Brancucci A, Tommasi L. The "consonance effect" and the hemispheres: a study on a split-brain patient. *Laterality*. 2015;20(3):257–269. Available from: https://doi.org/10.1080/1357650X.2014.959525.
52. Ioannidis JPA. Why most published research findings are false. *PLoS Med*. 2005;2(8):e124. Available from: https://doi.org/10.1371/journal.pmed.0020124.
53. Mancuso L, Uddin LQ, Nani A, Costa T, Cauda F. Brain functional connectivity in individuals with callosotomy and agenesis of the corpus callosum: a systematic review. *Neurosci Biobehav Rev*. 2019;105:231–248. Available from: https://doi.org/10.1016/j.neubiorev.2019.07.004.
54. Genç E, Ocklenburg S, Singer W, Güntürkün O. Abnormal interhemispheric motor interactions in patients with callosal agenesis. *Behav Brain Res*. 2015;293:1–9. Available from: https://doi.org/10.1016/j.bbr.2015.07.016.
55. Lassonde MC, Sauerwein HC, Lepore F. Agenesis of the corpus callosum. In: Zaidel E, Iacoboni M, eds. *The Parallel Brain: The Cognitive Neuroscience of the Corpus Callosum*. MIT Press; 2003.
56. Paul LK, Brown WS, Adolphs R, et al. Agenesis of the corpus callosum: genetic, developmental and functional aspects of connectivity. *Nat Rev Neurosci*. 2007;8(4):287–299. Available from: https://doi.org/10.1038/nrn2107.
57. Pânzaru M-C, Popa S, Lupu A, Gavrilovici C, Lupu VV, Gorduza EV. Genetic heterogeneity in corpus callosum agenesis. *Front Genet*. 2022;13:958570. Available from: https://doi.org/10.3389/fgene.2022.958570.
58. Filley C. *The Behavioral Neurology of White Matter*. 2nd ed. United States: Oxford University Press; 2012.
59. Palmer EE, Mowat D. Agenesis of the corpus callosum: a clinical approach to diagnosis. *Am J Med Genet Part C: SemMed Genet*. 2014;166C(2):184–197. Available from: https://doi.org/10.1002/ajmg.c.31405.
60. Lassonde MC, ed. *Callosal Agenesis: A Natural Split Brain?* Plenum Press; 1994.
61. Shevell MI. Clinical and diagnostic profile of agenesis of the corpus callosum. *J Child Neurol*. 2002;17(12):895–899. Available from: https://doi.org/10.1177/08830738020170122601.
62. Chiarello C. A house divided? Cognitive functioning with callosal agenesis. *Brain Lang*. 1980;11(1):128–158. Available from: https://doi.org/10.1016/0093-934x(80)90116-9.
63. Probst M. Ueber den Bau des vollständig balkenlosen Gross-hirnes sowie über Mikrogyrie und Heterotopie der grauen Substanz. *Arch f Psychiatrie*. 1901;34(3):709–786. Available from: https://doi.org/10.1007/BF02680175.
64. Bénézit A, Hertz-Pannier L, Dehaene-Lambertz G, et al. Organising white matter in a brain without corpus callosum fibres. *Cortex*. 2015;63:155–171. Available from: https://doi.org/10.1016/j.cortex.2014.08.022.
65. Lee S-K, Mori S, Kim DJ, Kim SY, Kim SY, Kim DI. Diffusion tensor MR imaging visualizes the altered hemispheric fiber connection in callosal dysgenesis. *AJNR Am J Neuroradiol*. 2004;25(1):25–28. Available from: http://www.ajnr.org/content/25/1/25.short.
66. Tovar-Moll F, Monteiro M, Andrade J, et al. Structural and functional brain rewiring clarifies preserved interhemispheric transfer in humans born without the corpus callosum. *PNAS*. 2014;111(21):7843–7848. Available from: https://doi.org/10.1073/pnas.1400806111.
67. Berlucchi G. Corpus callosum and simple visuomotor integration. *Neuropsychologia*. 1995;33(8):923–936. Available from: https://doi.org/10.1016/0028-3932(95)00031-w.
68. Jeeves MA. A comparison of interhemispheric transmission times in acallosals and normals. *Psychon Sci*. 1969;16(5):245–246. Available from: https://doi.org/10.3758/BF03332667.
69. Ocklenburg S, Ball A, Wolf CC, Genç E, Güntürkün O. Functional cerebral lateralization and interhemispheric interaction in patients with callosal agenesis. *Neuropsychology*. 2015;29(5):806–815. Available from: https://doi.org/10.1037/neu0000193.
70. Manhães AC, Schmidt SL, Filgueiras CC. Callosal agenesis affects consistency of laterality in a paw preference task in BALB/cCF mice. *Behav Brain Res*. 2005;159(1):43–49. Available from: https://doi.org/10.1016/j.bbr.2004.09.023.
71. Lassonde MC, Lortie J, Ptito M, Geoffroy G. Hemispheric asymmetry in callosal agenesis as revealed by dichotic listening performance. *Neuropsychologia*. 1981;19(3):455–458. Available from: https://doi.org/10.1016/0028-3932(81)90075-0.
72. Riecker A, Ackermann H, Schmitz B, Kassubek J, Herrnberger B, Steinbrink C. Bilateral language function in callosal agenesis: an fMRI and DTI study. *J Neurol*. 2007;254(4):528–530. Available from: https://doi.org/10.1007/s00415-006-0152-9.
73. Pelletier I, Paquette N, Lepore F, et al. Language lateralization in individuals with callosal agenesis: an fMRI study. *Neuropsychologia*. 2011;49(7):1987–1995. Available from: https://doi.org/10.1016/j.neuropsychologia.2011.03.028.

74. Jäncke L, Wunderlich G, Schlaug G, Steinmetz H. A case of callosal agenesis with strong anatomical and functional asymmetries. *Neuropsychologia*. 1997;35(10):1389–1394. Available from: https://doi.org/10.1016/s0028-3932(97)00068-7.
75. Letzner S, Simon A, Güntürkün O. Connectivity and neurochemistry of the commissura anterior of the pigeon (*Columba livia*). *J Comp Neurol*. 2016;524(2):343–361. Available from: https://doi.org/10.1002/cne.23858.
76. Giljov A, Karenina K, Ingram J, Malashichev Y. Parallel emergence of true handedness in the evolution of marsupials and placentals. *Curr Biol*. 2015;25(14):1878–1884. Available from: https://doi.org/10.1016/j.cub.2015.05.043.
77. Giljov A, Karenina K, Malashichev Y. Limb preferences in a marsupial, Macropus rufogriseus: evidence for postural effect. *Anim Behav*. 2012;83(2):525–534. Available from: https://doi.org/10.1016/j.anbehav.2011.11.031.
78. Suárez R, Paolino A, Fenlon LR, et al. A pan-mammalian map of interhemispheric brain connections predates the evolution of the corpus callosum. *PNAS*. 2018;115(38):9622–9627. Available from: https://doi.org/10.1073/pnas.1808262115.
79. Bisazza A, Rogers LJ, Vallortigara G. The origins of cerebral asymmetry: a review of evidence of behavioural and brain lateralization in fishes, reptiles and amphibians. *Neurosci Biobehav Rev*. 1998;22(3):411–426. Available from: https://doi.org/10.1016/s0149-7634(97)00050-x.
80. Manns M, Ströckens F. Functional and structural comparison of visual lateralization in birds—similar but still different. *Front Psychol*. 2014;5:206. Available from: https://doi.org/10.3389/fpsyg.2014.00206.
81. Ströckens F, Güntürkün O, Ocklenburg S. Limb preferences in non-human vertebrates. *Laterality*. 2013;18 (5):536–575. Available from: https://doi.org/10.1080/1357650X.2012.723008.
82. Frasnelli E. Brain and behavioral lateralization in invertebrates. *Front Psychol*. 2013;4:939. Available from: https://doi.org/10.3389/fpsyg.2013.00939.
83. Witelson SF. The brain connection: the corpus callosum is larger in left-handers. *Science*. 1985;229 (4714):665–668. Available from: https://doi.org/10.1126/science.4023705.
84. Moffat SD, Hampson E, Lee DH. Morphology of the planum temporale and corpus callosum in left handers with evidence of left and right hemisphere speech representation. *Brain*. 1998;121(Pt 12)(12):2369–2379. Available from: https://doi.org/10.1093/brain/121.12.2369.
85. Tuncer MC, Hatipoğlu ES, Ozateş M. Sexual dimorphism and handedness in the human corpus callosum based on magnetic resonance imaging. *Surg Radiol Anat*. 2005;27(3):254–259. Available from: https://doi.org/10.1007/s00276-004-0308-1.
86. Jäncke L, Staiger JF, Schlaug G, Huang Y, Steinmetz H. The relationship between corpus callosum size and forebrain volume. *Cereb Cortex*. 1997;7(1):48–56. Available from: https://doi.org/10.1093/cercor/7.1.48.
87. Westerhausen R, Kreuder F, Dos Santos Sequeira S, et al. Effects of handedness and gender on macro- and microstructure of the corpus callosum and its subregions: a combined high-resolution and diffusion-tensor MRI study. *Cognit Brain Res*. 2004;21(3):418–426. Available from: https://doi.org/10.1016/j.cogbrainres.2004.07.002.
88. Westerhausen R, Papadatou-Pastou M. Handedness and midsagittal corpus callosum morphology: a meta-analytic evaluation. *Brain Struct Funct*. 2022;227(2):545–559. Available from: https://doi.org/10.1007/s00429-021-02431-4.
89. Luders E, Cherbuin N, Thompson PM, et al. When more is less: associations between corpus callosum size and handedness lateralization. *NeuroImage*. 2010;52(1):43–49. Available from: https://doi.org/10.1016/j.neuroimage.2010.04.016.
90. Clarke JM, Lufkin RB, Zaidel E. Corpus callosum morphometry and dichotic listening performace: Individual differences in functional interhemispheric inhibition? *Neuropsychologia*. 1993;31(6):547–557. Available from: https://doi.org/10.1016/0028-3932(93)90051-z.
91. Hines M, Chiu L, McAdams LA, Bentler PM, Lipcamon J. Cognition and the corpus callosum: verbal fluency, visuospatial ability, and language lateralization related to midsagittal surface areas of callosal subregions. *Behav Neurosci*. 1992;106(1):3–14. Available from: https://doi.org/10.1037/0735-7044.106.1.3.
92. Musiek FE, Weihing J. Perspectives on dichotic listening and the corpus callosum. *Brain Cognition*. 2011;76 (2):225–232. Available from: https://doi.org/10.1016/j.bandc.2011.03.011.
93. Westerhausen R, Hugdahl K. The corpus callosum in dichotic listening studies of hemispheric asymmetry: a review of clinical and experimental evidence. *Neurosci Biobehav Rev*. 2008;32(5):1044–1054. Available from: https://doi.org/10.1016/j.neubiorev.2008.04.005.
94. Friedrich P, Ocklenburg S, Heins N, et al. Callosal microstructure affects the timing of electrophysiological left-right differences. *NeuroImage*. 2017;163:310–318. Available from: https://doi.org/10.1016/j.neuroimage.2017.09.048.
95. van der Knaap LJ, van der Ham IJM. How does the corpus callosum mediate interhemispheric transfer? A review. *Behav Brain Res*. 2011;223(1):211–221. Available from: https://doi.org/10.1016/j.bbr.2011.04.018.

96. Ringo JL, Doty RW, Demeter S, Simard PY. Time is of the essence: a conjecture that hemispheric specialization arises from interhemispheric conduction delay. *Cereb Cortex*. 1994;4(4):331–343. Available from: https://doi.org/10.1093/cercor/4.4.331.
97. Yazgan M, Wexler BE, Kinsbourne M, Peterson B, Leckman JF. Functional significance of individual variations in callosal area. *Neuropsychologia*. 1995;33(6):769–779. Available from: https://doi.org/10.1016/0028-3932(95)00018-x.
98. Karolis VR, Corbetta M, Thiebaut de Schotten M. The architecture of functional lateralisation and its relationship to callosal connectivity in the human brain. *Nat Commun*. 2019;10(1):1417. Available from: https://doi.org/10.1038/s41467-019-09344-1.
99. Bloom JS, Hynd GW. The role of the corpus callosum in interhemispheric transfer of information: excitation or inhibition? *Neuropsychol Rev*. 2005;15(2):59–71. Available from: https://doi.org/10.1007/s11065-005-6252-y.
100. Cook N. Homotopic callosal inhibition. *Brain Lang*. 1984;23(1):116–125. Available from: https://doi.org/10.1016/0093-934x(84)90010-5.
101. Josse G, Seghier ML, Kherif F, Price CJ. Explaining function with anatomy: language lateralization and corpus callosum size. *J Neurosci*. 2008;28(52):14132–14139. Available from: https://doi.org/10.1523/JNEUROSCI.4383-08.2008.
102. Nowicka A, Tacikowski P. Transcallosal transfer of information and functional asymmetry of the human brain. *Laterality*. 2011;16(1):35–74. Available from: https://doi.org/10.1080/13576500903154231.
103. Putnam MC, Wig GS, Grafton ST, Kelley WM, Gazzaniga MS. Structural organization of the corpus callosum predicts the extent and impact of cortical activity in the nondominant hemisphere. *J Neurosci*. 2008;28(11):2912–2918. Available from: https://doi.org/10.1523/JNEUROSCI.2295-07.2008.
104. Schulte T, Müller-Oehring EM, Rohlfing T, Pfefferbaum A, Sullivan EV. White matter fiber degradation attenuates hemispheric asymmetry when integrating visuomotor information. *J Neurosci*. 2010;30(36):12168–12178. Available from: https://doi.org/10.1523/JNEUROSCI.2160-10.2010.
105. Gootjes L, Bouma A, van Strien JW, van Schijndel R, Barkhof F, Scheltens P. Corpus callosum size correlates with asymmetric performance on a dichotic listening task in healthy aging but not in Alzheimer's disease. *Neuropsychologia*. 2006;44(2):208–217. Available from: https://doi.org/10.1016/j.neuropsychologia.2005.05.002.
106. Dimond S, Beaumont G. Use of two cerebral hemispheres to increase brain capacity. *Nature*. 1971;232(5308):270–271. Available from: https://doi.org/10.1038/232270a0.
107. Banich MT, Belger A. Interhemispheric interaction: how do the hemispheres divide and conquer a task? *Cortex*. 1990;26(1):77–94. Available from: https://doi.org/10.1016/s0010-9452(13)80076-7.
108. Levy J, Trevarthen C. Metacontrol of hemispheric function in human split-brain patients. *J Exp Psychol: Hum Percept Perform*. 1976;2(3):299–312. Available from: https://doi.org/10.1037/0096-1523.2.3.299.
109. Hellige JB, Jonsson JE, Michimata C. Processing from LVF, RVF and BILATERAL presentations: examinations of metacontrol and interhemispheric interaction. *Brain Cognition*. 1988;7(1):39–53. Available from: https://doi.org/10.1016/0278-2626(88)90020-6.
110. Urgesi C, Bricolo E, Aglioti SM. Hemispheric metacontrol and cerebral dominance in healthy individuals investigated by means of chimeric faces. *Cognit Brain Res*. 2005;24(3):513–525. Available from: https://doi.org/10.1016/j.cogbrainres.2005.03.005.
111. Kavcic V, Fei R, Hu S, Doty RW. Hemispheric interaction, metacontrol, and mnemonic processing in split-brain macaques. *Behav Brain Res*. 2000;111(1-2):71–82. Available from: https://doi.org/10.1016/s0166-4328(00)00141-8.
112. Adam R, Güntürkün O. When one hemisphere takes control: metacontrol in pigeons (Columba livia. *PLoS One*. 2009;4(4):e5307. Available from: https://doi.org/10.1371/journal.pone.0005307.
113. Freund N, Valencia-Alfonso CE, Kirsch J, Brodmann K, Manns M, Güntürkün O. Asymmetric top-down modulation of ascending visual pathways in pigeons. *Neuropsychologia*. 2016;83:37–47. Available from: https://doi.org/10.1016/j.neuropsychologia.2015.08.014.
114. Ünver E, Güntürkün O. Evidence for interhemispheric conflict during meta-control in pigeons. *Behav Brain Res*. 2014;270:146–150. Available from: https://doi.org/10.1016/j.bbr.2014.05.016.

CHAPTER

4

Structural hemispheric asymmetries

OUTLINE

The Lateralized Brain
DOI: https://doi.org/10.1016/B978-0-323-99737-9.00001-X

Introductory short story

The surgical instruments reflected the light emitted by the surgery room lamp. They had removed the dura mater and pia mater from the donated brain on the surgery table in front of them. "Now, divide the two hemispheres and cut them along the plane of the Sylvian fissure." The younger neurologist did as he was told. He revealed the upper surface of the left temporal lobe and then continued to do the same with the right hemisphere. He and his colleague were highly focused. Their project was extremely ambitious and could potentially revolutionize the understanding of language lateralization. Ever since the days of Pierre Paul Broca more than 100 years ago, it had been known that the left hemisphere was dominant for speech functions in the human brain. However, no researcher had yet shown convincing evidence for a neuroanatomical base for this remarkable feature of the cerebral cortex, and it was generally accepted that left-hemispheric language dominance was not associated with any significant structural differences between the left and the right hemisphere. Norman Geschwind and Walter Levitsky from the Department of Neurology at the Boston University School of Medicine, USA, conducted a study that was about to change this. During the 1960s, a period in which many researchers still conducted single-case patient studies or relied on small sample sizes to prove ambitious theories, they had gathered an impressive sample of 100 human post-mortem brain samples. They investigated the size of the left and right planum temporale, a temporal brain area containing Wernicke's area on the left. "It's again bigger on the left," whispered the younger neurologist. The older one knew what that meant. "This was the last one. That makes it bigger on the left in 65% of samples but bigger on the right in only 11% of samples. I think we found what we were looking for." They had finally found the neuroanatomical basis of human language lateralization, solving a riddle that had remained unsolved for more than 100 years.... Or had they?

Inspired by Geschwind and Levitsky.[1]

Introduction

What causes functional hemispheric asymmetries such as handedness or language lateralization? Since functional hemispheric asymmetries originate in the brain, it is an attractive hypothesis to assume that they are driven by asymmetries in brain structure.[2] The idea that brain function is guided by brain structure is not a recent one. The British neurologist John Hughlings Jackson, one of the fathers of modern epilepsy and aphasia research, stated as early as 1874:

> Difference of structure of necessity implies difference in function. ***Hughlings Jackson.***[3,4]

Given the large number of asymmetrically organized cognitive systems in the human brain, this statement would imply that structural hemispheric asymmetries are ubiquitous. Indeed, they have been reported for many different levels of brain structure.

Katrin Amunts from the Research Centre Jülich in Germany suggested that structural asymmetries can be differentiated into three larger categories[5]:

1. *Macroscopic asymmetries*: Asymmetries in volume, shape, and size of sulci, gyri, white matter tracts, brain lobes, and whole hemispheres.
2. *Microstructural asymmetries*: Asymmetries in the number and density of nerve cells in a specific brain region as well as asymmetries in other parameters of specific cytoarchitectonic areas (e.g., size, volume, surface, cortical thickness, number of spines, and degree or arborization of dendrites and axons).
3. *Molecular asymmetries*: Asymmetries in gene expression and protein distribution.

In the following sections of this chapter, we will describe the major types of structural hemispheric asymmetries for each of these categories. We will then continue to describe whether and how they are related to brain function.

Gray and white matter

Structural hemispheric asymmetries have been investigated in the two major components of the brain: Gray matter and white matter (see Box 3.1 in the previous chapter and Fig. 4.1). Gray matter mainly consists of neuronal cell bodies that control sensory, motor, and cognitive functions. White matter mainly consists of myelinated axons that are organized in fiber tracts that transport information from one area in the brain to another. Since myelin has a high lipid content, its appearance is lighter than that of gray matter. Its name originates from the fact that it looks almost white in post-mortem brain tissue samples that have been cleaned of blood. While the exact percentages differ as a function of age, their combined volume reaches about 80% of total brain volume in healthy human brains.[6] Overall, white matter volume represents 30%–40% of total brain volume, and gray matter volume represents 40%–50% of total brain volume.[7]

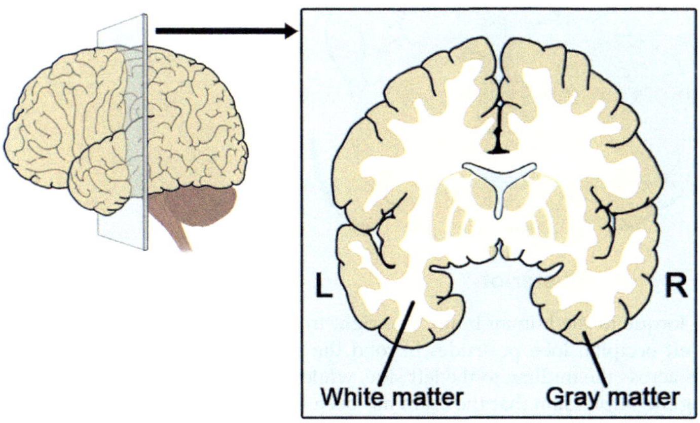

FIGURE 4.1 Gray and white matter in the human brain.

Macroscopic asymmetries

Gross anatomical differences between the hemispheres

While the average weight and overall size of the left and right hemispheres are very similar in healthy adults, two major forms of gross anatomical differences between the hemispheres have been described[8] (see Fig. 4.2).

1. *Petalia*: The petalia are the impressions that the two hemispheres' protrusions leave on the skull's inner surface. Analysis of petalia asymmetries has revealed that in most individuals, the right frontal lobe protrudes beyond the left in the anterior direction, while the left occipital lobe protrudes beyond the right in the posterior direction. Moreover, the right frontal cortex is wider than the left in most individuals, while the left occipital lobe is wider than the right in most individuals.[8]
2. *Yakovlevian torque*: Named after Paul Ivan Yakovlev, a neuroanatomist who first described this phenomenon, the Yakovlevian torque describes the fact that in most individuals, the right frontal cortex extends across the midline to the left side, while the left occipital cortex extends across the midline to the right side. Thus, the brain appears to be torqued in this direction.[8]

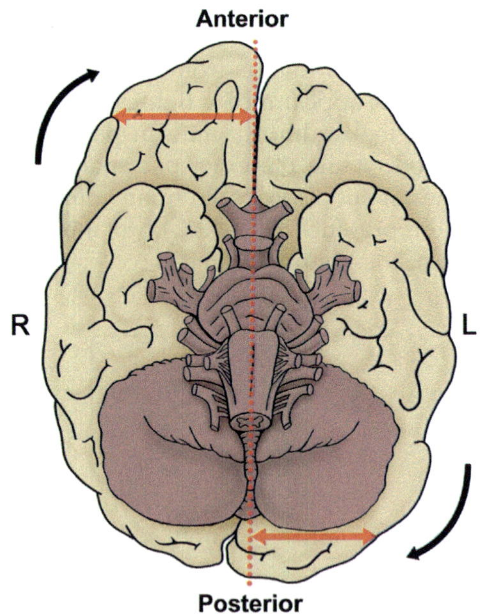

FIGURE 4.2 Petalia and Yakovlevian torque in the human brain. The right frontal lobe protrudes beyond the left in the anterior direction, while the left occipital lobe protrudes beyond the right in the posterior direction. Moreover, the right frontal cortex extends across the midline to the left side, while the left occipital cortex extends across the midline to the right side, giving the impression that the brain has been torqued in this direction.

Macroscopic asymmetries in gray matter

Macroscopic asymmetries in cortical structure

The largest study mapping macroscopic asymmetries in cortical structure in a non-clinical adult cohort was conducted by the ENIGMA (Enhancing Neuro Imaging Genetics through Meta-Analysis) Laterality Working Group (see Box 4.1). In the study, first-authored by Xiang-Zhen Kong from the Max Planck Institute for Psycholinguistics in Nijmegen, The Netherlands, asymmetries in macroscopic brain structure were assessed in 17,141 healthy individuals from 99 independent datasets.[13]

The authors found structural hemispheric asymmetries both on the level of hemispheres and the level of specific regions. In general, the cortex was thicker in the left hemisphere, but the surface area was larger in the right hemisphere. Moreover, hemispheric asymmetries were widespread on the regional level, and the authors showed that most brain areas in the human cortex displayed structural asymmetries. For cortical thickness, 26 out of 34 investigated brain regions (76.5%) showed statistically significant asymmetry after correction for multiple comparisons. While leftward asymmetries were more common in the anterior part of the brain, rightward asymmetries were more common in the posterior part of the brain (see Fig. 4.3 upper

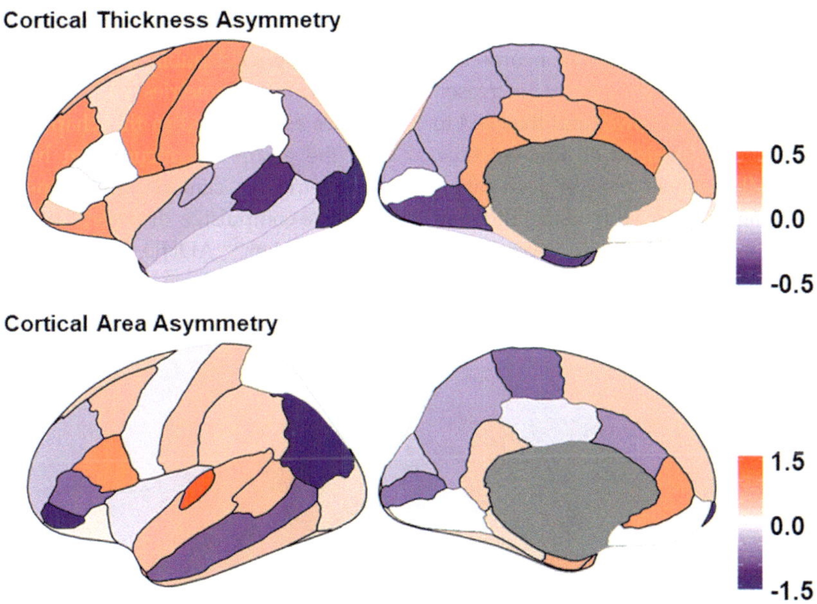

FIGURE 4.3 Macrostructural asymmetries in the human cortex based on a sample of more than 17,000 individuals.[13] The upper panel shows cortical thickness asymmetries, and the lower panel shows cortical area asymmetries. Colors indicate the effect size (Cohen's d) and direction of left-right differences in structure. Red indicates leftward asymmetry (greater cortical thickness or larger cortical area in the left hemisphere), and blue indicates rightward asymmetry. *Source: Reproduced from Kong X-Z, Postema MC, Guadalupe T, et al. Mapping brain asymmetry in health and disease through the ENIGMA consortium. Hum Brain Mapp. 2022;43(1):167–181. https://doi.org/10.1002/hbm.25033, with permission of the author. (Authors note that "This is an open access article distributed under the terms of the Creative Commons 4.0 International Attribution License, which permits unrestricted use, distribution, and reproduction in any medium, provided the original work is properly cited").*

BOX 4.1

The ENIGMA Laterality Working Group.

One of the major issues in neuroscientific research in the last decade is the so-called replication crisis.[9] In short, replication rates in psychology and neuroscience are low, partly because many studies are statistically underpowered due to small sample sizes leading to type-one errors (rejecting the null hypothesis when it is true) and spurious findings.[10] Moreover, low statistical power not only reduces the chance of detecting a true effect but also reduces the likelihood that a statistically significant result reflects a true effect. To combat this issue, more and more researchers organize themselves in multi-lab consortia with the aim of generating and analyzing large-scale data sets. Nevertheless, while data pooling and multi-site studies constitute modern solutions to well-known issues in the field, they also create new challenges for neuroscientists. These include the greatly increased demand for computing power and the need to harmonize datasets obtained at different sites, on different MRI scanner models, with different field strengths, and with using different scanning parameters. However, overall, the advantages seem to outweigh the disadvantages, so it is to be expected that consortium approaches will gain more and more popularity in the years to come, especially considering the surge in neuroimaging studies integrating whole-genome genetic data (where typical single-site studies would be severely underpowered). The ENIGMA Laterality Working Group (https://enigma.ini.usc.edu/ongoing/enigma-lateralization-working-group/), under the umbrella of ENIGMA (Enhancing Neuro Imaging Genetics Through Meta-Analysis) Consortium (https://enigma.ini.usc.edu), is a good example of what large-scale multinational research efforts could look like in the field of laterality research.[11,12] The logo of the ENIGMA consortium is shown in Fig. 4.4. Over the last years, this group has been very successful in implementing an international multi-site research strategy and contributing scientifically solid publications on cortical and subcortical asymmetries[11,13,14] in normative samples as discussed in this chapter. In addition, the group has focused on hemispheric asymmetries in clinical populations, such as obsessive-compulsive disorder,[15] autism spectrum disorders,[16] ADHD,[17] and major depressive disorder.[18] These will be discussed in Chapter 13.

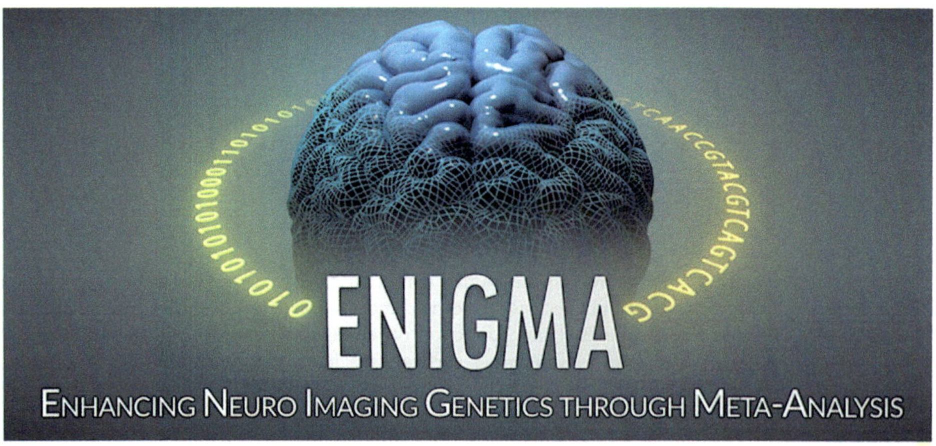

FIGURE 4.4 The logo of the ENIGMA consortium (Enhancing Neuro Imaging Genetics through Meta-Analysis). *Source: Created by Paul Thompson. Reprinted with permission of the creator.*

panel). For the surface area, even more regions showed asymmetry. For this measure, 31 out of 34 investigated regions (91.1%) showed statistically significant asymmetry after correction for multiple comparisons. For the surface area, no clear anterior–posterior distribution of asymmetries was observed (see Fig. 4.3 lower panel). Asymmetries of the surface area and cortical thickness were independent of each other and did not show any significant correlation. Taken together, the results of this study show that structural hemispheric asymmetries are the rule, not the exception, in the human cortex .

Macroscopic asymmetries in the planum temporale

While structural differences between the left and the right hemisphere of the human cortex certainly were noted by neuroanatomists before the 20th century,[19,20] the modern interest in structural asymmetries as a possible explanation for functional asymmetries was sparked by a seminal paper by Norman Geschwind and Walter Levitsky from the Department of Neurology at the Boston University School of Medicine, USA. Published in 1968 in the prestigious journal *Science*, the paper described the result of post-mortem analysis of 100 adult human brains.[1] The hemispheres of these brains were dissected along the plane of the Sylvian fissure, revealing the upper surfaces of the temporal lobes. Geschwind and Levitsky found that the planum temporale, the area posterior to the primary auditory cortex (Heschl's gyrus), was larger on the left in 65% of brains and larger on the right in only 11% of brains (see Fig. 4.5). On average, the left planum temporale was one-third bigger than the right.[1] What the authors did not know was that 32 years before this publication, Richard Arwed Pfeifer had mentioned this asymmetry in a chapter on the human cortical auditory system.[21] Pfeifer's study was not as meticulously conducted as the one by Geschwind and Levitzky and, most importantly, was in German—a language that many scientists outside Germany could not speak nor read.

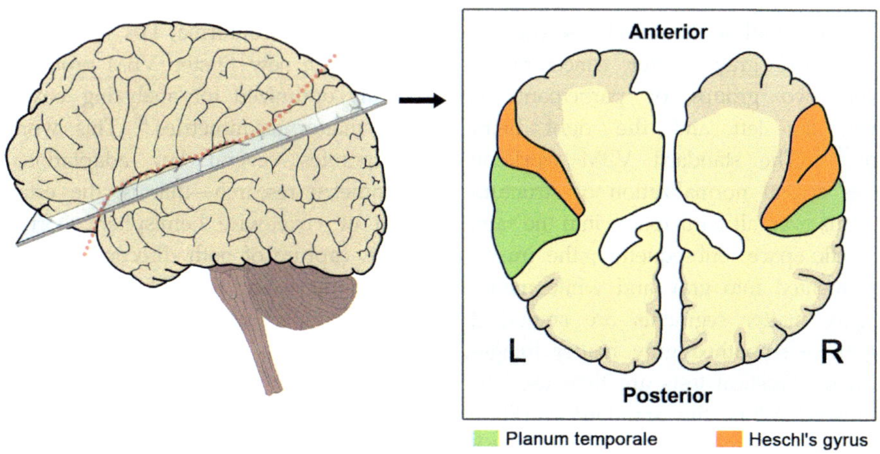

FIGURE 4.5 Structural asymmetries in the planum temporale and Heschl's gyri. The planum temporale has a larger volume in the left hemisphere than in the right hemisphere. In contrast, Heschl's gyri show a rightward asymmetry with many individuals having two gyri in the right hemisphere but only one in the left hemisphere.

The finding on asymmetry of the planum temporale was interesting since it contains Wernicke's area, one of the major language areas (see Chapter 5). Geschwind and Levitsky concluded that planum temporale asymmetry might be the structural basis of language lateralization. While Geschwind and Levitsky relied on the dissection of post-mortem brain tissue samples, other researchers have replicated the finding of leftward structural asymmetry of the planum temporale using modern in-vivo neuroimaging methods.

In the 1990s, Helmuth Steinmetz from the Department of Neurology at Heinrich-Heine-University in Düsseldorf, Germany, reported that in a sample of 154 healthy young adults studied with in vivo magnetic resonance morphometry, the planum temporale was lateralized to the left.[22] This finding was also confirmed by later meta-analysis of magnetic resonance data.[23] More recently, voxel-based morphometry (VBM, see Box 4.2) has been used to investigate macroscopic asymmetry in the planum temporale with higher regional specificity than traditional magnetic resonance approaches, further replicating pronounced leftward asymmetry.[26,27] Related to the leftward asymmetry in the planum temporale is the finding that Heschl's gyrus, located anterior to the planum temporale, shows pronounced rightward asymmetry. In about 54% of brains, there are two Heschl's gyri in the right hemisphere but only one in the left. The reverse pattern was observed in only 18% of brains[30] (see Fig. 4.5).

BOX 4.2

Key method: voxel-based morphometry (VBM)

VBM is one of the key neuroimaging techniques to assess macroscopic gray matter asymmetries with high regional specificity. VBM is an analysis workflow for structural MRI scans that allows for voxel-wise comparison of local gray matter concentrations between two groups of participants or between the left and the right hemisphere.[24,25] The standard VBM workflow involves spatial normalization of structural MRI scans from all participants into the same stereotactic space. Subsequently, the images are segmented into gray and white matter, and gray matter segments are smoothed. Using these smoothed gray matter images, voxel-wise statistical tests are then used for comparisons. While this standard workflow has been proven to reliably identify macroscopic gray matter asymmetries,[26,27] it was not specifically designed to investigate asymmetries. Recently, Florian Kurth from the Department of Neurology at the University of California, Los Angeles, USA, published a new 12-step VBM protocol specifically optimized for analyzing voxel-wise gray matter asymmetries.[28] This workflow incorporates several key adaptations for asymmetry research—that is, the establishment of voxel-wise hemispheric correspondence, capture of both direction and degree of asymmetry, and the prevention of blurring of information across hemispheres. An example of the results obtained with this toolbox—from a study investigating gray matter asymmetry differences between meditation practitioners and controls[29]—is shown in Fig. 4.6.

BOX 4.2 (*cont'd*)

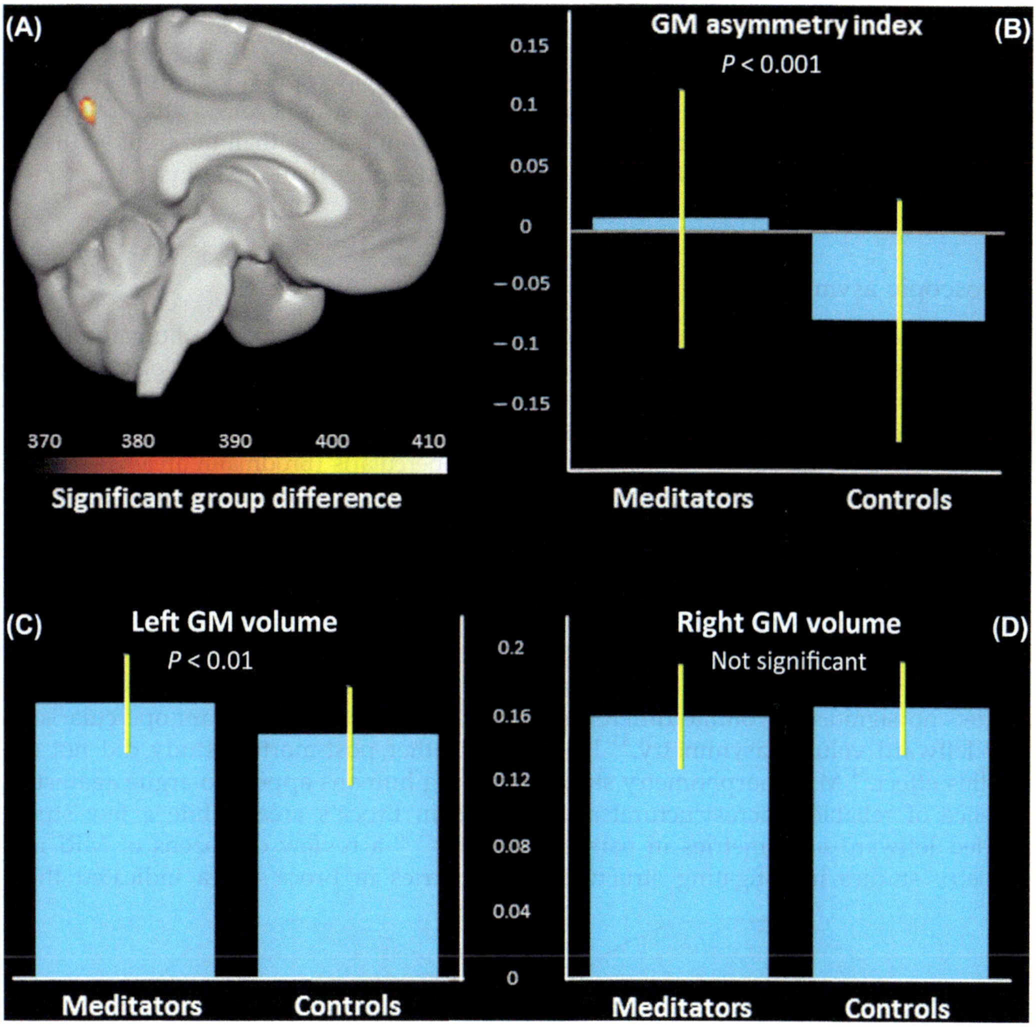

FIGURE 4.6 Exemplary voxel-based morphometry (VBM) results obtained with the VBM analysis workflow by Kurth and coworkers[28] from a study comparing gray matter asymmetries in meditators and nonmeditating controls.[29] (A) Significant gray matter asymmetry group differences between meditators and controls displayed on the medial view of the left hemisphere. (B) Cluster-specific gray matter asymmetry for both groups (positive values encode leftward asymmetry; negative values encode rightward asymmetry). (C) Cluster-specific gray matter volume (in ml) for the left hemisphere for both groups. (D) Cluster-specific gray matter volume (in ml) for the right hemisphere for both groups. *Source: Reproduced from Kurth F, MacKenzie-Graham A, Toga AW, Luders E. Shifting brain asymmetry: the link between meditation and structural lateralization. Soc Cogn Affect Neurosci. 2015;10(1):55–61. https://doi.org/10.1093/scan/nsu029, with permission from the author and publisher.*

Asymmetries of the Sylvian fissure

One of the first reliably found structural asymmetries in the brain was the asymmetry of the Sylvian fissure.[19] The Sylvian fissure—also sometimes called lateral sulcus or lateral fissure—is named after Franciscus Sylvius, a Dutch anatomist who first described it in the 17th century. It is a highly visible fissure that separates the temporal lobe from the frontal and parietal lobes. In most individuals, the left Sylvian fissure is longer than the right.[31] Moreover, the left Sylvian fissure runs at a more horizontal angle than the right so that the endpoint of the right Sylvian fissure is usually located higher than the left (see Fig. 4.7).

Macroscopic asymmetries in Broca's area

As the most pronounced structural asymmetry in the gray matter of the human brain has been observed for the planum temporale containing Wernicke's area, it was a logical next step for researchers to investigate gray matter asymmetry in the second major language region in the left hemisphere, Broca's area. Broca's area is the major speech production area within the language network and can be subdivided into pars opercularis (Brodmann area 44) and pars triangularis (Brodmann area 45) (see Fig. 4.8). Both areas are involved in semantic processing, but while pars triangularis shows particular involvement in lexical retrieval, pars opercularis shows increased relevance for phonological processing during word generation.[32]

Studies investigating macroscopic gray matter asymmetries in Broca's area often analyzed these two areas separately. In contrast to the well-replicated leftward gray matter asymmetry in the planum temporale, evidence for similar asymmetries in pars opercularis or pars triangularis is sparse and inconsistent. For example, one post-mortem study reported no significant volume difference in pars triangularis, while pars opercularis did show leftward volume asymmetry.[33] However, another post-mortem study did not confirm this effect.[34] MRI morphometry studies in living humans appear to argue against the existence of reliable macrostructural asymmetries in Broca's area. While a few studies reported leftward asymmetries in pars opercularis,[35,36] a review of dozens of MRI morphometry studies investigating structural asymmetries in Broca's area indicated that a

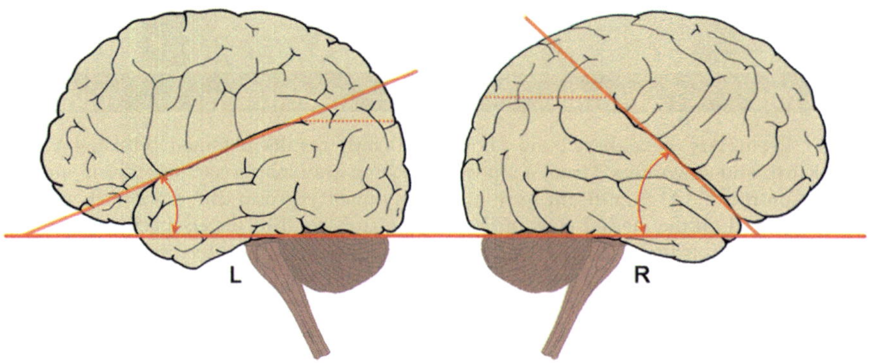

FIGURE 4.7 Structural asymmetries of the Sylvian fissure. The left Sylvian fissure is longer and runs at a more horizontal angle than the right.

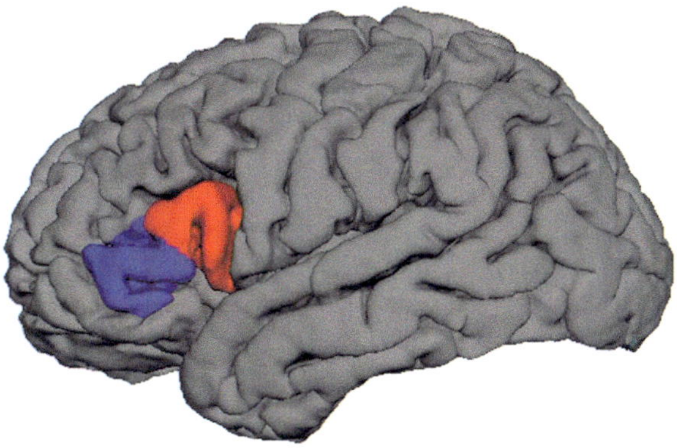

FIGURE 4.8 Broca's area can be subdivided into pars opercularis (Brodmann area 44, in red) and pars triangularis (Brodmann area 45, in blue).

large majority of these studies did not observe any statistically significant asymmetries in volume, surface area, or cortical thickness of pars triangularis or pars opercularis.[37]

Macroscopic asymmetries in the motor cortex

With a distribution of about 89.4% right-handers to 10.6% left-handers, handedness is the most pronounced functional asymmetry in humans.[38] Assuming a link between structural and functional hemispheric asymmetries, one could expect such a pronounced rightward behavioral preference to result from leftward structural asymmetries in the motor cortex.[8] The motor cortices are located in the precentral gyri, anterior to the central sulci. In each hemisphere, the motor cortex controls movements of the contralateral side of the body, with the representation of the hands located in the superior precentral gyrus.[39] Functionally, the motor cortex can be divided into three distinct areas (see Fig. 4.9):

1. *The primary motor cortex*: The primary motor cortex is located directly anterior to the central sulcus. It is the main brain area to control body movements and generates neural impulses that the spinal cord transfers to the peripheral nervous system to induce muscle activity.
2. *The premotor cortex*: The premotor cortex is located anterior to the primary motor cortex. It is involved in preparing movements and integrating sensory and spatial information into the planning of movements.
3. *The supplementary motor area*: The supplementary motor area is located anterior to the primary motor cortex on the midline surface of the respective hemisphere. It is involved in bimanual coordination and generating movement sequences.

A VBM study in right-handed subjects found a significant leftward asymmetrical cluster in an area of the precentral gyrus that corresponded to the region of hand representation,[39] a finding supported by other studies.[40,41] However, linking these structural asymmetries in the motor system to handedness has proven to be difficult, as the findings in right-handers do not seem to be completely reversed in left-handers. Two early magnetic resonance morphometry studies found that the depth of the central sulcus was related to handedness, with the left central sulcus being deeper than the right in right-handers and a reverse

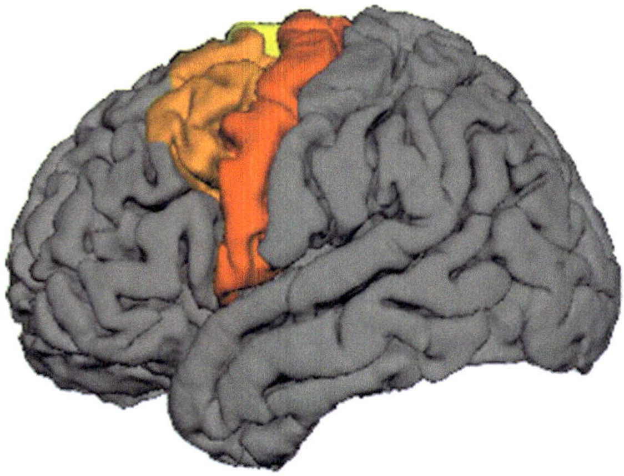

FIGURE 4.9 The motor cortex of the human brain. The primary motor cortex is shown in red, the premotor cortex is shown in orange, and the supplementary motor area is shown in yellow.

pattern in left-handers, at least for male participants.[42,43] These differences in sulcal depth imply an increased intrasulcal surface of the precentral gyrus in the dominant hemisphere.

A large-scale analysis of cortical surface area differences between 106 left-handed subjects and 1960 right-handed subjects in 74 different cortical areas in each hemisphere did not show a clear link between handedness and structural asymmetries in the motor cortex.[44] While the surface area of the left precentral sulcus showed a nominal association with left-handedness, this effect did not survive statistical correction for multiple comparisons. In contrast to this finding, an even larger study using neuroimaging data from 28,802 right-handers and 3062 left-handers from the UK Biobank population dataset did report significant effects.[45] In this study, differences between left-handers and right-handers were observed for surface area asymmetry of the precentral cortex, but also the fusiform cortex, the anterior insula, and the anterior middle cingulate cortex. In general, left-handers showed a less leftward/more rightward shift of surface area compared to right-handers in these regions. The authors concluded that this reflects a shift of cortical resources toward the right hemisphere that controls the dominant left hand in left-handers.

Macroscopic asymmetries in the planum parietale

Ever since Geschwind and Levitsky concluded that leftward planum temporale asymmetry might be the structural basis of leftward functional language lateralization,[1] researchers have been wondering whether a similar structural asymmetry could also be identified for right-dominant brain functions such as visuospatial processing.[46] One candidate region for this structure–function relationship is the so-called planum parietale, a region in the inferior parietal lobe encompassing the posterior wall of the left and right posterior ascending portion of the Sylvian fissure. A magnetic resonance morphometry study in 106 right-handers and 35 left-handers found that the planum parietale was lateralized to the right, with particularly strong rightward lateralization in right-handed men and left-handed women.[47] Rightward lateralization of the planum parietale was also observed in other studies in humans[48,49] and non-human primates.[50] While it has been proposed that rightward planum parietale asymmetries might relate to rightward functional asymmetries for visuospatial processing,[47] this relation has not yet been empirically tested.

Macroscopic asymmetries in subcortical structures

While structural brain asymmetries were mostly investigated in cortical structures, they have also been studied in subcortical structures—for example, in the limbic system or the basal ganglia. In 2016, Tulio Guadalupe and Clyde Francks from the Language and Genetics Department at the Max Planck Institute for Psycholinguistics in Nijmegen, The Netherlands, together with dozens of coauthors, presented a harmonized multi-site study analyzing subcortical brain asymmetries.[14] Using meta-analysis methods, they integrated volumetric asymmetry data of seven subcortical structures from more than 15,000 MRI scans from 52 different datasets. Based on volume estimates for the left and right nucleus accumbens, amygdala, caudate nucleus, globus pallidus, hippocampus, putamen, and thalamus, they calculated asymmetry indices for these bilaterally paired structures and found significant volume asymmetries for all of them. While the nucleus accumbens, putamen, globus pallidus, and thalamus had a larger volume on the left side, the caudate nucleus, amygdala, and hippocampus had a larger volume on the right side (see Fig. 4.10).

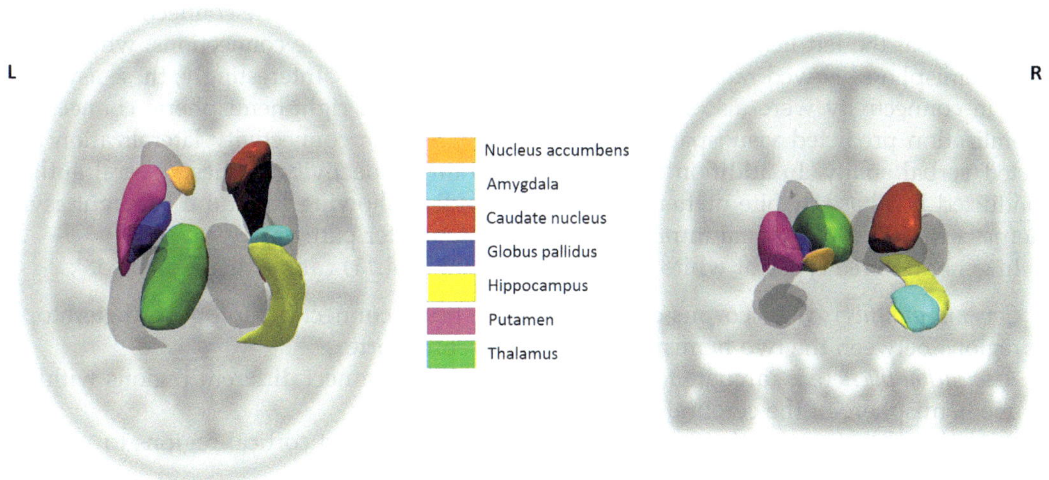

FIGURE 4.10 Visual representation of the seven bilaterally asymmetric subcortical structures from the study by Guadalupe and coworkers.[14] The structures are colored on the side of the relatively larger volume. *Source: Reproduced from Guadalupe T, Mathias SR, van Erp TGM, et al. Human subcortical brain asymmetries in 15,847 people worldwide reveal effects of age and sex. Brain Imaging Behav. 2017;11(5):1497–1514. https://doi.org/10.1007/s11682-016-9629-z, with permission of the author. (Authors note that "This is an open access article distributed under the terms of the Creative Commons Attribution License, which permits unrestricted use, distribution, and reproduction in any medium, provided the original work is properly cited.").*

Macroscopic asymmetries in white matter

While research on structural asymmetries has traditionally focused on gray matter, the wide availability of advanced in vivo white matter neuroimaging techniques such as diffusion tensor imaging (DTI, see Box 3.2) starting in the 2000s has led to an increased number of publications investigating white matter asymmetries. For example, in 2007, Thomas R. Barrick, from the Centre for Clinical Neuroscience at St. George's University of London, UK, and his coworkers published a DTI study investigating white matter asymmetries in 30 right-handed healthy young adults.[51] They identified a leftward asymmetric pathway connecting the parietal and frontal lobes to the temporal lobe as well as a rightward asymmetric pathway connecting the posterior temporal lobe to the superior parietal lobule. They speculated that the leftward asymmetric pathway may be related to functional language lateralization, while the rightward asymmetric pathway may be related to rightward functional asymmetries in auditory spatial attention and working memory. Support for the idea that intrahemispheric brain connections are organized in a fundamentally asymmetrical fashion also comes from whole-brain connectivity analysis. For example, a recent DTI and graph theory study indicated that the right hemisphere had greater intrahemispheric global efficiency and more interhemispheric interconnections, while the left hemisphere had more intrahemispheric local connections.[52] Moreover, right-hemispheric networks were characterized by increased efficiency in brain areas related to visuospatial attention and memory, while left-hemispheric networks show increased efficiency in language and motor areas.[53] Importantly, these asymmetries in structural whole-brain connectivity seem to be reflected by lateralization in functional connectivity. A recent fMRI resting-state connectivity analysis indicated that the left hemisphere showed more intrahemispheric connections, whereas the right hemisphere showed more interhemispheric connections.[46] Interestingly, these asymmetries in connectivity were also relevant for behavior: While right-hemispheric connectivity lateralization predicted visuospatial ability, left-hemispheric connectivity lateralization predicted verbal ability. In addition to such whole-brain connectivity studies, several authors conducted DTI-based white matter tractography studies, showing structural left-right asymmetries for several individual intrahemispheric white matter pathways.

In general, three types of white matter tracts (see Fig. 4.11) can be distinguished[54]:

1. *Commissural fibers* (e.g., corpus callosum): These tracts connect homotopic or heterotopic regions between the two hemispheres of the cerebral cortex.
2. *Association fibers* (e.g., arcuate fasciculus): These tracts connect different regions within one hemisphere of the cerebral cortex.
3. *Projection fibers* (e.g., corticospinal tract): These tracts connect cortical regions to subcortical regions and the spinal cord.

Macroscopic asymmetries in the arcuate fasciculus

Structural asymmetries have been reported for several white matter pathways, mostly concentrating on association fibers. The most widely investigated tract is probably the arcuate fasciculus (from the Latin meaning "curved bundle") (see Fig. 4.12). This lateral association tract bidirectionally connects the perisylvian regions of the parietal and

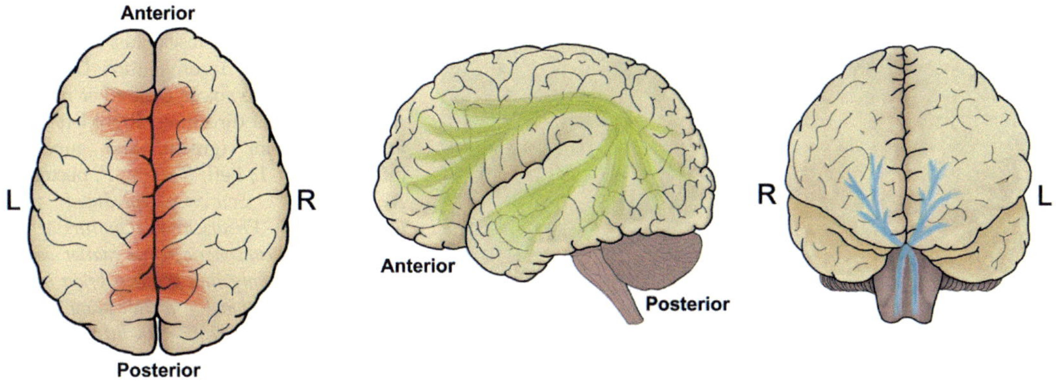

FIGURE 4.11 The three types of white matter tracts in the human brain. Commissural fibers (in red) connect regions between the two hemispheres of the cerebral cortex. Association fibers (in green) connect different regions within one hemisphere of the cerebral cortex. Projection fibers (in blue) connect cortical regions to subcortical regions and the spinal cord.

temporal cortex to the frontal lobe.[56] Originally, the arcuate fasciculus was thought to represent the connection between Broca's and Wernicke's area, but today, it is understood as part of a larger tract, the superior longitudinal fasciculus (SLF).[55,57]

Leftward asymmetry in the arcuate fasciculus has been reported for several types of structural measures. On the macrostructural level, it has been shown that tract volume is larger for the left arcuate fasciculus than for the right.[58–63] Moreover, leftward asymmetry has also been reported for tract length[61] and relative fiber density.[64,65] The two arcuate fasciculi have also been shown to differ in function. In the left hemisphere, the arcuate fasciculus has been shown to play a role in language[66,67] and in performing voluntary skilled movements.[68] In contrast, the right arcuate fasciculus is relevant for visuospatial processing[69] and prosodic aspects of language processing.[70]

Macroscopic asymmetries in the uncinate fasciculus

In addition to the arcuate fasciculus, another language-related white matter pathway exhibiting structural asymmetries is the uncinate fasciculus (from the Latin meaning "hooked bundle") (see Fig. 4.13). The uncinate fasciculus is a long-range association tract that bidirectionally connects the anterior temporal lobes with the anterior prefrontal and lateral orbitofrontal cortex.[71] The exact functional role of the uncinate fasciculus is not well understood, but empirical evidence indicates a role for language as well as episodic memory and social-emotional processing.[71]

For uncinate fasciculus tract volume, several authors have reported leftward asymmetry,[58,59,72] but two studies also reported rightward asymmetry.[73,74] How is it possible that some studies showed leftward tract volume asymmetry, while others showed rightward tract volume asymmetry for the same structure? To reconcile these conflicting asymmetry findings on uncinate fasciculus structure, a multinational research team led by Laurent

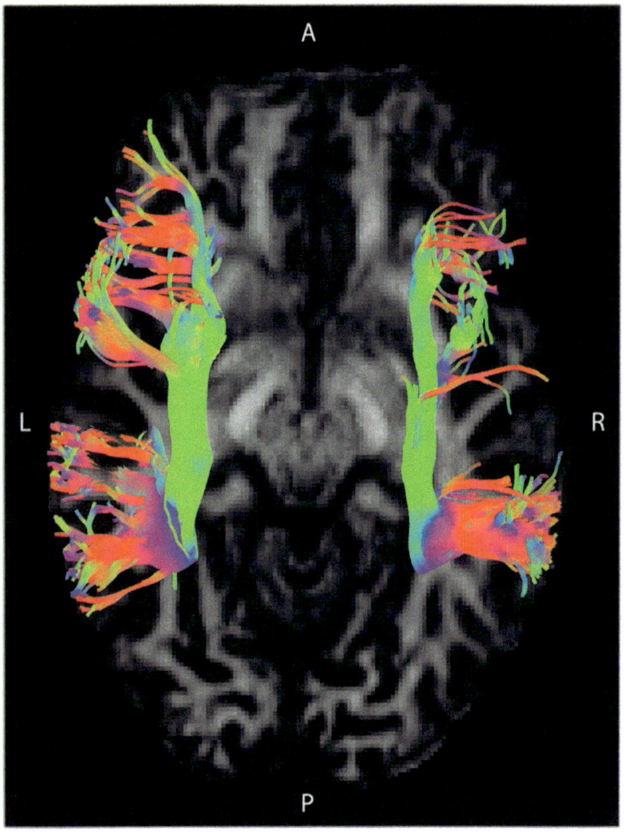

FIGURE 4.12 Axial view of the left and right human arcuate fasciculus, tracked by in vivo spherical deconvolution DTI fiber tractography. The colors code fiber orientations in different directions: Right-left (red), anterior-posterior (green), and superior-inferior (blue). *L*, Left; *R*, right; *P*, posterior; *A*, anterior.[55] *Source: Reproduced from Ocklenburg S, Friedrich P, Güntürkün O, Genç E. Intrahemispheric white matter asymmetries: the missing link between brain structure and functional lateralization? Rev Neurosci. 2016;27(5):465–480. https://doi.org/10.1515/revneuro-2015-0052, with permission from the authors and publisher.*

Petit from the University of Bordeaux in France performed an advanced tractography study.[75] Tractography was performed on the uncinate fasciculus, but instead of analyzing the whole tract, they subsequently applied an automatic clustering method to subdivide the structure into subcomponents. This analysis revealed five uncinate fasciculus subcomponents in each hemisphere. Importantly, these subcomponents showed distinct connectivity profiles and different structural asymmetries. While two subcomponents showed significant rightward asymmetries, one subcomponent showed significant leftward asymmetry. The two other subcomponents did not show any significant asymmetries. These findings were further supported by additional virtual microdissection analysis. This indicates that it can be beneficial to shift our understanding of certain white matter tracts from a unitary structure toward a multicomponent bundle to disambiguate conflicting asymmetry results. A specific way of analyzing the data could lead to the focus on one subcomponent of the overall tract that is lateralized in one direction in one study. If another study concentrates on a different subcomponent, puzzling discrepancies in results could be found.

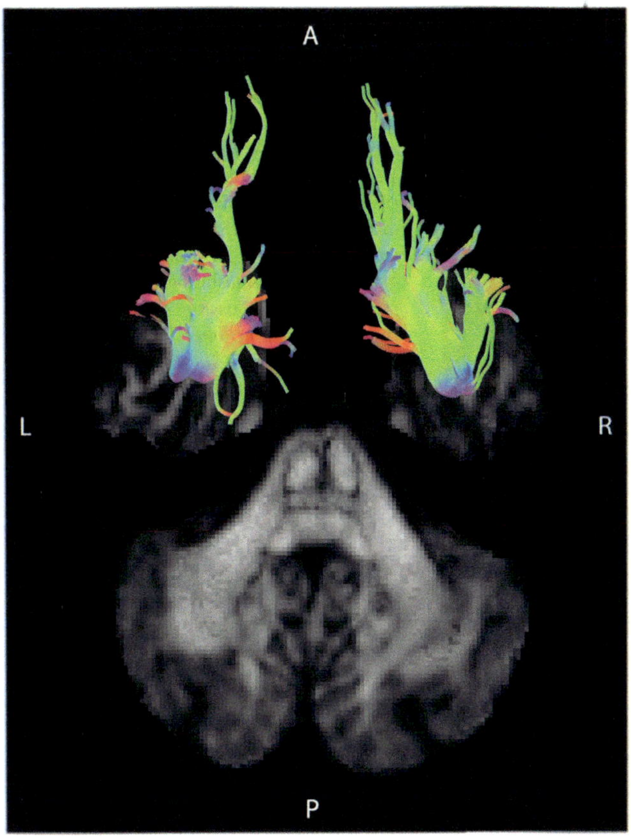

FIGURE 4.13 Axial view of the left and right human uncinate fasciculus, tracked by in vivo spherical deconvolution DTI fiber tractography. The colors code fiber orientations in different directions: Right-left (red), anterior-posterior (green), and superior-inferior (blue). *L*, Left; *R*, right; *P*, posterior; *A*, anterior.[55] *Source: Reproduced from Ocklenburg S, Friedrich P, Güntürkün O, Genç E. Intrahemispheric white matter asymmetries: the missing link between brain structure and functional lateralization?* Rev Neurosci. *2016;27(5):465–480. https://doi.org/10.1515/revneuro-2015-0052, with permission from the authors and publisher.*

Macroscopic asymmetries in the superior longitudinal fasciculus

Another multicomponent bundle for which macrostructural asymmetries have been reported for specific subcomponents is the SLF. The SLF is an association bundle that connects the parietal with the frontal lobe (see Fig. 4.14) and has functionally been implicated in visuospatial functions.[76] It can be separated into the dorsal SLF I, the middle SLF II, and the ventral SLF III. Interestingly, a recent tractography study revealed differential asymmetries in tract volume in these three subcomponents. While the tract volume of SLF III showed significant rightward asymmetry, SLF I was symmetric, and SLF II only showed a nonsignificant trend toward rightward lateralization.[76]

Microstructural asymmetries

Microstructural asymmetries in gray matter

Microstructural gray matter asymmetries refer to left-right differences in the number and density of nerve cells in specific brain regions as well as to asymmetries in other

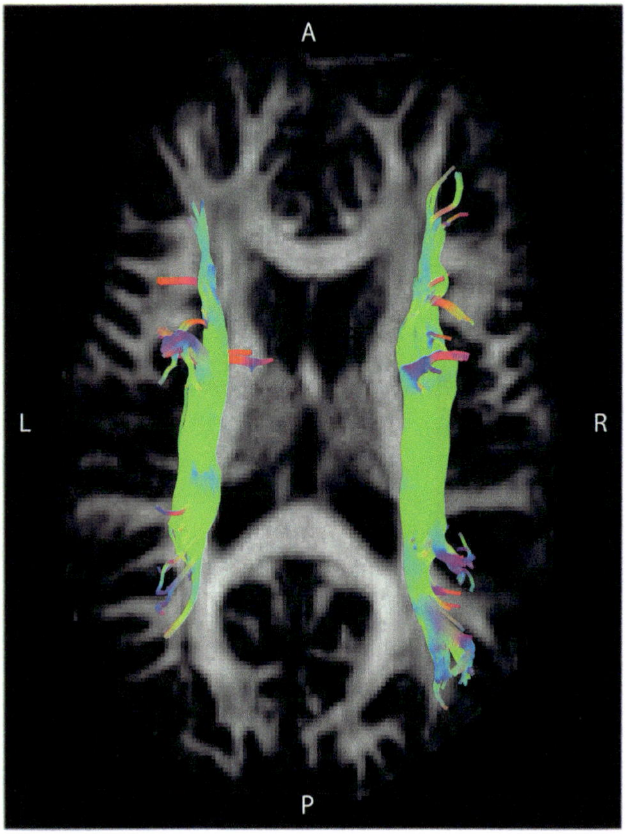

FIGURE 4.14 Axial view of the left and right human superior longitudinal fasciculus (SLF). For a clearer illustration, we only show SLF II and III, which are merged in this figure. The colors code fiber orientations in different directions: Right-left (red), anterior-posterior (green), and superior-inferior (blue). *L*, Left; *R*, right; *P*, posterior; *A*, anterior.[55] *Source: Reproduced from Ocklenburg S, Friedrich P, Güntürkün O, Genç E. Intrahemispheric white matter asymmetries: the missing link between brain structure and functional lateralization?* Rev Neurosci. *2016;27(5):465–480. https://doi.org/10.1515/revneuro-2015-0052, with permission from the authors and publisher.*

parameters of specific cytoarchitectonic areas—for example, size, volume, surface, cortical thickness, number of spines, and degree of arborization of dendrites and axons.[5] While research on structural asymmetries in gray matter has mainly focused on macroscopic asymmetries, microstructural asymmetries have been investigated in several brain areas. For example, Wernicke's area within the planum temporale and its right hemispheric counterpart contain vertical columns of pyramidal neurons that show strong leftward hemispheric asymmetries in terms of spacing. The columns were wider and farther apart from each other in the left hemisphere, implying that more subsystems can be included per surface unit in the left than in the right hemisphere. Overall, an increased number of differently specialized columns exists on the left side compared to the right, allowing for finer-grained analysis of the auditory signal in this hemisphere.[77–80] Interestingly, Broca's area seems to show more pronounced microstructural than macroscopic asymmetries. For example, it has been found that both pars opercularis and pars triangularis of Broca's area show leftward asymmetry in total neuron number.[81] Moreover, a left-larger-than-right asymmetry in the volume fraction of cell bodies was observed for both areas.[82]

Traditionally, the investigation of microstructural asymmetries in the human brain had the disadvantage that it was only possible in post-mortem brain tissue. However, with advanced diffusion neuroimaging sequences, it has recently become possible to estimate microstructural asymmetries in the human brain in vivo. For example, Judith Schmitz from the University of Göttingen, Germany, and coworkers used neurite orientation dispersion and density imaging (NODDI) to examine microstructural asymmetries in the human brain.[83]

NODDI is a diffusion MRI technique that allows for the estimation of the microstructural complexity of neurites (dendrites and axons) in vivo.[84] Schmitz and coworkers used NODDI to investigate microstructural asymmetries in three different parameters (see Fig. 4.15 for their workflow) in two independent adult cohorts (n = 269 and n = 251):

1. *Intraneurite volume fraction (INVF)*: A measure of neurite density. If INVF is high, the density of axons and dendrites in the investigated tissue is high. If INVF is low, the density of axons and dendrites in the investigated tissue is also low.
2. *Neurite orientation dispersion (ODI)*: A measure of neurite dispersion. If ODI is high, the dispersion of axons and dendrites in the investigated tissue is high. If ODI is low, axons and dendrites in the investigated tissue are largely aligned.
3. *Isotropic volume fraction (ISO)*: It is the proportion of free-moving water within each voxel. ISO is high in cerebrospinal fluid in the ventricles and lower in gray or white matter.

The results of the study indicated that microstructural asymmetries are quite common in the human brain. The authors reported only those asymmetries that were consistently found in both independent samples, to avoid spurious findings. The largest number of significant asymmetries were observed for INVF (see Fig. 4.16). For this NODDI parameter, pronounced leftward asymmetries were observed in early auditory, inferior parietal, and temporal-parietal-occipital brain areas. This is in line with a histological postmortem study reporting leftward microstructural asymmetries in neuron density in the planum temporale[85], and may reflect a structural underpinning of leftward functional language lateralization. Rightward asymmetries were observed in the inferior temporal gyrus and the fusiform gyrus, potentially reflecting a structural underpinning of rightward functional lateralization for face processing. For ODI (Fig. 4.17), leftward asymmetry was found in frontal areas and rightward asymmetry in early auditory areas. In contrast to INVF and ODI, ISO did show less pronounced asymmetries (see Fig. 4.18). This was to be expected, as ISO models the properties of cerebrospinal fluid, which is unlikely to be asymmetrically distributed. Taken together, the findings of this study suggest that microstructural asymmetries are widespread in the human brain. Future studies should investigate whether microstructural asymmetries may show stronger links to functional asymmetries than macrostructural asymmetries.

Microstructural asymmetries in white matter

In contrast to gray matter, for which the majority of studies on structural asymmetries have been performed using macroscopic asymmetry measures, research on structural asymmetries

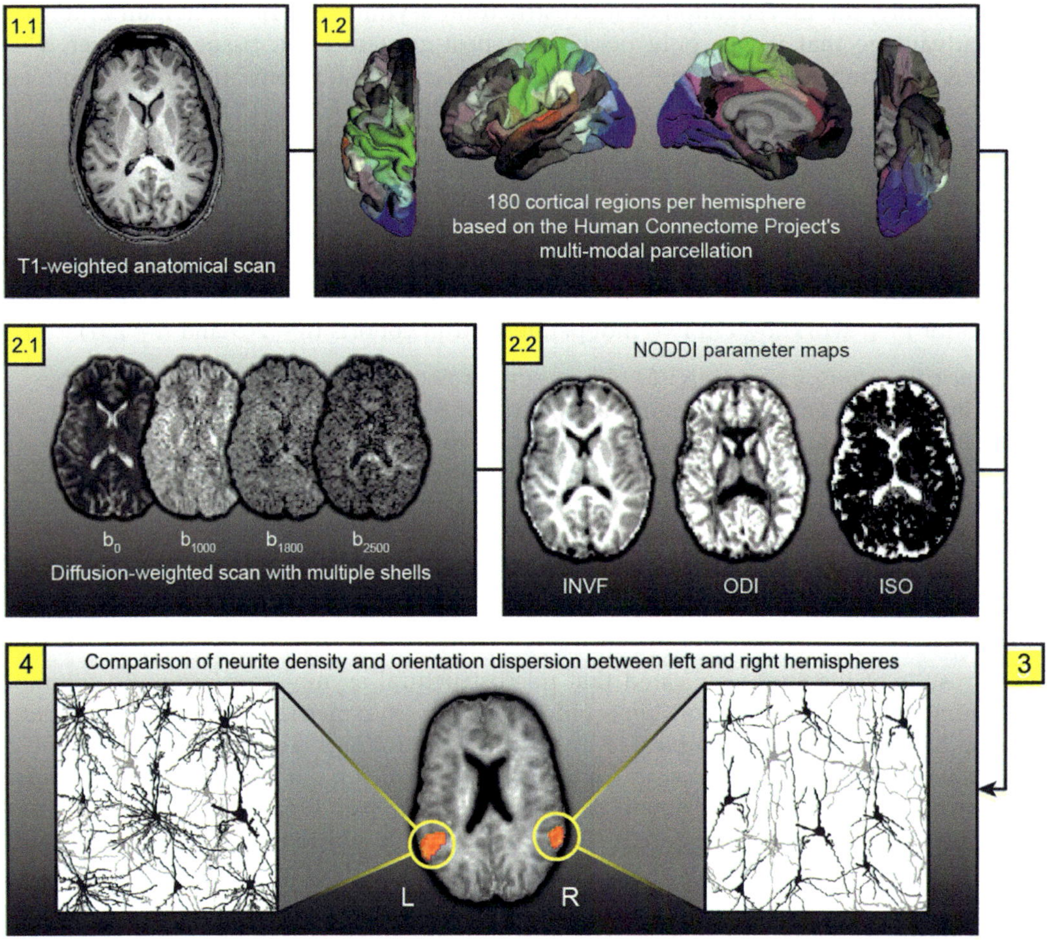

FIGURE 4.15 Schematic workflow for the use of NODDI to investigate microstructural asymmetries in the human brain.[83] Anatomical scans are delineated into 180 cortical regions in each hemisphere. Diffusion-weighted scans from the same individuals were used to create NODDI parameter maps for INVF, ODI, and ISO. NODDI parameters were than determined for each cortical region in the anatomical maps and compared between the left and the right hemisphere. *Source: Reproduced from Schmitz J, Fraenz C, Schlüter C, et al. Hemispheric asymmetries in cortical gray matter microstructure identified by neurite orientation dispersion and density imaging. NeuroImage. 2019;189:667–675. https://doi.org/10.1016/j.neuroimage.2019.01.079, with permission from the authors and publisher.*

in white matter has mainly focused on microstructural asymmetries. In particular, DTI (see Box 3.2) has been used to determine the so-called fractional anisotropy (FA), a measure of diffusion thought to reflect the microstructural integrity of white matter tracts.[86] FA is calculated from the three eigenvalues λ_1, λ_2, and λ_3 of each voxel, which indicate the diffusion of water molecules within that voxel along the *x*, *y*, and *z-axis*. FA has a value between 0 and 1, with higher values indicating stronger diffusion along one of the three axes compared to the other two. Thus, white matter within a voxel with high FA is thought to have higher microstructural integrity—that is, transporting more neuronal information in the same direction—than

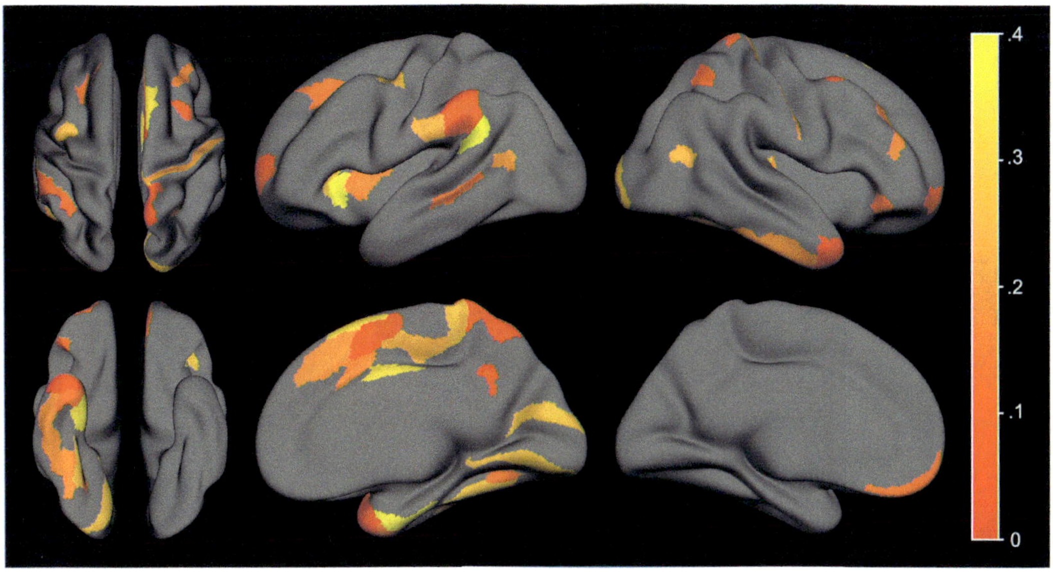

FIGURE 4.16 Microstructural asymmetries in INVF.[83] The color indicates the effect size of the main effect of hemisphere from small (red) to large (yellow). *Source: Reproduced from Schmitz J, Fraenz C, Schlüter C, et al. Hemispheric asymmetries in cortical gray matter microstructure identified by neurite orientation dispersion and density imaging.* NeuroImage. *2019;189:667–675. https://doi.org/10.1016/j.neuroimage.2019.01.079, with permission from the authors and publisher.*

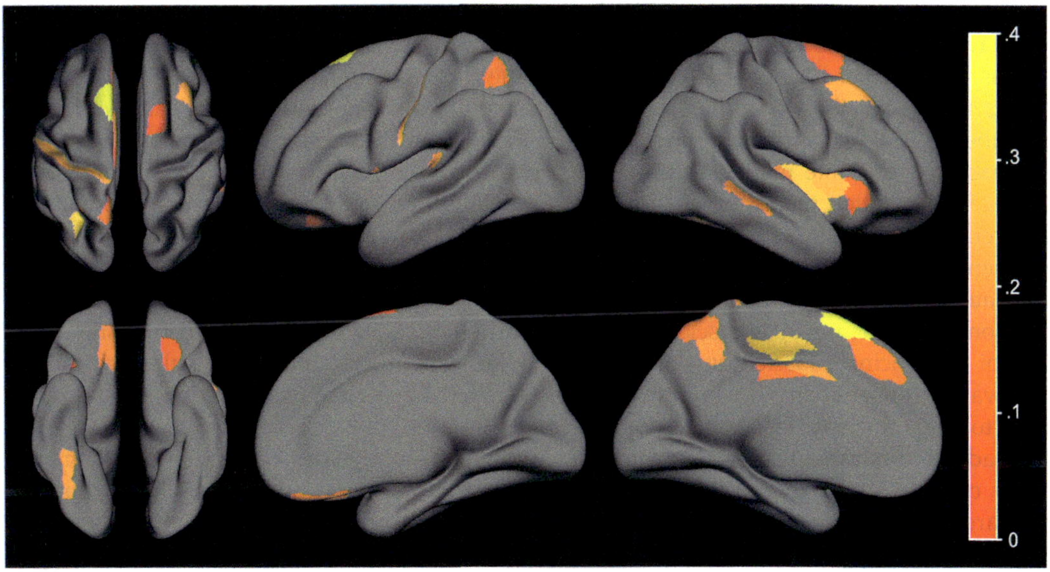

FIGURE 4.17 Microstructural asymmetries in ODI.[83] The color indicates the effect size of the main effect of hemisphere from small (red) to large (yellow). *Source: Reproduced from Schmitz J, Fraenz C, Schlüter C, et al. Hemispheric asymmetries in cortical gray matter microstructure identified by neurite orientation dispersion and density imaging.* NeuroImage. *2019;189:667–675. https://doi.org/10.1016/j.neuroimage.2019.01.079, with permission from the authors and publisher.*

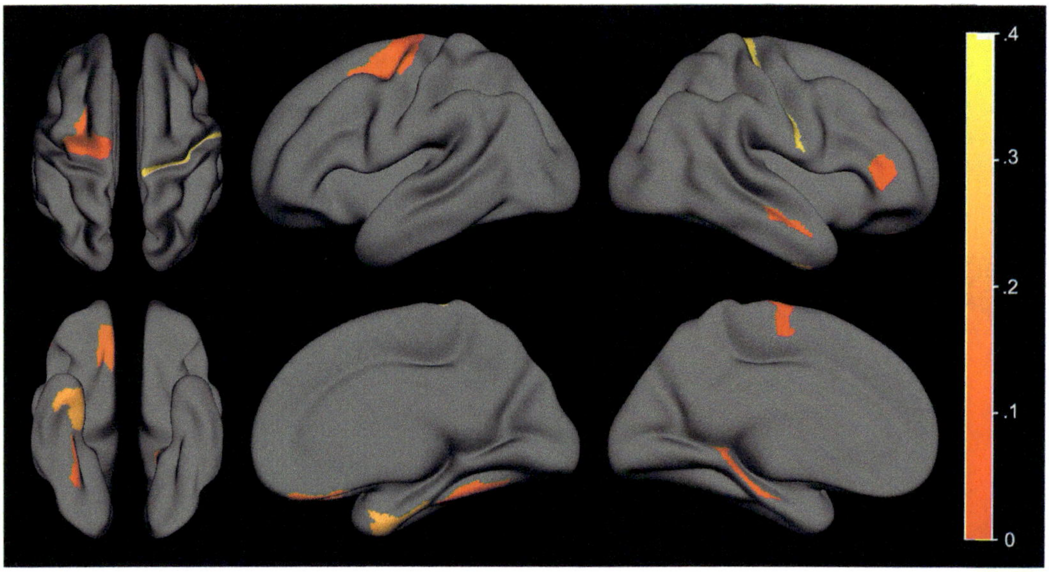

FIGURE 4.18 Microstructural asymmetries in ISO.[83] The color indicates the effect size of the main effect of hemisphere from small (red) to large (yellow). *Source: Reproduced from Schmitz J, Fraenz C, Schlüter C, et al. Hemispheric asymmetries in cortical gray matter microstructure identified by neurite orientation dispersion and density imaging.* NeuroImage. *2019;189:667–675. https://doi.org/10.1016/j.neuroimage.2019.01.079, with permission from the authors and publisher.*

white matter in a voxel with low FA. Comparable to what has been found for macroscopic white matter asymmetries, substantial leftward asymmetry has been found for the arcuate fasciculus with greater FA of the left compared to the right arcuate fasciculus.[54,58,60,74,87–96] These results imply that the left arcuate fasciculus transports neuronal information more efficiently than the right. For the uncinate fasciculus, the data pattern is similar to what has been found for macroscopic white matter asymmetries. For the whole tract, several authors have found higher FA in the left than the right uncinate fasciculus.[92,97–102] However, several groups have also reported rightward FA asymmetry for this tract.[74,95,103–106] Two studies did not observe any significant left-right differences in uncinate fasciculus FA.[73,107] Just like for macroscopic uncinate fasciculus asymmetries, such discrepant findings may be attributed to different subcomponents of the uncinate fasciculus showing differential asymmetries. While the subinsular part of the uncinate fasciculus shows leftward FA asymmetry, the extrainsular part shows rightward FA asymmetry.[94] Moreover, sample characteristics such as handedness[58] and sex[108] seem to affect uncinate fasciculus asymmetries.

In addition to arcuate and uncinate fasciculus, FA asymmetries have been discovered in many other white matter tracts.[95,99,104,105,109] For example, several authors have reported leftward FA asymmetry in the cingulum, a white matter tract in the limbic system that connects the cingulate cortex with the entorhinal cortex. Furthermore, leftward FA asymmetries have been observed in the corticospinal tract, optic radiation, inferior corona radiata, and superior cerebellar peduncle. Rightward FA asymmetries have been observed in the SLF, capsula interna, superior corona radiata, prefrontal white matter, and inferior longitudinal fasciculus.

Molecular asymmetries

Molecular asymmetries are differences in gene expression and protein distribution between the left and the right hemisphere.[5] Investigating these asymmetries is interesting, as they might constitute the molecular basis of both higher-level structural asymmetries and functional asymmetries. Evidence from gene expression studies in brain tissue obtained from human embryos indicates that several genes show consistent asymmetries in gene expression, that is, the amount to which a functional gene product is generated in the left and right hemispheres. For example, Tao Sun from Harvard Medical School, USA, and his coworkers used serial analysis of gene expression to compare gene expression levels between the left and right hemispheres of human embryos. They found 27 genes that showed reliable expression asymmetries[110–112] (see Fig. 4.19).

The most pronounced leftward expression asymmetries (up to sevenfold relative gene expression) were observed for *RABL2B* (RAB, member of RAS oncogene family-like 2B), the transcriptional repressor *HEY1* (hes-related family bHLH transcription factor with YRPW motif 1), and *DAPPER1* (disheveled binding antagonist of beta catenin 1). The most pronounced rightward expression asymmetries (up to 2.5-fold relative gene expression) were observed for the transcriptional regulator *LMO4* (LIM domain only 4), the stathmin-like 4 gene *STMN4* (stathmin 4), and *BAI1/ADGRB1* (adhesion G protein-coupled receptor B1). Of these genes, *LMO4* in particular gained the interest of the asymmetry research community, as it seems to influence behavioral asymmetries in vertebrates.[113]

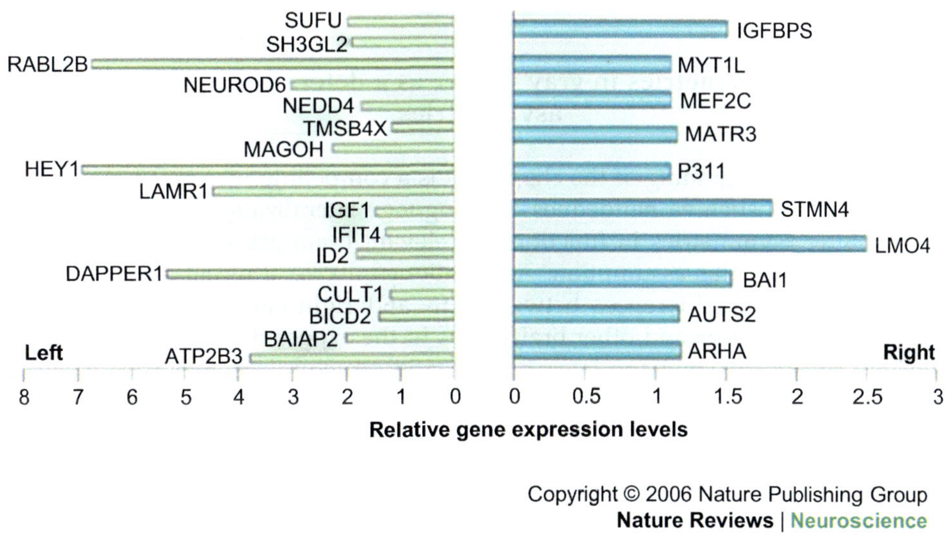

FIGURE 4.19 Gene expression asymmetries in human fetal brains at week 12.[112] *Source: Reproduced from Sun T, Walsh CA. Molecular approaches to brain asymmetry and handedness. Nat Rev Neurosci. 2006;7(8):655-662. https://doi.org/10.1038/nrn1930, with permission from the authors and publisher.*

In addition to studies identifying single genes that are asymmetrically expressed, a recent reanalysis of two transcriptomic datasets derived from post-mortem samples from the human language cortex reported left-right differences in gene expression not only for individual genes but also for so-called gene ontology (GO) groups.[114] GO groups classify genes together based on their function. There are three main groups—molecular function, cellular component, and biological process—as well as many more specific groups—for example, left/right axis specification—and one gene can belong to more than one GO group. GO groups can be assessed using AmiGO2 (http://amigo.geneontology.org/amigo/landing).[115] The GO approach has the advantage that it combines findings from single genes and integrates them, potentially leading to stronger effects than single gene analysis. Interestingly, analysis has revealed lateralization for several GO groups involved in the development of activity of the cortex. The four GO groups that showed the strongest effects all showed leftward asymmetry and included synaptic transmission, signal transduction, glutamate receptor activity, and nervous system development. The authors concluded that these GO groups might form the molecular base of leftward language lateralization and that the relationship between genes involved in these GO groups and functional lateralization would certainly be interesting to study.

In addition to this gene expression study, a recent developmental study also directly assessed hemispheric asymmetries in the proteome, the set of proteins produced by an organism.[116] By analyzing brain tissue samples from fetuses discarded following induced pregnancy termination, the authors found that at gestational weeks 9/11/13, 2969/2956/2766 proteins were expressed in the left frontal lobe, and 2932/2910/2784 proteins were expressed in the right frontal lobe. 2789/2778/2617 of these proteins were shared between the two hemispheres. Thus, asymmetries in the proteome exist, but are rather subtle. Most proteins are present in both hemispheres during early development.

Structural asymmetries in gray matter as a determinant of functional asymmetries

As outlined at the beginning of the chapter, it is a compelling idea that functional hemispheric asymmetries such as handedness or language lateralization are guided by structural gray matter asymmetries in functionally relevant brain areas. There is indeed some empirical evidence for this assumption. As mentioned above, a recent large-scale analysis of cortical surface area differences between left- and right-handers revealed significant effects for the motor cortex and other brain areas in the expected direction.[45] Similarly, the link between functional language lateralization and structural asymmetries in language-related areas has been investigated. For example, there is a significant positive correlation between structural asymmetry of the planum temporale and performance in the dichotic listening task, a behavioral measure of language lateralization,[117] in right-handed males. In this group, stronger structural lateralization of the planum temporal predicted stronger leftward language lateralization on the functional level. However, no such relation was found for women or left-handed men, indicating that factors other than asymmetry of the planum temporale affect language lateralization. Moreover, a recent comparison of structural asymmetries in individuals that were previously determined as functionally left- or right-dominant for language revealed significant leftward volume asymmetries in the planum temporale, Heschl's gyrus, pars opercularis of Broca's area, and the insula—a part of the cortex folded within the Sylvian fissure—in both groups of participants.[118] Thus, there was no rightward lateralization of these structures in participants with rightward functional language lateralization. The only area in which a significant group difference was observed was the insula, with larger insula volume in functionally leftward-dominant individuals compared to functionally rightward-dominant individuals. Moreover, the insula was the only brain area that showed a significant correlation with functional lateralization. While the extent of insula volume asymmetry correlated with dichotic listening performance and fMRI activation asymmetries during word generation, structural asymmetries in the planum temporale and Broca's area failed to show an association with functional language lateralization, a finding that has been replicated in another study.[119] Taken together, these findings support the idea that while structural gray matter asymmetries are associated with functional lateralization, they certainly are not the only influencing factor.

Structural asymmetries in white matter as a determinant of functional asymmetries

As for structural asymmetries in gray matter, structural asymmetries in white matter have been linked to functional asymmetries. For example, a combined fMRI/DTI study revealed a significant correlation between FA lateralization in the arcuate fasciculus with fMRI activation asymmetries during verb generation and reading comprehension.[60] Individuals with stronger FA asymmetry in the arcuate fasciculus also showed more pronounced activation asymmetries in the fMRI. Similar findings were reported in an fMRI/DTI study using tract-based spatial statistics, a method to statistically analyze DTI on the voxel instead of the tract level.[92] Here, several leftward asymmetric clusters were observed in the arcuate fasciculus. These structural asymmetries showed positive correlations with brain activation asymmetries during silent word generation, again indicating that stronger structural asymmetries in the arcuate fasciculus were related to stronger functional language lateralization. Moreover, asymmetries in the relative fiber density of the arcuate fasciculus have also been related to functional language lateralization. A DTI study showed that in right-handers, leftward asymmetry in arcuate fasciculus fiber density was positively correlated with brain activation asymmetries during verb generation.[65] However, no such relation was found in left-handers. This implies that handedness might affect the relation of asymmetries in white matter structure and functional lateralization, an idea that was supported by another recent study investigating asymmetries in arcuate fasciculus volume and length.[61] In this study, structural asymmetries in the arcuate fasciculus correlated with brain activation asymmetries during antonym (an antonym is a word that is the exact opposite of another word) generation only in left- but not in right-handers. For the uncinate fasciculus, a recent study combining DTI and dichotic listening indicated a positive correlation between the volume of the right uncinate fasciculus and dichotic listening performance.[92] Taken together, the empirical evidence supports the idea that white matter asymmetry, in particular in the arcuate fasciculus, might be one of the factors underlying functional language lateralization. As for gray matter asymmetries, this relation seems to be modulated by handedness.

An integrative view on structural asymmetries

Structural hemispheric asymmetries are abundant in the human brain and can be observed on the macroscopic and microstructural level in both gray and white matter. Additionally, left-right differences have been observed on the level of gene expression. Despite this abundance of structural asymmetries, it has proven surprisingly difficult to reliably link them to functional asymmetries. One problem in this regard seems to be that

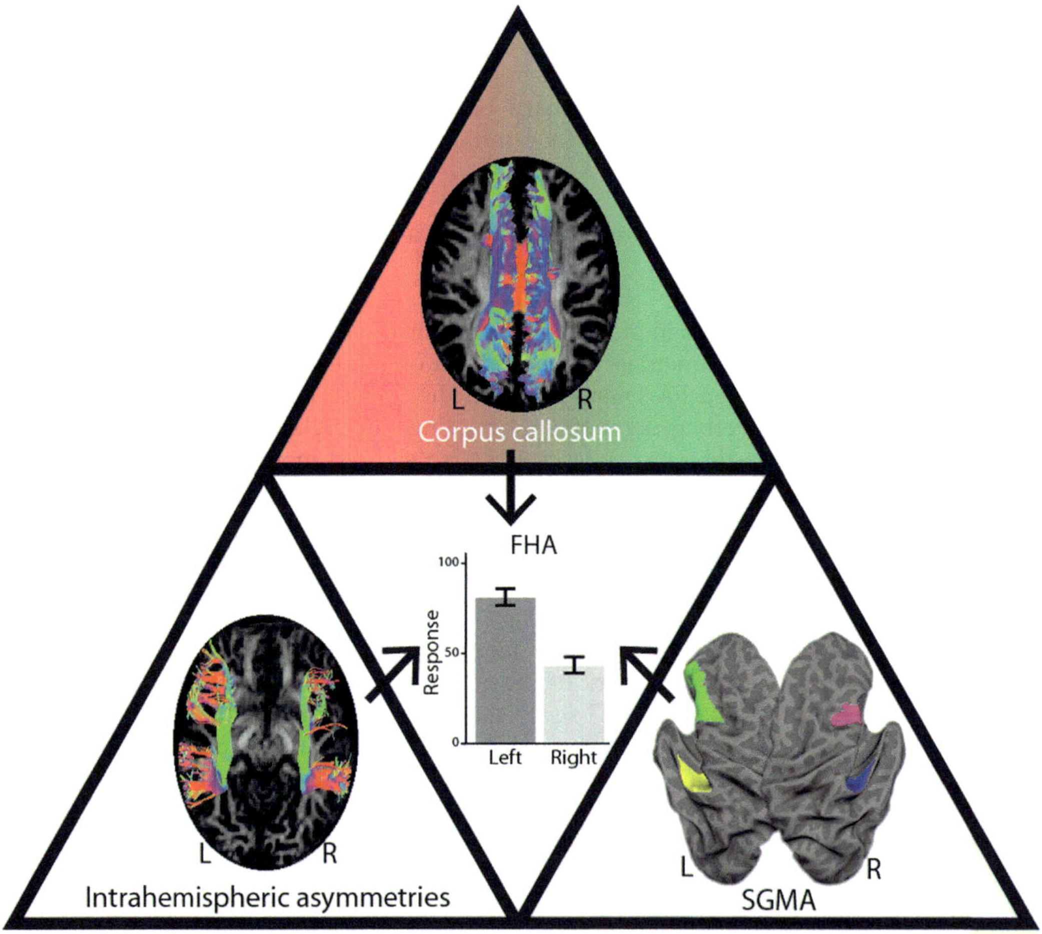

FIGURE 4.20 The triadic model: Structural factors that predict functional hemispheric asymmetries in the human brain. Corpus callosum structure (top), structural gray matter asymmetries (bottom right), and structural white matter asymmetries (bottom left) interact to determine functional lateralization.[55] *Source: Reproduced from Ocklenburg S, Friedrich P, Güntürkün O, Genç E. Intrahemispheric white matter asymmetries: the missing link between brain structure and functional lateralization?* Rev Neurosci. *2016;27(5):465–480. https://doi.org/10.1515/revneuro-2015-0052, with permission from the publisher and the authors.*

most studies that link functional lateralization—for example, in the language domain—to structure have focused on one or only a few key structures. Also, most studies have investigated structural asymmetries either in gray or in white matter. This approach does not fully take the network character of the human brain into account. Thus, it has recently been suggested that our understanding of asymmetrical structure–function relationships can only be advanced if all relevant structural determinants are included in the analysis.[55] This would not only include structural gray matter areas in relevant parts of the brain but also structural white matter asymmetries in the tracts connecting these areas, and structural properties of the corpus callosum which connects the hemispheres (see Chapter 3). By integrating these three factors into the analysis, interactions and inhibitory effects can be understood more accurately than by analyzing them separately (see Fig. 4.20).

Sample questions

- Describe the three classes of structural asymmetries in the brain. Give examples for each class.
- What is the ENIGMA consortium and why are such initiatives important for laterality research?
- What is known about macrostructural asymmetries in the human cortex?
- What is known about asymmetries in the planum temporale and their functional relevance?
- What is VBM?
- Describe the three types of white matter fiber tracts in the human brain.
- Describe two examples of structural gray matter asymmetries and structural white matter asymmetries in the human brain.
- What is NODDI and what are the major findings on structural asymmetries that were obtained with NODDI?
- What is known about hemispheric asymmetries at the level of gene expression?
- What is known about hemispheric asymmetries in the proteome?
- What is known about the functional relevance of structural asymmetries in the arcuate fasciculus?
- What are the brain structural correlates of human handedness?

References

1. Geschwind N, Levitsky W. Human brain: left-right asymmetries in temporal speech region. *Science*. 1968;161 (3837):186–187. Available from: https://doi.org/10.1126/science.161.3837.186.
2. Wada JA. Is functional hemispheric lateralization guided by structural cerebral asymmetry? *Can J Neurol Sci*. 2009;36(Suppl 2):25–31.
3. Hughlings Jackson J. On the nature of the duality of the brain. *Med Press Circular*. 1874;1:80–86.
4. Hughlings Jackson J. On the nature of the duality of the brain. *Brain*. 1915;38(1–2):80–86. Available from: https://doi.org/10.1093/brain/38.1-2.80.
5. Amunts K. Structural indices of asymmetry. In: Hugdahl K, Westerhausen R, eds. *The Two Halves of the Brain: Information Processing in the Cerebral Hemispheres*. MIT Press; 2010:145–176.
6. Zhang K, Sejnowski TJ. A universal scaling law between gray matter and white matter of cerebral cortex. *Proc Natl Acad Sci U S A*. 2000;97(10):5621–5626. Available from: https://doi.org/10.1073/pnas.090504197.

7. Ge Y, Grossman RI, Babb JS, Rabin ML, Mannon LJ, Kolson DL. Age-related total gray matter and white matter changes in normal adult brain. Part I: Volumetric MR imaging analysis. *AJNR Am J Neuroradiol*. 2002;23(8):1327–1333.
8. Toga AW, Thompson PM. Mapping brain asymmetry. *Nat Rev Neurosci*. 2003;4(1):37–48. Available from: https://doi.org/10.1038/nrn1009.
9. Ioannidis JPA. Why most published research findings are false. *PLoS Med*. 2005;2(8):e124. Available from: https://doi.org/10.1371/journal.pmed.0020124.
10. Button KS, Ioannidis JPA, Mokrysz C, et al. Power failure: why small sample size undermines the reliability of neuroscience. *Nat Rev Neurosci*. 2013;14(5):365–376. Available from: https://doi.org/10.1038/nrn3475.
11. Kong X-Z, Postema MC, Guadalupe T, et al. Mapping brain asymmetry in health and disease through the ENIGMA consortium. *Hum Brain Mapp*. 2022;43(1):167–181. Available from: https://doi.org/10.1002/hbm.25033.
12. Thompson PM, Stein JL, Medland SE, et al. The ENIGMA Consortium: large-scale collaborative analyses of neuroimaging and genetic data. *Brain Imaging Behav*. 2014;8(2):153–182. Available from: https://doi.org/10.1007/s11682-013-9269-5.
13. Kong X-Z, Mathias SR, Guadalupe T, et al. Mapping cortical brain asymmetry in 17,141 healthy individuals worldwide via the ENIGMA Consortium. *PNAS*. 2018;115(22):E5154–E5163. Available from: https://doi.org/10.1073/pnas.1718418115.
14. Guadalupe T, Mathias SR, van Erp TGM, et al. Human subcortical brain asymmetries in 15,847 people worldwide reveal effects of age and sex. *Brain Imaging Behav*. 2017;11(5):1497–1514. Available from: https://doi.org/10.1007/s11682-016-9629-z.
15. Kong X-Z, Boedhoe PSW, Abe Y, et al. Mapping cortical and subcortical asymmetry in obsessive-compulsive disorder: findings from the ENIGMA Consortium. *Biol Psychiatry*. 2020;87(12):1022–1034. Available from: https://doi.org/10.1016/j.biopsych.2019.04.022.
16. Postema MC, van Rooij D, Anagnostou E, et al. Altered structural brain asymmetry in autism spectrum disorder in a study of 54 datasets. *Nat Commun*. 2019;10(1):4958. Available from: https://doi.org/10.1038/s41467-019-13005-8.
17. Postema MC, Hoogman M, Ambrosino S, et al. Analysis of structural brain asymmetries in attention-deficit/hyperactivity disorder in 39 datasets. *J Child Psychol Psychiatry*. 2021;62(10):1202–1219. Available from: https://doi.org/10.1111/jcpp.13396.
18. Kovel CGF, de, Aftanas L, Aleman A, et al. No alterations of brain structural asymmetry in major depressive disorder: an ENIGMA Consortium analysis. *Am J Psychiatry*. 2019;176(12):1039–1049. Available from: https://doi.org/10.1176/appi.ajp.2019.18101144.
19. Eberstaller O. Zur Oberflächenanatomie der Grosshirnhemisphären. *Wien Med Blätter*. 1884;7:479–482.
20. Lichtheim L. On aphasia. *Brain*. 1885;7:433–484.
21. Pfeifer RA. Pathologie der Hörstrahlung und der corticalen Hörsphäre. In: Bumke O, Förster O, eds. *Handbuch der Neurologie, Band 6*. Springer; 1936:533–626.
22. Steinmetz H. Structure, function and cerebral asymmetry: in vivo morphometry of the planum temporale. *Neurosci & Biobehav Rev*. 1996;20(4):587–591. Available from: https://doi.org/10.1016/0149-7634(95)00071-2.
23. Shapleske J, Rossell S, Woodruff P, David A. The planum temporale: a systematic, quantitative review of its structural, functional and clinical significance. *Brain Res Rev*. 1999;29(1):26–49. Available from: https://doi.org/10.1016/s0165-0173(98)00047-2.
24. Ashburner J, Friston KJ. Voxel-based morphometry—the methods. *NeuroImage*. 2000;11(6 Pt 1):805–821. Available from: https://doi.org/10.1006/nimg.2000.0582.
25. Ashburner J, Friston KJ. Why voxel-based morphometry should be used. *NeuroImage*. 2001;14(6):1238–1243. Available from: https://doi.org/10.1006/nimg.2001.0961.
26. Good CD, Johnsrude I, Ashburner J, Henson RN, Friston KJ, Frackowiak RS. Cerebral asymmetry and the effects of sex and handedness on brain structure: a voxel-based morphometric analysis of 465 normal adult human brains. *NeuroImage*. 2001;14(3):685–700. Available from: https://doi.org/10.1006/nimg.2001.0857.
27. Luders E, Gaser C, Jancke L, Schlaug G. A voxel-based approach to gray matter asymmetries. *NeuroImage*. 2004;22(2):656–664. Available from: https://doi.org/10.1016/j.neuroimage.2004.01.032.
28. Kurth F, Gaser C, Luders E. A 12-step user guide for analyzing voxel-wise gray matter asymmetries in statistical parametric mapping (SPM. *Nat Protoc*. 2015;10(2):293–304. Available from: https://doi.org/10.1038/nprot.2015.014.

29. Kurth F, MacKenzie-Graham A, Toga AW, Luders E. Shifting brain asymmetry: the link between meditation and structural lateralization. *Soc Cogn Affect Neurosci.* 2015;10(1):55–61. Available from: https://doi.org/10.1093/scan/nsu029.
30. Chi JG, Dooling EC, Gilles FH. Left-right asymmetries of the temporal speech areas of the human fetus. *Arch Neurol.* 1977;34(6):346–348. Available from: https://doi.org/10.1001/archneur.1977.00500180040008.
31. Geschwind N, Galaburda AM. Cerebral lateralization. Biological mechanisms, associations, and pathology: I. A hypothesis and a program for research. *Arch Neurol.* 1985;42(5):428–459. Available from: https://doi.org/10.1001/archneur.1985.04060050026008.
32. Heim S, Eickhoff SB, Amunts K. Different roles of cytoarchitectonic BA 44 and BA 45 in phonological and semantic verbal fluency as revealed by dynamic causal modelling. *NeuroImage.* 2009;48(3):616–624. Available from: https://doi.org/10.1016/j.neuroimage.2009.06.044.
33. Amunts K, Schleicher A, Brgel U, Mohlberg H, Uylings HB, Zilles K. Broca's region revisited: cytoarchitecture and intersubject variability. *J Comp Neurol.* 1999;412(2):319–341. Available from: https://doi.org/10.1002/(SICI)1096-9861(19990920)412:2 < 319:AID-CNE10 > 3.0.CO;2-7.
34. Falzi G, Perrone P, Vignolo LA. Right-left asymmetry in anterior speech region. *Arch Neurol.* 1982;39(4):239–240. Available from: https://doi.org/10.1001/archneur.1982.00510160045009.
35. Foundas AL, Eure KF, Luevano LF, Weinberger DR. MRI asymmetries of Broca's area: the pars triangularis and pars opercularis. *Brain Lang.* 1998;64(3):282–296. Available from: https://doi.org/10.1006/brln.1998.1974.
36. Foundas AL, Bollich AM, Corey DM, Hurley M, Heilman KM. Anomalous anatomy of speech-language areas in adults with persistent developmental stuttering. *Neurology.* 2001;57(2):207–215. Available from: https://doi.org/10.1212/WNL.57.2.207.
37. Keller SS, Crow T, Foundas A, Amunts K, Roberts N. Broca's area: nomenclature, anatomy, typology and asymmetry. *Brain Lang.* 2009;109(1):29–48. Available from: https://doi.org/10.1016/j.bandl.2008.11.005.
38. Papadatou-Pastou M, Ntolka E, Schmitz J, et al. Human handedness: a meta-analysis. *Psychol Bull.* 2020;146(6):481–524. Available from: https://doi.org/10.1037/bul0000229.
39. Rose S, Rowland T, Pannek K, et al. Structural hemispheric asymmetries in the human precentral gyrus hand representation. *Neuroscience.* 2012;210:211–221. Available from: https://doi.org/10.1016/j.neuroscience.2012.02.044.
40. Goldberg E, Roediger D, Kucukboyaci NE, et al. Hemispheric asymmetries of cortical volume in the human brain. *Cortex.* 2013;49(1):200–210. Available from: https://doi.org/10.1016/j.cortex.2011.11.002.
41. Zilles K, Dabringhaus A, Geyer S, et al. Structural asymmetries in the human forebrain and the forebrain of non-human primates and rats. *Neurosci & Biobehav Rev.* 1996;20(4):593–605. Available from: https://doi.org/10.1016/0149-7634(95)00072-0.
42. Amunts K, Schlaug G, Schleicher A, et al. Asymmetry in the human motor cortex and handedness. *NeuroImage.* 1996;4(3 Pt 1):216–222. Available from: https://doi.org/10.1006/nimg.1996.0073.
43. Amunts K, Jäncke L, Mohlberg H, Steinmetz H, Zilles K. Interhemispheric asymmetry of the human motor cortex related to handedness and gender. *Neuropsychologia.* 2000;38(3):304–312. Available from: https://doi.org/10.1016/s0028-3932(99)00075-5.
44. Guadalupe T, Willems RM, Zwiers MP, et al. Differences in cerebral cortical anatomy of left- and right-handers. *Front Psychol.* 2014;5:261. Available from: https://doi.org/10.3389/fpsyg.2014.00261.
45. Sha Z, Pepe A, Schijven D, et al. Handedness and its genetic influences are associated with structural asymmetries of the cerebral cortex in 31,864 individuals. *PNAS.* 2021;118(47). Available from: https://doi.org/10.1073/pnas.2113095118.
46. Gotts SJ, Jo HJ, Wallace GL, Saad ZS, Cox RW, Martin A. Two distinct forms of functional lateralization in the human brain. *PNAS.* 2013;110(36):E3435–E3444. Available from: https://doi.org/10.1073/pnas.1302581110.
47. Jäncke L, Schlaug G, Huang Y, Steinmetz H. Asymmetry of the planum parietale. *NeuroReport.* 1994;5(9):1161–1163. Available from: https://doi.org/10.1097/00001756-199405000-00035.
48. Chiarello C, Welcome SE, Halderman LK, et al. A large-scale investigation of lateralization in cortical anatomy and word reading: are there sex differences. *Neuropsychology.* 2009;23(2):210–222. Available from: https://doi.org/10.1037/a0014265.
49. Heiervang E, Hugdahl K, Steinmetz H, et al. Planum temporale, planum parietale and dichotic listening in dyslexia. *Neuropsychologia.* 2000;38(13):1704–1713. Available from: https://doi.org/10.1016/s0028-3932(00)00085-3.

50. Gannon PJ, Kheck NM, Braun AR, Holloway RL. Planum parietale of chimpanzees and orangutans: a comparative resonance of human-like planum temporale asymmetry. *Anat Rec A Discov Mol Cell Evol Biol*. 2005;287(1):1128–1141. Available from: https://doi.org/10.1002/ar.a.20256.
51. Barrick TR, Lawes IN, Mackay CE, Clark CA. White matter pathway asymmetry underlies functional lateralization. *Cereb Cortex*. 2007;17(3):591–598. Available from: https://doi.org/10.1093/cercor/bhk004.
52. Iturria-Medina Y, Pérez Fernández A, Morris DM, et al. Brain hemispheric structural efficiency and interconnectivity rightward asymmetry in human and nonhuman primates. *Cereb Cortex*. 2011;21(1):56–67. Available from: https://doi.org/10.1093/cercor/bhq058.
53. Caeyenberghs K, Leemans A. Hemispheric lateralization of topological organization in structural brain networks. *Hum Brain Mapp*. 2014;35(9):4944–4957. Available from: https://doi.org/10.1002/hbm.22524.
54. Catani M, Forkel S, de Schotten MT. Asymmetry of white matter pathways. In: Hugdahl K, Westerhausen R, eds. *The Two Halves of the Brain: Information Processing in the Cerebral Hemispheres*. MIT Press; 2010:177–209.
55. Ocklenburg S, Friedrich P, Güntürkün O, Genç E. Intrahemispheric white matter asymmetries: the missing link between brain structure and functional lateralization? *Rev Neurosci*. 2016;27(5):465–480. Available from: https://doi.org/10.1515/revneuro-2015-0052.
56. Catani M, de Schotten MT. A diffusion tensor imaging tractography atlas for virtual in vivo dissections. *Cortex*. 2008;44(8):1105–1132. Available from: https://doi.org/10.1016/j.cortex.2008.05.004.
57. Ardila A. A review of conduction aphasia. *Curr Neurol Neurosci Rep*. 2010;10(6):499–503. Available from: https://doi.org/10.1007/s11910-010-0142-2.
58. James JS, Kumari SR, Sreedharan RM, Thomas B, Radhkrishnan A, Kesavadas C. Analyzing functional, structural, and anatomical correlation of hemispheric language lateralization in healthy subjects using functional MRI, diffusion tensor imaging, and voxel-based morphometry. *Neurol India*. 2015;63(1):49–57. Available from: https://doi.org/10.4103/0028-3886.152634.
59. Ocklenburg S, Schlaffke L, Hugdahl K, Westerhausen R. From structure to function in the lateralized brain: how structural properties of the arcuate and uncinate fasciculus are associated with dichotic listening performance. *Neurosci Lett*. 2014;580:32–36. Available from: https://doi.org/10.1016/j.neulet.2014.07.044.
60. Powell HWR, Parker GJM, Alexander DC, et al. Hemispheric asymmetries in language-related pathways: a combined functional MRI and tractography study. *NeuroImage*. 2006;32(1):388–399. Available from: https://doi.org/10.1016/j.neuroimage.2006.03.011.
61. Propper RE, O'Donnell LJ, Whalen S, et al. A combined fMRI and DTI examination of functional language lateralization and arcuate fasciculus structure: effects of degree versus direction of hand preference. *Brain Cognition*. 2010;73(2):85–92. Available from: https://doi.org/10.1016/j.bandc.2010.03.004.
62. Schlaug G, Marchina S, Norton A. Evidence for plasticity in white-matter tracts of patients with chronic Broca's aphasia undergoing intense intonation-based speech therapy. *Ann N Y Acad Sci*. 2009;1169:385–394. Available from: https://doi.org/10.1111/j.1749-6632.2009.04587.x.
63. Sreedharan RM, Menon AC, James JS, Kesavadas C, Thomas SV. Arcuate fasciculus laterality by diffusion tensor imaging correlates with language laterality by functional MRI in preadolescent children. *Neuroradiology*. 2015;57(3):291–297. Available from: https://doi.org/10.1007/s00234-014-1469-1.
64. Nucifora PGP, Verma R, Melhem ER, Gur RE, Gur RC. Leftward asymmetry in relative fiber density of the arcuate fasciculus. *NeuroReport*. 2005;16(8):791. Available from: https://doi.org/10.1097/00001756-200505310-00002.
65. Vernooij MW, Smits M, Wielopolski PA, Houston GC, Krestin GP, van der Lugt A. Fiber density asymmetry of the arcuate fasciculus in relation to functional hemispheric language lateralization in both right- and left-handed healthy subjects: a combined fMRI and DTI study. *NeuroImage*. 2007;35(3):1064–1076. Available from: https://doi.org/10.1016/j.neuroimage.2006.12.041.
66. Friederici AD. White-matter pathways for speech and language processing. In: Swaab DF, Boller F, Aminoff MJ, eds. *The Human Auditory System: Fundamental Organization and Clinical Disorders*. Elsevier; 2015:177–186.
67. Friederici AD, Gierhan SME. The language network. *Curr Opin Neurobiol*. 2013;23(2):250–254. Available from: https://doi.org/10.1016/j.conb.2012.10.002.
68. Heilman KM, Watson RT. The disconnection apraxias. *Cortex*. 2008;44(8):975–982. Available from: https://doi.org/10.1016/j.cortex.2007.10.010.
69. Doricchi F, de Schotten MT, Tomaiuolo F, Bartolomeo P. White matter (dis)connections and gray matter (dys) functions in visual neglect: gaining insights into the brain networks of spatial awareness. *Cortex*. 2008;44(8):983–995. Available from: https://doi.org/10.1016/j.cortex.2008.03.006.

70. Glasser MF, Rilling JK. DTI tractography of the human brain's language pathways. *Cereb Cortex*. 2008;18 (11):2471–2482. Available from: https://doi.org/10.1093/cercor/bhn011.
71. von der Heide RJ, Skipper LM, Klobusicky E, Olson IR. Dissecting the uncinate fasciculus: disorders, controversies and a hypothesis. *Brain*. 2013;136(Pt 6):1692–1707. Available from: https://doi.org/10.1093/brain/awt094.
72. Wakana S, Caprihan A, Panzenboeck MM, et al. Reproducibility of quantitative tractography methods applied to cerebral white matter. *NeuroImage*. 2007;36(3):630–644. Available from: https://doi.org/10.1016/j.neuroimage.2007.02.049.
73. Malykhin N, Concha L, Seres P, Beaulieu C, Coupland NJ. Diffusion tensor imaging tractography and reliability analysis for limbic and paralimbic white matter tracts. *Psychiatry Res*. 2008;164(2):132–142. Available from: https://doi.org/10.1016/j.pscychresns.2007.11.007.
74. Takao H, Hayashi N, Ohtomo K. White matter microstructure asymmetry: effects of volume asymmetry on fractional anisotropy asymmetry. *Neuroscience*. 2013;231:1–12. Available from: https://doi.org/10.1016/j.neuroscience.2012.11.038.
75. Hau J, Sarubbo S, Houde JC, et al. Revisiting the human uncinate fasciculus, its subcomponents and asymmetries with stem-based tractography and microdissection validation. *Brain Struct Funct*. 2017;222(4):1645–1662. Available from: https://doi.org/10.1007/s00429-016-1298-6.
76. de Schotten MT, Dell'Acqua F, Forkel SJ, et al. A lateralized brain network for visuospatial attention. *Nat Neurosci*. 2011;14(10):1245–1246. Available from: https://doi.org/10.1038/nn.2905.
77. Buxhoeveden DP, Switala AE, Litaker M, Roy E, Casanova MF. Lateralization of minicolumns in human planum temporale is absent in nonhuman primate cortex. *BBE*. 2001;57(6):349–358. Available from: https://doi.org/10.1159/000047253.
78. Galuske RA, Schlote W, Bratzke H, Singer W. Interhemispheric asymmetries of the modular structure in human temporal cortex. *Science*. 2000;289(5486):1946–1949. Available from: https://doi.org/10.1126/science.289.5486.1946.
79. Hutsler J. The specialized structure of human language cortex: pyramidal cell size asymmetries within auditory and language-associated regions of the temporal lobes. *Brain Lang*. 2003;86(2):226–242. Available from: https://doi.org/10.1016/s0093-934x(02)00531-x.
80. Hutsler J, Galuske RAW. Hemispheric asymmetries in cerebral cortical networks. *Trends Neurosci*. 2003;26 (8):429–435. Available from: https://doi.org/10.1016/S0166-2236(03)00198-X.
81. Uylings HB, Jacobsen AM, Zilles K, Amunts K. Left-right asymmetry in volume and number of neurons in adult Broca's area. *Cortex*. 2006;42(4):652–658. Available from: https://doi.org/10.1016/S0010-9452(08)70401-5.
82. Amunts K, Schleicher A, Ditterich A, Zilles K. Broca's region: cytoarchitectonic asymmetry and developmental changes. *J Comp Neurol*. 2003;465(1):72–89. Available from: https://doi.org/10.1002/cne.10829.
83. Schmitz J, Fraenz C, Schlüter C, et al. Hemispheric asymmetries in cortical gray matter microstructure identified by neurite orientation dispersion and density imaging. *NeuroImage*. 2019;189:667–675. Available from: https://doi.org/10.1016/j.neuroimage.2019.01.079.
84. Zhang H, Schneider T, Wheeler-Kingshott CA, Alexander DC. NODDI: practical in vivo neurite orientation dispersion and density imaging of the human brain. *NeuroImage*. 2012;61(4):1000–1016. Available from: https://doi.org/10.1016/j.neuroimage.2012.03.072.
85. Smiley JF, Rosoklija G, Mancevski B, et al. Hemispheric comparisons of neuron density in the planum temporale of schizophrenia and nonpsychiatric brains. *Psychiatry Res*. 2011;192(1):1–11. Available from: https://doi.org/10.1016/j.pscychresns.2010.11.007.
86. Alexander AL, Lee JE, Lazar M, Field AS. Diffusion tensor imaging of the brain. *Neurotherapeutics: J Am Soc Exp NeuroTherapeutics*. 2007;4(3):316–329. Available from: https://doi.org/10.1016/j.nurt.2007.05.011.
87. Büchel C, Raedler T, Sommer M, Sach M, Weiller C, Koch MA. White matter asymmetry in the human brain: a diffusion tensor MRI study. *Cereb Cortex*. 2004;14(9):945–951. Available from: https://doi.org/10.1093/cercor/bhh055.
88. Catani M, Allin MPG, Husain M, et al. Symmetries in human brain language pathways correlate with verbal recall. *Proc Natl Acad Sci U S A*. 2007;104(43):17163–17168. Available from: https://doi.org/10.1073/pnas.0702116104.
89. Choi C-H, Lee J-M, Koo B-B, et al. Sex differences in the temporal lobe white matter and the corpus callosum: a diffusion tensor tractography study. *NeuroReport*. 2010;21(1):73–77. Available from: https://doi.org/10.1097/WNR.0b013e3283345eb0.

90. Häberling IS, Badzakova-Trajkov G, Corballis MC. Asymmetries of the arcuate fasciculus in monozygotic twins: genetic and nongenetic influences. *PLoS One*. 2013;8(1):e52315. Available from: https://doi.org/10.1371/journal.pone.0052315.
91. Lebel C, Beaulieu C. Lateralization of the arcuate fasciculus from childhood to adulthood and its relation to cognitive abilities in children. *Hum Brain Mapp*. 2009;30(11):3563–3573. Available from: https://doi.org/10.1002/hbm.20779.
92. Ocklenburg S, Hugdahl K, Westerhausen R. Structural white matter asymmetries in relation to functional asymmetries during speech perception and production. *NeuroImage*. 2013;83:1088–1097. Available from: https://doi.org/10.1016/j.neuroimage.2013.07.076.
93. O'Donnell LJ, Westin C-F, Golby AJ. Tract-based morphometry for white matter group analysis. *NeuroImage*. 2009;45(3):832–844. Available from: https://doi.org/10.1016/j.neuroimage.2008.12.023.
94. Rodrigo S, Naggara O, Oppenheim C, et al. Human subinsular asymmetry studied by diffusion tensor imaging and fiber tracking. *AJNR Am J Neuroradiol*. 2007;28(8):1526–1531. Available from: https://doi.org/10.3174/ajnr.A0584.
95. Takao H, Abe O, Yamasue H, et al. Gray and white matter asymmetries in healthy individuals aged 21-29 years: a voxel-based morphometry and diffusion tensor imaging study. *Hum Brain Mapp*. 2011;32(10):1762–1773. Available from: https://doi.org/10.1002/hbm.21145.
96. Upadhyay J, Hallock K, Ducros M, Kim D-S, Ronen I. Diffusion tensor spectroscopy and imaging of the arcuate fasciculus. *NeuroImage*. 2008;39(1):1–9. Available from: https://doi.org/10.1016/j.neuroimage.2007.08.046.
97. Danielian LE, Iwata NK, Thomasson DM, Floeter MK. Reliability of fiber tracking measurements in diffusion tensor imaging for longitudinal study. *NeuroImage*. 2010;49(2):1572–1580. Available from: https://doi.org/10.1016/j.neuroimage.2009.08.062.
98. Hasan KM, Iftikhar A, Kamali A, et al. Development and aging of the healthy human brain uncinate fasciculus across the lifespan using diffusion tensor tractography. *Brain Res*. 2009;1276:67–76. Available from: https://doi.org/10.1016/j.brainres.2009.04.025.
99. Hasan KM, Kamali A, Abid H, Kramer LA, Fletcher JM, Ewing-Cobbs L. Quantification of the spatiotemporal microstructural organization of the human brain association, projection and commissural pathways across the lifespan using diffusion tensor tractography. *Brain Struct Funct*. 2010;214(4):361–373. Available from: https://doi.org/10.1007/s00429-009-0238-0.
100. Jahanshad N, Lee AD, Barysheva M, et al. Genetic influences on brain asymmetry: a DTI study of 374 twins and siblings. *NeuroImage*. 2010;52(2):455–469. Available from: https://doi.org/10.1016/j.neuroimage.2010.04.236.
101. Kubicki M, Westin C-F, Maier SE, et al. Uncinate fasciculus findings in schizophrenia: a magnetic resonance diffusion tensor imaging study. *Am J Psychiatry*. 2002;159(5):813–820. Available from: https://doi.org/10.1176/appi.ajp.159.5.813.
102. Rocca MA, Valsasina P, Ceccarelli A, et al. Structural and functional MRI correlates of Stroop control in benign MS. *Hum Brain Mapp*. 2009;30(1):276–290. Available from: https://doi.org/10.1002/hbm.20504.
103. DeRosse P, Nitzburg GC, Ikuta T, Peters BD, Malhotra AK, Szeszko PR. Evidence from structural and diffusion tensor imaging for frontotemporal deficits in psychometric schizotypy. *Schizophrenia Bull*. 2015;41(1):104–114. Available from: https://doi.org/10.1093/schbul/sbu150.
104. Madsen KS, Jernigan TL, Iversen P, et al. Cortisol awakening response and negative emotionality linked to asymmetry in major limbic fibre bundle architecture. *Psychiatry Res*. 2012;201(1):63–72. Available from: https://doi.org/10.1016/j.pscychresns.2011.07.015.
105. Park H-J, Westin C-F, Kubicki M, et al. White matter hemisphere asymmetries in healthy subjects and in schizophrenia: a diffusion tensor MRI study. *NeuroImage*. 2004;23(1):213–223. Available from: https://doi.org/10.1016/j.neuroimage.2004.04.036.
106. Yasmin H, Aoki S, Abe O, et al. Tract-specific analysis of white matter pathways in healthy subjects: a pilot study using diffusion tensor MRI. *Neuroradiology*. 2009;51(12):831–840. Available from: https://doi.org/10.1007/s00234-009-0580-1.
107. Hernando KA, Szaflarski JP, Ver Hoef LW, Lee S, Allendorfer JB. Uncinate fasciculus connectivity in patients with psychogenic nonepileptic seizures: a preliminary diffusion tensor tractography study. *Epilepsy Behav*. 2015;45:68–73. Available from: https://doi.org/10.1016/j.yebeh.2015.02.022.
108. Kitamura S, Morikawa M, Kiuchi K, et al. Asymmetry, sex differences and age-related changes in the white matter in the healthy elderly: a tract-based study. *BMC Res Notes*. 2011;4(1):378. Available from: https://doi.org/10.1186/1756-0500-4-378.

109. Yin X, Han Y, Ge H, et al. Inferior frontal white matter asymmetry correlates with executive control of attention. *Hum Brain Mapp*. 2013;34(4):796–813. Available from: https://doi.org/10.1002/hbm.21477.
110. Sun T, Patoine C, Abu-Khalil A, et al. Early asymmetry of gene transcription in embryonic human left and right cerebral cortex. *Science*. 2005;308(5729):1794–1798. Available from: https://doi.org/10.1126/science.1110324.
111. Sun T, Collura RV, Ruvolo M, Walsh CA. Genomic and evolutionary analyses of asymmetrically expressed genes in human fetal left and right cerebral cortex. *Cereb Cortex*. 2006;16(Suppl 1):i18–i25. Available from: https://doi.org/10.1093/cercor/bhk026.
112. Sun T, Walsh CA. Molecular approaches to brain asymmetry and handedness. *Nat Rev Neurosci*. 2006;7(8):655–662. Available from: https://doi.org/10.1038/nrn1930.
113. Li Q, Bian S, Liu B, Hong J, Toth M, Sun T. Establishing brain functional laterality in adult mice through unilateral gene manipulation in the embryonic cortex. *Cell Res*. 2013;23(9):1147–1149. Available from: https://doi.org/10.1038/cr.2013.106.
114. Karlebach G, Francks C. Lateralization of gene expression in human language cortex. *Cortex*. 2015;67:30–36. Available from: https://doi.org/10.1016/j.cortex.2015.03.003.
115. Gene Ontology Consortium. AmiGO2: Welcome. Published April 8, 2021. http://amigo.geneontology.org/amigo/landing. Accessed.
116. Zhao X, Liang W, Wang W, et al. Changes in and asymmetry of the proteome in the human fetal frontal lobe during early development. *Commun Biol*. 2022;5(1):1031. Available from: https://doi.org/10.1038/s42003-022-04003-6.
117. Dos Santos Sequeira S, Woerner W, Walter C, et al. Handedness, dichotic-listening ear advantage, and gender effects on planum temporale asymmetry—a volumetric investigation using structural magnetic resonance imaging. *Neuropsychologia*. 2006;44(4):622–636. Available from: https://doi.org/10.1016/j.neuropsychologia.2005.06.014.
118. Greve DN, van der Haegen L, Cai Q, et al. A surface-based analysis of language lateralization and cortical asymmetry. *J Cogn Neurosci*. 2013;25(9):1477–1492. Available from: https://doi.org/10.1162/jocn_a_00405.
119. Chiarello C, Vazquez D, Felton A, Leonard CM. Structural asymmetry of anterior insula: behavioral correlates and individual differences. *Brain Lang*. 2013;126(2):109–122. Available from: https://doi.org/10.1016/j.bandl.2013.03.005.

CHAPTER

5

Language and the left hemisphere

OUTLINE

Introductory short story

He was drunk. Again. "I can hit anything I want. Anything. Just name it." He slurred while trying his best to stand upright. The other GIs burst into laughter as he could barely

The Lateralized Brain
DOI: https://doi.org/10.1016/B978-0-323-99737-9.00010-0

keep himself from falling over after spending most of last week's salary on beer in town. Now he had returned to the camp and was rummaging through the kitchen, trying to find something edible. When the other soldiers did not stop laughing at him, his face turned into a purple-reddish color. The boy knew that this was not good. "I can hit anything I want, you know. I am not drunk. That Japanese soap water is not even real beer." When the others laughed at him again, the GI drew his weapon, wildly pointing at things in the kitchen. "The cook's hat! I can hit that with a single bullet, no problem." The boy shook his head, but it was too late.

When he awoke in the hospital room, the boy's head was in bandages. The doctors told him that the GI's bullet had missed the hat, instead penetrating his skull, and damaging the left half of his brain. This damage to his brain was also the reason why he could not move his right foot anymore. After a few weeks in the hospital, the seizures started. At first, it was only his right leg, but soon his whole body would shake uncontrollably. When his epilepsy became worse, he was admitted to the neurologic ward at another hospital. When routine treatments failed to show any effects, a young doctor suggested trying out a new procedure: He would inject a powerful sedative into the artery supplying the left half of the brain with blood. This would effectively anesthetize the left hemisphere and thereby hopefully stop an ongoing seizure, as the damaged tissue in the left hemisphere was likely the cause of the epileptic seizures. After the boy had consented to the procedure, the doctor injected the sedative and the seizures stopped immediately. However, he was unable to move the right half of his body at all, and he could not speak a single meaningful word. It dawned upon the young doctor that not only had he discovered a possible new treatment for epilepsy, but he had also found a reliable way to determine the language-dominant half of the brain of his patient. By putting one hemisphere to sleep while leaving the other one awake, he could test which half of the brain was controlling the patient's ability to speak. This would allow neurosurgeons to determine the potential impact of brain surgery in one hemisphere on the patient and to adjust their intervention to spare the patient's ability to speak, if possible. To this day, the procedure he invented is carried out by medical professionals all around the world, helping patients to retain their ability to speak.

Inspired by Juhn Atsushi Wada.[1]

Introduction

The procedure described in this story is called the Wada test, named after Juhn Atsushi Wada, a Japanese-Canadian neurologist who developed the test as a medical resident in Japan after World War II.[1–3] During the Wada test, one half of the brain is anesthetized for $2-5$ min by injecting sodium amobarbital, a powerful sedative, into the left or right internal carotid artery, the main artery supplying the respective hemisphere with blood (see Fig. 5.1).

The patient is then asked to perform a cognitive task (e.g., overtly counting backward from 100). If the patient fails to perform the task correctly, it can be concluded that the anesthetized hemisphere is dominant for the tested cognitive function in this individual. In about 87% of right-handers and 65% of left-handers, anesthesia of the left hemisphere

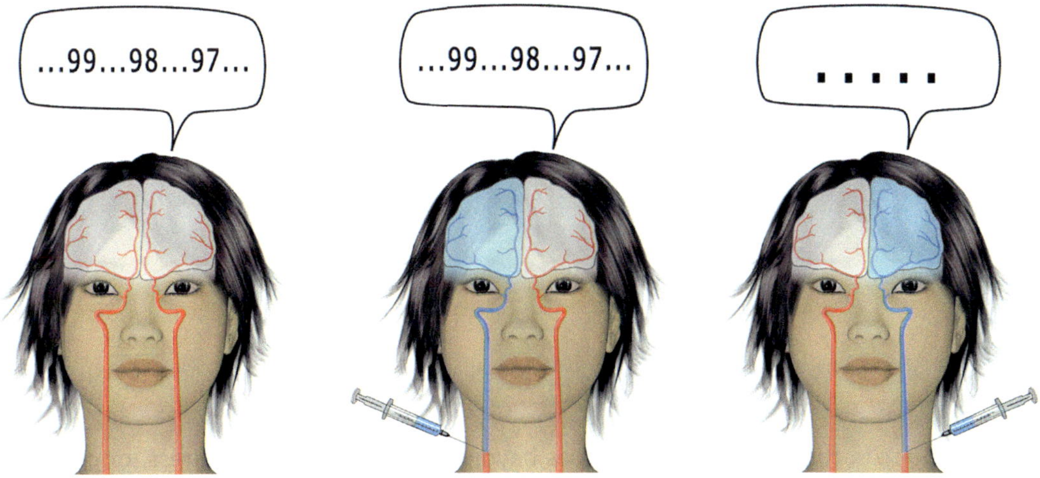

FIGURE 5.1 The Wada Test: In order to determine the language-dominant hemisphere, the patient is asked to count loudly backward from 100 (first panel). When the hemisphere nondominant for language is anesthetized, the patient's ability to speak is unimpaired (second panel). When the language-dominant hemisphere is anesthetized by injecting sodium amobarbital into the internal carotid artery, the patient loses the ability to speak (third panel).

leads to severe impairments in language production.[4] The possibility to reliably determine the language-dominant hemisphere in patients was a boon in particular for neurosurgeons planning ablative surgery as a treatment to stop seizures in otherwise untreatable epilepsy. Although evidence has been accumulating showing that noninvasive neuroimaging methods such as functional magnetic resonance imaging (fMRI) are starting to allow equally reliable determination of cognitive functions in the brain, the Wada test is still used to this day.[5,6]

While both research in split-brain patients and the results from the Wada test indicate that speech processing is impossible without the left hemisphere in most individuals, the notion that one hemisphere completely controls a complex cognitive function such as language processing is grossly oversimplified. Language processing is not one unified brain function but consists of several functional modules, each showing its own typical pattern of hemispheric asymmetries. For example, when we communicate with another human being, we participate in both speech perception—such as hearing and understanding the words—and speech production—such as choosing and speaking the words. In addition to spoken speech, language processing can also refer to written material—such as reading or writing a sentence. Moreover, there are also nonverbal aspects of language—such as gesturing or facial movements—and some languages—such as sign language—that are based completely on nonverbal means of communication.

Motor aspects of language lateralization

Human language is the most complex system of communication any species on earth has ever developed. While many species use vocal calls to communicate, only humans

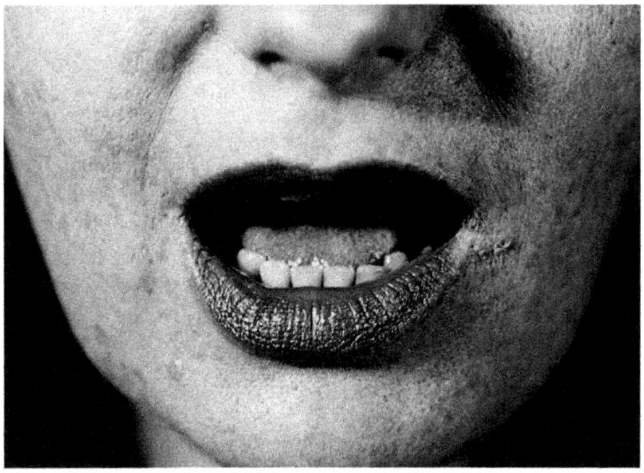

FIGURE 5.2 During speech most people open the right half of their mouth to a larger extent than the left. *Source: From Hausmann M, Behrendt-Körbitz S, Kautz H, Lamm C, Radelt F, Güntürkün O. Sex differences in oral asymmetries during wordrepetition. Neuropsychologia. 1998;36(12):1397–1402. https://doi.org/10.1016/s0028-3932(98)00027-x. Used with permission of the authors and the publisher.*

have developed articulate speech that goes beyond cries of warning, distress, or courtship behavior. One of the major theories regarding the evolution of human language is the hypothesis that it evolved from ancient gestural systems of communication, with vocalizations being gradually incorporated.[7] This implies that left-hemispheric language lateralization might already be evident on the level of motor gestures carried out, for example, with the arms or the face. Indeed, it was shown that people gesture more with their right hand than with their left during speaking, while no differences between the hands were observed for silent activities.[8,9] Other motor aspects of language are also lateralized. During speech, the right side of the mouth opens faster and reaches higher maximal opening values in most individuals, with about 76% showing a greater right-side opening[10,11] (Fig. 5.2). This motor asymmetry reflects left-hemispheric control of articulation in these subjects, resulting in greater activity of the right-side lip opening muscles.

Lateralization of language processing: the methods

While the Wada test has been used successfully to determine the language-dominant hemisphere before epilepsy surgery in clinical settings, its high level of invasiveness renders it an inapplicable tool for basic research on functional hemispheric asymmetries. Therefore, several noninvasive techniques were developed to determine individual language hemispheric dominance.

Dichotic listening

The most commonly used performance-based test to determine language lateralization is the dichotic listening task.[12,13] In this task, pairs of auditory stimuli (e.g., consonant-vowel syllables like /BA/, /DA/, /GA/, /KA/, /PA/, /TA/) are presented to the subject simultaneously via headphones, one to each ear (see Fig. 5.3).

Participants have to indicate which syllable they hear best by pressing the appropriate response button or telling this syllable to the experimenter. Unbeknown to the subject,

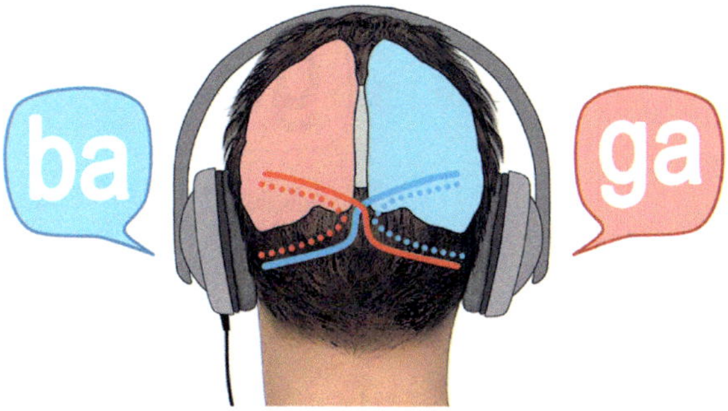

FIGURE 5.3 The dichotic listening task: Participants wear headphones, and two different consonant-vowel syllables are presented to the left and right ear simultaneously. Left ear input is mainly processed by the right hemisphere and vice versa, as contralateral auditory connections (bold lines) are stronger than ipsilateral auditory connections (dotted lines).

many stimulus pairs are presented "dichotically" so that a different syllable is played to each ear (e.g., /BA/ to the left ear and /GA/ to the right ear). Typically, subjects report more syllables presented to the right ear and tend to neglect left ear input. For this phenomenon, the term "right-ear advantage" was coined. Overall, about 74% of the population show a right-ear advantage on this test, while 20% show a left-ear advantage, and 6% do not show a clear ear advantage.[14] A more recent study of 1554 individuals found that 82.1% of the participants showed a right-ear advantage.[15] Better recall of syllables that have been presented to the right ear is interpreted as a behavioral indicator for left-hemispheric dominance in the processing of verbal information, as information from the right ear is projected mostly to the left hemisphere.[13] For researchers interested in using the dichotic listening task in their work, René Westerhausen from the University of Oslo has written an excellent primer on the methodological aspects of using this task to assess hemispheric asymmetries.[16]

The dichotic listening task is a highly valid measure to determine the language-dominant hemisphere and matches the results of the Wada test in more than 90% of epileptic patients.[17] Two major theoretical models have been developed to explain the emergence of the right-ear advantage during dichotic listening[18] (see Fig. 5.4):

1. The structural model: This model assumes that both the left and the right auditory cortex receive projections from each ear. However, contralateral projections are more numerous and faster, whereas information from ipsilateral projections is relatively suppressed. During dichotic listening, the syllable presented to the right ear has preferential access to the language-dominant left hemisphere, while the syllable presented to the left ear is processed by the right hemisphere first and then transferred to the left via the corpus callosum. As a result, the syllable presented to the right ear is usually processed faster by the language-dominant left hemisphere and therefore more likely to be reported by the participant. Following this logic, a left-ear advantage indicates right-hemispheric language dominance.[19]
2. The attentional model: This model assumes that the right-ear advantage is not based on structural features of the auditory system but on an attentional bias toward the right ear caused by preponderant activation of the left hemisphere during language-based tasks. Basically, the participant is thought to anticipate the incoming verbal stimulus during dichotic listening. This anticipation then activates the left hemisphere before the stimulus is perceived. This priming process biases attention toward the right ear and

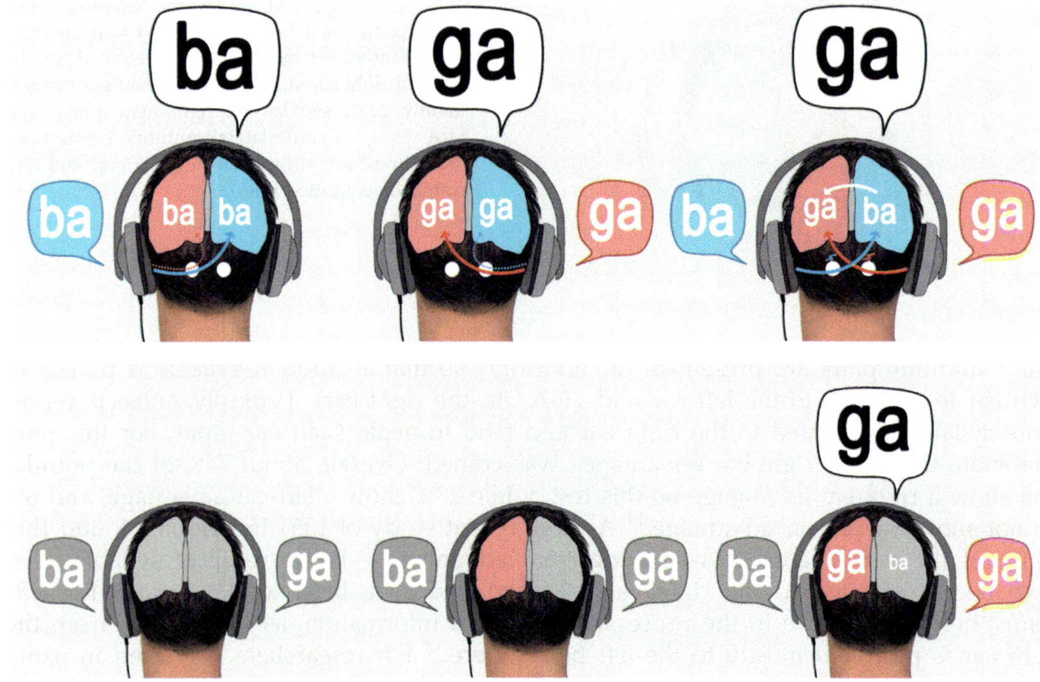

FIGURE 5.4 The two main models of dichotic listening. Upper panel: The structural model: Both the left and the right auditory cortex receive projections from each ear. When a syllable is only presented to the left or right ear (1 and 2), the syllable is represented in both hemispheres. During dichotic listening (3), the syllable presented to the right ear has preferential access to the language-dominant left hemisphere, while the syllable presented to the left ear is processed by the right hemisphere first and then transferred to the left via the corpus callosum. Thus, the syllable presented to the right ear (GA) is processed faster and is reported by the participant. Lower panel: The attentional model: Before the beginning of the dichotic listening task, there is no anticipation for a specific type of incoming stimuli (1). After a few trials, the participant starts to anticipate the incoming of verbal stimuli, resulting in an activation of the left hemisphere even before the next stimulus is perceived (2). This priming process then biases attention toward the right ear and gives the left hemisphere an advantage for processing the stimulus when it is presented so that participants report the GA presented to the right ear (3).

gives the left hemisphere an advantage for processing the stimulus when it is presented, resulting in the right-ear advantage.[20]

Both models have been supported by empirical evidence,[18] and it is likely that both the structure of the auditory system and attentional processes influence dichotic listening performance. One commonly used variation of the classic dichotic listening paradigm allowing the investigation of the role of attention and cognitive control processes during dichotic listening is the so-called forced attention dichotic listening paradigm.[21,22] In this version of the task, participants are tested in three conditions:

1. Nonforced condition: Participants are instructed to report the syllable they heard best, irrespective of the ear it was presented to. This condition parallels the classic dichotic listening task.

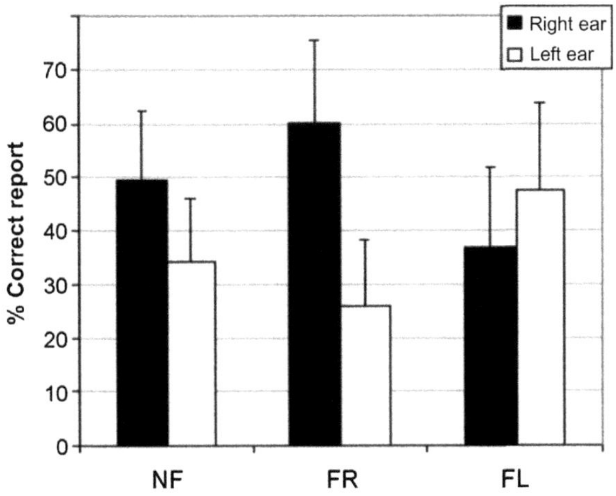

FIGURE 5.5 Typical results for the forced attention dichotic listening task. Mean percentage and standard deviation for left and right ear report for the nonforced condition (NF), for the forced-right condition (FR), and the forced-left condition (FL). *Source: From Westerhausen R, Hugdahl K. The corpus callosum in dichotic listening studies of hemispheric asymmetry: a review of clinical and experimental evidence. Neurosci Biobehav Rev. 2008;32(5): 1044–1054. https://doi.org/10.1016/j.neubiorev.2008.04.005. Used with permission of the authors and publisher.*

2. Forced-right condition: Participants are instructed to concentrate only on syllables presented to the right ear.
3. Forced-left condition: Participants are instructed to attend only to syllables presented to the left ear.

Typically, subjects show a right-ear advantage in the nonforced condition, and this advantage is substantially increased in the forced-right condition. In contrast, instructing the subjects to only attend to the left ear stimulus in the forced-left condition reverses the right-ear advantage to a small left-ear advantage[18] (Fig. 5.5). Systematic comparison of the three conditions allows for investigation of the effects of attention on dichotic listening performance. By comparing the extent of the individual bottom-up driven right-ear advantage in the nonforced condition with performance in the two forced conditions, the test can determine how strongly performance is modulated by the instructions to attend to just one side.

The forced attention dichotic listening paradigm has been implemented in a smartphone app by the research laboratory of Kenneth Hugdahl from the University of Bergen, Norway.[23,24] The dichotic listening task has also been implemented in an online behavioral laterality battery with good test-retest reliability.[25]

Visual half-field experiments in healthy subjects

In addition to the dichotic listening task, another commonly used behavioral paradigm to determine functional hemispheric asymmetries is the visual half-field technique (see Box 3.3). While this technique has been established in split-brain patients, it can also be used in healthy participants. To recap, the visual half-field technique is based on the principle that input from the left visual field is processed by the right hemisphere, while input from the right visual field is processed by the left hemisphere.[26] This neural organization principle is utilized by presenting stimuli tachistoscopically (very briefly, usually less than 200 ms) in either the left or right visual field. Doing this ensures that the stimulus is only processed by the contralateral hemisphere and that no interhemispheric transfer via the corpus callosum occurs during sensory stimulus processing.[27]

One commonly used paradigm applying this technique is the so-called lexical decision task. At the beginning of each trial, one word is presented centrally, followed by another word presented either in the left or right visual field. Participants have to indicate whether the two words were the same or different. Generally, a right visual field/left hemisphere advantage has been reported for this task.[28–30] A similar behavioral technique used to assess language lateralization is the tachistoscopic version of the classic Go-NoGo Task with verbal stimuli. In this type of task, two verbal stimuli (e.g., the words "press" and "stop") are presented tachistoscopically in either the left or right visual half-field. Typically, participants have to press a key if the more frequent Go-stimulus is presented and are required to refrain from responding if the less frequent NoGo-stimulus is presented. It was shown that participants were more accurate when stimuli were presented in the right visual field, reflecting left-hemispheric dominance for processing verbal stimuli.[31] The validity of the visual half-field technique has been corroborated by neuroimaging, as the direction of the visual field advantage for verbal stimuli shows a high correlation with activation in Broca's area.[32]

Functional transcranial Doppler sonography

While behavioral techniques like dichotic listening and visual half-field paradigms provide reliable measures to identify the language-dominant hemisphere, they only provide very indirect estimates of brain activity in the two hemispheres. A more direct way to assess which hemisphere is more active while speaking or hearing words is provided by functional transcranial Doppler sonography. This technique uses ultrasound to assess how fast blood flows through the blood vessels supplying the two hemispheres with blood. When subjects speak, their language-dominant hemisphere's oxygen demand increases. This results in faster blood flow. By comparing blood flow to the left and right hemisphere during speaking, the language-dominant hemisphere can be identified. For example, hemispheric perfusion differences between the left and right middle cerebral artery during a word generation task in healthy right-handers indicated that 92.5% of individuals showed left-hemispheric language dominance.[33] Similar results have also been obtained in a study measuring language lateralization with functional transcranial Doppler sonography during a word generation task.[34] In this research, the incidence of left-hemispheric language dominance was 96% in consistent right-handers. Interestingly, the authors found that the degree of left hemispheric language dominance decreased linearly with the degree of left-handedness, the percentage dropping to 85% in ambidextrous individuals and to only 73% in consistent left-handers—more on handedness in Chapter 6. These findings were also corroborated by a more recent functional transcranial Doppler sonography study which used a similar paradigm.[35] In this study, left-hemispheric language dominance was observed in 97% of right-handed and 74% of left-handed participants. Functional transcranial Doppler sonography has also been used for assessing lateralization during writing in left-handers and right-handers.[36] In this study, right-handers showed stronger leftward language lateralization than left-handers.

Using functional magnetic resonance imaging to assess language lateralization

Functional transcranial Doppler sonography allows for identification of the language-dominant hemisphere, but beyond that, it is impossible to determine which areas in that hemisphere are activated during speech perception or production. In order to determine not only whether the left or the right hemisphere is more active during a specific task but also which specific areas show activation differences between the left and the right half of the brain, a number of neuroimaging techniques can be used to visualize task-dependent activity in the human brain. Initially, this was mostly done with positron emission tomography (PET), a technique in which a radioactive tracer is injected into the blood flow and its concentration in the brain is measured. As functional activity in a brain area leads to regional changes in the distribution of cerebral blood flow, activity in a certain region leads to an increased tracer concentration in this area. For example, Kenneth Hugdahl and colleagues from the University of Bergen, Norway, used this technique to assess language lateralization in a study with dichotically presented consonant-vowel syllables.[37] They showed stronger activation in the left than in the right temporal lobe during perception of syllables. Today, PET is still routinely used in clinical settings, for example, in oncology. However, the need to inject a radioactive tracer limits its usefulness for cognitive neuroscience, rendering less or noninvasive alternatives such as fMRI (Box 5.1) the preferred methods of choice. Determining the language dominant hemisphere using fMRI has high overlap with the results of the Wada test, and a recent study showed that the same results can be found for 91.4% of individuals for reading and 96.9% of individuals for auditory speech perception.[40]

BOX 5.1

Key technique: functional magnetic resonance imaging (fMRI).

fMRI is the main neuroimaging technique to visualize task-related activity in the human brain in vivo.[26,38] fMRI is based on MRI, and fMRI pictures are acquired by scanning participants' brains in an MRI scanner (see Fig. 5.6) using a special fMRI recording sequence. MRI is a neuroimaging technique that generates a series of sectional images of structures inside the body. It is based on the fact that the hydrogen atoms inside our bodies are magnetic. The MRI scanner generates a strong magnetic field. If a participant is inside the MRI scanner, the hydrogen atoms in their body align themselves with the direction of this magnetic field. The MRI scanner can also generate a second magnetic field with a different direction. When this magnetic field is switched on, the hydrogen atoms start to spin like a spinning top about to tip over. This spinning movement is measured by the MRI scanner. When the second magnetic field is switched off again, the hydrogen atoms again align themselves with the first

BOX 5.1 (*cont'd*)

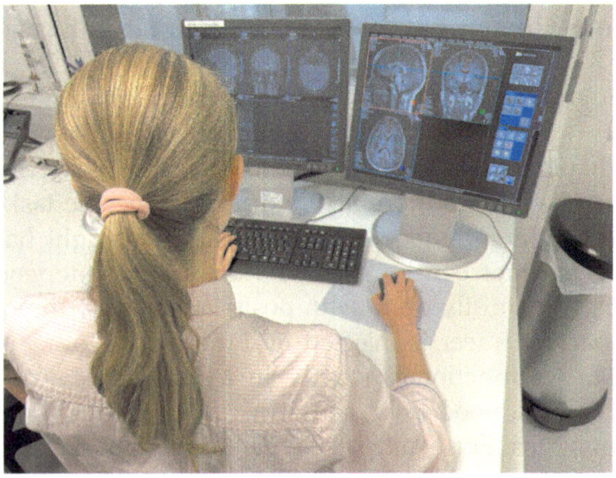

FIGURE 5.6 Researcher operating an MRI scanner.

magnetic field. This looks like when a spinning top has almost tipped over and stabilizes vertically again. The time that the hydrogen atoms need to go from the "almost tipped over" state to being aligned with the first magnetic field again is called relaxation time (TR). Importantly, the TR of a hydrogen atom depends on the molecule which it is part of. Since different tissue types (e.g., gray and white matter in the brain) are built of different molecules, their TRs are different, too. This leads to differences in signal strength (brightness of different areas in the MRI picture), thus allowing researchers to differentiate between different types of tissue. MRI is widely used in medical diagnostics (e.g., to detect tumors) and also in neuroscientific research to get anatomical pictures of the brain, which enable detection of anatomical asymmetries in different brain areas (more on that in Chapter 4).

fMRI is an extension of MRI and allows for visualization of changes in blood flow in different brain areas. It is based on the principle that oxygen-rich blood has different magnetic properties than oxygen-poor blood. Within the blood, hemoglobin molecules in red blood cells carry the oxygen, and oxygenated hemoglobin is less magnetic than deoxygenated hemoglobin. When functional activity in a brain area increases, that area's need for oxygen also increases. This results in stronger blood flow to that area, providing it with oxygen-rich blood. The MRI scanner measures these changes in the magnetic properties of the blood and maps it onto specific locations in the brain, allowing for brain maps of functional activity during a task. In a typical fMRI experiment, two or more conditions, such as one experimental condition—for example, perception of speech sounds—and one control condition—for example, perception of white noise—are presented either in separate blocks—block design—or as intermixed events—event-related design. The idea is that neurons in language-associated areas are active only in the experimental condition, resulting in an increased blood flow to

BOX 5.1 (*cont'd*)

language-relevant areas. By statistically comparing differences in blood flow between the experimental and control condition, brain areas associated with cognitive processing that differ between the conditions can be identified. Therefore, choosing the right control condition is very important in fMRI research.

fMRI is not without drawbacks. MRI scanners are expensive to purchase and maintain, and many researchers in cognitive neuroscience only have limited access to scanners in hospitals. Also, for some subjects, for example, individuals with claustrophobia or patients with pacemakers and other ferromagnetic implants, fMRI scans might not be possible solely for research purposes. Moreover, concerns regarding the validity of functional activation in small structures like the amygdala have been expressed, as blood flow in nearby blood vessels might confound results.[39] Nevertheless, since fMRI provides brain activation images with high spatial resolution without the need to inject any radioactive tracers or to perform any other invasive procedures, it is the main workhorse technique of cognitive neuroscience and one of the most important tools in lateralization research.

In principle, language lateralization can be assessed by comparing left and right hemisphere activity in any type of language-based task. However, while the neuronal networks involved in producing and perceiving spoken language and reading written words overlap to a certain extent, they are, of course, not identical.[41] fMRI-based language lateralization tasks can roughly be categorized into four different types:

1. Speech generation
2. Auditory speech perception
3. Writing
4. Visual word perception

Asymmetrical brain activation during speech generation is typically assessed using word generation tasks.[42] For example, participants are shown a probe letter (e.g., "B") while in the fMRI scanner and are asked to think of as many words starting with the probe letter as possible within a timeframe of 30 s (e.g., "beach, breakfast, birthday"). Participants are asked to merely think of the words, rather than saying them aloud, since jaw movements can affect the fMRI data. In order to assess whether the participants are actually able to perform the task, it is typically performed outside the scanner again after the scan has been performed. Data from this task (see Fig. 5.7) suggests that speech generation leads to stronger left-hemispheric activations in the inferior frontal gyrus, the precentral gyrus, the supplementary motor area, and the superior and inferior parietal lobules as well as in the inferior occipital gyrus.[42]

Asymmetrical brain activation during auditory speech perception can be assessed using a simple task in which participants have to listen to a series of spoken words using headphones while in the fMRI scanner.[43–45] In this task, activation is typically observed

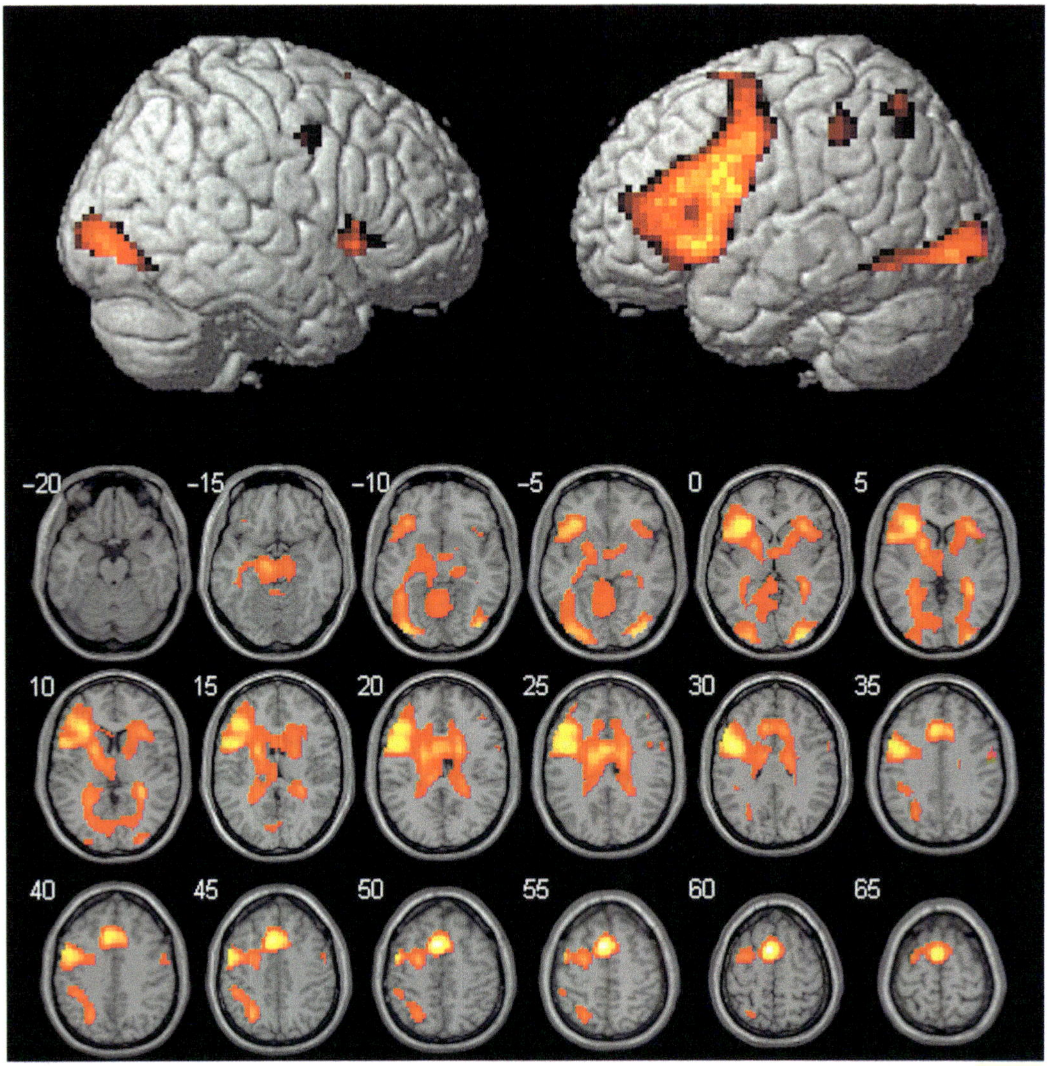

FIGURE 5.7 Asymmetrical brain activation generated by a word generation task. *Source: From Badzakova-Trajkov G, Häberling IS, Roberts RP, Corballis MC. Cerebral asymmetries: complementary and independent processes. PLoS One. 2010;5(3):e9682. https://doi.org/10.1371/journal.pone.0009682. Used with permission of the author.*

bilaterally in auditory sensory areas in the temporal lobe, with more extensive activation in the left temporal gyrus (see Fig. 5.8). Also, frontal areas such as the inferior and superior frontal gyrus, the frontal operculum, the precentral gyrus, and the supplementary motor area show leftward asymmetry.[43]

One important question that can be asked regarding listening tasks to determine language lateralization is whether there is a difference between language families. Several thousand languages are spoken across the globe, yet most fMRI tasks typically use only very few of

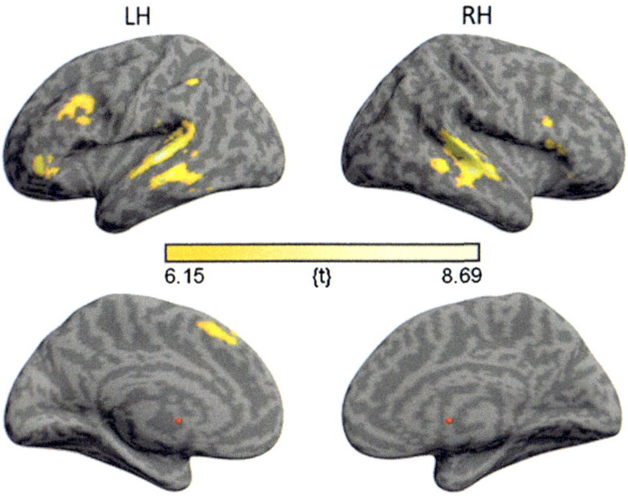

FIGURE 5.8 fMRI results for passive listening of verbal stimuli. Color-coded regions showed a significant effect of experimental versus rest condition. Color indicates *t* values, ranging from a minimum of 6.15 (orange) to a maximum of 8.69 (white). *Source: From Ocklenburg S, Hugdahl K, Westerhausen R. Structural white matter asymmetries in relation to functional asymmetries during speech perception and production. NeuroImage. 2013;83:1088–1097. https://doi.org/10.1016/j.neuroimage.2013.07.076, with permission of the author and the publisher.*

those such as English, French, or German. To answer this question, researchers Saima Malik-Moraleda and Dima Ayyash together with an international team of scientists assessed language lateralization across 45 languages and 12 language families.[46] Among other paradigms, the researchers used the "Alice" language localizer in which participants listened to short passages from Alice in Wonderland in their native language. Acoustically degraded versions of the native language passages and passages in an unfamiliar language were used as control conditions. The results of the "Alice" localizer are shown in Fig. 5.9. The results revealed that the general pattern of activation in the left hemisphere was very consistent across the investigated languages. Both the strength and the extent of activation were significantly higher in left compared to right brain areas. The researchers suggested that there is a universal language network covering several areas in the left temporal, frontal, and parietal cortices, and that leftward lateralization is one of the defining characteristics of this network.

In addition to passive listening, lateralization in auditory speech perception can also be assessed using the dichotic listening task in the MRI scanner. While attending only to the stimulus presented to the right ear leads to activation in Broca's area and the left superior temporal gyrus,[47] left lateralization of hemodynamic responses to dichotic stimuli has been observed mainly in the planum temporale and the superior temporal gyrus.[48,49] Interestingly, an fMRI study by René Westerhausen[50] from the University of Oslo, Norway, revealed that dichotic listening leads to largely symmetric activations in temporal, parietal, inferior frontal, and medial superior frontal regions. Leftward asymmetries were mainly found in the planum temporale and in the postcentral and medial superior frontal gyri, while rightward asymmetries were found in middle temporal and middle frontal gyri (see Fig. 5.10). Thus, even if this task yields clearly asymmetrical behavioral output, it is controlled by bilateral cortical networks with only few strongly leftward asymmetrical activation centers.

The forced attention version of the dichotic listening task can be used in the MRI scanner just like the nonforced version. Kristiina Kompus from the University of Bergen,

FIGURE 5.9 Left hemisphere brain activation contrast maps (native language vs degraded language) for the Alice language localizer for one sample participant for each language. Red and yellow indicate the 80th and 99th percentiles of positive-contrast activation for each participant, respectively. *Source: From Malik-Moraleda S, Ayyash D, Gallée J, et al. An investigation across 45 languages and 12 language families reveals a universal language network. Nat Neurosci. 2022;25(8):1014–1019. https://doi.org/10.1038/s41593-022-01114-5. Used with permission of the authors and the publisher.*

Norway, scanned 113 participants, while they performed the forced attention dichotic listening task. She wanted to test the hypothesis that attending to the left-ear stimulus in the forced-left condition induces a cognitive conflict, requiring cognitive control processes, which does not happen in the forced-right condition, because the right-ear stimulus has a higher perceptual salience during dichotic listening.[51] This idea was supported by the results of the study, as activation differences in the left inferior frontal gyrus and left striatum, areas likely to be involved in cognitive control processes, were observed between the forced-left and the other two conditions.

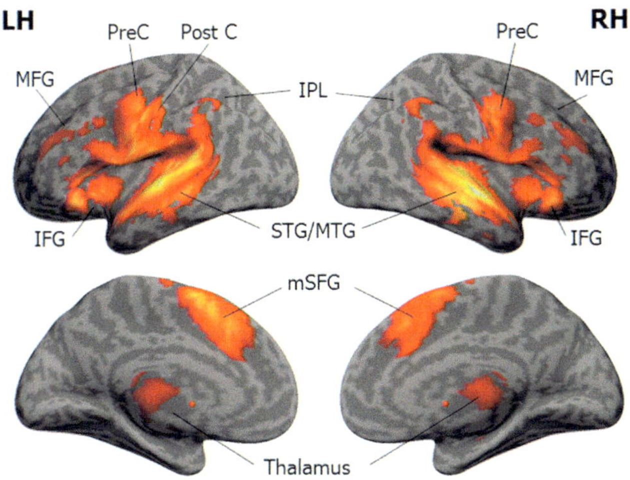

FIGURE 5.10 Contrast between dichotic listening and a resting baseline. *LH*, Left hemisphere; *RH*, right hemisphere; *IFG*, inferior frontal gyrus; *MFG*, middle frontal gyrus; *PreC*, precentral gyrus; *Post C*, postcentral gyrus; *IPL*, inferior parietal lobe; *STG*, superior temporal gyrus; *MTG*, middle temporal gyrus; *mSFG*, medial superior frontal gyrus. *Source: From Westerhausen R, Kompus K, Hugdahl K. Mapping hemispheric symmetries, relative asymmetries, and absolute asymmetries underlying the auditory laterality effect.* NeuroImage. *2014;84:962–970, with permission of the author and the publisher.*

In addition to auditory paradigms, both language production and perception can also be assessed in the visual modality, for example, by using reading or writing tasks.[41] For example, asymmetrical brain activation during visual word perception can be assessed by performing tasks such as deciding whether two visually presented words have the same meaning—for example, "thorn" and "sting"—or are semantically unrelated, for example, "monk" and "oak."[52] Data from this task suggests that visual word perception leads to greater left- than right-hemispheric activation in the inferior frontal sulcus, the inferior part of the inferior frontal gyrus, the posterior part of the superior temporal sulcus, and the ascending branch of the superior temporal sulcus. Language production in the visual domain can be assessed using a writing task in the MRI scanner. Writing activates different neural networks than speaking, with activation foci in several regions of the posterior parietal cortex which are connected to language processing areas, motor, and visual sensory areas.[53]

Using electroencephalography to investigate hemispheric asymmetries

While fMRI has a very good spatial resolution, allowing for an exact localization of differential activations between the hemispheres, its temporal resolution is in the seconds range. This can be problematic in research on hemispheric asymmetries, as the corpus callosum can exchange information between the two hemispheres much faster than 1 s after stimulus presentation, making it possible that hemispheric asymmetries that are observed shortly after stimulus presentation disappear quickly due to interhemispheric information transfer. Most lateralization researchers interested in timing-sensitive processing use electroencephalography (EEG) to track fast changes in neuronal activity. Compared to fMRI, EEG allows for a much higher temporal resolution—in the milliseconds range—albeit sacrificing spatial resolution (see Box 5.2).

BOX 5.2

Key technique: electroencephalography (EEG).

In the brain, hundred thousands of active neurons generate electrical activity at any given moment. EEG is a noninvasive technique to estimate this electrical activity in the brain by recording it from the skull.[54] As electrical activity in the brain fluctuates based on neuronal activity, changes in EEG parameters are thought to reflect cognitive processing—or its absence. A typical EEG setup (see Fig. 5.11A) consists of 32, 64, or 128 recording electrodes as well as a reference and a ground electrode. The recording electrodes are placed on the scalp at predefined positions using an elastic EEG cap like the one shown in the picture. As the electric signals recorded by the electrodes are very small (in the micro-volt range), they are amplified using an EEG amplifier (also shown in Fig. 5.11A). The amplifier box is connected to a recording computer with software to record the EEG data. Typically, a second computer is used to present a task to the participant. An EEG recording yields a raw EEG signal for each electrode (Fig. 5.11B) that can be analyzed in several different ways.

Analysis of asymmetries in EEG data has focused on two different aspects of the raw EEG:

Frequency bands

The raw EEG of one electrode is composed of signals from millions of neurons firing with different frequencies. Using a specific algorithm (the Fourier transformation), different frequency bands can be identified within the raw EEG signal. These frequency bands are differentially modulated by sleeping/waking cycles and cognitive activity. The most commonly distinguished ones are:

Delta-band (0.1–4 Hz): This frequency band is dominant in deep sleep.
Theta-band (4–8 Hz): This frequency band is commonly observed during the change from sleeping to waking.
Alpha-band (8–12 Hz): This frequency band is commonly observed when eyes are closed and in a relaxed state.
Beta-band (14–25 Hz): This frequency band is dominant in a state of mental activity.
Gamma-band (>30 Hz): This frequency band is observed during focused concentration on a task.
By comparing the power of a frequency band (how much electrical energy was produced during a given time period) between electrodes located on the left and right side of the head, hemispheric asymmetries can be determined.

Event-related potentials

When a participant perceives a stimulus or makes a reaction, there is a slight change in the raw EEG, often too minimal to detect in a single trial and confounded by other activity in the EEG. EEG researchers have developed a simple trick to detect these changes. They present a stimulus repeatedly and average out the EEG signal over many (often hundreds) time segments that each start with the presentation of the stimulus. This way, random activity from other sources in the EEG raw signal can be filtered out and so-called event-related potential components (ERPs) can be detected reliably. Researchers have identified a large number of ERP components, the N1 (see the next section) being the most

BOX 5.2 (*cont'd*)

(A)

(B)

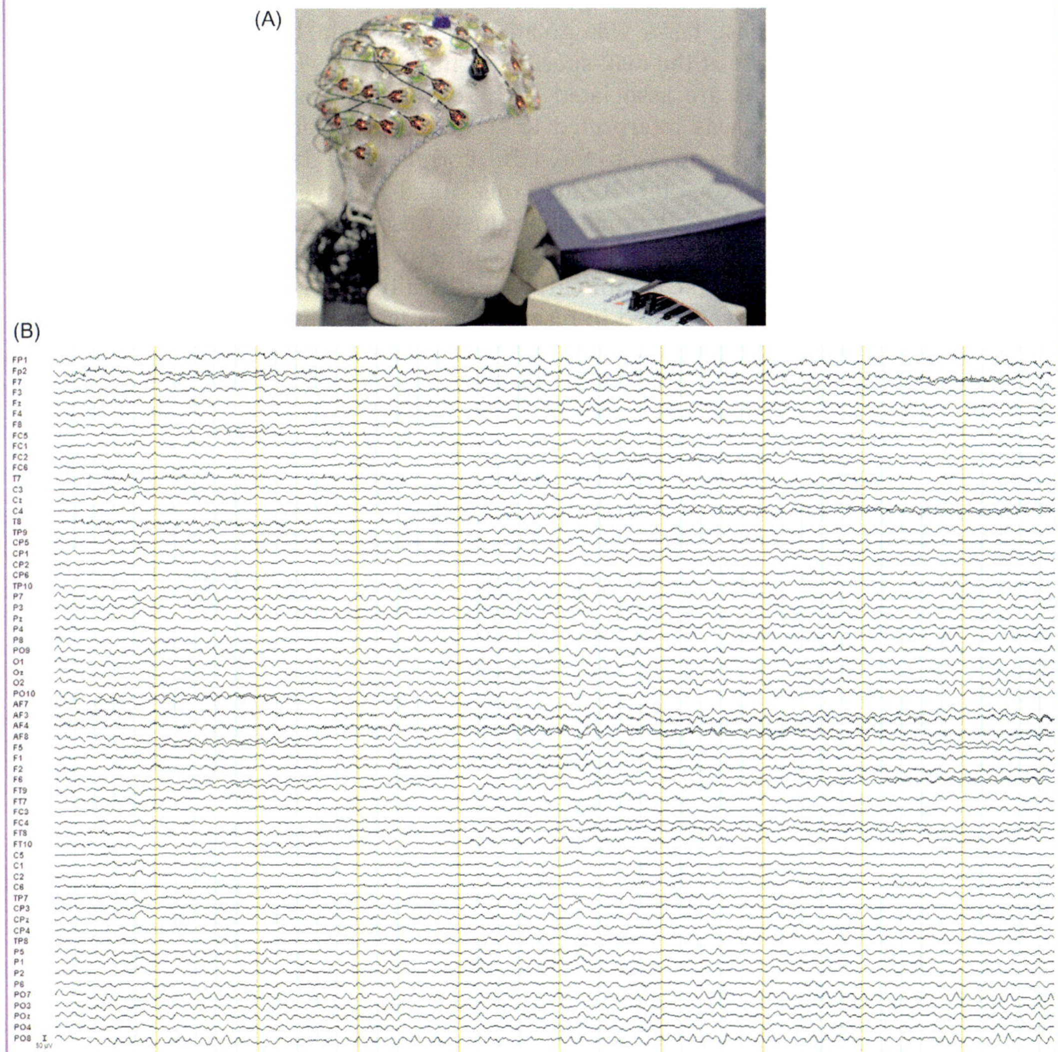

FIGURE 5.11 (A) A typical EEG-setup with amplifier and active electrodes mounted on an electrode cap. (B) Raw EEG signal for 64 electrodes.

widely investigated ERP component in language lateralization research. By comparing the amplitude and latency of ERP components at right- and left-sided electrodes, hemispheric asymmetries can be detected.

Language-related EEG asymmetries were first investigated by David Galin and Robert Ornstein from the Langley Porter Neuropsychiatric Institute in San Francisco, USA, who continuously recorded EEG from their subjects while they wrote a letter or mentally composed a letter while fixating their gaze on a central spot on the table.[55] They could show that EEG power during these tasks was greater over the right hemisphere compared to the left hemisphere. As most of the task-specific asymmetry was found in the Alpha-band, and since alpha oscillations are associated with relaxing and are reduced by cognitive activity or excitement, this was interpreted to show left-hemispheric dominance for language production. While research on the EEG correlates of hemispheric asymmetries in emotion processing (see Chapter 9) since this initial discovery has continued to use asymmetries in Alpha-band oscillations, subsequent research on language asymmetries has been more focused on ERPs, presumably because they offer a more direct assessment of cognitive activity than alpha asymmetries. Especially the N1 ERP component (see Fig. 5.12) has been widely investigated in research on language lateralization.

The N1 is an early negative ERP component (occurring 140–200 ms after stimulus onset) which is modulated by physical stimulus features.[56] Functionally, it is related to orientation of attention toward visual stimuli[57–60] or with mechanisms involved in the categorization of visual stimuli.[61] The generators of the N1 lie within the posterior part of the cortex, involving the lateral extrastriate cortex as well as dorsal occipitoparietal and ventral occipitotemporal areas.[62,63] The N1 is the earliest ERP component reflecting recognition of verbal stimuli, and it has been linked to the automatic lexical classification of words.[64] For tasks involving visually presented words—for example, semantic tasks such as indicating whether a presented word could complete a sentence in a meaningful way or not—a larger amplitude of the N1 has been observed for left occipitotemporal brain areas than for the same areas in the right hemisphere.[65,66] Interestingly, N1 asymmetries have been shown to correlate with the right-ear advantage in an EEG variant of the dichotic listening task.[67] Participants with greater differences in N1 latency between the left and the right temporal cortex also show a stronger right-ear advantage, indicating that it might be caused by the processing speed advantage of the language-dominant left hemisphere.

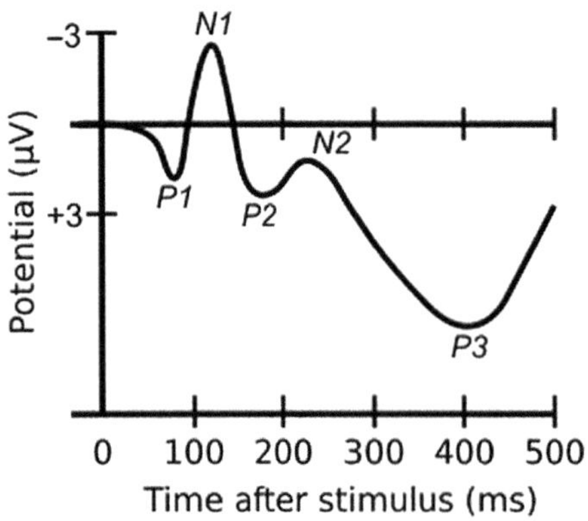

FIGURE 5.12 The N1 ERP component and other stimulus-locked ERP components (P1, P2, N2, and P3). *Source: Wikimedia commons, licensed under CC BY-SA 3.0 license.*

The language system: Broca's and Wernicke's area

The techniques described in the previous section allow for identification of an individual's language-dominant hemisphere and specific asymmetries in left- and right-sided brain areas, but they do not necessarily tell us why the left hemisphere is language-dominant in most people. To understand the emergence of language lateralization, it is important to know the different brain areas involved in the language system and its individual properties.

As described in Chapter 1, in 1861, Pierre Paul Broca, a French surgeon, published a scientific paper describing a patient of his who was called Monsieur Tan because the only word he was able to utter was "tan."[68] Post-mortem analysis of the brain of this severely speech-impaired patient revealed massive damage to the left frontal lobe, indicating that the left hemisphere and especially the area now known as Broca's area—the left posterior inferior frontal gyrus, Brodmann area 44/45—is highly relevant for language production. Today we know that damage to Broca's area leads to Broca's aphasia. The term aphasia describes a family of acquired speech problems that (mostly) result from left-hemispheric damage. Impairment may manifest during the production, comprehension, reading, or writing of language. Broca's aphasia (also called expressive aphasia) is defined by three core symptoms:

1. Agrammatism: Severe problems in using grammar as a means to structure spoken information.
2. Anomia: Inability to produce the correct words to communicate meaning.
3. Articulation problems.

Speech comprehension is largely preserved in patients with Broca's aphasia. A typical response by a patient with Broca's aphasia could look like this:

Psychologist: "Have you visited a doctor before?"
Patient: "No ... tills ... eh, eh... pills I took ... Kavergot ... 2 years ago ... eh ... this ... doctor came and I ... eh ... l.l ... little needles."

In 1874, a few years after Broca published his findings on the role of the left hemisphere for speech production, Carl Wernicke (1848–1905), a German anatomist, published a book including the first description of a new type of language disorder.[69] He could show that lesions within the left superior temporal gyrus resulted in deficits in language comprehension. This region is now referred to as Wernicke's area, and the associated syndrome is known as Wernicke's aphasia or receptive aphasia. Specifically, patients suffering from Wernicke's aphasia are able to speak fluently, but their language is meaningless. This is in stark contrast to patients suffering from Broca's aphasia—resulting from lesions of the inferior frontal cortex—who can understand language but are impaired in their ability to form or articulate words.

Wernicke assumed that Wernicke's area contained the memory for the associations between auditory sequences and words. Lesions of this area should render the patient unable to "hear" a word when listening to its auditory pattern. Indeed, patients with Wernicke's aphasia are defined by comprehension problems while at the same time

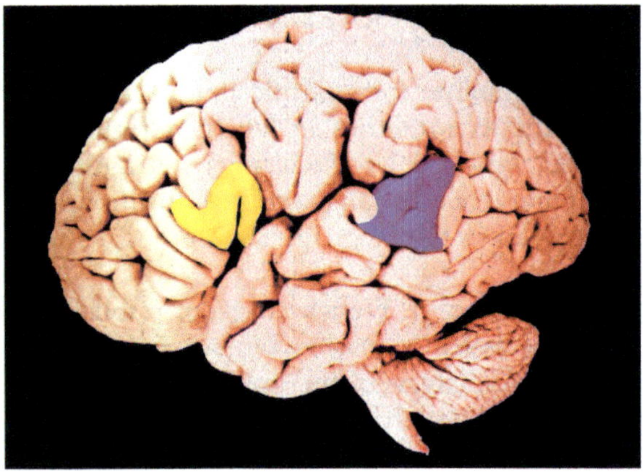

FIGURE 5.13 Broca's area (in yellow) and Wernicke's area (in purple). *Source: The figure of the human brain has been adapted with permission from http://brainmuseum.org and http://www.brains.rad.msu.edu, supported by the US National Science Foundation.*

having a largely preserved ability to speak. This condition is also called pure word deafness. These patients hear everything, and, of course, they also hear spoken words. They also understand the meaning of sounds like doorbells and birdsong, but they have lost their comprehension of words. For example, NB, a patient with Wernicke's aphasia, described his condition as follows:

> I hear everything. I also know what to do with sounds that are not words. But when you start speaking, it is like listening to a foreign language.

Today we know that Wernicke's and Broca's area (see Fig. 5.13) are the core areas of the human language system. However, language production and perception of language are not solely controlled by Broca's and Wernicke's area, respectively.[70] Instead, an extended frontal-temporal network is active during both language production and perception—let us have a look at its components.

Auditory information processing in the temporal lobe: Wernicke's area and beyond

Generally, the task of the superior temporal gyri is to act as computational hubs that disambiguate complex sounds by isolating different properties of acoustic information—for example, temporal and spectral variation—and match them to stored templates. However, what makes the left superior temporal gyrus so special that lesions here can cause Wernicke's aphasia, while homologous right-sided lesions cannot? Different processing properties of the left and right superior temporal gyrus were analyzed in a seminal fMRI study by Marc Schönwiesner from the Auditory Cognitive Neuroscience Lab at the University of Montreal, Canada.[71] In this study, subjects listened to sounds that either varied in their spectral or temporal complexity. Spectral complexity of an auditory stimulus refers to the frequency domain, for example, the pitch—high or low—of the stimulus.

Temporal complexity, on the other hand, refers to the time domain—for example, how many modulations of the sound occur within a given time window (see Fig. 5.14).

Schönwiesner and his colleagues analyzed brain activations in Heschl's gyrus (HG) and also in the anterolateral portions of the planum temporale (abbreviated AL for anterolateral). HG is located within the lateral sulcus and contains the primary auditory cortex. For this area, no brain asymmetries were observed, as the left and right HG both showed a linear increase in activation with increasing spectral complexity. In contrast, no effect was observed for temporal modulation, showing that HG is only sensitive to spectral modulations. Contrary to that, AL, containing Wernicke's area on the left, was sensitive to temporal variations in the left hemisphere and spectral variations in the right hemisphere (see Fig. 5.15).

Thus, the left hemisphere seems to be specialized in analyzing fast temporal changes of the auditory input.[72] This feature is critical for speech comprehension, as it is needed for the correct identification of consonants. The left hemisphere excels in hearing the

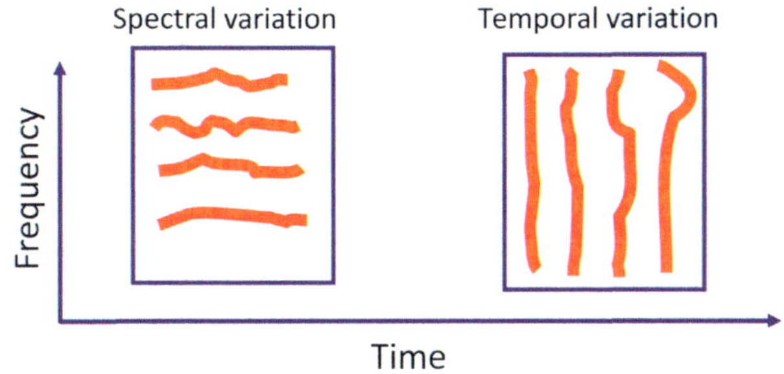

FIGURE 5.14 Spectral and temporal variations in sound.

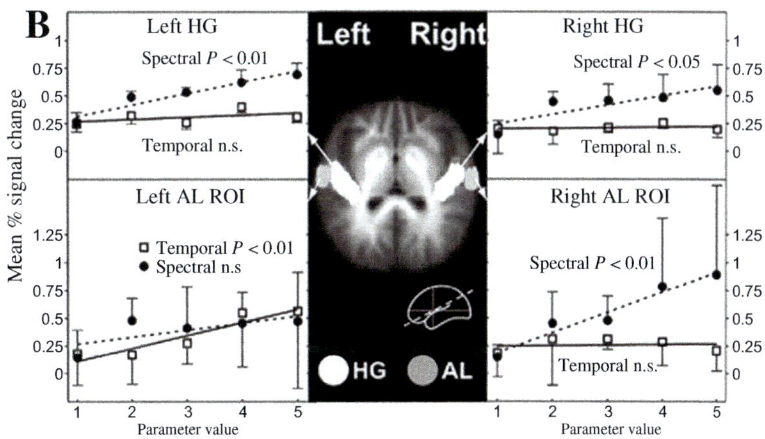

FIGURE 5.15 fMRI signal changes in four regions of interest (left and right HG (Heschl's gyrus) and AL (anterolateral belt area on the superior temporal gyrus)) as a function of spectral and temporal input parameters. Parameter value indicates stimulus complexity. *Source: From Schönwiesner M, Rübsamen R, von Cramon DY. Spectral and temporal processing in the human auditory cortex revisited. Ann N Y Acad Sci. 2005;1060(1):89-92. https://doi.org/10.1196/annals.1360.051. Used with permission of the authors and the publisher.*

minute time differences that are critical to differentiate consonants due to its better temporal resolution.[73] Accordingly, patients with Wernicke's aphasia often particularly struggle to comprehend consonants with fast start/stop-sequences—for example, /t/, /d/, /k/, /p/.

Interestingly, there seems to be an anatomical foundation for the left hemisphere's specialization in detecting temporal variations in sounds. Wernicke's area in the left hemisphere and the homotopic area in the right hemisphere are organized in cortical modules. These modules are represented by vertical columns of pyramidal neurons.[74] There are two types of such columns (see Fig. 5.16):

- Microcolumns have a width of about 60−90 μm. They are defined anatomically by the vertical arrangement of pyramidal neurons.
- Macrocolumns are formed by several microcolumns and have a width of about 600−800 μm. They are defined by their zones of axonal termination—for example, thalamocortical or corticocortical.

Importantly, these microstructural features of Wernicke's area show strong hemispheric asymmetries. In general, areas in the auditory processing stream show asymmetry in pyramidal cell sizes, with the left hemisphere containing a greater number of large pyramidal cells.[75] Moreover, in Wernicke's area—but not the primary auditory cortex—the columns consisting of these pyramidal cells are wider and farther apart from each other in the left hemisphere.[74−77] This wider spacing implies that more subsystems can be included per surface unit in the left than in the right hemisphere, as individual pyramidal cells contact fewer adjacent cell columns. This results in a smaller number of interconnected columns that potentially influence each other's activity and thus a less redundant processing

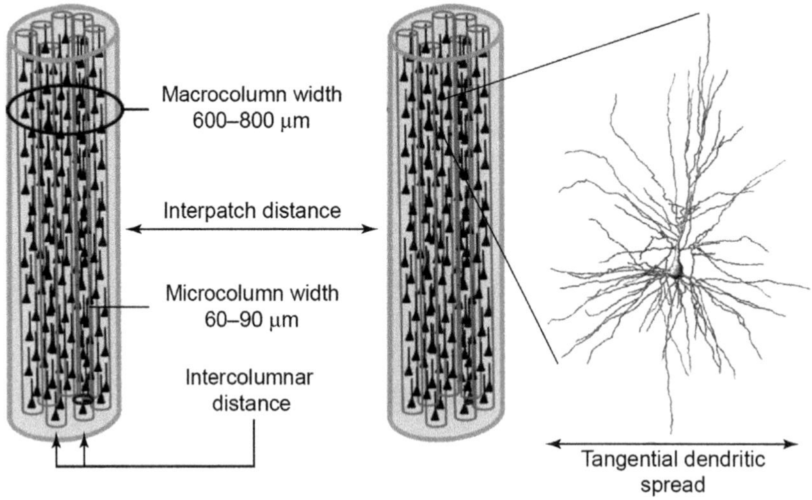

FIGURE 5.16 Columnar organization of the cerebral cortex. *Source: From Hutsler J, Galuske RA. Hemispheric asymmetries in cerebral cortical networks.* Trends Neurosci. *2003;26(8):429−435. https://doi.org/10.1016/s0166-2236(03)00198-x. Used with permission of the publisher.*

architecture[74] (see Fig. 5.17). Overall, more differently specialized columns exist on the left side as compared to the right, and the auditory signal can probably be analyzed in greater detail and with higher temporal resolution in the left hemisphere.

In a recent study, this assumption was tested using a combination of NODDI (see Chapter 4 for details) and EEG.[78] To assess leftward lateralization of syllable recognition speed, an EEG version of the dichotic listening paradigm was used and the latency of the N1 was assessed. Moreover, macrostructural volume asymmetries and macrostructural asymmetries in the density of axons and dendrites in the planum temporale were determined using NODDI. Strong leftward asymmetries were found both for volume of the planum temporale and the density of axons and dendrites. Subsequently, lateralization indices reflecting the extent of physiological and structural asymmetries were correlated. A higher density of axons and dendrites in the planum temporale was associated with faster neurophysiological processing of auditory speech. This implies that a higher density and higher number of synaptic contacts in the left planum temporale increases temporal precision and decreases latency of language processes.

Interestingly, an MRI and fMRI study also found that while 80% of individuals show a leftward planum temporale asymmetry in surface area, this macrostructural asymmetry was not correlated to functional hemispheric asymmetries while listening to speech.[79] Taken together, these studies suggest a high relevance of microstructural but not macrostructural asymmetries for functional language lateralization.

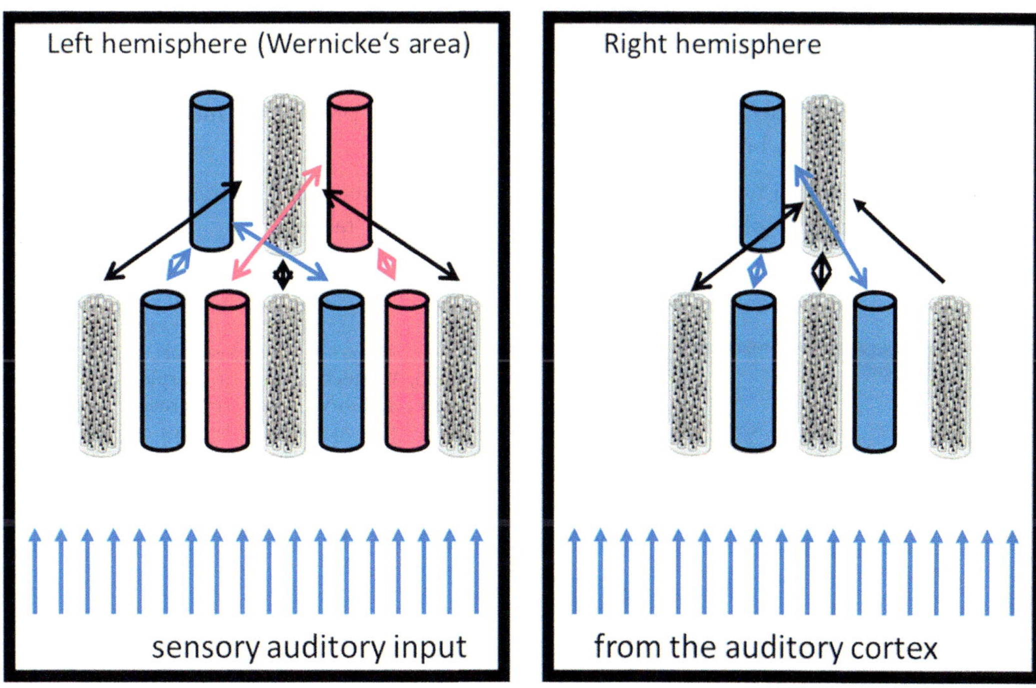

FIGURE 5.17 Hemispheric asymmetries in the columnar organization of Wernicke's area and its right hemispheric homolog.

The role of the posterior speech area

As mentioned above, Wernicke's aphasia is a syndrome that is characterized by a drastic reduction of speech comprehension and mostly intact speech production. Additionally, problems in finding the correct word, phonetic mistakes, and semantic and phonematic paraphasias—production of unintended words or syllables during speech—can occur. However, patients diagnosed with Wernicke's aphasia sometimes show deficits after lesions that go beyond the classic boundaries of Wernicke's area, for example, into the so-called posterior speech area surrounding Wernicke's area (see Fig. 5.18).

While reduction of speech comprehension with fluent speech production is the key symptom for damage to Wernicke's area itself, the other three symptoms described above—problems in word finding, phonetic problems, and semantic and phonematic paraphasia—have been linked to damage to the posterior speech area.

Problems in word finding can become manifest in different ways:

1. Long pauses during conversations.
2. Escape to catchwords (e.g., "Well, you know what I want to tell you. You know, these things, do you.")
3. Repetitions (e.g., "...later with the car...directly with the car...directly with the car I came to you with.")
4. Abruptly stopping sentences followed by unrelated content (e.g., "I was then for 6 years...other countries...I organized for 12 years...for 6 years I did...car construction... no...6 years I worked...")
5. Using pantomimic acts and gestures instead of speech.

Phonetic problems become manifest in difficulties in producing meaningful speech sounds. Paraphasias can be:

1. Semantic: Replacing a word with a different word that has a similar meaning (e.g., "luggage" instead of "plastic bag," or "chair" instead of "table").

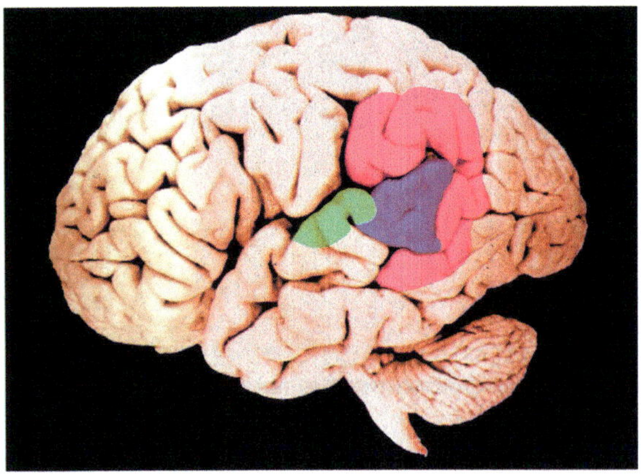

FIGURE 5.18 The posterior speech area (in pink), Wernicke's area (in purple), and the primary auditory cortex (in green). *Source: The figure of the human brain has been adapted with permission from http://brainmuseum.org and http://www.brains.rad.msu.edu, supported by the US National Science Foundation.*

2. Phonematic: Replacing a word with another word or non-word that has a similar sound structure (e.g., "Kiev" instead of "kiwi," "brass" instead of "glass," or "chakalote" instead of "chocolate").

These deficits indicate that within the posterior speech area, the auditory representations of words from Wernicke's area are associated with their meaning. It is important to stress that memories for objects or events are distributed over many cortical areas within the memory system and are thus not exclusively stored in the posterior speech area. The posterior speech area, therefore, is the place where the associations between words and objects happen. If only the posterior speech area but not Wernicke's area is damaged, patients develop transcortical sensory aphasia. Patients with this type of aphasia can repeat speech, indicating that translation from the auditory stream into a word-based acoustic output is functional. However, they are unable to understand what was said and cannot generate meaningful speech.

Information transfer in the language system

As mentioned above, an extended fronto-temporal network is active during both language production and perception. However, language function is not only based on these gray matter areas but also relies on the white matter connections between them. Thus, information transfer between the areas compromising this network is critical for successful language processing.[80] The most important white matter pathway in the language system is the arcuate fasciculus, an extended structure connecting Broca's and Wernicke's area dorsally. The arcuate fasciculus transfers linguistic information relevant for both the phonological and syntactic processing of speech.[81] Importantly, the organization of the left- and right-hemispheric networks via the arcuate fasciculus differs profoundly. While a direct connection between Wernicke's and Broca's area (or their right hemispheric homologs) only exists in the left hemisphere, an indirect pathway via the posterior speech area is present in both hemispheres[82] (see Fig. 5.19).

The functions of the direct and the indirect pathway can be determined from cases of conduction aphasia in which they are damaged. The direct connection between Broca's area and Wernicke's area transfers phonological information but not semantic meaning. If the direct connection is transected, a typical patient would behave like this:

Psychologist: "Repeat my words: Car."
GE: "Car."
Psychologist: "The sun is shining."
GE: "The sun is shining."
Psychologist: "Kalinopa."
GE: "What? What did you say."
Psychologist: "For Mother's Day, I bought flowers."
GE: "Mother's day I brought flowers home."
Psychologist: "Seat."
GE: "Sofa."
Psychologist: "Lestin."

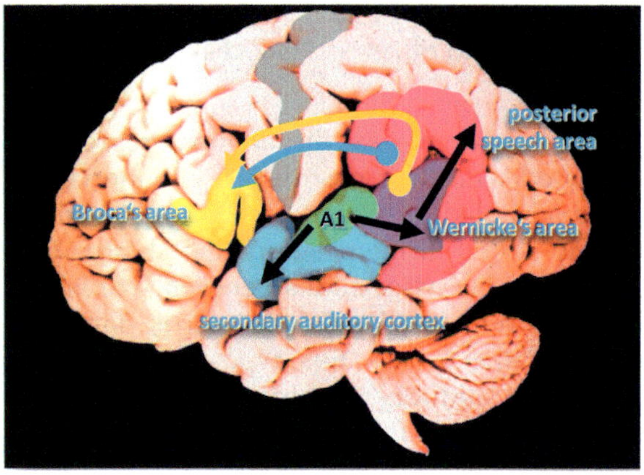

FIGURE 5.19 The direct (in orange) and indirect (in blue) connection between Wernicke's and Broca's area. *Source: The figure of the human brain has been adapted with permission from http://brainmuseum.org and http://www.brains.rad.msu.edu, supported by the US National Science Foundation.*

GE: "What's that?"

The indirect connection between Broca's area and Wernicke's area over the posterior speech area transfers words with their semantic meaning. If the indirect connection is transected but the direct connection is still intact, a typical patient would behave like this:

Psychologist: "Repeat my words: Car."
GE: "Repeat my words: Car."
Psychologist: "The sun is shining."
GE: "The sun is shining."
Psychologist: "Kalinopa."
GE: "Kalinopa."
Psychologist: "For Mother's day I bought flowers."
GE: "For Mother's day I bought floffs."
Psychologist: "Seat."
GE: "Seat."
Psychologist: "Lestin."
GE: "Lestin."

As the direct pathway is only present in the left hemisphere but the indirect pathway is present in both hemispheres, direct transmission of phonological content is mainly confined to the left hemisphere, while semantic content can also be transferred within the right hemisphere.

Interestingly, alterations in connectivity of the language network have been suggested to contribute to the pathogenesis of auditory hallucinations, one of the core symptoms of schizophrenia. Schizophrenic patients with auditory hallucinations often display reduced white matter integrity of the arcuate fasciculus compared to patients without acoustic hallucinations or healthy controls.[83] Disorganized reciprocal connections between perceptual and motor language areas may affect differentiation between overt and inner speech, and a reduced synchronization of activity may account for the patients' inability to distinguish self-generated thoughts from external stimulation. Thus, apart from proper functioning of singular brain regions, organized connectivity of intra- and interhemispheric fronto-temporal networks is critical for normal language processing.

Right-hemispheric contributions to language

While the left hemisphere is superior in language processing in the majority of individuals, this does not imply that the right hemisphere completely lacks any linguistic capabilities. Indeed, in recent years, a growing body of empirical evidence has supported the idea that the right hemisphere plays a crucial role in certain aspects of language processing.[84] There are three main aspects of spoken language that the right hemisphere is critically involved in:

1. Prosody: Wernicke's homologous area in the right hemisphere specializes in comprehending slow tonal changes that are typical for prosodic information, for example, intonation, tone, or rhythm of speech. The right hemisphere specializes in analyzing the direction of pitch differences, timbre, and many aspects of music.
2. Pragmatics: The right hemisphere is important for the integration of contextual information into the meaning of speech. For example, when figural speech is used or jokes are made, patients with right-hemispheric lesions often do not understand the punchline and take these figures of speech literally.
3. Paralinguistics: The right hemisphere is critical for the integration of paralinguistic aspects (e.g., emotional tone of a sentence, meaning of gasps or sighs during a conversation) into the meaning of speech. Patients with right-hemispheric lesions often fail to comprehend these aspects of communication.

These findings show that the right hemisphere is far from being nonverbal and that it substantially contributes to normal language functions.

Atypical language lateralization

While the right hemisphere contributes to language processing in everyone, there are a few individuals in whom the right hemisphere is language-dominant—for example, it shows greater activation than the left. The exact prevalence of these individuals greatly depends on the size of the sample used to assess language lateralization. It has been estimated that about 1%–5% of right-handers show right-hemispheric language dominance, as do 10% – 25% of left-handers.[85–88] While most authors distinguish between typical and atypical language lateralization, recently, a differentiation into three groups has been proposed based on Gaussian modeling[89]:

1. Typical lateralization (clear left hemisphere dominance): 88% of right-handers and 78% of left-handers.
2. Ambilateralization (no dominant hemisphere): 12% of right-handers and 15% of left-handers.
3. Strongly atypical lateralization (clear right hemisphere dominance): 0% of right-handers and 7% of left-handers.

Irrespective of the classification system that is used, the reason for the existence of atypical language lateralization to date is largely a scientific mystery. One hypothesis that has been put forward is the idea that atypical language lateralization is caused by anatomical asymmetries in the language system (more on structural asymmetries in Chapter 4). However, it has been shown that left- and right-lateralized individuals show almost identical structural asymmetries—for example, in the planum temporale and HG.[90,91] Moreover, individuals with situs inversus, a medical condition in which all asymmetric visceral organs are mirror-image reversed, show no alterations of language lateralization.[92,93] Another factor that has been suggested to determine language lateralization is asymmetrical sensory input. Along these lines, individuals who are deaf in the right ear would receive relatively less auditory input to the left than to the right hemisphere and thus might develop atypical right hemispheric language asymmetries. This hypothesis was tested by Lise van der Haegen from the University of Ghent in Belgium.[94] She tested language dominance with three different tests in congenitally deaf subjects and healthy control subjects. In contrast to the aforementioned hypothesis, her results clearly showed that a lack of sensory auditory input to the left hemisphere did not cause atypical language lateralization as all but one patient showed typical left-sided language lateralization. While these results suggest that genetic factors possibly override the role of sensory input for the ontogenesis of language lateralization, no such gene(s) has/have yet been identified in molecular genetic studies.

Recently, the consortium on language asymmetry published a large-scale registered report to address the question whether language lateralization can be different for different components of language and whether such dissociations are more common in left-handers.[95] To this end, 621 individuals were tested with an online laterality battery containing the dichotic listening task, as well as a rhyme decision task and a word comprehension task. In the second step, 108 left-handers and 101 right-handers were tested with different language tasks while undergoing functional transcranial Doppler sonography. In general, strong leftward lateralization was found for speech generation tasks, but only weak lateralization for language perception tasks. Overall, the results suggested that there seem to be at least two distinct forms of language lateralization. Online test results and functional transcranial Doppler sonography results showed only weak correlations. In general, right-handers showed a stronger left hemisphere bias than left-handers. However, there was no stronger dissociation between tasks in left-handers compared to right-handers. In line with the findings of earlier functional transcranial Doppler sonography studies, these findings suggest that it is important to use several different language lateralization tasks containing both speech perception tasks and speech generation tasks when assessing atypical or inconsistent language lateralization.[96,97] Moreover, it is crucial to always test both left-handed and right-handed participants.

Language lateralization in bilinguals

A question related to atypical lateralization is whether a second language somebody learns is lateralized the same way the first is, or whether different brain areas are involved in processing this language.[98] Two main hypotheses have been developed regarding language lateralization in bilingual individuals:

1. *The age of language acquisition hypothesis* assumes that differences in language lateralization between the first and the second language depend on whether the two languages are learned at a similar age or not. Early, simultaneous acquisition of the two languages will lead to a more left-hemispheric pattern of lateralization, while later acquisition will lead to a stronger involvement of the right hemisphere.[99]
2. *The stage of language acquisition hypothesis* assumes that language lateralization for the second language is not determined by age at time of learning but by relative proficiency in that language. Increasing proficiency in the second language will lead to a more left-hemispheric dominance for that language, as the speaker depends less on pragmatic cues processed by the right hemisphere.[100]

Rachel Hull and Jyotsna Vaid from the Department of Psychology of the Texas A&M University, USA, conducted two meta-analyses—a statistical approach to combine the results of different empirical studies in order to gain higher statistical power—to clarify which of these hypotheses is supported by empirical data.[98] Integrating the results of 66 research studies, they found that the age at onset of bilingualism was the most important factor to influence language lateralization. Bilinguals who had learned both their first and their second language before their sixth birthday showed a more bilateral pattern for both languages than bilinguals who had learned their second language after the age of 6. In this group, both languages showed a stronger left-dominant pattern. However, language proficiency also affected lateralization. Among bilinguals who had learned their second language after age 6, those who showed lower proficiency in the second language also showed more left-hemispheric dominance.

In a more recent functional transcranial Doppler study on language lateralization of first and second languages in bilinguals, participants were significantly left-lateralized for both languages during a verbal fluency task, with high correlation between them.[101]

Hemispheric asymmetries in nonvocal languages: the case of whistle language

One factor that might critically influence to what extent linguistic processing activates the left and the right hemisphere of the brain is the nature of the linguistic stimulus. While left-hemispheric dominance has been demonstrated for tonal languages like Cantonese as well as nontonal languages like English, and also for African click languages,[102–104] it is conceivable that the right hemisphere would be dominant for a language in which the listener would need to encode acoustic properties for which the right hemisphere is specialized. This hypothesis was tested by Onur Güntürkün and his colleagues,[105] who investigated language lateralization in native whistle-speaking people from a mountainous area in Northeast Turkey. Whistle languages are an extremely rare form of human communication. Only three groups of people use them as a means of communication:

- The Mazatecs, an indigenous tribe from the Oaxaca region in southern Mexico
- The Spanish inhabitants of La Gomera on the Canary Islands.

- People around the village of Kuşköy in the Turkish Black Sea Province Giresun.

What makes whistle languages unique is that whistles largely consist of changes in pitch and timbre, classic right-hemispheric functions. It uses slow melodic modulations to code meaning. Also, time-sensitive voice-onset-time computations are not necessary. Thus, the physical structure of a whistle language does not necessarily require the left hemisphere for comprehension. However, whistle language processing should not be completely independent of the left hemisphere. Whistled Turkish is based on vocal Turkish and uses the full lexical and syntactic information of vocal Turkish (see Box 5.3 for a translation of an exemplary dialogue). It is usually learned between the ages of 6 and 12, when vocal Turkish is already fully established. Since vocabulary and grammar of vocal Turkish are largely left-lateralized, whistled Turkish should also use this left-lateralized cognitive system.

BOX 5.3

A typical dialogue in whistled Turkish.

A wide variety of intentions and instructions can be communicated using whistle languages as exemplified by the dialogue in whistled Turkish. A video showing this conversation can be found here: https://www.bio.psy.ruhr-uni-bochum.de/bpsy/mam/images/tuerkische_pfeifsprache.mp4

Halil: Küçük, iftara gel bu akşam ("Küçük, come for iftar [breaking the feast in the evening during Ramadan] this evening").
Küçük: Gelirim ("I will come").
Halil: İftara gel bu aksam, kahvede yaparız ("Come for iftar this evening, we will do it in the café").
Küçük: Tamam, tamam ("OK, OK").
Halil: Dinle ("Listen").
Küçük: Evet ("Yes").
Halil: Güntürkün'ün sana selamı var. Güntürkün'ün sana selamı var ("Güntürkün sends his best regards. Güntürkün sends his best regards").
Küçük: Aleykümselam. Ne zaman gidiyor? ("Aleykümselam [traditional counter-greeting]. When does he leave?").
Halil: Hoca yarın gidiyor; yarın gidecek ("Hoca [traditional word for a scholar] leaves tomorrow; his departure is tomorrow").
Küçük: Tamam. Benimde selamımı söyle ("OK. Tell him my best regards").
Halil: Küçük! ("Küçük").
Küçük: Evet ("Yes").
Halil: Osman'ın da selamı var. Osman'ın da selamı var ("Osman greets you too. Osman greets you too").
Küçük: Aleykümselam. ("Aleykümselam [traditional counter-greeting]").
Halil: Hepimizden sana iyi günler ("We all wish you a good day").
Küçük: Osman oradamı? ("Is Osman there?").
Halil: Evet, Nazmiye'nin yanında; Nazmiye'nin yanında ("Yes, beneath Nazmiye; beneath Nazmiye").
Küçük: Tamam, tamam ("OK, OK").

Language lateralization of Turkish whistle speakers for both whistled and vocal Turkish was tested using the dichotic listening task. Compared to vocal Turkish, hearing whistle language sounds increased left-ear/right-hemispheric comprehension. At the same time, it decreased right-ear/left-hemispheric comprehension (see Fig. 5.20).

Therefore, there was no significant right-ear advantage for whistled Turkish. Possibly, symmetrical performance in whistle language dichotic listening reflects involvement of both hemispheres in the analysis of whistled language input. Being Turkish, whistled Turkish activates left-hemispheric language resources. The fact that right-ear comprehension was relatively equal in vocal and whistled dichotic listening tasks shows that whistled Turkish resulted in a similar left-hemispheric activation as dichotic listening tasks with vocal Turkish. Thus, even without the requirement for voice-onset-time analysis, a left-hemispheric superiority emerged. However, since whistled Turkish also requires melodic analysis patterns of the right hemisphere, dichotic listening with whistled Turkish resulted in an increase of left ear (right hemisphere) activations, resulting in a bilateral activity pattern. A similar decrease in left and a relative increase in right-hemispheric processing was

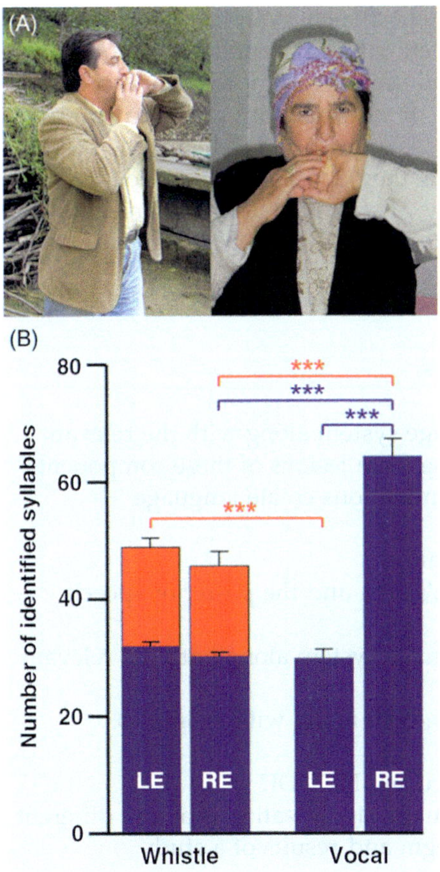

FIGURE 5.20 Hemispheric asymmetries of whistle language. (A) Two whistle language speakers from Northeastern Turkey. (B) Left ear (LE) and right ear (RE) responses for whistled and vocal dichotic syllables. Blue bars represent the number of correctly identified syllables and red bars represent whistle responses adjusted for recognition performance. All differences marked with *** were significant at $P < .001$. *Source: From Güntürkün O, Güntürkün M, Hahn C. Whistled Turkish alters language asymmetries.* Curr Biol. *2015;25(16):R706–R708. https://doi.org/10.1016/j.cub.2015.06.067. Used with permission of the author and the publisher.*

also found in a subsequent dichotic listening study in Silbo Gomero, a form of whistled Spanish.[106] This suggests that bihemispheric processing of whistled languages may occur independent of language.

Hemispheric asymmetries in nonvocal languages: the case of sign language

Another form of nonvocal language that is particularly interesting regarding hemispheric asymmetries is sign language. In contrast to whistle language, there is no auditory input. Instead, spatial information (e.g., position of the fingers and hands of the speaker) is much more relevant for sign processing than for any other form of language. Since spatial processing is mainly a right-hemispheric function, one could assume that sign processing involves the right hemisphere to a greater extent than speech. However, is this really the case? Probably not. In general, spoken and signed languages seem to rely largely on similar neuronal networks. Lesion studies showed that damage to the left hemisphere leads to impaired language processing in both spoken and signed language users, whereas damage to the right hemisphere does not elicit such symptoms.[107] Moreover, neuroimaging studies indicate a crucial role of the left inferior frontal gyrus for signed language processing, just as for spoken languages, and sentence comprehension predominantly activated left-hemispheric areas independent of the modality of verbal input.[108] Thus, all natural human languages seem to rely on a core set of left-hemispheric brain regions independent of the modality through which they are communicated. It is the linguistic nature of the information, not the modality, that drives brain organization for language.

Sample questions

- Specify the major components of the human language system along with the relevant connections. What are the major syndromes resulting from lesions of these components?
- How can asymmetries in time and frequency discriminations create language asymmetries in humans?
- Outline the anatomical asymmetries of Wernicke's area.
- What are the lead symptoms of lesions to Wernicke's area and the posterior speech region? Provide examples for each of your points.
- Draw and label the components of the human language system along with the relevant connections (on a lateral view of the human brain).
- There are two types of conduction aphasias. Explain both along with examples.
- What are the lead symptoms of Broca's aphasia?
- What has been found about language lateralization using NODDI?
- What is known about how different or similar language lateralization is across different languages? Please describe the experimental paradigm and results of a study investigating this question.

References

1. Wada JA. Youthful season revisited. *Brain Cognition.* 1997;33(1):7–10. Available from: https://doi.org/10.1006/brcg.1997.0879.
2. Wada JA. A new method of determining the side of cerebral speech dominance: a preliminary report on the intracarotid injection of sodium amytal in man. *Igaku Seibutsugaku.* 1949;14:221–222 [Paper in Japanese]. Available from: https://ci.nii.ac.jp/naid/10010374848/.
3. Perria L, Rosadini G, Rossi GF. Determination of side of cerebral dominance with amobarbital. *Arch Neurol.* 1961;4(2):173–181. Available from: https://doi.org/10.1001/archneur.1961.00450080055006.
4. Carey DP, Johnstone LT. Quantifying cerebral asymmetries for language in dextrals and adextrals with random-effects meta analysis. *Front Psychol.* 2014;5:1128. Available from: https://doi.org/10.3389/fpsyg.2014.01128.
5. Papanicolaou AC, Rezaie R, Narayana S, et al. Is it time to replace the Wada test and put awake craniotomy to sleep? *Epilepsia.* 2014;55(5):629–632. Available from: https://doi.org/10.1111/epi.12569.
6. Bruzsa A-K, Walther K, Kasper BS, Gollwitzer S, Hamer H, Schwarz M. WADA test for postoperative memory prediction in left TLE. Is it still useful in the 21st century? *Clin Neurol Neurosurg.* 2023;225:107580. Available from: https://doi.org/10.1016/j.clineuro.2022.107580.
7. Corballis MC. From mouth to hand: gesture, speech, and the evolution of right-handedness. *Behav Brain Sci.* 2003;26(2):199–208. Available from: https://doi.org/10.1017/s0140525x03000062. discussion 208-60.
8. Kimura D. Manual activity during speaking—I. Right-handers. *Neuropsychologia.* 1973;11(1):45–50. Available from: https://doi.org/10.1016/0028-3932(73)90063-8.
9. Lavergne J, Kimura D. Hand movement asymmetry during speech: no effect of speaking topic. *Neuropsychologia.* 1987;25(4):689–693. Available from: https://doi.org/10.1016/0028-3932(87)90060-1.
10. Graves R, Goodglass H, Landis T. Mouth asymmetry during spontaneous speech. *Neuropsychologia.* 1982;20(4):371–381. Available from: https://doi.org/10.1016/0028-3932(82)90037-9.
11. Hausmann M, Behrendt-Körbitz S, Kautz H, Lamm C, Radelt F, Güntürkün O. Sex differences in oral asymmetries during wordrepetition. *Neuropsychologia.* 1998;36(12):1397–1402. Available from: https://doi.org/10.1016/s0028-3932(98)00027-x.
12. Kimura D. Cerebral dominance and the perception of verbal stimuli. *Can J Psychology/Revue Canadienne de Psychologie.* 1961;15(3):166–171. Available from: https://doi.org/10.1037/h0083219.
13. Tervaniemi M, Hugdahl K. Lateralization of auditory-cortex functions. *Brain Res Brain Res Rev.* 2003;43(3):231–246. Available from: https://doi.org/10.1016/j.brainresrev.2003.08.004.
14. Hugdahl K. Dichotic listening in the study of auditory laterality. In: Hugdahl K, Davidson RJ, eds. *The Ssymmetrical Brain.* MIT Press; 2004.
15. Packheiser J, Schmitz J, Arning L, Beste C, Güntürkün O, Ocklenburg S. A large-scale estimate on the relationship between language and motor lateralization. *Sci Rep.* 2020;10(1):13027. Available from: https://doi.org/10.1038/s41598-020-70057-3.
16. Westerhausen R. A primer on dichotic listening as a paradigm for the assessment of hemispheric asymmetry. *Laterality.* 2019;24(6):740–771. Available from: https://doi.org/10.1080/1357650X.2019.1598426.
17. Hugdahl K, Carlsson G, Uvebrant P, Lundervold AJ. Dichotic-listening performance and intracarotid injections of amobarbital in children and adolescents. Preoperative and postoperative comparisons. *Arch Neurol.* 1997;54(12):1494–1500. Available from: https://doi.org/10.1001/archneur.1997.00550240046011.
18. Westerhausen R, Hugdahl K. The corpus callosum in dichotic listening studies of hemispheric asymmetry: a review of clinical and experimental evidence. *Neurosci Biobehav Rev.* 2008;32(5):1044–1054. Available from: https://doi.org/10.1016/j.neubiorev.2008.04.005.
19. Kimura D. Functional asymmetry of the brain in dichotic listening. *Cortex.* 1967;3(2):163–178. Available from: https://doi.org/10.1016/s0010-9452(67)80010-8.
20. Kinsbourne M. The cerebral basis of lateral asymmetries in attention. *Acta Psychologica.* 1970;33:193–201. Available from: https://doi.org/10.1016/0001-6918(70)90132-0.
21. Bryden M, Munhall K, Allard F. Attentional biases and the right-ear effect in dichotic listening. *Brain Lang.* 1983;18(2):236–248. Available from: https://doi.org/10.1016/0093-934x(83)90018-4.
22. Hugdahl K, Andersson L. The "forced-attention paradigm" in dichotic listening to CV-syllables: a comparison between adults and children. *Cortex.* 1986;22(3):417–432. Available from: https://doi.org/10.1016/s0010-9452(86)80005-3.

23. Bless JJ, Westerhausen R, Arciuli J, Kompus K, Gudmundsen M, Hugdahl K. "Right on all occasions?"—On the feasibility of laterality research using a smartphone dichotic listening application. *Front Psychol.* 2013;4:42. Available from: https://doi.org/10.3389/fpsyg.2013.00042.
24. Bless JJ, Westerhausen R, von Koss Torkildsen J, Gudmundsen M, Kompus K, Hugdahl K. Laterality across languages: results from a global dichotic listening study using a smartphone application. *Laterality.* 2015;20(4):434–452. Available from: https://doi.org/10.1080/1357650X.2014.997245.
25. Parker AJ, Woodhead ZVJ, Thompson PA, Bishop DVM. Assessing the reliability of an online behavioural laterality battery: a pre-registered study. *Laterality.* 2021;26(4):359–397. Available from: https://doi.org/10.1080/1357650X.2020.1859526.
26. Kolb B, Whishaw IQ. 3. printing *Fundamentals of Human Neuropsychology.* 5th ed. Worth Publ; 2003.
27. Bourne VJ. The divided visual field paradigm: methodological considerations. *Laterality.* 2006;11(4):373–393. Available from: https://doi.org/10.1080/13576500600633982.
28. Hausmann M. Functional cerebral asymmetries during the menstrual cycle: a cross-sectional and longitudinal analysis. *Neuropsychologia.* 2002;40(7):808–816. Available from: https://doi.org/10.1016/s0028-3932(01)00179-8.
29. Hausmann M, Güntürkün O. Steroid fluctuations modify functional cerebral asymmetries: the hypothesis of progesterone-mediated interhemispheric decoupling. *Neuropsychologia.* 2000;38(10):1362–1374. Available from: https://doi.org/10.1016/s0028-3932(00)00045-2.
30. Hausmann M, Güntürkün O. Sex differences in functional cerebral asymmetries in a repeated measures design. *Brain Cognition.* 1998;41(3):263–275. Available from: https://doi.org/10.1006/brcg.1999.1126.
31. Measso G, Zaidel E. Effect of response programming on hemispheric differences in lexical decision. *Neuropsychologia.* 1990;28(7):635–646. Available from: https://doi.org/10.1016/0028-3932(90)90118-8.
32. van der Haegen L, Cai Q, Seurinck R, Brysbaert M. Further fMRI validation of the visual half field technique as an indicator of language laterality: a large-group analysis. *Neuropsychologia.* 2011;49(10):2879–2888. Available from: https://doi.org/10.1016/j.neuropsychologia.2011.06.014.
33. Knecht S, Deppe M, Dräger B, et al. Language lateralization in healthy right-handers. *Brain.* 2000;123(Pt 1):74–81. Available from: https://doi.org/10.1093/brain/123.1.74.
34. Knecht S, Dräger B, Deppe M, et al. Handedness and hemispheric language dominance in healthy humans. *Brain.* 2000;123(Pt 12):2512–2518. Available from: https://doi.org/10.1093/brain/123.12.2512.
35. Flöel A, Buyx A, Breitenstein C, Lohmann H, Knecht S. Hemispheric lateralization of spatial attention in right- and left-hemispheric language dominance. *Behav Brain Res.* 2005;158(2):269–275. Available from: https://doi.org/10.1016/j.bbr.2004.09.016.
36. Papadatou-Pastou M, Sampanis P, Koumzis I, et al. Cerebral laterality of writing in right- and left-handers: a functional transcranial Doppler ultrasound study. *Eur J Neurosci.* 2022;56(2):3921–3937. Available from: https://doi.org/10.1111/ejn.15723.
37. Hugdahl K, Brønnick K, Kyllingsbrk S, Law I, Gade A, Paulson OB. Brain activation during dichotic presentations of consonant-vowel and musical instrument stimuli: a 15 O-PET study. *Neuropsychologia.* 1999;37(4):431–440. Available from: https://doi.org/10.1016/S0028-3932(98)00101-8.
38. Huettel SA, Song AW, McCarthy G. *Functional Magnetic Resonance Imaging.* 3rd ed. Sinauer Associates. Inc. Publishers; 2014.
39. Boubela RN, Kalcher K, Huf W, et al. fMRI measurements of amygdala activation are confounded by stimulus correlated signal fluctuation in nearby veins draining distant brain regions. *Sci Rep.* 2015;5(1)10499. Available from: https://doi.org/10.1038/srep10499.
40. Campbell JM, Kundu B, Lee JN, et al. Evaluating the concordance of functional MRI-based language lateralization and Wada testing in epilepsy patients: a single-center analysis. *Interv Neuroradiol.* 2022;. Available from: https://doi.org/10.1177/15910199221121384.
41. Price CJ. A review and synthesis of the first 20 years of PET and fMRI studies of heard speech, spoken language and reading. *NeuroImage.* 2012;62(2):816–847. Available from: https://doi.org/10.1016/j.neuroimage.2012.04.062.
42. Badzakova-Trajkov G, Häberling IS, Roberts RP, Corballis MC. Cerebral asymmetries: complementary and independent processes. *PLoS One.* 2010;5(3):e9682. Available from: https://doi.org/10.1371/journal.pone.0009682.
43. Ocklenburg S, Hugdahl K, Westerhausen R. Structural white matter asymmetries in relation to functional asymmetries during speech perception and production. *NeuroImage.* 2013;83:1088–1097. Available from: https://doi.org/10.1016/j.neuroimage.2013.07.076.

44. Binder JR, Swanson SJ, Hammeke TA, Sabsevitz DS. A comparison of five fMRI protocols for mapping speech comprehension systems. *Epilepsia*. 2008;49(12):1980–1997. Available from: https://doi.org/10.1111/j.1528-1167.2008.01683.x.
45. Specht K, Reul J. Functional segregation of the temporal lobes into highly differentiated subsystems for auditory perception: an auditory rapid event-related fMRI-task. *NeuroImage*. 2003;20(4):1944–1954. Available from: https://doi.org/10.1016/j.neuroimage.2003.07.034.
46. Malik-Moraleda S, Ayyash D, Gallée J, et al. An investigation across 45 languages and 12 language families reveals a universal language network. *Nat Neurosci*. 2022;25(8):1014–1019. Available from: https://doi.org/10.1038/s41593-022-01114-5.
47. Jäncke L, Shah NJ. Does dichotic listening probe temporal lobe functions? *Neurology*. 2002;58(5):736–743. Available from: https://doi.org/10.1212/WNL.58.5.736.
48. Jäncke L, Buchanan TW, Lutz K, Shah NJ. Focused and nonfocused attention in verbal and emotional dichotic listening: an FMRI study. *Brain Lang*. 2001;78(3):349–363. Available from: https://doi.org/10.1006/brln.2000.2476.
49. Sandmann P, Eichele T, Specht K, et al. Hemispheric asymmetries in the processing of temporal acoustic cues in consonant-vowel syllables. *Restor Neurol Neurosci*. 2007;25(3–4):227–240. Available from: https://content.iospress.com/articles/restorative-neurology-and-neuroscience/rnn253405.
50. Westerhausen R, Kompus K, Hugdahl K. Mapping hemispheric symmetries, relative asymmetries, and absolute asymmetries underlying the auditory laterality effect. *NeuroImage*. 2014;84:962–970. Available from: https://doi.org/10.1016/j.neuroimage.2013.09.074.
51. Kompus K, Specht K, Ersland L, et al. A forced-attention dichotic listening fMRI study on 113 subjects. *Brain Lang*. 2012;121(3):240–247. Available from: https://doi.org/10.1016/j.bandl.2012.03.004.
52. Bethmann A, Tempelmann C, de Bleser R, Scheich H, Brechmann A. Determining language laterality by fMRI and dichotic listening. *Brain Res*. 2007;1133(1):145–157. Available from: https://doi.org/10.1016/j.brainres.2006.11.057.
53. Segal E, Petrides M. The anterior superior parietal lobule and its interactions with language and motor areas during writing. *Eur J Neurosci*. 2012;35(2):309–322. Available from: https://doi.org/10.1111/j.1460-9568.2011.07937.x.
54. Luck SJ. *An Introduction to the Event-Related Potential Technique. 2nd ed*. MIT Press; 2014.
55. Galin D, Ornstein R. Lateral specialization of cognitive mode: an EEG study. *Psychophysiology*. 1972;9(4):412–418. Available from: https://doi.org/10.1111/j.1469-8986.1972.tb01788.x.
56. Herrmann CS, Knight RT. Mechanisms of human attention: event-related potentials and oscillations. *Neurosci Biobehav Rev*. 2001;25(6):465–476. Available from: https://doi.org/10.1016/s0149-7634(01)00027-6.
57. Beste C, Baune BT, Falkenstein M, Konrad C. Variations in the TNF-α gene (TNF-α -308G → A) affect attention and action selection mechanisms in a dissociated fashion. *J Neurophysiol*. 2010;104(5):2523–2531. Available from: https://doi.org/10.1152/jn.00561.2010.
58. Beste C, Saft C, Andrich J, Gold R, Falkenstein M. Stimulus-response compatibility in Huntington's disease: a cognitive-neurophysiological analysis. *J Neurophysiol*. 2008;99(3):1213–1223. Available from: https://doi.org/10.1152/jn.01152.2007.
59. Hillyard SA, Anllo-Vento L. Event-related brain potentials in the study of visual selective attention. *Proc Natl Acad Sci U S A*. 1998;95(3):781–787. Available from: https://doi.org/10.1073/pnas.95.3.781.
60. Wascher E, Beste C. Tuning perceptual competition. *J Neurophysiol*. 2010;103(2):1057–1065. Available from: https://doi.org/10.1152/jn.00376.2009.
61. Grossi G, Savill N, Thomas E, Thierry G. Posterior N1 asymmetry to English and Welsh words in Early and Late English-Welsh bilinguals. *Biol Psychol*. 2010;85(1):124–133. Available from: https://doi.org/10.1016/j.biopsycho.2010.06.003.
62. Gomez Gonzalez CM, Clark VP, Fan S, Luck SJ, Hillyard SA. Sources of attention-sensitive visual event-related potentials. *Brain Topogr*. 1994;7(1):41–51. Available from: https://doi.org/10.1007/BF01184836.
63. Yamazaki T, Kamijo K, Kenmochi A, et al. Multiple equivalent current dipole source localization of visual event-related potentials during oddball paradigm with motor response. *Brain Topogr*. 2000;12(3):159–175. Available from: https://doi.org/10.1023/A:1023467806268.
64. Spironelli C, Angrilli A. Developmental aspects of automatic word processing: language lateralization of early ERP components in children, young adults and middle-aged subjects. *Biol Psychol*. 2009;80(1):35–45. Available from: https://doi.org/10.1016/j.biopsycho.2008.01.012.

65. Spironelli C, Angrilli A. Influence of Phonological, Semantic and Orthographic tasks on the early linguistic components N150 and N350. *Int J Psychophysiol*. 2007;64(2):190–198. Available from: https://doi.org/10.1016/j.ijpsycho.2007.02.002.
66. Proverbio AM, Cok B, Zani A. Electrophysiological measures of language processing in bilinguals. *J Cognit Neurosci*. 2002;14(7):994–1017. Available from: https://doi.org/10.1162/089892902320474463.
67. Eichele T, Nordby H, Rimol LM, Hugdahl K. Asymmetry of evoked potential latency to speech sounds predicts the ear advantage in dichotic listening. *Cognit Brain Res*. 2005;24(3):405–412. Available from: https://doi.org/10.1016/j.cogbrainres.2005.02.017.
68. Broca P. Remarques sur le siège de la faculté du langage articulé, suivies d'une observation d'aphémie (perte de la parole). *Bull et Memoires de la Societe anatomique de Paris*. 1861;6:330–357.
69. Wernicke C. *Der aphasische Symptomencomplex: Eine psychologische Studie auf anatomischer Basis*. M. Cohn & Weigert; 1874.
70. Hickok G, Venezia J, Teghipco A. Beyond Broca: neural architecture and evolution of a dual motor speech coordination system. *Brain*. 2022;. Available from: https://doi.org/10.1093/brain/awac454.
71. Schönwiesner M, Rübsamen R, von Cramon DY. Spectral and temporal processing in the human auditory cortex–revisited. *Ann N Y Acad Sci*. 2005;1060(1):89–92. Available from: https://doi.org/10.1196/annals.1360.051.
72. Zatorre RJ. Hemispheric asymmetries for music and speech: spectrotemporal modulations and top-down influences. *Front Neurosci*. 2022;161075511. Available from: https://doi.org/10.3389/fnins.2022.1075511.
73. Zatorre RJ, Belin P, Penhune VB. Structure and function of auditory cortex: music and speech. *Trends Cognit Sci*. 2002;6(1):37–46. Available from: https://doi.org/10.1016/s1364-6613(00)01816-7.
74. Hutsler J, Galuske RA. Hemispheric asymmetries in cerebral cortical networks. *Trends Neurosci*. 2003;26(8):429–435. Available from: https://doi.org/10.1016/s0166-2236(03)00198-x.
75. Hutsler J. The specialized structure of human language cortex: pyramidal cell size asymmetries within auditory and language-associated regions of the temporal lobes. *Brain Lang*. 2003;86(2):226–242. Available from: https://doi.org/10.1016/s0093-934x(02)00531-x.
76. Buxhoeveden DP, Switala AE, Litaker M, Roy E, Casanova MF. Lateralization of minicolumns in human planum temporale is absent in nonhuman primate cortex. *BBE*. 2001;57(6):349–358. Available from: https://doi.org/10.1159/000047253.
77. Galuske RA, Schlote W, Bratzke H, Singer W. Interhemispheric asymmetries of the modular structure in human temporal cortex. *Science*. 2000;289(5486):1946–1949. Available from: https://doi.org/10.1126/science.289.5486.1946.
78. Ocklenburg S, Friedrich P, Fraenz C, et al. Neurite architecture of the planum temporale predicts neurophysiological processing of auditory speech. *Sci Adv*. 2018;4(7)eaar6830. Available from: https://doi.org/10.1126/sciadv.aar6830.
79. Tzourio-Mazoyer N, Crivello F, Mazoyer B. Is the planum temporale surface area a marker of hemispheric or regional language lateralization? *Brain Struct Funct*. 2018;223(3):1217–1228. Available from: https://doi.org/10.1007/s00429-017-1551-7.
80. Friederici AD. Pathways to language: fiber tracts in the human brain. *Trends Cognit Sci*. 2009;13(4):175–181. Available from: https://doi.org/10.1016/j.tics.2009.01.001.
81. Häberling IS, Badzakova-Trajkov G, Corballis MC. Asymmetries of the arcuate fasciculus in monozygotic twins: genetic and nongenetic influences. *PLoS One*. 2013;8(1):e52315. Available from: https://doi.org/10.1371/journal.pone.0052315.
82. Catani M, Allin MPG, Husain M, et al. Symmetries in human brain language pathways correlate with verbal recall. *Proc Natl Acad Sci U S A*. 2007;104(43):17163–17168. Available from: https://doi.org/10.1073/pnas.0702116104.
83. Hubl D, Koenig T, Strik W, et al. Pathways that make voices: white matter changes in auditory hallucinations. *Arch Gen Psychiatry*. 2004;61(7):658–668. Available from: https://doi.org/10.1001/archpsyc.61.7.658.
84. Lindell AK. In your right mind: right hemisphere contributions to language processing and production. *Neuropsychol Rev*. 2006;16(3):131–148. Available from: https://doi.org/10.1007/s11065-006-9011-9.
85. Brysbaert M, Cai Q, van der Haegen L. Brain asymmetry and visual word recognition. In: Adelman JS, ed. *Visual Word Recognition: Models and Methods, Orthography and Phonology*. Psychology Press; 2012:139–158.
86. Cai Q, van der Haegen L. What can atypical language hemispheric specialization tell us about cognitive functions? *Neurosci Bull*. 2015;31(2):220–226. Available from: https://doi.org/10.1007/s12264-014-1505-5.
87. Knecht S, Flöel A, Dräger B, et al. Degree of language lateralization determines susceptibility to unilateral brain lesions. *Nat Neurosci*. 2002;5(7):695–699. Available from: https://doi.org/10.1038/nn868.

88. Knecht S, Dräger B, Flöel A, et al. Behavioural relevance of atypical language lateralization in healthy subjects. *Brain*. 2001;124(Pt 8):1657–1665. Available from: https://doi.org/10.1093/brain/124.8.1657.
89. Mazoyer B, Zago L, Jobard G, et al. Gaussian mixture modeling of hemispheric lateralization for language in a large sample of healthy individuals balanced for handedness. *PLoS One*. 2014;9(6):e101165. Available from: https://doi.org/10.1371/journal.pone.0101165.
90. Greve DN, van der Haegen L, Cai Q, et al. A surface-based analysis of language lateralization and cortical asymmetry. *J Cognit Neurosci*. 2013;25(9):1477–1492. Available from: https://doi.org/10.1162/jocn_a_00405.
91. Leroy F, Cai Q, Bogart SL, et al. New human-specific brain landmark: the depth asymmetry of superior temporal sulcus. *PNAS*. 2015;112(4):1208–1213. Available from: https://doi.org/10.1073/pnas.1412389112.
92. Kennedy DN, O'Craven KM, Ticho BS, Goldstein AM, Makris N, Henson JW. Structural and functional brain asymmetries in human situs inversus totalis. *Neurology*. 1999;53(6):1260–1265. Available from: https://doi.org/10.1212/WNL.53.6.1260.
93. Schuler A-L, Kasprian G, Schwartz E, et al. Mens inversus in corpore inverso? Language lateralization in a boy with situs inversus totalis. *Brain Lang*. 2017;174:9–15. Available from: https://doi.org/10.1016/j.bandl.2017.06.005.
94. van der Haegen L, Acke F, Vingerhoets G, et al. Laterality and unilateral deafness: patients with congenital right ear deafness do not develop atypical language dominance. *Neuropsychologia*. 2016;93(Pt B):482–492. Available from: https://doi.org/10.1016/j.neuropsychologia.2015.10.032.
95. Parker AJ, Woodhead ZVJ, Carey DP, et al. Inconsistent language lateralisation—testing the dissociable language laterality hypothesis using behaviour and lateralised cerebral blood flow. *Cortex*. 2022;154:105–134. Available from: https://doi.org/10.1016/j.cortex.2022.05.013.
96. Woodhead ZVJ, Bradshaw AR, Wilson AC, Thompson PA, Bishop DVM. Testing the unitary theory of language lateralization using functional transcranial Doppler sonography in adults. *R Soc Open Sci*. 2019;6(3):181801. Available from: https://doi.org/10.1098/rsos.181801.
97. Woodhead ZVJ, Thompson PA, Karlsson EM, Bishop DVM. An updated investigation of the multidimensional structure of language lateralization in left- and right-handed adults: a test-retest functional transcranial Doppler sonography study with six language tasks. *R Soc Open Sci*. 2021;8(2):200696. Available from: https://doi.org/10.1098/rsos.200696.
98. Hull R, Vaid J. Bilingual language lateralization: a meta-analytic tale of two hemispheres. *Neuropsychologia*. 2007;45(9):1987–2008. Available from: https://doi.org/10.1016/j.neuropsychologia.2007.03.002.
99. Genesee F, Hamers J, Lambert WE, Mononen L, Seitz M, Starck R. Language processing in bilinguals. *Brain Lang*. 1978;5(1):1–12. Available from: https://doi.org/10.1016/0093-934x(78)90002-0.
100. Galloway LM, Scarcella R. Cerebral organization in adult second language acquisition: Is the right hemisphere more involved? *Brain Lang*. 1982;16(1):56–60. Available from: https://doi.org/10.1016/0093-934x(82)90071-2.
101. Bishop DVM, Grabitz CR, Harte SC, et al. Cerebral lateralisation of first and second languages in bilinguals assessed using functional transcranial Doppler ultrasound. *Wellcome Open Res*. 2016;1:15. Available from: https://doi.org/10.12688/wellcomeopenres.9869.2.
102. Gu F, Zhang C, Hu A, Zhao G. Left hemisphere lateralization for lexical and acoustic pitch processing in Cantonese speakers as revealed by mismatch negativity. *NeuroImage*. 2013;83:637–645. Available from: https://doi.org/10.1016/j.neuroimage.2013.02.080.
103. Best CT, Avery RA. Left-hemisphere advantage for click consonants is determined by linguistic significance and experience. *Psychol Sci*. 1999;10(1):65–70. Available from: https://doi.org/10.1111/1467-9280.00108.
104. Marsolek CJ, Deason RG. Hemispheric asymmetries in visual word-form processing: progress, conflict, and evaluating theories. *Brain Lang*. 2007;103(3):304–307. Available from: https://doi.org/10.1016/j.bandl.2007.02.009. author reply 308-12.
105. Güntürkün O, Güntürkün M, Hahn C. Whistled Turkish alters language asymmetries. *Curr Biol*. 2015;25(16):R706–R708. Available from: https://doi.org/10.1016/j.cub.2015.06.067.
106. Villar González P, Güntürkün O, Ocklenburg S. Lateralization of auditory processing of silbo gomero. *Symmetry*. 2020;12(7):1183. Available from: https://doi.org/10.3390/sym12071183.
107. Hickok G, Bellugi U, Klima ES. The neurobiology of sign language and its implications for the neural basis of language. *Nature*. 1996;381(6584):699–702. Available from: https://doi.org/10.1038/381699a0.
108. MacSweeney M, Capek CM, Campbell R, Woll B. The signing brain: the neurobiology of sign language. *Trends Cognit Sci*. 2008;12(11):432–440. Available from: https://doi.org/10.1016/j.tics.2008.07.010.

CHAPTER

6

Handedness and other behavioral asymmetries

OUTLINE

The Lateralized Brain
DOI: https://doi.org/10.1016/B978-0-323-99737-9.00012-4

Introductory short story

Growing up in Germany in the 1950s, Anneliese was a bright and cheerful child. In kindergarten, she started drawing pictures earlier than most other children, always using her left hand to hold the pencil or paintbrush. No one objected to her left-handedness until she got to primary school and started writing exercises. To avoid smudging the ink with her left hand when writing, she would hold her hand in a different position than her teacher Miss Walkenhorst had instructed her. Miss Walkenhorst did not like this. Sometimes, when Anneliese was concentrating on writing a complicated word correctly, the teacher would hit the back of her head with her wooden ruler. "Ouch" cried Anneliese and turned her head toward the teacher, tears in her eyes. Miss Walkenhorst smiled at her. "How many times do I need to tell you? Write with your GOOD hand." Anneliese would then try to write with her right hand for a while, but after a few minutes, her right hand would start cramping terribly, making her writing unreadable. The situation became even more serious at the end of second grade when Miss Walkenhorst asked Anneliese's parents to visit her at school and told them that she would not allow their daughter to progress to third grade unless they managed to get her to write with the proper hand. Anneliese's parents were not happy, but what could they do? They paid a private tutor for private extra writing exercises with the right hand at home. After a while, Anneliese was able to force herself to write with the right hand without as many cramps as before. While she would continue to use her right hand for writing as an adult, she never much enjoyed it. She continued to use the left hand for drawing and all other fine motor tasks. Even decades later, years after she had retired from her position as an emergency room nurse, Anneliese would sometimes wake in the middle of the night, hearing those all-too-familiar words: Write with your GOOD hand, Anneliese!

Based on actual events.

Measuring handedness

When Barack Obama was elected as the 44th president of the United States of America, several magazines, newspapers, and websites published articles about the fact that he, like many US presidents before him, is left-handed. Like Barack Obama, roughly 1 person out of 10 is left-handed.[1] The exact prevalence of left-handedness was determined in a set of meta-analyses based on data from more than 2.3 million individuals from 200 different studies.[2] The research team, led by Marietta Papadatou-Pastou from the National and Kapodistrian University of Athens, Greece, found that the best overall estimate for left-handedness in the population was 10.6%, with a range between 9.3% and 18.3%. Interestingly, this clear asymmetry in behavioral preference is not paralleled by any distinct asymmetries in biomechanical features of the two hands—for example, in muscle strength. Handedness thus is a manifestation of varying skills between the left and right hand, reflecting hemispheric asymmetry in motor areas of the brain.[3,4] Since handedness affects every single person in the world, there has been great interest in the topic as reflected by several books that have been published on the psychology and neuroscience of left-handedness.[4–7] Recently, also a philosophical examination of left-handedness was published, highlighting how handedness shapes experience and inequality.[8]

While the assessment of handedness seems trivial at first glance, there are several methodological approaches to determining an individual's handedness. First, there is the important distinction between hand preference and hand performance[9]:

- *Hand preference* refers to the individual's subjectively preferred hand for manual activities and is typically assessed using questionnaires.
- *Hand performance* refers to the individual's objectively measured skills in manual activities.

Hand preference

The easiest way to determine subjective hand preference is a dichotomous differentiation between left- and right-handers which can, for example, be based on an assessment of the hand preference for writing or drawing (see Fig. 6.1).

This approach is most commonly used in studies on handedness in which time constraints limit the amount of time available for data collection from each participant.[10,11]

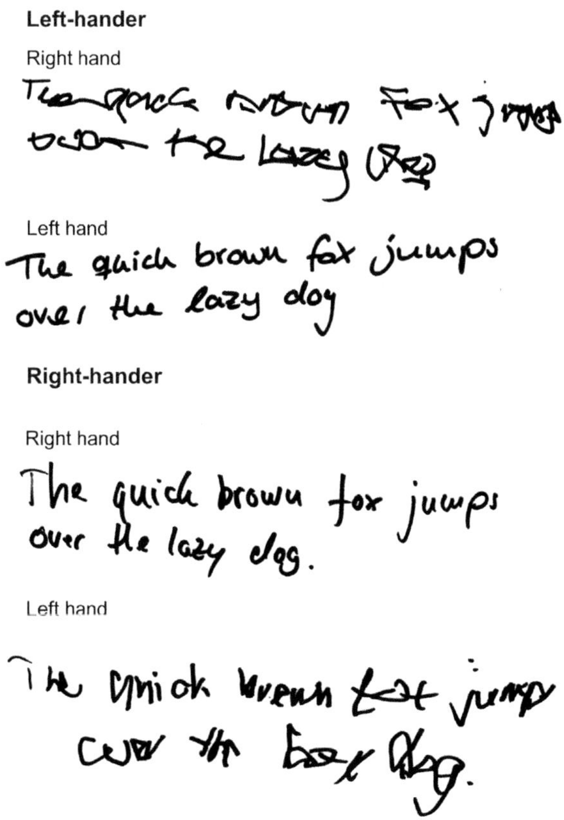

FIGURE 6.1 Right- and left-handed handwriting of a left-hander (upper panel) and a right-hander (lower panel).

However, use of this method can lead to biased results under some circumstances, for example, if left-handers who have been forced to learn to write with their right hand (see introductory short story) are included in the sample.[12] Aside from this qualitative measure, several quantitative measures of hand preference have been developed in order to allow for assessment not only of the direction but also of the strength of hand preference. The most widely used tool is the Edinburgh Handedness Inventory (EHI) (see Box 6.1), published by R.C. Oldfield from the University of Edinburgh, Scotland, in 1971.[13] To date, variants of this 10-item questionnaire have been used in several hundred different studies, and James M. Edlin from the University of Louisville, USA, reported that there is a substantial diversity between handedness studies regarding the addition or omission of test items, the response format, and the criteria applied to classify individuals as right-handed or left-handed.[14] Since such inconsistent use of the EHI may lead to conflicting results in the scientific community, we show the original version of the questionnaire from 1971 despite several more recent suggestions on how to optimize it.[15–17]

The LQ you have just calculated in Box 6.1 has a range between −100 and +100.

BOX 6.1

Test your own hand preference!

The EHI[13] (Table 6.1) is the handedness questionnaire most widely used in scientific research. Here, you can try it out yourself and calculate your laterality quotient (LQ) to determine your own hand preference.

TABLE 6.1 The Edinburgh handedness questionnaire.[13]

		Left	Right
1	Writing		
2	Drawing		
3	Throwing		
4	Scissors		
5	Toothbrush		
6	Knife (without fork)		
7	Spoon		
8	Broom (upper hand)		
9	Striking match (match)		
10	Opening box (lid)		
Sum left (L):	**Sum right (R):**		
R − L =	**R + L =**		
Lateralization quotient (LQ): $LQ = [(R - L)/(R + L)] \times 100 =$			

Reprinted with permission of the publisher.

BOX 6.1 (*cont'd*)

Instructions:

Please indicate your preferences in the use of hands in the following activities by putting + in the appropriate column. Where the preference is so strong that you would never try to use the other hand unless absolutely forced to, put + +. If in any case you are really indifferent, put + in both columns. Some of the activities require both hands. In these cases, the part of the task or object for which hand preference is wanted is indicated in brackets. Please try to answer all the questions, and only leave a blank if you have no experience at all of the object or task.

From.[13]

When you have filled in all boxes, calculate the sum of pluses you made for the left hand (L) and the sum of pluses you made for the right hand (R). Now you can calculate your LQ using the formula:

$$LQ = [(R - L)]/[(R + L)] \times 100.$$

First, calculate the dividend of the formula by subtracting L from R and then the divisor by adding R and L. Calculate the quotient and multiply it by 100 and you have calculated your LQ.

Positive values indicate right-sided preference and negative values indicate left-sided preference. Higher values indicate a more consistent handedness. Individuals with an LQ of +100 always use their right and never their left hand for all tested activities, while individuals with an LQ of −100 always use their left and never their right hand for all tested activities. Individuals with an LQ of 0 are completely ambidextrous and are equally likely to use both hands. Typically, LQ data show a J-shaped distribution that is skewed to the right, with most participants showing high positive values, some showing high negative values, and only a few showing values around 0—see Fig. 6.2 for an exemplary distribution from 1050 participants tested by the authors. Typically, right-handers are more often extremely consistent—that is, they do everything with the right hand—than left-handers.

Based on their LQ, individuals can be grouped into different categories (see Fig. 6.3), with the simplest model being that participants with an LQ larger than zero are considered right-handed and participants with an LQ smaller than zero are considered left-handed. Compared to assessment of handedness based on writing preference, this method has the advantage that left-handers who have been forced to write with the right hand are considered left-handers instead of right-handers. Typically, they show negative LQs because they tend to use their left hands for most nonwriting activities. As their genetic predisposition is likely toward left-handedness even though they use their right hand for writing, this method is preferable over the assessment of the writing hand for genetic studies of handedness.

Aside from this dichotomous differentiation between left- and right-handers, several finer categorization methods for LQ data have been proposed. For example, it has been suggested to distinguish so-called *consistent*, or strong, left- and right-handers—individuals, respectively, with an LQ of −100 and +100—from *moderate*—individuals with an absolute LQ between 60 and 99—and *weak*—individuals with an absolute LQ below 60—left- and right-handers.[18]

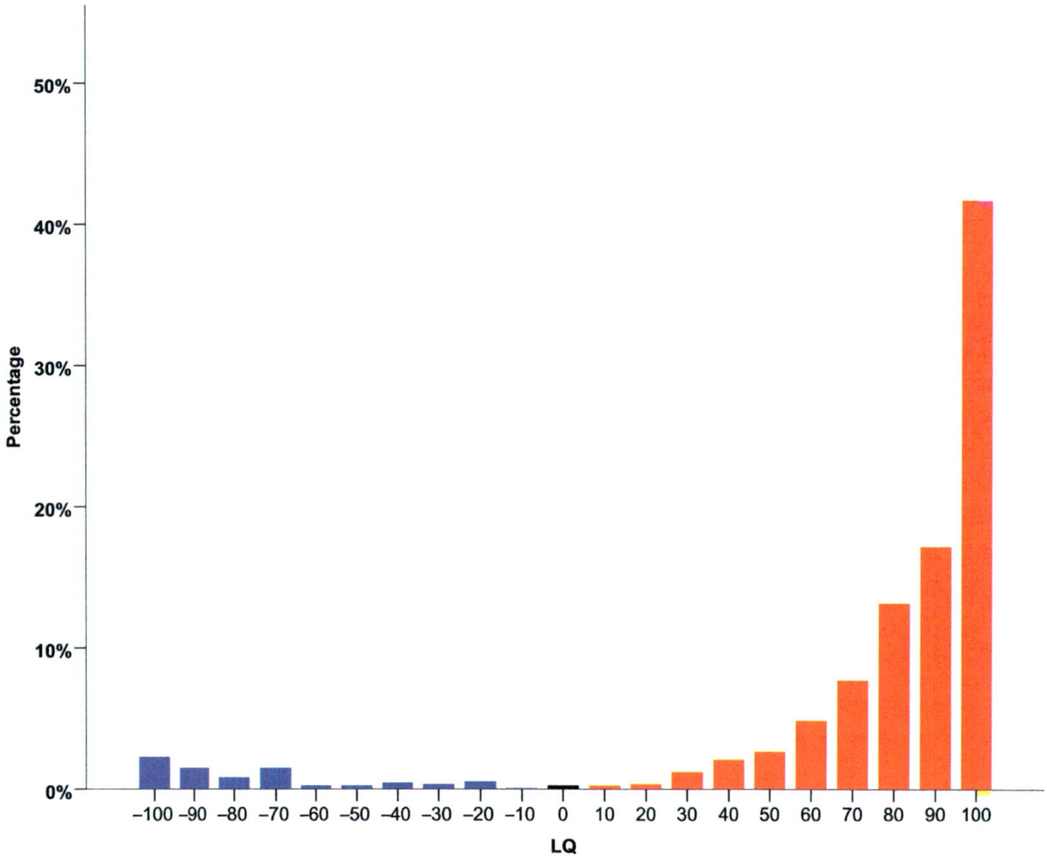

FIGURE 6.2 The typical J-shaped distribution of LQ data obtained from the EHI exemplified in a dataset of 1050 participants tested by the authors. Blue bars show left-handers, and red bars show right-handers.

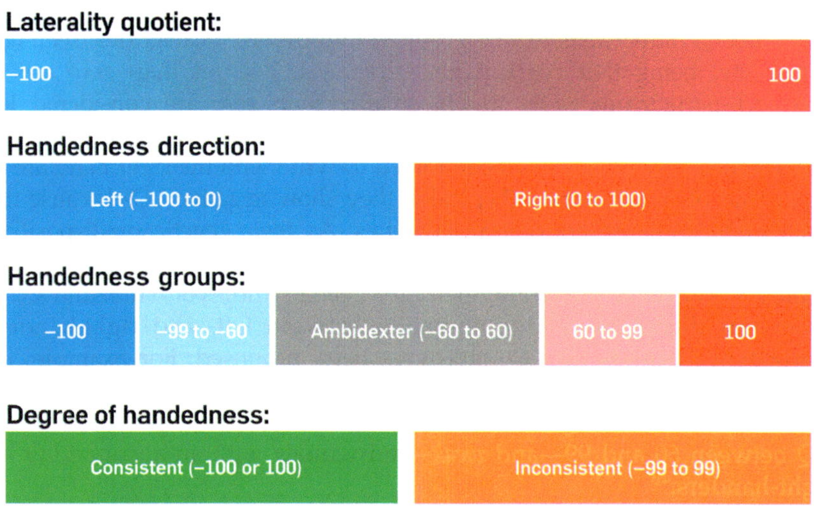

FIGURE 6.3 Typical categorization schemes for LQ data.

It has also been proposed to classify individuals with a low absolute LQ score—for example, less than 60—not as weak left- or right-handers but as mixed-handers or ambidexters.[19] In the abovementioned meta-analysis, all 72 studies in which participants were classified as left-handers, right-handers, or mixed-handers were analyzed in a separate analysis to determine the prevalence of mixed-handedness.[2] In this analysis, the prevalence of mixed-handedness was 9.33%. Mixed-handedness is particularly interesting in the context of clinical laterality research, as for several mental or neurodevelopmental disorders, patients show elevated rates of mixed-handedness.[20,21] These findings will be discussed in detail in the last chapter of this book.

Moreover, it has been suggested to differentiate consistent handedness—for example, individuals who always use one hand for all activities and thus have an LQ of either −100 or +100—from inconsistent handedness (LQ between −99 and +99). This classification is supported by studies showing that especially consistent and inconsistent left-handers show distinct performance patterns on a number of cognitive tests.[22,23]

Aside from the EHI, several other handedness questionnaires have been developed to assess hand preference. These include (but are certainly not limited to):

- The Flinders Handedness survey (FLANDERS)[24]
- The Crovitz and Zener questionnaire[25,26]
- The Waterloo Handedness Questionnaire[27]
- The Coren Lateral Preference Inventory[28,29]
- The Annett hand preference questionnaire[30]

Moreover, under some circumstances, it might not be feasible to assess hand preference using a questionnaire—for example, in children who are not yet able to read—so that behavioral methods have to be applied. For example, the WatHand Cabinet Test[31] involves several different activities that have to be performed—for example, using a toy hammer or tossing a ball to a target. Based on the hand used for each activity, a skill score similar to the LQ can be calculated for each participant. Similarly, some authors asked their participants to perform the tasks described in the EHI.[32]

Hand performance

In contrast to hand preference, hand performance refers to objectively measurable differences in fine motor skills between the two hands. Several different tests have been used to test left-right differences in hand performance. One of the most commonly used tests is the so-called Pegboard Task.[33] In this task, participants are asked several times to move a row of 10 pegs from 10 holes on one side of a board to 10 holes on the other side of that board (see Fig. 6.4).

Hand performance is measured by comparing reaction times for the left and right hand. Often the so-called PegQ is calculated in order to gain a quantitative measure of hand performance. This is done using the formula $\text{PegQ} = [2 \times (L - R)/(L + R)]$, with L indicating the average time the participant needed to move all 10 pegs with the left hand and R indicating the average time the participant needed to move all 10 pegs with the right hand. Negative values indicate that the participant was faster with their left hand, and positive values indicate that the participant was faster with their right hand. Interestingly, the EHI and the Pegboard Task yield significantly different distributions. While hand preference

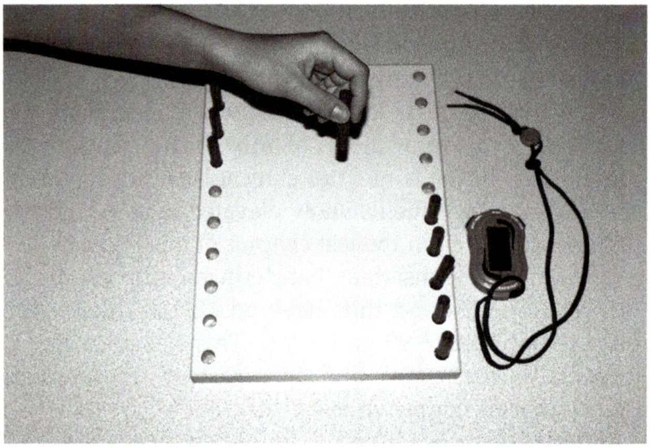

FIGURE 6.4 The Pegboard task.

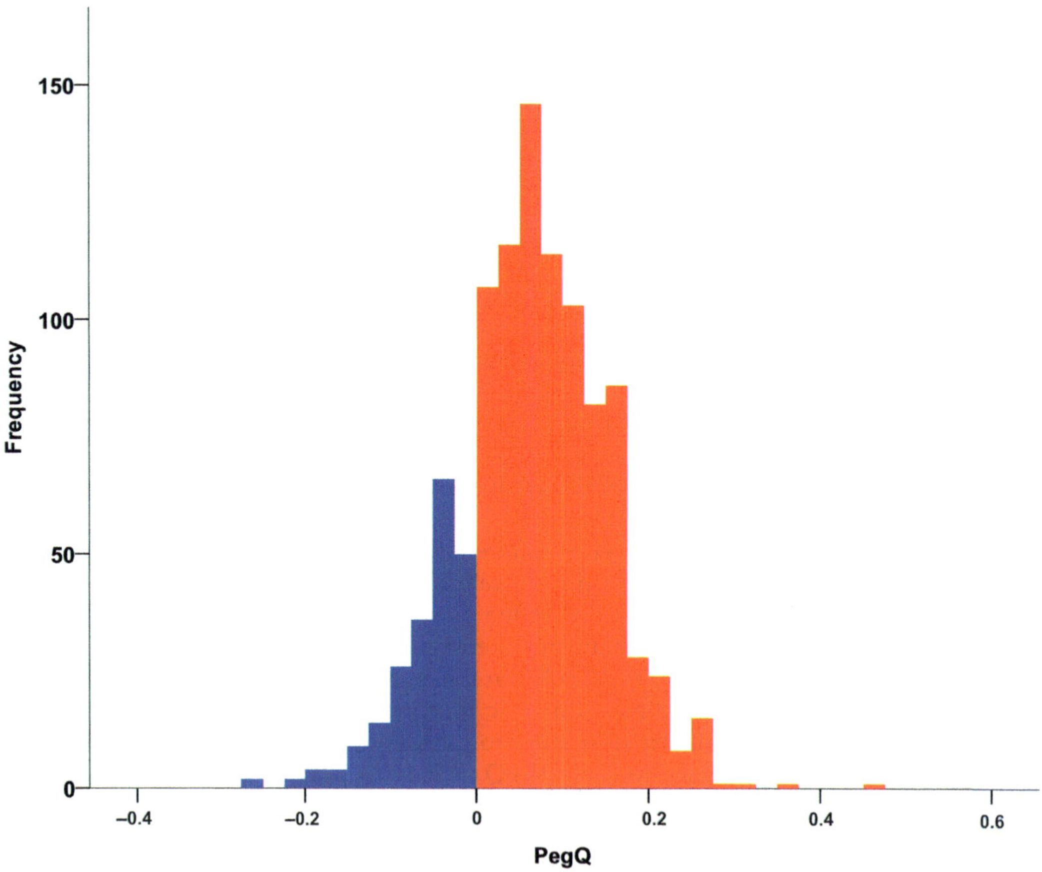

FIGURE 6.5 The typical bell-shaped distribution of PegQ-data obtained from the Pegboard task exemplified in a dataset of 1050 participants tested by the authors. Blue bars show left-handers, and red bars show right-handers.

typically has a J-shaped distribution, hand performance measured with the Pegboard Task typically shows a more bell-shaped distribution with a shift to the right side (see Fig. 6.5). While most individuals actually show better hand performance with their preferred hand, this difference in distribution makes it clear that the correlation between the two variables typically is significant but not very high.[34]

In addition to the Pegboard Task, several different methods have been used to measure hand performance:

- Finger tapping: Subjects are asked to tap with their index fingers on a tapping apparatus as fast as possible. Performance is compared between the left and the right hand.[35]
- The Tapley and Bryden test of hand performance: Subjects are presented with a series of circles (see Fig. 6.6) and are asked to draw a dot in each circle, following the pattern of circles. The task is performed twice with each hand with a 20 s time limit in place for each trial. A laterality index can be calculated using the formula (R − L)/(R + L), with R and L as the average numbers of circles correctly marked with the right and left hand.[36,37] There are also other versions of the dot marking test in which the dots follow different patterns.[38]

FIGURE 6.6 The Tapley and Bryden test of hand performance.

- Handwriting speed: Subjects are asked to write the alphabet as one word as quickly as possible. This is done three times with each hand, and reaction times are compared between the left and right hand.[39]
- Grip strength: Subjects are asked to make three maximum squeezes of a hand-held dynamometer, and average squeeze strength is compared between the two hands.[39]

The ontogenesis of handedness

As outlined in Chapter 1, quotas of left- and right-handers are stable across historical eras and geographical locations.[40] As mentioned above, 10.6% of the population are left-handed.[2] While it has been suggested that there is some geographical variation in the exact percentages,[41] there is no culture with a higher percentage of left- than right-handers. This remarkable stability of the trait has led to several partly bizarre theories explaining the prevalence of right-handedness as outlined in Chapter 1.

An intriguing idea was that handedness may be caused by the orientation of the inner organs. While individuals with situs solitus, the normal configuration of inner organs, were thought to be right-handed, individuals with situs inversus, a mirrored configuration of inner organs, were thought to be left-handed. Situs inversus is a rare condition, with about 1 in 10,000 people affected,[42] and it can be caused by several different defects during left-right differentiation of the body. Early anatomical evidence clearly speaks against the view that the asymmetry of the inner organs affects handedness. In 1950, Johan Torgersen from the Anatomical Institute of the University of Oslo in Norway analyzed 998,862 X-rays of Norwegian citizens.[43] This way, he identified 122 cases of situs inversus (about 1 in 10,000). In addition, Torgersen was able to identify 38 further cases in the published literature. Of these 160 cases of situs inversus, only 11 were left-handers (6.9%, a percentage lower than the population average). The results of this study, therefore, suggest that situs inversus does not affect handedness. A more recent genetic study, however, suggests that the specific subtype of situs inversus needs to be taken into account.[44] Situs inversus can occur with or without recessive primary ciliary dyskinesia. Primary ciliary dyskinesia is a rare disorder in which the cilia, hair-like cell organelles that are important for left-right distribution of proteins during embryogenesis, do not move properly. In the study, patients that had situs inversus but not primary ciliary dyskinesia showed an elevated rate of left-handedness, with five out of nine patients (about 56%) being left-handed. This shows that there may be a link between asymmetries of the inner organs and handedness, but that more research into subtypes of situs inversus is needed.

Aside from these and other early theories of handedness, one notion put forward by many authors is the idea that handedness is, at least to some extent, a heritable trait.[45] This notion appears plausible, given the geographical and historical predominance of right-handedness. Moreover, several other lines of research also strongly suggest genetic influences on handedness:

- Family studies: There is a positive correlation between children's handedness and that of their biological parents. In one study, two right-handed parents had a chance of 7% to raise a left-handed child. This percentage was increased to 21% in couples with one right- and one left-handed parent.[46] Another study found that two right-handed parents had a chance of 9.5% to raise a left-handed child, while couples with one left-handed parent had a chance of 19.5% to raise a left-handed child. Two left-handed parents had the highest chance of raising a left-handed child (about 26.1%).[33]

- Adoption studies: Louise Carter-Saltzman from the Department of Psychology at the University of Washington, USA, found that left-handedness in children was significantly related to left-handedness in their biological but not their adoptive parents. For example, if both biological parents were left-handed, the chance of their child being left-handed was 27%, but if both adoptive parents were left-handed, this chance was only 5%.[47]
- Twin studies: Identical twins are more likely to be concordant for hand preference than nonidentical twins,[48] a finding that has recently been confirmed by meta-analysis.[49]

Single gene models

These phenomena led several researchers to develop genetic models of handedness and hemispheric asymmetries. Before the widespread use of modern genetic techniques, development of these models was largely based on fitting the behavioral data into a classification scheme, without taking any molecular evidence into account. Monogenic models of handedness have been clearly refuted by recent molecular genetic evidence[50–52] and should not be used as an explanation for handedness ontogenesis anymore. However, given their historical significance, we will briefly describe them in this section.

There were two major single gene models, one of them the so-called Right Shift Theory developed by Marian Annett in the 1960s.[33] According to this model, right-handedness and left-hemispheric language dominance are under the control of a single hypothetical gene with two alleles (see Box 6.2 for a definition).

This claim is supported by the finding that more than 95% of right-handers also show left-hemispheric language dominance but fails to explain why at least 70% of left-handers also show left-hemispheric language dominance.[54,55] In Annett's[33] terminology, one allele called RS + (for right-shift +) leads to a relative advantage of the left over the right hemisphere, resulting in an increased chance of being right-handed and having left-hemispheric

BOX 6.2

Genes and alleles

A gene is the molecular unit of inheritance and comprises regulatory, transcribed, and other functional sequence regions within the DNA. Overall, there are about 22,000 genes in the human genome.[53] Alleles represent sequential variations of single genes. Alleles can differ only minimally in their nucleotide sequence but can nevertheless lead to different gene products and hence to different feature characteristics. Importantly, genes never directly impact behavioral phenotypes such as handedness; they are just code for proteins. These proteins, however, may affect development of specific parts of the brain—for example, organization of gray and white matter—which in turn control neuronal processing and ultimately behavior.

language dominance. In contrast, the other allele, called RS −, leads to handedness and language dominance being determined at random. This yields three possible genotypes:

- Homozygous RS− − individuals have a 50% chance of either being left- or right-handed with either left- or right-hemispheric language dominance.
- Heterozygous RS+ − individuals have an increased chance of being right-handed and left-hemispheric language dominance is increased.
- Homozygous RS+ + individuals have the highest chance of being right-handed and left-hemispheric language-dominant and only have a very slight chance of being left-handed or right-hemispheric language-dominant.

According to Annett,[33] the heterozygous RS+ − genotype is the most common genotype with a prevalence of about 49%. In comparison, the homozygous RS+ + genotype is supposed to have a prevalence of about 32%, and the homozygous RS− − genotype is supposed to have a prevalence of about 19%.

The other major single gene model of handedness was developed by Chris McManus from University College London, UK.[4] Similar to Annett, the initial version of this theory assumed that handedness and language lateralization are under the control of a single hypothetical gene with two alleles. In his terminology, one is called the C allele—for chance—while the other is called the D allele—for dextral. While the D allele increases the chance of being right-handed, the C allele leaves the direction of handedness up to chance. Along these lines, DD individuals are supposed to be right-handed in 100% of cases, while CC individuals have a 50% chance of being left- and a 50% chance of being right-handed. Since D and C alleles are assumed to be codominant, they both are believed to contribute to the phenotype, resulting in DC individuals having a 25% chance of being left-handed and a 75% chance of being right-handed. The differences between the two models are rather subtle, with the main discrepancy being that Annett predicts homozygous RS+ + individuals to have a slight chance of being left-handed, while McManus' model predicts that DD individuals are always right-handed. It is thus not surprising that both authors could show that their respective models were able to statistically fit the distributions of handedness and language lateralization in the population reasonably well.[56,57]

Aside from these two models, several other authors have developed genetic theories of handedness and hemispheric asymmetries. These include but are not limited to:

- The X-linked three-allele model for handedness.[58]
- The recessive X-linkage theory of handedness and language dominance.[59]
- The so-called "big bang" theory of handedness, hemispheric asymmetries in general, and schizophrenia.[60]

These models, however, have been criticized for a number of reasons and were never as well-established as the models proposed by Annett and McManus.

Molecular genetic research on handedness

While single-gene models reflect the statistical distribution of handedness data in the population reasonably well, they have always had a major drawback: At the time these theories were conceived, it was technically impossible to determine which actual gene in the human genome represented the elusive right-shift gene. After the human genome had finally been sequenced in the early 2000s[61,62] and actual molecular genetic research on handedness became feasible, handedness researchers got very excited. Surely, a single gene controlling a clearly defined phenotype such as handedness would be very easy to identify. They were bitterly disappointed. Despite unwavering efforts, a single gene that controls handedness or other forms of lateralization has not yet been identified, and it is very unlikely that it exists.

In the early days of molecular genetic research on handedness, scientists mainly used the so-called candidate gene approach. In a candidate gene study, a gene is chosen based on its theoretical significance for a phenotype. Subsequently, one or more single nucleotide polymorphisms (see Box 6.3) in this gene are identified. Then, participants are tested and

BOX 6.3

Genetic variation and genome-wide association studies (GWAS)

While the DNA sequences of two random human genomes are about 99.5% similar,[63] there are still millions of structural genetic variations between individuals.[64] These include:

- Single nucleotide polymorphisms (SNPs): SNPs are the most common type of genetic polymorphism and involve a change in a single nucleotide—for example, from adenine to cytosine.
- Microsatellites: Repetition of short DNA sequences—2–5 base pairs—for a variable number of times.
- Copy number variations: Duplication of long stretches of DNA.
- Other mutations, for example, deletion of a base pair in the sequence.

By comparing the relation between genetic variation (typically SNPs) and phenotypic variation (e.g., left- or right-handedness) in large numbers of individuals, it can be determined whether variation in a certain gene is associated with changes in the phenotype. This principle is used in GWAS. GWAS investigate DNA variations throughout the whole genome and associate them to the phenotype.[64] To this end, DNA samples and phenotypic data are collected from a large number of participants—typically thousands up to hundreds of thousands. In each DNA sample, millions of SNPs are determined using so-called SNP microarrays. Typically, two phenotypic groups—for example, left- and right-handers or patients and controls—are then compared statistically regarding the frequency of different alleles within each SNP. A significant overrepresentation of one allele in one group implies an association of this SNP with the phenotype. As millions of statistical comparisons are calculated, GWAS need drastic corrections for multiple comparisons in order to avoid false positive results. Most often, a p-value below 5×10^{-8} is thought to indicate genome-wide significance. This generates the issue that most GWAS need massive sample sizes in order to achieve enough statistical power to detect existing effects, in particular when multiple genetic influences with small amounts of explained variance interact. Typically, results from GWAS are depicted in a Manhattan plot (see Fig. 6.7), a special type of scatterplot named after its similarity to the Manhattan skyline with a few tall skyscrapers and a large number of lower-level buildings.

the phenotype (e.g., left-handed or right-handed) and genotype (individual genetic variants) are determined. By comparing participants with different genetic variants with each other, it can be determined whether the investigated polymorphisms affect the phenotype. Using these or similar methods, several genetic loci potentially involved in handedness have been identified. These include:

- *LRRTM1* (leucine rich repeat transmembrane neuronal 1): This gene possibly has an influence on neuronal differentiation and connectivity.[65]
- *PCSK6* (proprotein convertase subtilisin/kexin type 6): This gene encodes a protein that influences functionality of the left-right axis determining protein NODAL (see Chapter 10).[66–70]
- *AR* (androgen receptor): This gene is relevant for testosterone functions.[10,71]

However, none of these associations was replicated in later large-scale genome-wide association studies (GWAS), and the candidate gene approach has been criticized for low replication rates. Therefore, modern molecular genetic largely relies on GWAS (see Box 6.3).

The first handedness GWAS was a web-based study that incorporated more than 500,000 different SNPs[72] and looked at handedness and a number of other traits in a sample of 4268 participants. No single SNP reached genome-wide significance for handedness, and in fact, the top hit for handedness did not even represent a trend toward significance. Similar results were also found in a more recent GWAS in 3940 twins that also failed to identify any locus associated with handedness at a genome-wide level of significance[73] (see Fig. 6.7). Based on an analysis of these early GWAS data, Chris McManus estimated the number of genetic loci involved in handedness to be at least 30–40 that should be detected by larger GWAS with higher statistical power.[74] This is largely in line with the findings of recent large-scale GWAS, but future GWAS with larger sample sizes and more statistical power may detect even more relevant loci.

Using data from the UK Biobank cohort (N = 331,037), Carolien de Kovel and Clyde Francks discovered three SNPs that reached genome-wide significance, most importantly one implicating that the microtubule-associated gene *MAP2* may play a role in handedness ontogenesis.[51] Microtubules are an important part of the cytoskeleton and give cells their structure and shape. They also form the internal structure of cilia, a type of cell organelle that is important for left-right axis differentiation in the developing embryo (see Chapter 10). Similar results regarding *MAP2* were also reported in a second handedness GWAS using the UK Biobank Cohort.[52] The largest handedness GWAS was performed by the International Handedness Consortium and included data from more than 1.7 million individual participants.[50] In this analysis, 41 genetic loci reached genome-wide significance for left-handedness and 7 genetic loci reached genome-wide significance for ambidexterity (it can be assumed that self-identified ambidexterity in this study largely overlaps with mixed-handedness in other studies). Fig. 6.8 shows the Manhattan plot for the left-handedness GWAS.

This GWAS contained data from more than 13 million SNPs and revealed several important insights. First, it replicated and strengthened the association between handedness and microtubule that was reported by the first two handedness GWAS. In this study, several of the 41 genetic loci that reached significance for left-handedness were close to genes that have a function in the formation or regulation of microtubules (e.g., *MAP2*, *TUBB*, *TUBB3*,

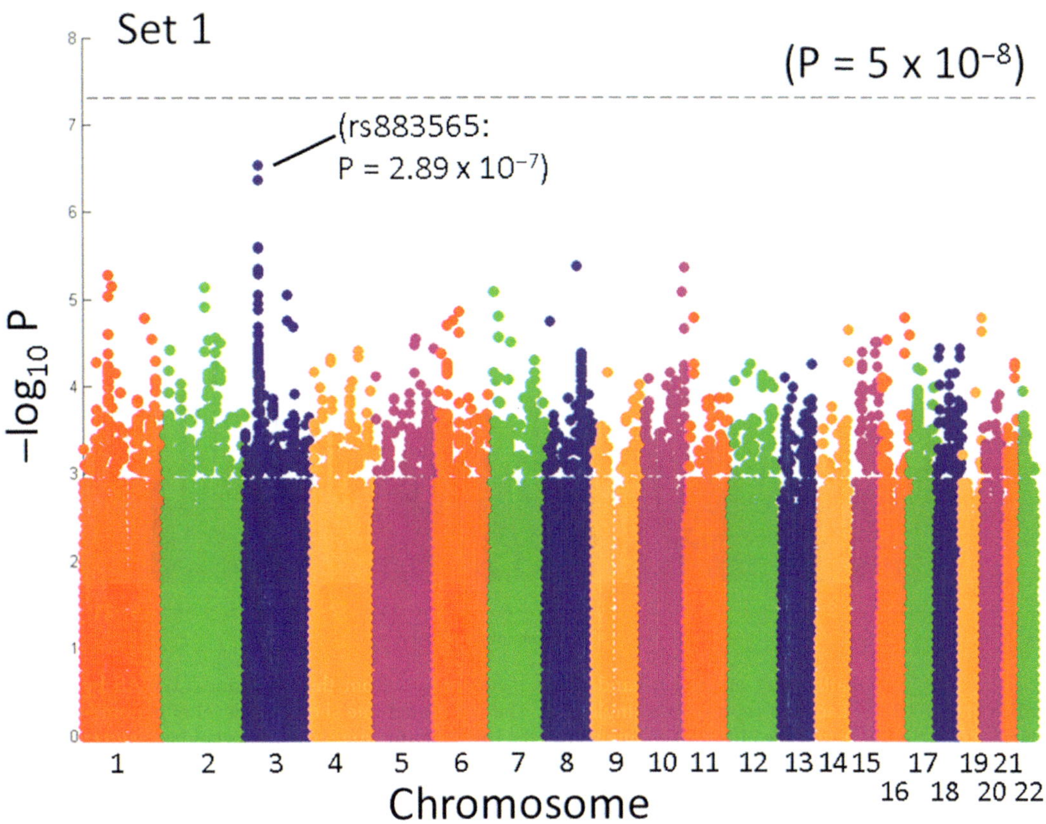

FIGURE 6.7 GWAS results for handedness from the study by Armour et al.[73] shown as Manhattan plot. Association between genetic variation in 2,535,688 different SNPs across the human genome and handedness is shown as $-\log_{10}$ P-values. Each dot represents a single SNP. No association reached the level of genome-wide significance ($P < 5 \times 10^{-8}$) indicated by the dotted line. *Source: Reprinted with the permission of the author and publisher. Armour, J. A. L., Davison, A., & McManus, C. Genome-wide association study of handedness excludes simple genetic models.* Heredity. *2014;112:221–225.*

NDRG1, *TUBB4A*, *TUBA1B*, *BUB3*, and *TTC28*). This strongly suggests that any biologically informed model of handedness ontogenesis needs to integrate microtubule function. Moreover, three loci had previously been associated with schizophrenia (rs6224, rs13107325, and rs45527431). Interestingly, those variants that were associated with an increased prevalence of left-handedness were also related to an increased schizophrenia risk, suggesting a molecular link between handedness and mental disorders (see Chapter 13).

Tissue enrichment analysis indicated that the identified genes were mostly relevant for the central nervous system. Moreover, pathway analysis suggested that their functional relevance included regulation of microtubules and axons, but also neurogenesis and regulation of the morphology of the cerebral cortex and hippocampus. In addition to these findings, the results of the study indicated that additive genetic effects explained 11.9% of

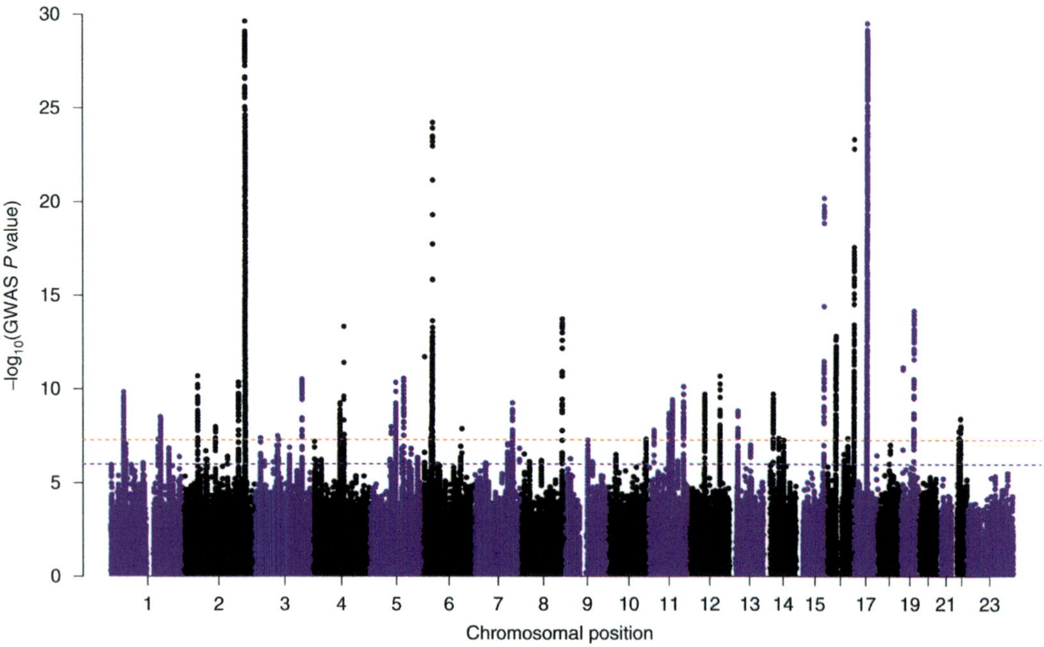

FIGURE 6.8 Manhattan plot of the left-handedness meta-analysis from the handedness GWAS by Cuellar-Partida et al. (2021).[50] Each dot represents a single SNP. The red dotted line shows the level of genome-wide significance ($P < 5 \times 10^{-8}$), and the blue broken line shows the threshold for suggestive associations. *Source: Reprinted with the permission of the author and publisher. Cuellar-Partida et al. Genome-wide association study identifies 48 common genetic variants associated with handedness.* Nat Hum Behav. *2021;5:59–70.*

the variance in handedness, while shared environmental effects explained 4.6% of the variance, and individual environment effects explained 83.6%. Thus, the largest genetic study on handedness to date clearly suggests that handedness is a polygenic trait, but also that it is largely determined by nongenetic influences.

While GWAS are called "genome-wide," they do not cover all possible genetic variation. They have the drawback that they typically only assess so-called common genetic variants that are present in a certain percentage of the population. This implies that rare genetic variants or novel mutations that affect handedness may not be detected by GWAS. Therefore, some researchers investigated genetic bottleneck populations with increased left-handedness with the aim to detect genes that affect handedness that may not have been found by GWAS. For example, the group of Iris E. Sommer from the University of Utrecht in the Netherlands investigated handedness in 37 families from an isolated Dutch town that had been an island for many centuries and only connected to the Dutch mainland 75 years ago.[75] Due to this isolation and a genetic bottleneck event in the 17th century (a sharp reduction of the population size due to environmental factors such as famine and human action such as war) this population showed a lower genetic heterogeneity than a random sample from the Dutch population. The sample showed increased prevalence of left-handedness, with 202 subjects being right-handed and 166 left-handed. Sommer and colleagues used parametric linkage analysis, a technique that allows for investigation of the relationship between genetic and phenotypic similarities in samples of individuals with known familial relationships, to investigate the genetics of handedness in their sample. However, no gene or genetic region reached genome-wide significance, thus providing a strong indication against a single-gene model.

Similar results were obtained by Clyde Francks' group at the Max Planck Institute for Psycholinguistics in Nijmegen, the Netherlands.[76] They performed whole exome sequencing—the exome is the 1% of the DNA that codes proteins—in a family from Pakistan with a history of consanguineous marriages and overrepresentation of left-handedness. Like the group of Sommer, they did not find any gene or mutation that could qualify as a single-gene explanation for handedness.

Asymmetrically expressed genes: the *LMO4* story

In order to identify the remaining genes that may be involved in the ontogenesis of handedness, several different strategies can be used. One approach is to identify genes that show differential transcription in the left and right hemisphere.[77,78] For example, the relative expression level of *LMO4* (LIM domain only 4), a gene that possibly plays a role as a transcriptional regulator, is about 2.5 times higher in the right compared to the left hemisphere, whereas the relative expression level of *HEY1* (hairy/enhancer-of-split related with YRPW motif 1), a gene that is possibly involved in neurogenesis, is about seven times higher in the left compared to the right hemisphere.[79] Interestingly, a genetic knock-out study in mice showed that knocking out *Lmo4* affects pawedness.[80] Wildtype mice did not show a clear preference for the left or the right paw. Mice in which *Lmo4* was knocked out in the left hemisphere during embryonic development showed a preference to use the left paw. In contrast, mice in which *Lmo4* was knocked out in the right hemisphere during embryonic development showed a preference to use the right paw. This suggests that while mice normally do not show a clear preference for one paw, unilateral knock-out of *Lmo4* in one hemisphere leads to a contralateral paw bias, suggesting a possible relevance of this asymmetrically expressed gene for functional lateralization.

Multilocus genetic models

Taken together, the evidence from GWAS clearly indicates that handedness is a polygenic trait and is not controlled by a single gene.[74,81] Based on these findings, McManus revised his original single-gene model to include multiple genetic loci while still being conceptually equivalent to the assumptions of the single gene version. This, however, still does not solve the problem that actual molecular evidence is needed in order to make a functional biological model of handedness ontogenesis. For example, as all three large-scale GWAS on handedness identified microtubule function as a relevant factor for handedness ontogenesis, a theory on handedness genetic should integrate the molecular pathways involved in microtubule formation and function.

Nongenetic influences on handedness

While it is undisputed that handedness is, to some extent, genetically determined, there is substantial evidence that a purely genetic model is not able to fully explain how handedness develops. The authors of a large-scale twin study reported that based on behavioral data, additive genetic effects only account for 23.64% of the variance in handedness, while the remaining 76.36% are accounted for by nonshared environmental influences.[11] Moreover, based on the genetic data mentioned above, the largest handedness GWAS reported that only 11.9% of the variance in handedness data can be explained by genetic effects.[50]

Interestingly, there are several lines of more direct evidence that suggest that nongenetic factors play an important role in the ontogenesis of handedness. One of these research lines comes from studies on twins. Identical twins are more likely to be concordant for hand preference than nonidentical twins,[48,49] but the concordance rate is not 100%. Discordant handedness (one twin is left-handed and the other twin is right-handed) is observed in about 20%–25% of identical twins.[82] As identical twins are genetically identical, it is not possible to explain this finding by a model that proposes solely genetic inheritance of handedness,[83] emphasizing the role of environmental factors for handedness development.[84]

Pawedness in mice: the Collins experiments

Strong evidence for the role of environmental factors in the development of handedness in nonhuman mammals comes from a classic series of experiments conducted by Robert L. Collins at the Jackson Laboratory in Bar Harbor, Maine, USA.[85–87] Just like humans show handedness, mice show pawedness—for example, a preference to use one paw over the other when handling food or manipulating objects. To test pawedness in these research animals, Collins invented a special feeding chamber (see Fig. 6.9). On the front wall of each chamber, a narrow glass feeding tube was positioned exactly equidistant between the left and the right wall of the chamber. At the end of the tube, maple-flavored rolled wheat, a highly desirable food for mice, was placed. Since the feeding tube was so narrow, the food-deprived mice could withdraw food by using only a single paw. By repeatedly testing an animal in such a chamber, its pawedness could be determined.

In the first experiment, Collins used these chambers to analyze the pawedness of mice and found about 50% to be left- and about 50% to be right-pawed. In the second experiment, he created two biased versions of the testing chamber, one in which the feeding tube was flush with the left wall—encouraging left paw use—and one in which the feeding tube was flush with the right wall—encouraging right paw use. The rationale behind this experiment was that if pawedness was determined genetically, the mice should use the paw they preferred in the unbiased chamber, but if it was determined by environmental pressures, they would follow the chamber's bias. Surprisingly, about 90% of the animals followed the environmental pressures and only 10% resisted it. Collins then transferred mice tested in the "left-world" chamber to the "right-world" chamber and vice versa. Again, only about 10% of the mice showed pawedness inconsistent with the demands of their environment, while 90%

FIGURE 6.9 The basic experimental apparatus used in the classic Collins experiments.[85–87] The mouse has to retrieve a food flake from a feeding tube that is so narrow that the mouse can only insert one paw.

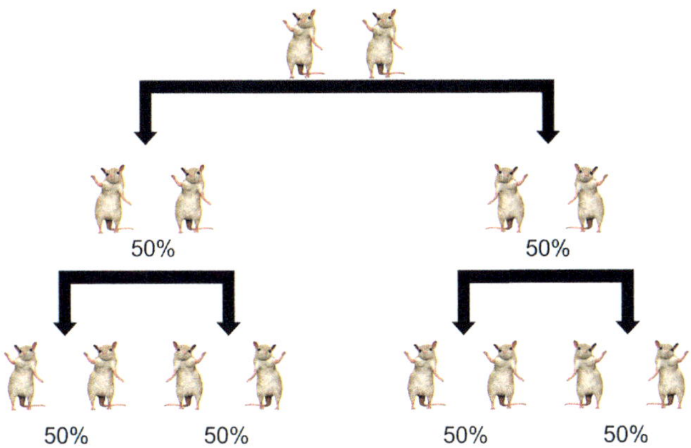

FIGURE 6.10 The breeding scheme used in the classic Collins experiments.[85–87]

followed it. Thus, the environment seems to be a crucial determinant of pawedness in mice. Only in a small number of mice did endogenous, possibly genetic, factors induce insensitivity to biased environmental experience. To further analyze the inheritance of pawedness, Collins started to separately breed left- and right-pawed mice for several generations and tested their pawedness. In each generation, there were again 50% left- and 50% right-pawed mice (see Fig. 6.10). Thus, Collins concluded that the side of paw preferences in mice was not determined by genetic means and that the direction of asymmetry is a matter of chance in mice.

Cultural threat

The big question is this: To what extent do humans show comparable effects? While humans rarely extend their arms into narrow feeding tubes, there are certain environments that promote right-handedness. For example, until the early 1970s, many German primary school teachers actively discouraged using the left hand for writing. Left-handed children were forced, sometimes with corporal punishment—for example, being hit on the fingers or the head with a ruler—to switch their writing hand when they started school. Here is how Anneliese Meier, a German retiree, remembers the practice:

> *Take the nice hand! my grandmother said and meant: Take the right hand, not the left. But I had done everything with the left: Writing, drawing, throwing. If I did not obey, I was hit slightly on my fingers. My right hand was always cramping. So, I sometimes placed a book on my desk so that the teacher would not see which hand I was using.*

Judging only by the written pages, the teachers would have been unable to say which hand had been used. After some time, handwriting in converted left-handers was almost as accurate with their right as with their left hand. Interestingly, other activities conducted with the hands were usually not affected, as converted left-handers were only forced to switch hands for writing. As a result, most nonwriting actions were predominantly executed with the left hand in converted left-handers.[88,89]

Handedness conversion can have negative effects on wellbeing, particularly in individuals who struggle with right-handed writing. A large scale study (n = 1277 adults) reported that in a subset of older adults who were forced to switch from left-handed to right-handed writing but were unsuccessful in doing so, several markers of psychological and physiological wellbeing were lower than in other participants in the sample.[90]

While the practice of forcing left-handers to write with the right hand has been abandoned in most countries, there are still some cultures in which the use of the left hand for many activities is actively discouraged. Typically, societies in which use of the left hand is associated with uncleanliness—for example, if toilet paper is not used in a society, the left hand is traditionally used for personal hygiene—or bad luck have a lower prevalence of left-handedness than most Western countries. This is the case, for example, in several African countries such as Sudan or the Ivory Coast. A recent study investigated handedness in two African student samples, reporting incidence of left-handedness in 7.9% of Ivorian and only 5% of Sudanese students, both rates being considerably lower than the 10% found in Western societies.[91] A reduction of left-handedness due to cultural pressures has also been observed in two studies investigating handedness in students from Malawi, a country in which more than 90% of teachers explicitly discourage the use of the left hand for writing in school.[92,93] Zverev investigated handedness in Malawian school children between the ages of 6 and 17 years and found that the prevalence of left-handedness was only 3.9%, with 5.7% of the children being mixed-handed and 90.4% being right-handed.[92] In a similar study, Zverev investigated handedness in Malawian students, guardians, and teachers and found that only 4.3% of participants were left-handed, while 4.7% were mixed- and 91% were right-handed.[93] Similar results have also been found in some Asian countries. For example, it was reported that only 3.5% of Taiwanese school children and adults are left-handed.[94] Handedness conversion is not irreversible, but it does have long-lasting effects.

A recent study investigated the handwriting kinematics of left-handers who had been converted to write with the right hand and participated in a training to learn writing with the left hand again.[95] Even after two years of training, participants did not reach the performance of innate left-handers in complex tasks, despite making substantial progress.

Taken together, the findings of cultural threat on handedness in humans are not comparable to the results Collins reported for mice. While forcing children to use their right hand for writing reduces the number of left-handers in the population, this does not seem to work the other way around. Even under very liberal cultural conditions, the prevalence of left-handers does not seem to exceed 13%. Thus, in contrast to mice, humans may have multiple genes that are not "direction-blind."

The early insult theory

One theory on the impact of environmental factors on handedness that generated substantial interest in the media and general public is the idea that left-handedness might have a pathological origin.[96–99] In particular, birth complications have been suggested to be a causative factor for left-handedness. The first widely recognized study testing this idea was conducted by Paul Bakan from Simon Fraser University in Burnaby, Canada. This researcher and his colleagues asked 510 university students to complete a handedness and a birth stress evaluation questionnaire.[96] In this questionnaire, the students had to indicate stressful conditions associated with birth, such as birth of multiples, premature birth, prolonged labor, Caesarian birth, breech birth, blue baby, or breathing difficulty at birth. All students reporting at least one of these conditions were assigned to the "stress" group, while all other students were assigned to the "normal" group. The researchers found a clear, significant relationship between birth stress and handedness: Birth stress was reported by almost twice as many left-handers (40%) than right-handers (22%). They concluded that left-handedness is a neuropathological condition caused by adverse birth circumstances.

Based on these and other findings, Paul Satz from the Neuropsychology Laboratory at the University of Florida, USA, developed the so-called early insult theory of left-handedness.[100] This model assumed that birth complications could lead to unilateral brain damage—for example, due to an oxygen shortage—resulting in a shift in handedness from the genetically predisposed hemisphere to the other. Fig. 6.11 illustrates the model.

Out of 100 healthy babies, about 10 would be left- and 90 would be right-handed. It was assumed that 1 out of 10 babies suffers from birth complications resulting in unilateral brain damage. This means that out of 10 left-handers, 9 would remain left-handers and 1 would become a pathological right-hander. On the other hand, out of the 90 right-handers, 81 would remain right-handers and 9 would become pathological left-handers. Thus, overall, there would be 18 left-handers. In 50% of them, left-handedness would have been caused by the early insult, while the other 50% would be healthy left-handers. In the right-handers, however, the incidence of pathological handedness would be much lower. Here, there would be 82 right-handers, with only 1.2% of them having pathological handedness, while the other 98.8% would have nonpathological handedness.

It is important to note that the early insult theory does not require major brain damage with massive cell death. Mostly, minute effects of the fetal environment and experience of

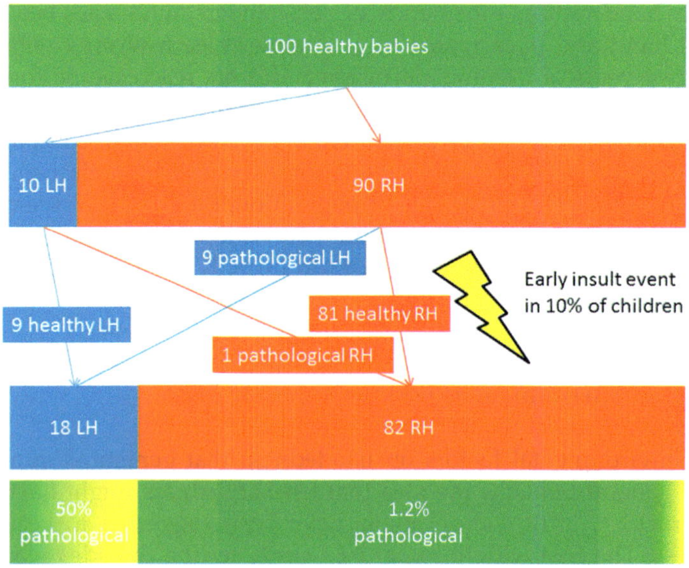

FIGURE 6.11 The early insult theory of pathological left-handedness. *LH*, Left-handed; *RH*, Right-handed.

birth stress are considered significant in explaining handedness. A proportion of individuals may suffer some minor brain insult either prenatally or perinatally, and this causes a cognitive decline as well as a shift toward right-hemispheric dominance, leading to left-handedness. Indeed, some studies have shown that left-handedness is associated with birth stress, low birth weight, and prenatal exposure to maternal depressive symptoms or critical life events, but the results were often markedly weaker than the initial findings by Bakan.[101–107] In contrast, a considerable number of studies have completely failed to find a relationship between birth stress and handedness, disproving the early insult theory.[108–117]

As these differences between studies might have been caused by a combination of small sample sizes and differential sample characteristics, it is important to look at the relation of handedness and birth stress in large samples with several thousand participants. David W. Johnston from Queensland University of Technology, Australia, and his coworkers conducted a study with 4942 Australian children (4–5 years) from the Longitudinal Study of Australian Children (LSAC) and found no evidence for an association of birth complications or problematic health scores and left-handedness.[98] More recently, Michael E.R. Nicholls from Flinders University in Adelaide, Australia, and his coworkers assessed birth stress in relation to handedness and cognitive ability in a large sample of about 10,000 children from the British Cohort Study. While birth stress affected cognitive ability, it did not have any effect on direction or degree of hand preference.[118] Taken together, the evidence from statistics in large samples thus refutes the early insult theory.

Seasonal anisotropy in handedness

Another finding that cannot be explained with a purely genetic model is the observation of seasonal anisotropy in handedness, that is, differences in the direction of handedness between individuals born in different parts of the year. In the northern hemisphere, there is a significantly higher incidence of left-handers among persons born in spring and ensuing months (March to July) than among persons born during the rest of the year.[119] This pattern is reversed in the southern hemisphere, with a large-scale study in Australian students finding a higher incidence of left-handedness among persons born in winter and autumn.[120] While there are some individual studies which failed to find a significant association between month of birth and handedness,[121] a meta-analysis combining the data from 13 different studies found a significant increase in left-handedness for spring births, with the ratio of left-handed to right-handed people being about 10% higher for individuals born during spring months than for individuals born during the rest of the year.[119] While the cause for this phenomenon is still not entirely clear, different theories have been proposed, including seasonal differences in prenatal exposure to testosterone as well as seasonal differences in patterns of nutrition and disease.[119,120]

Parental influences

Another form of a culturally transmitted bias proposed to affect handedness is familial influence.[122] Parental handedness may influence the direction of handedness in a child either by direct instructions or via observational learning processes. This assumption is supported by the finding that children's and parents' handedness are correlated[46], but at the same time is somewhat contradicted by the finding that left-handedness in children is not related to left-handedness in adoptive parents.[47]

The role of early experience

Interestingly, learning processes due to parental care are not the only forms of experience allegedly influencing handedness. One major line of research on the ontogenesis of handedness focuses on the role of early perceptual asymmetries for the consolidation of handedness and other forms of lateralization. As pointed out in Chapter 10, Ontogenesis of Hemispheric Asymmetries, in detail, in birds, visual lateralization is induced by asymmetrical light stimulation of the embryo in the egg. Bird embryos turn their head to the right side during embryogenesis due to the asymmetric expression of a complex gene cascade. This way, the right eye is closer to the translucent eggshell and gets stimulated by light, while the left eye is not. This light stimulation is a prerequisite for the development of visual asymmetries, as birds raised in the dark do not show lateralized behavior. Thus, genetic factors can lead to asymmetrical body positions or morphology that always direct the action of an environmental factor in the same direction. However, they alone may not be sufficient to induce functional asymmetries, as dark-incubated animals do not show asymmetries despite showing a rightward head turn.[123–125]

In birds, lateralized experience shapes asymmetry formation. But the interesting question is: Are there similar gene-environment interactions during human development that might shape handedness? Obviously, human embryos do not grow inside a hard-shelled egg like bird embryos, but nevertheless there is growing evidence that early experience may shape

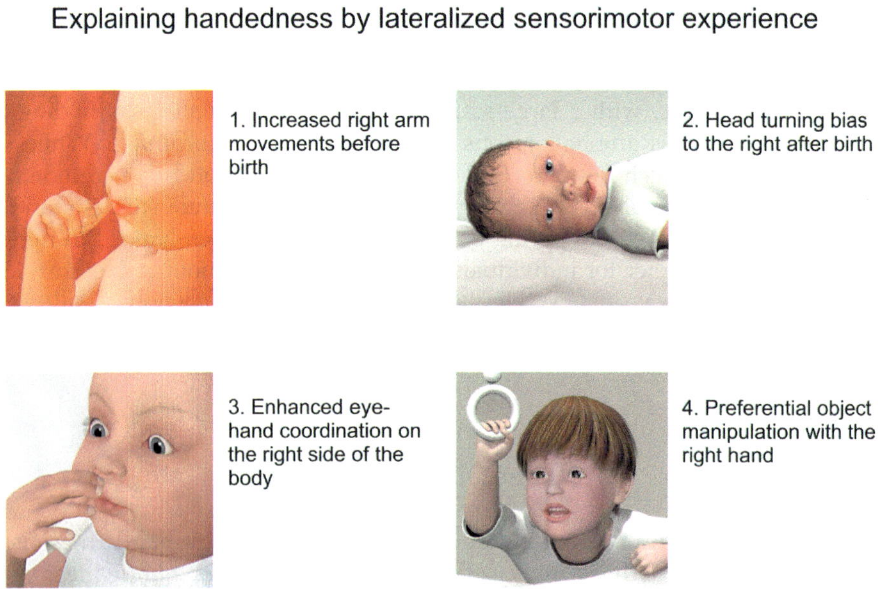

FIGURE 6.12 The role of experience for handedness development.

human handedness. In particular, perinatal asymmetrical perception of the environment induced by head-turning asymmetries has been suggested to affect handedness[83] (Fig. 6.12).

Human embryos and newborns preferentially turn their head to the right side. About 83% of newborn children show a tendency to turn their head to the right side when brought into a supine position, and this tendency is highly correlated with the hand that is used when the mouth is contacted.[126] A similar, albeit less-pronounced rightward bias in relation to the own body can be observed in human fetuses around the 38th week of gestation by using ultrasound.[127,128] Since preferred head position influences the amount of spontaneous visual experience of the left and right hand, the right-sided bias in head turning preference has been suggested to contribute to handedness development.[129] This assumption is supported by the finding that the direction of neonatal head-turning preference significantly predicts preferential hand use.[129,130] Moreover, it has been shown that the rightward neonatal head-turning preference leads to enhanced eye-hand-coordination on the right body side[131] and preferential right-hand object manipulation during the first year of life.[132]

Adult head-turning asymmetries and handedness

One important question regarding the influence of head-turning asymmetries on handedness is whether they disappear completely during development or whether they persist. Interestingly, the neonatal right-sided head-turning preference makes a surprisingly romantic reappearance later in life. One activity during which adults turn their head to the left or right is kissing, as kissers have to lean to one side in order to avoid bumping their noses. Thus, individuals can be categorized into left-kissers and right-kissers (see Fig. 6.13).

FIGURE 6.13 Head-turning asymmetries while kissing.

Güntürkün[133] assessed the turning direction in 124 kissing couples. To minimize the possible influence of cultural factors, kissing observations were conducted on international airports in the United States, Germany, and Turkey. Kissers had to face each other, and the kiss had to be clearly visible. None of the participants carried heavy bags, and no cheek kisses were recorded. Overall, 64.5% of the kissing pairs turned their heads to the right and 35.5% turned their heads to the left, indicating that the rightward head-motor bias persists into adulthood. A follow-up study tested whether the head-turning bias actually affected handedness.[134] In this study, the participants' head-turning preferences were assessed by asking them to kiss a plastic head (see Fig. 6.14), and their handedness was assessed using the EHI. Right-kissers had a significantly higher LQ (about 80) than left-kissers (about 60), indicating that head-turning preferences might indeed influence handedness.

While these findings clearly suggest a correlation between head-turning bias and handedness, it is not clear whether this relationship is a causal one. Head-turning bias may merely be caused by the same genetic factors that drive handedness, without a causal link between the two variables. In order to causally test the influence of a skewed visual experience of the hands on the development of handedness, it would be necessary to test children born with a permanent motor bias of the head which would increase the probability of visual experiences of one hand. This was done in a study investigating handedness in children with congenital muscular torticollis.[32] These children display a permanently tilted asymmetric head posture to the left or to the right in combination with a contralateral rotation of face and chin. This leads to an increased visual experience of the hand contralateral to the head tilt (see Fig. 6.15).

Relative to controls, children with torticollis had a higher probability of right- or left-handedness when having a head tilt to the opposite side. While 100% of children with

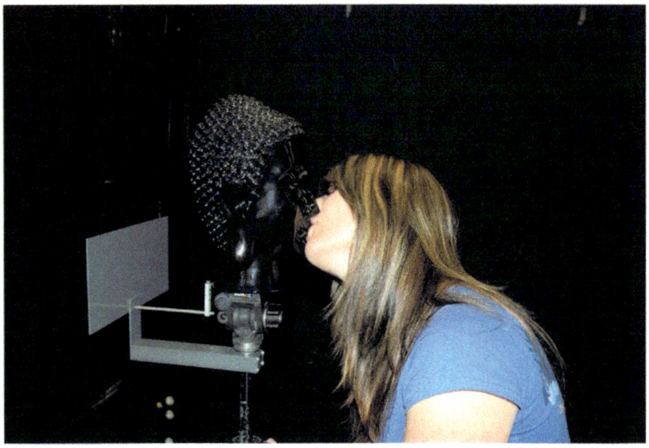

FIGURE 6.14 The experimental apparatus used in the study by Ocklenburg et al. [134] Participants' head-turning preferences were assessed by asking them to kiss a plastic head.

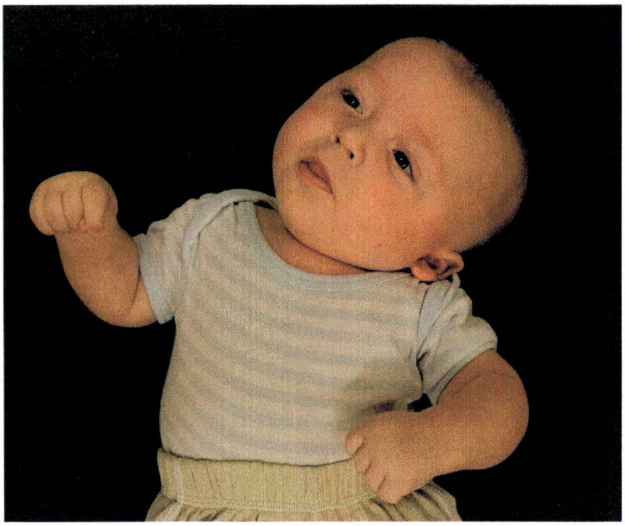

FIGURE 6.15 A patient with congenital muscular torticollis.[32] The patient displays a permanently tilted asymmetric head posture to the left or to the right in combination with a contralateral rotation of face and chin. This leads to an increased visual experience of the hand contralateral to the head tilt. *Source: Reprinted with permission of the author and publisher. Ocklenburg S, Bürger C, Westermann C, Schneider D, Biedermann H, Güntürkün O. Visual experience affects handedness.* Behav Brain Res. *2010;207:447–451.*

left-sided torticollis—and more visual experience of the right hand—were right-handed, only 78% of the children with right-sided torticollis—and more visual experience of the left hand—were right-handed. These findings indicate that increased visual experience of one hand affects human handedness.

Large-scale studies on nongenetic factors in handedness ontogenesis

Comparable to recent trends for handedness genetics studies, nongenetic factors involved in handedness ontogenesis have also been investigated in large-scale population samples.

For example, Carolien de Kovel and Clyde Francks used data from the UK Biobank (n = 501,730 individuals) to investigate the role of early life factors for handedness in a large scale population study.[135] They found out that the probability of being left-handed was significantly affected by several different factors, including the year and location of birth, which likely reflects the cultural threat effects described earlier in this chapter. In general, people had a 0.7% higher chance of being left-handed per decade. Moreover, handedness was associated with birth weight, with left-handers on average being 26g lighter at birth than right-handers. Left-handers were also more likely to be part of a multiple birth and to be born in summer.

In addition, left-handers were more likely to not have been breastfed than right-handers. This finding was also confirmed by a recent meta-analysis.[136] Also, men were more likely than women to be left-handed, a finding that also reached significance in the abovementioned meta-analysis for handedness.[2]

Taken together, the analysis clearly showed that several early life factors significantly affect handedness. Importantly, even when taken all together, these factors only explained a small fraction of the variance in phenotypic handedness data. The authors reported that for all factors together, the pseudo R^2 according to MacFadden was only 0.005. As this measure can range from 0 (no predictive value) to 1 (high predictive value), the authors concluded that the early life factors included in the UK Biobank have minimal predictive value for handedness.

The epigenetics of handedness

For all nongenetic influences on handedness, it is crucial to understand how they affect brain structure and function on the molecular level. This led several authors to investigate epigenetic mechanisms such as DNA methylation in the context of handedness research[137] (see Fig. 6.16). Epigenetics refers to molecular processes that alter gene expression but do not lead to mutations in the DNA itself.[138] Comparable to genetic candidate gene studies for handedness, early attempts at investigating the epigenetics of handedness focused on epigenetic processes affecting the expression of specific candidate genes.[139,140] More recent attempts have instead focused on covering the whole epigenome. Similar to GWAS for genetic influences, epigenome-wide association study (EWAS) can be performed for epigenetic influences such as methylation. The largest handedness EWAS to date included analysis of whole-blood DNA methylation in 3914 adults.[141] Overall, 409,562 CpGs were analyzed in the study. A CpG is a specific segment of the DNA in which a cytosine nucleotide is followed by a guanine nucleotide. Methylation of the cytosine in a CpG can alter the expression of the gene, thus making CpGs a prime target for epigenetic research. In the study, no association reached epigenome-wide significance. When the CpGs with the lowest p-values in the sample were assessed, the methylation differences between left-handers and right-handers were very small (from 0.06% to 0.8%). The authors concluded that methylation in peripheral tissue such as blood (which was investigated in this study) captures only very little variance in handedness. Instead, future studies should focus on biologically relevant tissue such as brain tissue. This shows a principal problem in epigenetic research: Unlike DNA which is stable across cell types and developmental stages, epigenetic processes often differ between different cell types and also between different developmental stages.

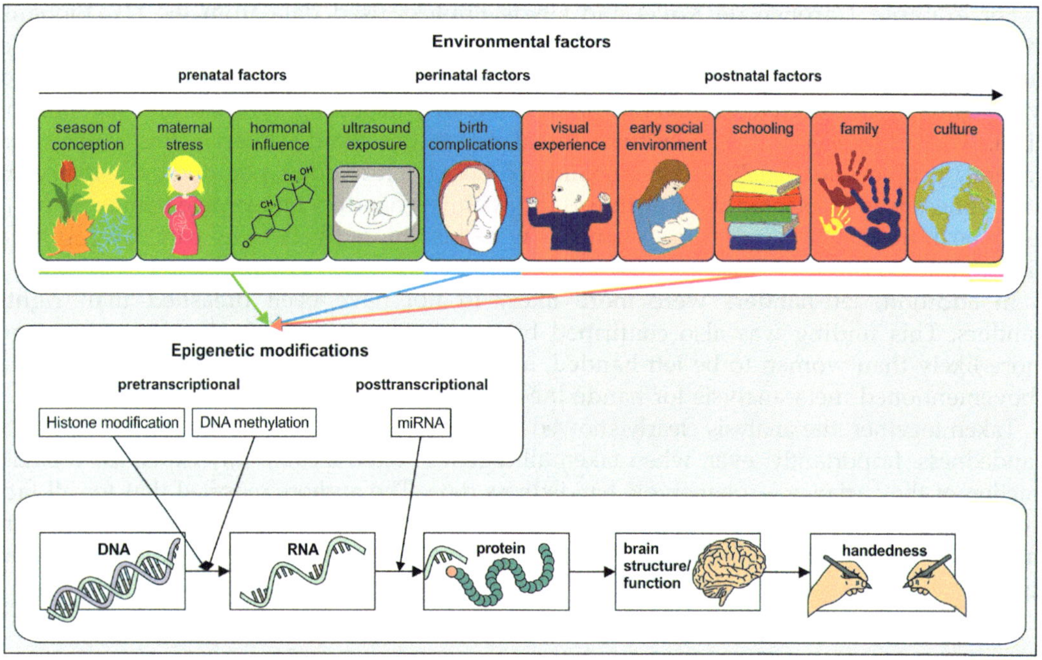

FIGURE 6.16 A model integrating epigenetic processes in handedness ontogenesis.[137] Environmental factors induce epigenetic modifications, which influence protein synthesis via pre- or posttranscriptional mechanisms. This in turn affects brain structure and function and ultimately behavioral preferences like handedness. *Source: Reprinted with permission of the author and publisher. Schmitz J, Metz GAS, Güntürkün O, Ocklenburg S. Beyond the genome-Towards an epigenetic understanding of handedness ontogenesis.* Prog Neurobiol. *2017;159:69–89.*

Thus, even if strong epigenetic effects occur in the motor cortex of the brain during development, they may not be captured if methylation of bloods cells is investigated in adulthood. Thus, more research on the epigenetics of handedness is needed.

Evidence for a multifactorial determination of handedness

Taken together, research on the role of early experience for the ontogenesis of handedness indicates that, despite the presence of genetic factors, handedness displays some degree of plasticity in response to environmental experience. Humans have an embryonic and postdelivery tendency to turn their heads toward the right side. This tendency never disappears completely but is present throughout lifetime in a subtle way. This lateralized bias seems to modify handedness. However, this modification explains only a small part of the variance of handedness. These findings illustrate that it does not make sense to try to understand handedness from a nature versus nurture standpoint. Like other complex traits, it is not determined by either genes or environment. Instead, multiple genetic and environmental factors that potentially show complex gene–environment interactions influence its ontogenesis. Moreover, it has recently been suggested that besides nature and nurture, another factor affects handedness ontogenesis: (developmental) noise.[142] Comparable to similar "Nature, nurture, noise" models that have been suggested for bodily asymmetries,[143] Chris McManus recently suggested that randomness in brain development may be an important factor in handedness ontogenesis that is captured by neither genetic nor epigenetic techniques.[81,142] Taken together, even after decades of research, we have only begun to understand the ontogenesis of handedness.[144] Clearly, more research on epigenetic processes is needed, but also finding ways to empirically assess the role of randomness in brain development for handedness is critical for moving forward. Uncovering this complex cascade will be a great challenge in handedness research over the next decades.

Handedness and the brain

As pointed out above, genes are unlikely to directly impact behavioral phenotypes such as handedness. However, the proteins they are coding for affect the development of brain regions that control neuronal processing and behavior. Understanding the neuronal correlates of handedness is, therefore, an essential step in understanding the development of handedness.

In principle, the left hand is controlled by the right motor cortex and vice versa. Typically, the neuronal correlates of hand use are assessed by asking participants to perform different finger and hand movements and measuring their brain activity while doing so (see Fig. 6.17).

For example, Klöppel and his colleagues used functional magnetic resonance imaging (fMRI) to investigate the impact of handedness on neuronal activation during simple

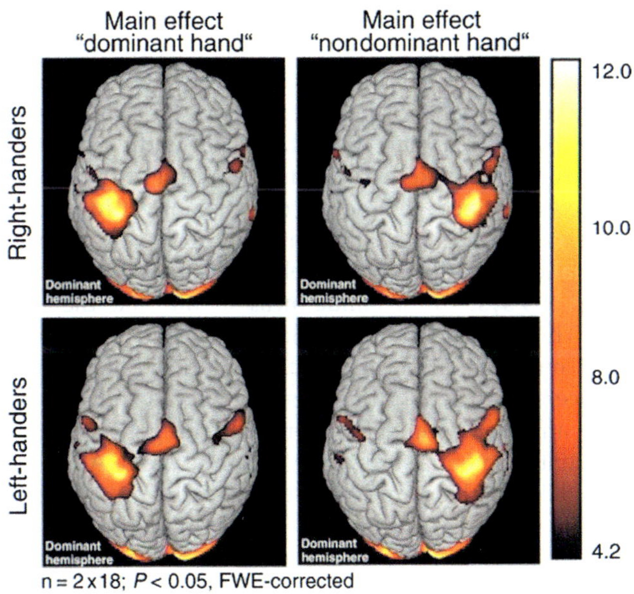

FIGURE 6.17 Neural activity for visually paced fist closures in left- and right-handers from the study by Pool et al.[145] *Reprinted with permission of the author and publisher. Pool EM, Rehme AK, Fink GR, Eickhoff SB, Grefkes C. Handedness and effective connectivity of the motor system.* Neuroimage. *2014;99:451–460.*

unilateral and bilateral finger movements.[146] They found a relative increase of activity in dorsal premotor and right primary sensorimotor cortices in right-handers, but not left-handers, during simultaneous bilateral button presses. Moreover, right-handers showed general predominance of left-hemispheric activation relative to right-hemispheric activation, while in left-handers this pattern was reversed. In general, movements of the dominant hand seem to be controlled by the contralateral hemisphere, while movements of the non-dominant hand lead to a more bilateral activation pattern.[147,148] In general, right- and left-handers show very similar brain activation patterns during simple hand movements, with differential brain activations only observed for more complex or sequential movements.[149–151] For example, Verstynen and coworkers could show that complex movements with the left hand lead to stronger activations in the left than in the right dorsal premotor cortex.[152] While both left- and right-handers showed this pattern, the effect was far less pronounced in left-handers. Left- and right-handers seemed to not only differ with regard to activations of specific areas during motor tasks but also in effective connectivity of areas within the motor network. A recent dynamic causal modeling study investigating effective connectivity between key motor areas during fist closures of both hands in right- and left-handers revealed that in right-handers, the dominant supplementary motor area played a more prominent role within the network than in left-handers. Also, left-handers showed less asymmetry of effective connectivity, implicating more bilateral motor control of the hands.[145] In addition to these functional correlates of handedness, several researchers have also investigated structural correlates of handedness, which are discussed in Chapter 4.

Handedness and the spinal cord

While the brain clearly is relevant for handedness after birth, it has been suggested that during handedness ontogenesis before birth, the spinal cord may play an important role.

Peter G. Hepper and coworkers used ultrasound to show that human fetuses show a clear rightward preference to suck the right thumb in-utero at as early as 15 weeks of gestation.[153] A later study showed that already at 10 weeks of gestation, about 85% of fetuses show a preference to move the right arm more often than the left.[154] This finding left the authors to conclude that these arm movements are unlikely to be under control of the brain, but more likely to be caused by the spinal cord. Indeed, the motor cortex is not yet functionally linked to the spinal cord at 10 weeks of gestation and the outgrowth of corticospinal projections does not enter the anterior spinal cord until several weeks later.[155] This suggests that research on the role of the central nervous system for handedness should take the spinal cord into account, not only the brain. This was, for example, done in a study in which genome-wide mRNA expression and DNA methylation in cervical and anterior thoracal spinal cord segments of five human fetuses were analyzed.[156] In this study, it was shown that asymmetric methylation of CpG islands affects asymmetric expression of genes in the human fetal spinal cord, which may represent a potential precursor of handedness. A recent computational modeling study also suggested that the connections between the motor cortices and the spinal cord are relevant for handedness.[157] In the study network, modeling was performed using simultaneous fMRI data from brain and spinal cord. The authors found a greater leftward asymmetry in the direct connection between motor cortex and spinal cord was linked to rightward hand preference. In addition, the indirect pathway between motor cortex and spinal cord also played a role for handedness through the control of the nondominant hand. Taken together, these findings suggest that the spinal cord and its connections to the brain should be considered in future handedness research.

The neuroscience of handedness conversion

Converted left-handers like Anneliese in the introductory story constitute a particularly interesting population for the investigation of the neuronal correlates of handedness. Using positron emission tomography, Siebner and coworkers assessed regional cerebral blood flow in adult converted left-handers and right-handed controls who repetitively wrote the German verb "bellen"—"to bark"—with their right hand on a digital graphics tablet.[89] In right-handers, the handwriting task caused mostly left-hemispheric activation in the parietal cortex and in premotor association areas. In converted left-handers, however, the task caused more bilateral brain activation with activation foci in the right lateral premotor, parietal, and temporal cortex. A more bilateral activation pattern in converted left-handers was also reported in a subsequent fMRI study.[88] Here, converted left-handers, right-handers, and left-handers performed simple movements with their index fingers while their brains were scanned. Compared to right-handers, converted left-handers retained two areas of higher-level motor control in the right hemisphere (their naturally dominant motor hemisphere): the dorsolateral premotor cortex and the inferior parietal cortex. These two areas serve motor planning and action selection. Taken together, these findings indicate that in converted left-handers, the left hemisphere controls the right hand with regard to actual movements, while the right hemisphere organizes actions at a higher level. Conversion thus increases the need for cross-hemispheric communication, which requires additional resources and possibly slows down processing.

Cognitive correlates of handedness

Since handedness seems to be related to differential brain organization, several researchers have investigated its relation to cognitive abilities. Unfortunately, this is an area full of exaggerations. For example, one common myth about left-handers' cognitive abilities is that there are more left-handed geniuses (for another common myth about left-handedness, see Box 6.4).

However, what are the facts? Is handedness really related to cognitive abilities? In general, results from single studies with small sample sizes on this topic often seem conflicting, emphasizing the need to look at results from large-scale studies or meta-analyses. Marietta Papadatou-Pastou from the University of Athens, Greece, conducted several meta-analyses to find out if left-handers were indeed more intelligent than right-handers.[161] She integrated the results of studies that assessed handedness prevalence in individuals with intellectual giftedness and compared them to typically developing individuals. Moreover, she also integrated the findings of studies that assessed handedness in

BOX 6.4

Myths and facts about left-handedness: do left-handers die younger?

In 1988, the prestigious scientific journal *Nature* published an article that asked a simple question: Do right-handers live longer? In this short article by Diane Halpern and Stanley Coren,[158] the authors explored the relationship between handedness and age at death by identifying these parameters for more than 1000 baseball players listed in a commonly used baseball encyclopedia. They found that for 1472 right-handed baseball players, the mean age at death was 64.64 years, while for the 236 left-handed baseball players, the mean age at death 63.97 years—a significant effect. The conclusion that left-handers die younger than right-handers was also supported by a follow-up study by the same authors in 1991.[159] Published in the *New England Journal of Medicine*, one of the most respected medical journals in the world, this article focused on death certificates from two counties in southern California. To the astonishment of not only the medical community but also left-handers all around the world, Halpern and Coren could show that left-handers in these counties died on average at age 66, while right-handers on average lived to the age of 75—a significant 9-year age difference in life span. These and other results culminated in a 1991 review article by Coren and Halpern[97] that claimed that left-handedness is selectively associated with earlier mortality because left-handers have an elevated risk for accidents in a right-handed world and are also more prone to birth stress which in turn is supposed to be related to neuropathies, developmental delays, and immune deficiencies. These are bleak prospects for left-handers, but are they really doomed to an early death? Probably not. In his book "Right hand, left hand," Chris McManus, a highly-respected authority on all things concerning handedness, explained the statistical fallacy underlying the results reported by Coren and Halpern.[4] They essentially conducted a "death cohort" analysis, taking people who died at a given time and looking back at their lives. However, depending on the birth date of these people, the socioeconomic circumstances in which they grew up could be strikingly different. With regard to handedness, there was much more conversion of left-handers to right-handers in the early 20th century than after 1950. According to McManus, the percentage of left-handers hovered between 3% and 4% between 1900 and 1910, while being at 10% or higher since 1940.[4] Thus, the left-handers in the studies by Coren and Halpern did not die younger, there were just less of them present in the older cohorts due to forced relearning. This explanation of the data pattern was supported by other researchers, among them Clare Porac from Penn State University, USA, who did not find any evidence whatsoever to support the idea that left-handers were more prone to health problems.[1,160] Today, the laterality research community has completely rejected the idea that left-handers die younger.

individuals with intellectual disabilities and compared them to individuals with typical intelligence. Her analysis comprised over 16,000 individual participants and revealed no evidence for an elevated rate of left-handedness in particularly intelligent individuals. However, there was a significant effect in the meta-analysis comparing individuals with intellectual disabilities to typically developed individuals. Here, in individuals with intellectual disabilities, prevalence of left-handedness was almost twice as large as in typically developing individuals—19.6%—indicating a possible link between developmental disorders and handedness—more on the relation of hemispheric asymmetries and pathology will be discussed in Chapter 13. Similar results were also obtained in a large study investigating handedness and general cognitive ability in 4942 Australian children between the ages of 4–5 years.[98] On average, left-handers scored significantly lower than right-handers in different mental aptitude tests. Interestingly, the lowest scores were found for mixed-handers, indicating that absence of asymmetry might reduce efficacy of cognitive processing. However, there is also empirical evidence supporting the notion that mixed-handedness might have some benefits for specific cognitive domains, given reports of greater semantic fluency or cognitive flexibility in mixed-handers compared to strong left- or right-handers.[162,163] The relation of handedness and specific cognitive domains has also been investigated in several other studies. For example, a recent meta-analysis investigated whether left- or right-handers had an advantage in spatial or verbal abilities.[164] While no difference was found for verbal abilities, a small but significant cognitive advantage of right-handers was found for spatial abilities—for example, mental rotation. Left-handers and right-handers also have been compared regarding their artistic creativity, since it is a common idea that left-handers may be more creative than right-handers. Interestingly, a large-scale study found that while left-handers rate themselves as more artistically creative than right-handers, both groups were in fact equally artistically creative.[165]

Footedness

Handedness is not the only limb-use preference in humans. For example, it is well known that Lionel Messi, one of the best soccer players in the history of the game and part of the Argentinian team that won the soccer world cup in 2022, prefers to use his left foot when kicking a ball. David Carey from the School of Psychology at Bangor University, Wales, systematically investigated footedness in soccer players and found that, unlike Messi, about 79% of professional soccer players show a preference to use the right foot when kicking a ball. Only 21% showed a preference for the left foot, indicating that the majority of individuals exhibit a right-sided bias for foot use, albeit at a somewhat lower rate than for handedness.[166] A rate of about 80% right-footers was also reported by the same group in a later paper on amateur soccer players.[167] Similar to what has been done for handedness, a comprehensive meta-analysis has been conducted for footedness.[168] By integrating data from 164 studies, it was determined that 12.1% of individuals were left-footed. Overall, 23.7% of individuals were either left-footed or mixed-footed.

Quantitatively, footedness can be assessed with the Waterloo Footedness Questionnaire[169,170] (see Box 6.5). In this short questionnaire, participants have to indicate their preferred foot for 10 different activities—for example, "If you wanted to pick up a marble with your toes, which foot would you use?" Similar to the EHI, an LQ ranging between −100 and +100 can be calculated, with positive values indicating a preference for the right foot and negative values a preference for the left foot.

BOX 6.5

Test your footedness!

The Waterloo Footedness Questionnaire[169,170] (Table 6.2) is the footedness questionnaire most widely used in scientific research. Here, you can try it out yourself and calculate your LQ to determine your footedness.

Instructions:

Answer each of the following questions as best as you can. If you always use one foot to perform the described activity, circle Ra or La (for right always or left always). If you usually use one foot, circle Ru or Lu as appropriate. If you use both feet equally often, circle Eq. Please do not simply circle one answer for all questions, but imagine yourself performing each activity in turn, and then mark the appropriate answer. If necessary, stop and pantomime the activity.

From[169,170]

When you have filled out all boxes, calculate the sum of pluses you made for the left foot (L) and the sum of pluses you made for the right foot (R). Now you can calculate your own footedness LQ using the formula:

$$LQ = [(R-L)/(R+L)] \times 100.$$

First, calculate the dividend of the formula by subtracting L from R and then the divisor by adding R and L. Calculate the quotient and multiply it by 100, and you have calculated your footedness LQ.

TABLE 6.2 The Waterloo Footedness Questionnaire.[170]

1	Which foot would you use to kick a stationary ball at a target straight in front of you?	La Lu Eq Ru Ra
2	If you had to stand on one foot, which foot would it be?	La Lu Eq Ru Ra
3	Which foot would you use to smooth sand at the beach?	La Lu Eq Ru Ra
4	If you had to step up onto a chair, which foot would you place on the chair first?	La Lu Eq Ru Ra
5	Which foot would you use to stomp on a fast-moving bug?	La Lu Eq Ru Ra
6	If you were to balance on one foot on a railway track, which foot would you use?	La Lu Eq Ru Ra
7	If you wanted to pick up a marble with your toes, which foot would you use?	La Lu Eq Ru Ra
8	If you had to hop on one foot, which foot would you use?	La Lu Eq Ru Ra
9	Which foot would you use to help push a shovel into the ground?	La Lu Eq Ru Ra
10	During relaxed standing, people initially put most of their weight on one foot, leaving the other leg slightly bent. Which foot do you put most of your weight on first?	La Lu Eq Ru Ra
11	Is there any reason (i.e., injury) why you have changed your foot preference for any of the abovementioned activities?	Yes No (circle one)
12	Have you ever been given special training or encouragement to use a particular foot for certain activities?	Yes No (circle one)
13	If you have answered Yes for either question 11 or 12, please explain.	

Reprinted with permission of the publisher.

Eyedness and earedness and further forms of lateral preferences

In addition to handedness and footedness which can be characterized as motor preferences, preferences in the usage of lateralized sensory organs have been observed for eyes and ears. Eye preference can be assessed by asking participants to indicate which eye they prefer—for example, when looking through a telescope, microscope, or keyhole—and it was reported that about 66% of the population prefer the right eye for these activities.[171] Interestingly, it has been suggested that eye preference may be more plastic than handedness and footedness and may change depending on the angle of gaze.[172] A similar distribution has also been observed for ear preference, with about 65% of the population preferably using the right ear when listening to a conversation going on behind a closed door.[46]

Importantly, there are many more forms of lateral preferences and side biases in human behavior and laterality researchers continue to discover new ones. In his recent book "Side effects—How left-brain right-brain differences shape everyday behavior," Lorin J. Elias from the University of Saskatchewan, Canada, gives an overview of research trends regarding lateral preferences in everyday life. In addition to handedness, footedness, eyedness, and earedness, other interesting forms of lateral preference as seating biases, turning biases, lateral biases during gesturing, posing biases, and cradling biases are described.[173] It is truly fascinating how many aspects of everyday life are determined by lateral preferences.

Associations between different forms of lateral preferences

The question whether and to which extent lateral preferences for hand, foot, eye, and ear use are interrelated has yet to be answered unequivocally. While some authors[174] report highly significant positive correlations between all four measures, suggesting a common ontogenetic background, others confirm that handedness and footedness are highly correlated but do not have a significant association with the lateral preferences for eye or ear use.[175] Moreover, a study on the association between eye preference and handedness found no association between the strength of hand preference and the strength of eye preference.[176] These results suggest that the ontogenesis of lateral preferences in sensory and motor systems may at least partly rely on independent processes. For the association between handedness and footedness, the results of the abovementioned meta-analysis[168] indicate that about 60.1% of left-handers were left-footed, but only 3.2% of right-handers were left-footed, indicating a substantial link between handedness and footedness.

Aside from footedness, handedness has also been related to language lateralization. In a widely cited study, Knecht et al.[55] measured language lateralization using functional transcranial Doppler sonography and linked it to handedness. They found that the incidence of atypical right-hemispheric language dominance increased with the degree of left-handedness. While strong right-handers only had a prevalence of 4% for right-hemispheric language dominance, this value increased to 15% in mixed-handers and 27% in strong left-handers. This shows that the ontogenesis of handedness and language lateralization is influenced by at least partly overlapping factors.

Sample questions

- What is the difference between hand preference and hand performance, and which techniques have been used to test them?
- What is *situs inversus*? In which way does it affect the organization of the body? Does it have an effect on handedness?
- Please describe three lines of evidence which support the idea that handedness has a genetic component.
- Collins started a famous series of studies on the genetics of pawedness in mice. What was the experiment? What did he find out? Can his insight be applied to human genetics of handedness?
- Please outline the present knowledge on the genetic background of handedness.
- What is a GWAS and what are the main insights about the ontogenesis of handedness that have been obtained from GWAS?
- A cultural pressure to change handedness in school children creates a characteristic behavioral and neural pattern. What is this pattern?
- What is the essence of the early insult theory? What kind of evidence speaks against it?
- Could head-turning asymmetries explain human handedness? Please review the experimental evidence for your answer.
- What is known about the epigenetics of handedness?

References

1. Porac C. *Laterality: Exploring the Enigma of Left-Handedness*. Elsevier Science; 2016.
2. Papadatou-Pastou M, Ntolka E, Schmitz J, et al. Human handedness: a meta-analysis. *Psychological Bull*. 2020;146(6):481–524. Available from: https://doi.org/10.1037/bul0000229.
3. Corballis MC. The evolution and genetics of cerebral asymmetry. *Philos Trans R Soc Lond B Biol Sci*. 2009;364 (1519):867–879. Available from: https://doi.org/10.1098/rstb.2008.0232.
4. McManus C. *Right Hand, Left Hand: The Origins of Asymmetry in Brains, Bodies, Atoms, and Cultures*. Harvard University Press; 2002.
5. de Kay JT. *The Natural Superiority of the Left-Hander*. Large print edition. Published by M. Evans, an imprint of the Rowman & Littlefield Publishing Group; 1979.
6 Ocklenburg S. *Linkshändigkeit und Hirnasymmetrien: Eine Einführung*. Springer; 2022.
7. Smits R, Waters L. *The Puzzle of Left-Handedness*. Reaktion Books; 2011.
8. Westmoreland P. *How Handedness Shapes Lived Experience, Intersectionality, and Inequality: Hand and World*. Palgrave Macmillan; 2023.
9. Scharoun SM, Bryden PJ. Hand preference, performance abilities, and hand selection in children. *Front Psychol*. 2014;5:82. Available from: https://doi.org/10.3389/fpsyg.2014.00082.
10. Medland SE, Duffy DL, Spurdle AB, et al. Opposite effects of androgen receptor CAG repeat length on increased risk of left-handedness in males and females. *Behav Genet*. 2005;35(6):735–744. Available from: https://doi.org/10.1007/s10519-005-6187-3.
11. Medland SE, Duffy DL, Wright MJ, et al. Genetic influences on handedness: data from 25,732 Australian and Dutch twin families. *Neuropsychologia*. 2009;47(2):330–337. Available from: https://doi.org/10.1016/j.neuropsychologia.2008.09.005.
12. Klöppel S, Mangin J-F, Vongerichten A, Frackowiak RSJ, Siebner HR. Nurture versus nature: long-term impact of forced right-handedness on structure of pericentral cortex and basal ganglia. *J Neurosci*. 2010;30 (9):3271–3275. Available from: https://doi.org/10.1523/JNEUROSCI.4394-09.2010.

13. Oldfield RC. The assessment and analysis of handedness: the Edinburgh inventory. *Neuropsychologia*. 1971;9(1):97–113. Available from: https://doi.org/10.1016/0028-3932(71)90067-4.
14. Edlin JM, Leppanen ML, Fain RJ, Hackländer RP, Hanaver-Torrez SD, Lyle KB. On the use (and misuse?) of the Edinburgh Handedness Inventory. *Brain Cognition*. 2015;94:44–51. Available from: https://doi.org/10.1016/j.bandc.2015.01.003.
15. Fazio R, Coenen C, Denney RL. The original instructions for the Edinburgh Handedness Inventory are misunderstood by a majority of participants. *Laterality*. 2012;17(1):70–77. Available from: https://doi.org/10.1080/1357650X.2010.532801.
16. Milenkovic S, Dragovic M. Modification of the Edinburgh Handedness Inventory: a replication study. *Laterality*. 2013;18(3):340–348. Available from: https://doi.org/10.1080/1357650X.2012.683196.
17. Veale JF. Edinburgh Handedness Inventory—Short Form: a revised version based on confirmatory factor analysis. *Laterality*. 2014;19(2):164–177. Available from: https://doi.org/10.1080/1357650X.2013.783045.
18. Tzourio-Mazoyer N, Simon G, Crivello F, et al. Effect of familial sinistrality on planum temporale surface and brain tissue asymmetries. *Cereb Cortex*. 2010;20(6):1476–1485. Available from: https://doi.org/10.1093/cercor/bhp209.
19. Isaacs KL, Barr WB, Nelson PK, Devinsky O. Degree of handedness and cerebral dominance. *Neurology*. 2006;66(12):1855–1858. Available from: https://doi.org/10.1212/01.wnl.0000219623.28769.74.
20. Borawski J, Papadatou-Pastou M, Packheiser J, Ocklenburg S. Handedness in post-traumatic stress disorder: a meta-analysis. *Neurosci Biobehav Rev*. 2023;145:105009. Available from: https://doi.org/10.1016/j.neubiorev.2022.105009.
21. Nastou E, Ocklenburg S, Hoogman M, Papadatou-Pastou M. Handedness in ADHD: meta-analyses. *Neuropsychol Rev*. 2022;32(4):877–892. Available from: https://doi.org/10.1007/s11065-021-09530-3.
22. Gardner MR, Potts R. Hand dominance influences the processing of observed bodies. *Brain Cognition*. 2010;73(1):35–40. Available from: https://doi.org/10.1016/j.bandc.2010.02.002.
23. Peters M, Servos P. Performance of subgroups of left-handers and right-handers. *Can J Psychol*. 1989;43(3):341–358. Available from: https://doi.org/10.1037/h0084226.
24. Nicholls MER, Thomas NA, Loetscher T, Grimshaw GM. The Flinders Handedness survey (FLANDERS): a brief measure of skilled hand preference. *Cortex*. 2013;49(10):2914–2926. Available from: https://doi.org/10.1016/j.cortex.2013.02.002.
25. Bryden MP. Measuring handedness with questionnaires. *Neuropsychologia*. 1977;15(4–5):617–624. Available from: https://doi.org/10.1016/0028-3932(77)90067-7.
26. Crovitz HF, Zener K. A group-test for assessing hand- and eye-dominance. *Am J Psychol*. 1962;75(2):271. Available from: https://doi.org/10.2307/1419611.
27. Steenhuis RE, Bryden MP. Different dimensions of hand preference that relate to skilled and unskilled activities. *Cortex*. 1989;25(2):289–304. Available from: https://doi.org/10.1016/s0010-9452(89)80044-9.
28. Coren S. The lateral preference inventory for measurement of handedness, footedness, eyedness, and earedness: norms for young adults. *Bull Psychon Soc*. 1993;31(1):1–3. Available from: https://doi.org/10.3758/BF03334122.
29. Siefer A, Ehrenstein WH, Arnold-Schulz-Gahmen BE, Sokeland J, Luttmann A. Population statistics and association analysis of sensorimotor lateral preference and its relevance for various fields of professional activity. *Zentralblatt fur Arbeitsmedizin Arbeitsschutz und Ergonomie*. 2003;53(6):346–353.
30. Annett M. A classification of hand preference by association analysis. *Br J Psychol*. 1970;61(3):303–321. Available from: https://doi.org/10.1111/j.2044-8295.1970.tb01248.x.
31. Bryden PJ, Roy EA, Spence J. An observational method of assessing handedness in children and adults. *Devel Neuropsychol*. 2007;32(3):825–846. Available from: https://doi.org/10.1080/87565640701539667.
32. Ocklenburg S, Bürger C, Westermann C, Schneider D, Biedermann H, Güntürkün O. Visual experience affects handedness. *Behav Brain Res*. 2010;207(2):447–451. Available from: https://doi.org/10.1016/j.bbr.2009.10.036.
33. Annett M. *Handedness and Brain Asymmetry: The Right Shift Theory*. Psychology Press; 2002.
34. Nicholls MER, Chapman HL, Loetscher T, Grimshaw GM. The relationship between hand preference, hand performance, and general cognitive ability. *J Int Neuropsychol Soc*. 2010;16(4):585–592. Available from: https://doi.org/10.1017/S1355617710000184.
35. Peters M, Durding BM. Handedness measured by finger tapping: a continuous variable. *Can J Psychol*. 1978;32(4):257–261. Available from: https://doi.org/10.1037/h0081694.
36. McManus C, van Horn JD, Bryden PJ. The Tapley and Bryden test of performance differences between the hands: the original data, newer data, and the relation to pegboard and other tasks. *Laterality*. 2016;21(4-6):371–396. Available from: https://doi.org/10.1080/1357650X.2016.1141916.

37. Tapley SM, Bryden MP. A group test for the assessment of performance between the hands. *Neuropsychologia*. 1985;23(2):215–221. Available from: https://doi.org/10.1016/0028-3932(85)90105-8.
38. Porac C. Hand preference and skilled hand performance among individuals with successful rightward conversions of the writing hand. *Laterality*. 2009;14(2):105–121. Available from: https://doi.org/10.1080/13576500802299418.
39. Provins KA, Magliaro J. The measurement of handedness by preference and performance tests. *Brain Cognition*. 1993;22(2):171–181. Available from: https://doi.org/10.1006/brcg.1993.1032.
40. Coren S, Porac C. Fifty centuries of right-handedness: the historical record. *Science*. 1977;198(4317):631–632. Available from: https://doi.org/10.1126/science.335510.
41. Raymond M, Pontier D. Is there geographical variation in human handedness? *Laterality*. 2004;9(1):35–51. Available from: https://doi.org/10.1080/13576500244000274.
42. Eitler K, Bibok A, Telkes G. Situs inversus totalis: a clinical review. *Int J Gen Med*. 2022;15:2437–2449. Available from: https://doi.org/10.2147/IJGM.S295444.
43. Torgersen J. Situs inversus, asymmetry, and twinning. *Am J Hum Genet*. 1950;2(4):361–370.
44. Postema MC, Carrion-Castillo A, Fisher SE, Vingerhoets G, Francks C. The genetics of situs inversus without primary ciliary dyskinesia. *Sci Rep*. 2020;10(1):3677. Available from: https://doi.org/10.1038/s41598-020-60589-z.
45. Jordan HE. The inheritance of left-handedness. *J Hered*. 1911;2(1):19–29. Available from: https://doi.org/10.1093/jhered/2.1.19.
46. Reiss M, Reiss G. Earedness and handedness: distribution in a German sample with some Family Data. *Cortex*. 1999;35(3):403–412. Available from: https://doi.org/10.1016/S0010-9452(08)70808-6.
47. Carter-Saltzman L. Biological and sociocultural effects on handedness: comparison between biological and adoptive families. *Science*. 1980;209(4462):1263–1265. Available from: https://doi.org/10.1126/science.7403887.
48. Sicotte NL, Woods RP, Mazziotta JC. Handedness in twins: a meta-analysis. *Laterality*. 1999;4(3):265–286. Available from: https://doi.org/10.1080/713754339.
49. Pfeifer LS, Schmitz J, Papadatou-Pastou M, Peterburs J, Paracchini S, Ocklenburg S. Handedness in twins: meta-analyses. *BMC Psychol*. 2022;10(1):11. Available from: https://doi.org/10.1186/s40359-021-00695-3.
50. Cuellar-Partida G, Tung JY, Eriksson N, et al. Genome-wide association study identifies 48 common genetic variants associated with handedness. *Nat Hum Behav*. 2021;5(1):59–70. Available from: https://doi.org/10.1038/s41562-020-00956-y.
51. Kovel CGFde, Francks C. The molecular genetics of hand preference revisited. *Sci Rep*. 2019;9(1):5986. Available from: https://doi.org/10.1038/s41598-019-42515-0.
52. Wiberg A, Ng M, Al Omran Y, et al. Handedness, language areas and neuropsychiatric diseases: insights from brain imaging and genetics. *Brain*. 2019;142(10):2938–2947. Available from: https://doi.org/10.1093/brain/awz257.
53. Pertea M, Salzberg SL. Between a chicken and a grape: estimating the number of human genes. *Genome Biol*. 2010;11(5):206. Available from: https://doi.org/10.1186/gb-2010-11-5-206.
54. Flöel A, Buyx A, Breitenstein C, Lohmann H, Knecht S. Hemispheric lateralization of spatial attention in right- and left-hemispheric language dominance. *Behav Brain Res*. 2005;158(2):269–275. Available from: https://doi.org/10.1016/j.bbr.2004.09.016.
55. Knecht S, Dräger B, Deppe M, et al. Handedness and hemispheric language dominance in healthy humans. *Brain*. 2000;123(12):2512–2518. Available from: https://doi.org/10.1093/brain/123.12.2512. Pt 12.
56. Annett M. A Single Gene Explanation of Right and Left Handedness and Brainedness. Coventry, UK: Lanchester Polytechnic; 1978.
57. McManus C. Handedness, language dominance and aphasia: A genetic model. *Psychol Med Monogr Suppl*. 1985;8:3–40. Available from: https://doi.org/10.1017/S0264180100001879.
58. McKeever WF. An X-linked three allele model of hand preference and hand posture for writing. *Laterality*. 2004;9(2):149–173. Available from: https://doi.org/10.1080/13576500244000292.
59. Jones GV, Martin M. Language dominance, handedness and sex: recessive X-linkage theory and test. *Cortex*. 2010;46(6):781–786. Available from: https://doi.org/10.1016/j.cortex.2009.07.009.
60. Crow TJ. The 'big bang' theory of the origin of psychosis and the faculty of language. *Schizophrenia Res*. 2008;102(1-3):31–52. Available from: https://doi.org/10.1016/j.schres.2008.03.010.

61. Lander ES, Linton LM, Birren B, et al. Initial sequencing and analysis of the human genome. *Nature*. 2001;409 (6822):860–921. Available from: https://doi.org/10.1038/35057062.
62. Venter JC, Adams MD, Myers EW, et al. The sequence of the human genome. *Science*. 2001;291 (5507):1304–1351. Available from: https://doi.org/10.1126/science.1058040.
63. Levy S, Sutton G, Ng PC, et al. The diploid genome sequence of an individual human. *PLoS Biol*. 2007;5(10): e254. Available from: https://doi.org/10.1371/journal.pbio.0050254.
64. Plomin R, DeFries JC, Knopik VS, Neiderhiser JM. *Behavioral Genetics*. 6th ed. Worth Publ; 2013.
65. Francks C, Maegawa S, Laurén J, et al. LRRTM1 on chromosome 2p12 is a maternally suppressed gene that is associated paternally with handedness and schizophrenia. *Mol Psychiatry*. 2007;12(12):1129–1139, 1057. Available from: https://doi.org/10.1038/sj.mp.4002053.
66. Arning L, Ocklenburg S, Schulz S, et al. PCSK6 VNTR polymorphism is associated with degree of handedness but not direction of handedness. *PLoS One*. 2013;8(6):e67251. Available from: https://doi.org/10.1371/journal.pone.0067251.
67. Brandler WM, Morris AP, Evans DM, et al. Common variants in left/right asymmetry genes and pathways are associated with relative hand skill. *PLoS Genet*. 2013;9(9):e1003751. Available from: https://doi.org/10.1371/journal.pgen.1003751.
68. Robinson KJ, Hurd PL, Read S, Crespi BJ. The PCSK6 gene is associated with handedness, the autism spectrum, and magical ideation in a non-clinical population. *Neuropsychologia*. 2016;84:205–212. Available from: https://doi.org/10.1016/j.neuropsychologia.2016.02.020.
69. Scerri TS, Brandler WM, Paracchini S, et al. PCSK6 is associated with handedness in individuals with dyslexia. *Hum Mol Genet*. 2011;20(3):608–614. Available from: https://doi.org/10.1093/hmg/ddq475.
70. Shore R, Covill L, Pettigrew KA, et al. The handedness-associated PCSK6 locus spans an intronic promoter regulating novel transcripts. *Hum Mol Genet*. 2016;25(9):1771–1779. Available from: https://doi.org/10.1093/hmg/ddw047.
71. Arning L, Ocklenburg S, Schulz S, et al. Handedness and the X chromosome: the role of androgen receptor CAG-repeat length. *Sci Rep*. 2015;5(1):8325. Available from: https://doi.org/10.1038/srep08325.
72. Eriksson N, Macpherson JM, Tung JY, et al. Web-based, participant-driven studies yield novel genetic associations for common traits. *PLoS Genet*. 2010;6(6):e1000993. Available from: https://doi.org/10.1371/journal.pgen.1000993.
73. Armour JAL, Davison A, McManus C. Genome-wide association study of handedness excludes simple genetic models. *Heredity*. 2014;112(3):221–225. Available from: https://doi.org/10.1038/hdy.2013.93.
74. McManus C, Davison A, Armour JAL. Multilocus genetic models of handedness closely resemble single-locus models in explaining family data and are compatible with genome-wide association studies. *Ann N Y Acad Sci*. 2013;1288:48–58. Available from: https://doi.org/10.1111/nyas.12102.
75. Somers M, Ophoff RA, Aukes MF, et al. Linkage analysis in a Dutch population isolate shows no major gene for left-handedness or atypical language lateralization. *J Neurosci*. 2015;35(23):8730–8736. Available from: https://doi.org/10.1523/JNEUROSCI.3287-14.2015.
76. Kavaklioglu T, Ajmal M, Hameed A, Francks C. Whole exome sequencing for handedness in a large and highly consanguineous family. *Neuropsychologia*. 2016;93(Pt B):342–349. Available from: https://doi.org/10.1016/j.neuropsychologia.2015.11.010.
77. Karlebach G, Francks C. Lateralization of gene expression in human language cortex. *Cortex*. 2015;67:30–36. Available from: https://doi.org/10.1016/j.cortex.2015.03.003.
78. Sun T. Differential gene transcription in the left and right cerebral cortex. In: Hugdahl K, Westerhausen R, eds. *The Two Halves of the Brain: Information Processing in the Cerebral Hemispheres*. MIT Press; 2010.
79. Sun T, Walsh CA. Molecular approaches to brain asymmetry and handedness. *Nat Rev Neurosci*. 2006;7 (8):655–662. Available from: https://doi.org/10.1038/nrn1930.
80. Li Q, Bian S, Liu B, Hong J, Toth M, Sun T. Establishing brain functional laterality in adult mice through unilateral gene manipulation in the embryonic cortex. *Cell Res*. 2013;23(9):1147–1149. Available from: https://doi.org/10.1038/cr.2013.106.
81. McManus C. Cerebral polymorphisms for lateralisation: modelling the genetic and phenotypic architectures of multiple functional modules. *Symmetry*. 2022;14(4):814. Available from: https://doi.org/10.3390/sym14040814.
82. Gurd JM, Schulz J, Cherkas L, Ebers GC. Hand preference and performance in 20 pairs of monozygotic twins with discordant handedness. *Cortex*. 2006;42(6):934–945. Available from: https://doi.org/10.1016/s0010-9452 (08)70438-6.

83. Schaafsma SM, Riedstra BJ, Pfannkuche KA, Bouma A, Groothuis TGG. Epigenesis of behavioural lateralization in humans and other animals. *Philos Trans R Soc Lond B Biol Sci*. 2009;364(1519):915–927. Available from: https://doi.org/10.1098/rstb.2008.0244.
84. Previc FH. A general theory concerning the prenatal origins of cerebral lateralization in humans. *Psychol Rev*. 1991;98(3):299–334. Available from: https://doi.org/10.1037/0033-295x.98.3.299.
85. Collins RL. On the inheritance of handedness. I. Laterality in inbred mice. *J Hered*. 1968;59(1):9–12. Available from: https://doi.org/10.1093/oxfordjournals.jhered.a107656.
86. Collins RL. On the inheritance of handedness. II. Selection for sinistrality in mice. *J Hered*. 1969;60(3):117–119. Available from: https://doi.org/10.1093/oxfordjournals.jhered.a107951.
87. Collins RL. When left-handed mice live in right-handed worlds. *Science*. 1975;187(4172):181–184. Available from: http://www.jstor.org/stable/1739756.
88. Klöppel S, Vongerichten A, van Eimeren T, Frackowiak RSJ, Siebner HR. Can left-handedness be switched? Insights from an early switch of handwriting. *J Neurosci*. 2007;27(29):7847–7853. Available from: https://doi.org/10.1523/JNEUROSCI.1299-07.2007.
89. Siebner HR, Limmer C, Peinemann A, et al. Long-term consequences of switching handedness: a positron emission tomography study on handwriting in "converted" left-handers. *J Neurosci*. 2002;22(7):2816–2825. Available from: https://doi.org/10.1523/JNEUROSCI.22-07-02816.2002.
90. Porac C, Searleman A. The effects of hand preference side and hand preference switch history on measures of psychological and physical well-being and cognitive performance in a sample of older adult right-and left-handers. *Neuropsychologia*. 2002;40(12):2074–2083. Available from: https://doi.org/10.1016/s0028-3932(02)00058-1.
91. de Agostini M, Khamis AH, Ahui AM, Dellatolas G. Environmental influences in hand preference: an African point of view. *Brain Cognition*. 1997;35(2):151–167. Available from: https://doi.org/10.1006/brcg.1997.0935.
92. Zverev YP. Prevalence of the three categories of handedness among Malawian school children. *Collegium Antropologicum*. 2004;28(2):755–760.
93. Zverev YP. Cultural and environmental pressure against left-hand preference in urban and semi-urban Malawi. *Brain Cognition*. 2006;60(3):295–303. Available from: https://doi.org/10.1016/j.bandc.2005.07.008.
94. Hung C-C, Tu Y-K, Chen S-H, Chen R-C. A study on handedness and cerebral speech dominance in right-handed Chinese. *J Neurolinguistics*. 1985;1(1):143–163. Available from: https://doi.org/10.1016/s0911-6044(85)80007-5.
95. Stetter L, Sattler JB, Marquardt C, Hermsdörfer J. Handwriting kinematics during learning to write with the dominant left hand in converted left-handers. *Sci Rep*. 2023;13(1):2171. Available from: https://doi.org/10.1038/s41598-023-28911-7.
96. Bakan P, Dibb G, Reed P. Handedness and birth stress. *Neuropsychologia*. 1973;11(3):363–366. Available from: https://doi.org/10.1016/0028-3932(73)90050-x.
97. Coren S, Halpern DF. Left-handedness: a marker for decreased survival fitness. *Psychol Bull*. 1991;109(1):90–106. Available from: https://doi.org/10.1037/0033-2909.109.1.90.
98. Johnston DW, Nicholls MER, Shah M, Shields MA. Nature's experiment? Handedness and early childhood development. *Demography*. 2009;46(2):281–301. Available from: https://doi.org/10.1353/dem.0.0053.
99. Satz P, Orsini DL, Saslow E, Henry R. The pathological left-handedness syndrome. *Brain Cognition*. 1985;4(1):27–46. Available from: https://doi.org/10.1016/0278-2626(85)90052-1.
100. Satz P. Left-handedness and early brain insult: an explanation. *Neuropsychologia*. 1973;11(1):115–117. Available from: https://doi.org/10.1016/0028-3932(73)90071-7.
101. Coren S. Family patterns in handedness: evidence for indirect inheritance mediated by birth stress. *Behav Genet*. 1995;25(6):517–524. Available from: https://doi.org/10.1007/BF02327575.
102. Coren S, Porac C. Birth factors and laterality: effects of birth order, parental age, and birth stress on four indices of lateral preference. *Behav Genet*. 1980;10(2):123–138. Available from: https://doi.org/10.1007/BF01066263.
103. Leiber L, Axelrod S. Not all sinistrality is pathological. *Cortex*. 1981;17(2):259–271. Available from: https://doi.org/10.1016/s0010-9452(81)80046-9.
104. Orlebeke JF, Knol DL, Koopmans JR, Boomsma DI, Bleker OP. Left-handedness in twins: genes or environment? *Cortex*. 1996;32(3):479–490. Available from: https://doi.org/10.1016/s0010-9452(96)80005-0.
105. Segal NL. Origins and implications of handedness and relative birth weight for IQ in monozygotic twin pairs. *Neuropsychologia*. 1989;27(4):549–561. Available from: https://doi.org/10.1016/0028-3932(89)90059-6.

106. van Strien JW, Bouma A, Bakker DJ. Birth stress, autoimmune diseases, and handedness. *J Clin Exp Neuropsychol*. 1987;9(6):775–780. Available from: https://doi.org/10.1080/01688638708405216.
107. Williams CS, Buss KA, Eskenazi B. Infant resuscitation is associated with an increased risk of left-handedness. *Am J Epidemiol*. 1992;136(3):277–286. Available from: https://doi.org/10.1093/oxfordjournals.aje.a116493.
108. Annett M, Ockwell A. Birth order, birth stress and handedness. *Cortex*. 1980;16(1):181–187. Available from: https://doi.org/10.1016/s0010-9452(80)80033-5.
109. Bailey LM, McKeever WF. A large-scale study of handedness and pregnancy/birth risk events: implications for genetic theories of handedness. *Laterality*. 2004;9(2):175–188. Available from: https://doi.org/10.1080/13576500342000013.
110. Ehrlichman H, Zoccolotti P, Owen D. Perinatal factors in hand and eye preference: data from the Collaborative Perinatal Project. *Int J Neurosci*. 1982;17(1):17–22. Available from: https://doi.org/10.3109/00207458208985084.
111. Levander M, Schalling D, Levander SE. Birth stress, handedness and cognitive performance. *Cortex*. 1989;25(4):673–681. Available from: https://doi.org/10.1016/s0010-9452(89)80027-9.
112. McKeever WF, Suter PJ, Rich DA. Maternal age and parity correlates of handedness: gender, but no parental handedness modulation of effects. *Cortex*. 1995;31(3):543–553. Available from: https://doi.org/10.1016/s0010-9452(13)80065-2.
113. McManus C. Handedness and birth stress. *Psychol Med*. 1981;11(3):485–496. Available from: https://doi.org/10.1017/s0033291700052806.
114. Ounsted M, Cockburn J, Moar VA. Hand preference: its provenance, development, and associations with intellectual ability at the age of 7.5 years. *J Dev Behav Pediatr*. 1985;6(2):76–80.
115. Reiss M. Ist Linkshändigkeit pathologisch bedingt? *Prax (Bern 1994)*. 1996;85(34):997–1000.
116. Tan LE, Nettleton NC. Left handedness, birth order and birth stress. *Cortex*. 1980;16(3):363–373. Available from: https://doi.org/10.1016/s0010-9452(80)80038-4.
117. van der Elst W, Hurks PPM, Wassenberg R, Meijs CJC, van Boxtel MPJ, Jolles J. On the association between lateral preferences and pregnancy/birth stress events in a nonclinical sample of school-aged children. *J Clin Exp Neuropsychol*. 2011;33(1):1–8. Available from: https://doi.org/10.1080/13803391003757825.
118. Nicholls MER, Johnston DW, Shields MA. Adverse birth factors predict cognitive ability, but not hand preference. *Neuropsychology*. 2012;26(5):578–587. Available from: https://doi.org/10.1037/a0029151.
119. Jones GV, Martin M. Seasonal anisotropy in handedness. *Cortex*. 2008;44(1):8–12. Available from: https://doi.org/10.1016/j.cortex.2006.05.001.
120. Nicholls MER. Seasonal trends in the birth of sinistrals. *Laterality*. 1998;3(3):241–253. Available from: https://doi.org/10.1080/713754306.
121. Karev GB. Season of birth and parental age in right, mixed and left handers. *Cortex*. 2008;44(1):79–81. Available from: https://doi.org/10.1016/j.cortex.2006.01.001.
122. Laland KN. Exploring gene-culture interactions: insights from handedness, sexual selection and niche-construction case studies. *Philos Trans R Soc Lond B Biol Sci*. 2008;363(1509):3577–3589. Available from: https://doi.org/10.1098/rstb.2008.0132.
123. Manns M, Freund N, Patzke N, Güntürkün O. Organization of telencephalotectal projections in pigeons: impact for lateralized top-down control. *Neuroscience*. 2007;144(2):645–653. Available from: https://doi.org/10.1016/j.neuroscience.2006.09.043.
124. Manns M, Güntürkün O. 'Natural' and artificial monocular deprivation effects on thalamic soma sizes in pigeons. *NeuroReport*. 1999;10(15):3223. Available from: https://journals.lww.com/neuroreport/fulltext/1999/10190/_natural__and_artificial_monocular_deprivation.18.aspx.
125. Manns M, Güntürkün O. Dual coding of visual asymmetries in the pigeon brain: the interaction of bottom-up and top-down systems. *Exp Brain Res*. 2009;199(3-4):323–332. Available from: https://doi.org/10.1007/s00221-009-1702-z.
126. Hopkins B, Lems W, Janssen B, Butterworth G. Postural and motor asymmetries in newlyborns. *Hum Neurobiol*. 1987;6(3):153–156. Available from: https://pascal-francis.inist.fr/vibad/index.php?action=getRecordDetail&lang=en&idt=7708426.
127. Ververs IA, Vries JIde, van Geijn HP, Hopkins B. Prenatal head position from 12–38 weeks. I. Developmental aspects. *Early Hum Dev*. 1994;39(2):83–91. Available from: https://doi.org/10.1016/0378-3782(94)90157-0.

128. Ververs IA, Vries JIde, van Geijn HP, Hopkins B. Prenatal head position from 12–38 weeks. II. The effects of fetal orientation and placental localization. *Early Hum Dev*. 1994;39(2):93–100. Available from: https://doi.org/10.1016/0378-3782(94)90158-9.
129. Michel GF. Right-handedness: a consequence of infant supine head-orientation preference? *Science*. 1981;212 (4495):685–687. Available from: https://doi.org/10.1126/science.7221558.
130. Konishi Y, Mikawa H, Suzuki J. Asymmetrical head-turning of preterm infants: some effects on later postural and functional lateralities. *Dev Med Child Neurol*. 1986;28(4):450–457. Available from: https://doi.org/10.1111/j.1469-8749.1986.tb14282.x.
131. Michel GF, Harkins DA. Postural and lateral asymmetries in the ontogeny of handedness during infancy. *Devel Psychobiol*. 1986;19(3):247–258. Available from: https://doi.org/10.1002/dev.420190310.
132. Hinojosa T, Sheu C-F, Michel GF. Infant hand-use preferences for grasping objects contributes to the development of a hand-use preference for manipulating objects. *Devel Psychobiol*. 2003;43(4):328–334. Available from: https://doi.org/10.1002/dev.10142.
133. Güntürkün O. Human behaviour: adult persistence of head-turning asymmetry. *Nature*. 2003;421(6924):711. Available from: https://doi.org/10.1038/421711a.
134. Ocklenburg S, Güntürkün O. Head-turning asymmetries during kissing and their association with lateral preference. *Laterality*. 2009;14(1):79–85. Available from: https://doi.org/10.1080/13576500802243689.
135. de Kovel CGF, Carrión-Castillo A, Francks C. A large-scale population study of early life factors influencing left-handedness. *Sci Rep*. 2019;9(1):584. Available from: https://doi.org/10.1038/s41598-018-37423-8.
136. Hujoel PP. Breastfeeding and handedness: a systematic review and meta-analysis of individual participant data. *Laterality*. 2019;24(5):582–599. Available from: https://doi.org/10.1080/1357650X.2018.1555254.
137. Schmitz J, Metz GAS, Güntürkün O, Ocklenburg S. Beyond the genome-towards an epigenetic understanding of handedness ontogenesis. *Prog Neurobiol*. 2017;159:69–89. Available from: https://doi.org/10.1016/j.pneurobio.2017.10.005.
138. Jaenisch R, Bird A. Epigenetic regulation of gene expression: how the genome integrates intrinsic and environmental signals. *Nat Genet*. 2003;33(Suppl):245–254. Available from: https://doi.org/10.1038/ng1089.
139. Leach EL, Prefontaine G, Hurd PL, Crespi BJ. The imprinted gene LRRTM1 mediates schizotypy and handedness in a nonclinical population. *J Hum Genet*. 2014;59(6):332–336. Available from: https://doi.org/10.1038/jhg.2014.30.
140. Schmitz J, Kumsta R, Moser D, Güntürkün O, Ocklenburg S. DNA methylation in candidate genes for handedness predicts handedness direction. *Laterality*. 2018;23(4):441–461. Available from: https://doi.org/10.1080/1357650X.2017.1377726.
141. Odintsova VV, Suderman M, Hagenbeek FA, et al. DNA methylation in peripheral tissues and left-handedness. *Sci Rep*. 2022;12(1):5606. Available from: https://doi.org/10.1038/s41598-022-08998-0.
142. McManus C. Is any but a tiny fraction of handedness variance likely to be due to the external environment? *Laterality*. 2021;26(3):310–314. Available from: https://doi.org/10.1080/1357650X.2021.1892126.
143. Graham JH. Nature, nurture, and noise: developmental instability, fluctuating asymmetry, and the causes of phenotypic variation. *Symmetry*. 2021;13(7):1204. Available from: https://doi.org/10.3390/sym13071204.
144. McManus C. Half a century of handedness research: myths, truths; fictions, facts; backwards, but mostly forwards. *Brain Neurosci Adv*. 2019;3. Available from: https://doi.org/10.1177/2398212818820513.
145. Pool E-M, Rehme AK, Fink GR, Eickhoff SB, Grefkes C. Handedness and effective connectivity of the motor system. *NeuroImage*. 2014;99:451–460. Available from: https://doi.org/10.1016/j.neuroimage.2014.05.048.
146. Klöppel S, van Eimeren T, Glauche V, et al. The effect of handedness on cortical motor activation during simple bilateral movements. *NeuroImage*. 2007;34(1):274–280. Available from: https://doi.org/10.1016/j.neuroimage.2006.08.038.
147. Grabowska A, Gut M, Binder M, Forsberg L, Rymarczyk K, Urbanik A. Switching handedness: fMRI study of hand motor control in right-handers, left-handers and converted left-handers. *Acta Neurobiologiae Experimentalis*. 2012;72(4):439–451. Available from: https://pubmed.ncbi.nlm.nih.gov/23377273/.
148. van den Berg FE, Swinnen SP, Wenderoth N. Involvement of the primary motor cortex in controlling movements executed with the ipsilateral hand differs between left- and right-handers. *J Cogn Neurosci*. 2011;23 (11):3456–3469. Available from: https://doi.org/10.1162/jocn_a_00018.
149. Dassonville P, Zhu XH, Uurbil K, Kim SG, Ashe J. Functional activation in motor cortex reflects the direction and the degree of handedness. *Proc Natl Acad Sci U S A*. 1997;94(25):14015–14018. Available from: https://doi.org/10.1073/pnas.94.25.14015.

150. Solodkin A, Hlustik P, Noll DC, Small SL. Lateralization of motor circuits and handedness during finger movements. *Eur J Neurol*. 2001;8(5):425–434. Available from: https://doi.org/10.1046/j.1468-1331.2001.00242.x.
151. Willems RM, van der Haegen L, Fisher SE, Francks C. On the other hand: including left-handers in cognitive neuroscience and neurogenetics. *Nat Rev Neurosci*. 2014;15(3):193–201. Available from: https://doi.org/10.1038/nrn3679.
152. Verstynen T, Diedrichsen J, Albert N, Aparicio P, Ivry RB. Ipsilateral motor cortex activity during unimanual hand movements relates to task complexity. *J Neurophysiol*. 2005;93(3):1209–1222. Available from: https://doi.org/10.1152/jn.00720.2004.
153. Hepper PG, Shahidullah S, White R. Origins of fetal handedness. *Nature*. 1990;347(6292):431. Available from: https://doi.org/10.1038/347431b0.
154. Hepper PG, McCartney GR, Shannon EA. Lateralised behaviour in first trimester human foetuses. *Neuropsychologia*. 1998;36(6):531–534. Available from: https://doi.org/10.1016/s0028-3932(97)00156-5.
155. Donkelaar HJten, Lammens M, Wesseling P, Hori A, Keyser A, Rotteveel J. Development and malformations of the human pyramidal tract. *J Neurol*. 2004;251(12):1429–1442. Available from: https://doi.org/10.1007/s00415-004-0653-3.
156. Ocklenburg S, Schmitz J, Moinfar Z, et al. Epigenetic regulation of lateralized fetal spinal gene expression underlies hemispheric asymmetries. *Elife*. 2017;6:e22784. Available from: https://doi.org/10.7554/eLife.22784.
157. Takasawa E, Abe M, Chikuda H, Hanakawa T. A computational model based on corticospinal functional MRI revealed asymmetrically organized motor corticospinal networks in humans. *Commun Biol*. 2022;5(1):664. Available from: https://doi.org/10.1038/s42003-022-03615-2.
158. Halpern DF, Coren S. Do right-handers live longer? *Nature*. 1988;333(6170):213. Available from: https://doi.org/10.1038/333213b0.
159. Halpern DF, Coren S. Handedness and life span. *N Engl J Med*. 1991;324(14):998. Available from: https://doi.org/10.1056/NEJM199104043241418.
160. Porac C, Friesen IC, Barnes MP, Gruppuso V. Illness and accidental injury in young and older adult left- and right-handers: implications for genetic theories of hand preference. *Devel Neuropsychol*. 1998;14(1):157–172. Available from: https://doi.org/10.1080/87565649809540705.
161. Papadatou-Pastou M, Tomprou D-M. Intelligence and handedness: meta-analyses of studies on intellectually disabled, typically developing, and gifted individuals. *Neurosci Biobehav Rev*. 2015;56:151–165. Available from: https://doi.org/10.1016/j.neubiorev.2015.06.017.
162. Sontam V, Christman SD. Semantic organisation and handedness: mixed-handedness is associated with more diffuse activation of ambiguous word associates. *Laterality*. 2012;17(1):38–50. Available from: https://doi.org/10.1080/1357650X.2010.529450.
163. Sontam V, Christman SD, Jasper JD. Individual differences in semantic switching flexibility: effects of handedness. *J Int Neuropsychol Soc*. 2009;15(6):1023–1027. Available from: https://doi.org/10.1017/S1355617709990440.
164. Somers M, Shields LS, Boks MP, Kahn RS, Sommer IE. Cognitive benefits of right-handedness: a meta-analysis. *Neurosci Biobehav Rev*. 2015;51:48–63. Available from: https://doi.org/10.1016/j.neubiorev.2015.01.003.
165. van der Feen FE, Zickert N, Groothuis TGG, Geuze RH. Does hand skill asymmetry relate to creativity, developmental and health issues and aggression as markers of fitness? *Laterality*. 2020;25(1):53–86. Available from: https://doi.org/10.1080/1357650X.2019.1619750.
166. Carey DP, Smith G, Smith DT, et al. Footedness in world soccer: an analysis of France '98. *J Sports Sci*. 2001;19(11):855–864. Available from: https://doi.org/10.1080/026404101753113804.
167. Carey DP, Smith DT, Martin D, et al. The bi-pedal ape: plasticity and asymmetry in footedness. *Cortex*. 2009;45(5):650–661. Available from: https://doi.org/10.1016/j.cortex.2008.05.011.
168. Packheiser J, Schmitz J, Berretz G, et al. Four meta-analyses across 164 studies on atypical footedness prevalence and its relation to handedness. *Sci Rep*. 2020;10(1):14501. Available from: https://doi.org/10.1038/s41598-020-71478-w.
169. Elias LJ, Bryden MP. Footedness is a better predictor of language lateralisation than handedness. *Laterality*. 1998;3(1):41–51. Available from: https://doi.org/10.1080/713754287.
170. Elias LJ, Bryden MP, Bulman-Fleming MB. Footedness is a better predictor than is handedness of emotional lateralization. *Neuropsychologia*. 1998;36(1):37–43. Available from: https://doi.org/10.1016/s0028-3932(97)00107-3.

171. Reiss M, Reiss G. Ocular dominance: some family data. *Laterality*. 1997;2(1):7–16. Available from: https://doi.org/10.1080/713754254.
172. Carey DP. Vision research: losing sight of eye dominance. *Curr Biol*. 2001;11(20):R828–R830. Available from: https://doi.org/10.1016/s0960-9822(01)00496-1.
173. Elias LJ. *Side Effects: How Left-Brain Right-Brain Differences Shape Everyday Behaviour*. Dundurn Press; 2022.
174. Dittmar M. Functional and postural lateral preferences in humans: interrelations and life-span age differences. *Hum Biol*. 2002;74(4):569–585. Available from: https://doi.org/10.1353/hub.2002.0040.
175. Reiss M, Tymnik G, Kögler P, Kögler W, Reiss G. Laterality of hand, foot, eye, and ear in twins. *Laterality*. 1999;4(3):287–297. Available from: https://doi.org/10.1080/713754337.
176. Carey DP, Hutchinson CV. Looking at eye dominance from a different angle: is sighting strength related to hand preference? *Cortex*. 2013;49(9):2542–2552. Available from: https://doi.org/10.1016/j.cortex.2012.11.011.

CHAPTER

7

Spatial attention, neglect, and the right hemisphere

OUTLINE

Introductory short story

The neuropsychologist browsed attentively through the patient's records. The patient, an 86-year-old woman called IG, had worked as a manager before retiring. On July 24th, 1977, she had a stroke and since then experienced neurological symptoms. The left half of her body was paralyzed, her vision in the left visual field was impaired, and she showed an almost complete loss of responses to sensory stimulation on the left side. Because of these symptoms, she had been referred to the Center of Neuropsychology at the University of Milan. It had been known for a long time that a stroke in the right hemisphere could lead to a condition called spatial hemineglect or just neglect. In this

The Lateralized Brain
DOI: https://doi.org/10.1016/B978-0-323-99737-9.00006-9

condition, patients are unaware of objects in the left visual field. Now, the neuropsychologist Edoardo Bisiach and his colleague Claudio Luzzatti wanted to test with this patient whether spatial hemineglect could only be observed in real life or also when patients were merely imagining a visual scene. To investigate this question, Bisiach and Luzzatti developed a clever experimental design. Being a lifelong resident of Milan, the patient was familiar with the Piazza del Duomo, the central city square of Milan, and home of the world-famous Duomo, the Cathedral of Milan (see Fig. 7.1).

The neuropsychologist asked the patient to describe the buildings surrounding the Piazza del Duomo while imagining herself looking at the front of the Duomo from the opposite side of the square. The patient nodded and started describing the buildings: "The Royal Palace, the stairs, the Arengario, the Archiepiscopal Palace, and the Via delle Ore." While she spoke, the neuropsychologist looked at a map of the Piazza del Duomo (see Fig. 7.2).

As expected, the patient had only reported buildings on the right side of her imagined position on the Piazza del Duomo and had completely omitted everything on the left side. Now, it was time to conduct the critical test. He asked the patient to take a different perspective. This time, she was asked to imagine standing at the front doors of the cathedral. Thus buildings that had been on her left side from her previous vantage point were now on her right side and vice versa. Without hesitation, IG started to describe the Piazza del Duomo from her new imagined perspective. "I see the arcades with the shops, the Via Dante, the Rinascente, the jewelers' shops, a shirt shop, and Motta." It was exactly like Bisiach and Luzzatti had predicted: Again, the patient only reported buildings located on her right side, completely omitting those on the left side, even though she had been perfectly able to describe them when imagining herself standing across from the cathedral. For the first time, Bisiach and Luzzatti had shown that spatial hemineglect indeed existed in representational space.

Inspired by Bisiach and Luzzatti.[1]

FIGURE 7.1 The Piazza del Duomo ("Cathedral Square") in Milan, Italy Steffen Schmitz (Carschten)/ Wikimedia Commons/published under a CC-BY-SA-4.0. *https://commons.wikimedia.org/wiki/File:Milano,_Duomo,_2016-06_CN-01.jpg.*

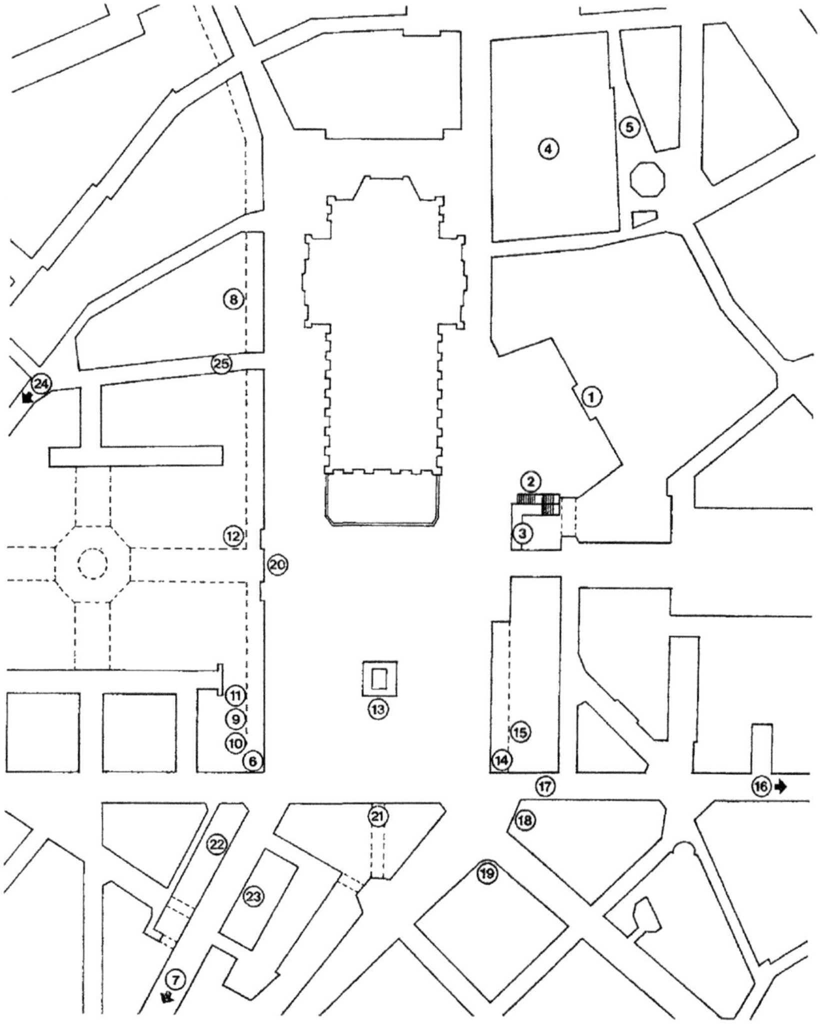

FIGURE 7.2 Plan of the Piazza del Duomo ("Cathedral Square") in Milan, Italy, from the study by Bisiach and Luzzatti.[1] The numbers indicate different buildings the patient mentioned in her report. The buildings described by IG when imagining standing opposite to the cathedral were numbers 1–5. The buildings described by IG when imaging looking at the square from the cathedral's doors are numbers 6–12. *Reproduced with permission from the publisher from Bisiach E, Luzzatti C. Unilateral neglect of representational space.* Cortex. *1978;14:129–133.*

Spatial hemineglect

If the left hemisphere largely controls verbal abilities, what are the complementing specializations of the right hemisphere? One answer to this question is given by research on spatial hemineglect (commonly just called neglect), a neurological condition that has consequences which Edoardo Bisiach and Claudio Luzzatti from the Centro di

Neuropsicologia dell'Università di Milano in Milan, Italy, strikingly demonstrated in their classic patient study described in the introductory short story of this chapter.[1]

Neglect is a common neurological condition in which patients are unaware of objects or people in the visual field contralateral to an acquired brain lesion—for example, after suffering a stroke.[2] Neglect patients typically are not aware of their deficits. The most common cause of neglect is a stroke in the territory of the middle cerebral artery, one of the three major arteries supplying the brain with blood.[3] Additional causes include tumors,[4] neurodegenerative processes (e.g., in the context of Alzheimer's dementia),[5,6] or traumatic brain injury.[7] Neglect is not a sensory (e.g., inability to see one visual field), motor (e.g., inability to move one half of the body), or cognitive (e.g., reduced attention to the environment due to depression) deficit, but a deficit in attention to and awareness of one side of space.[8] It is largely independent of sensory input or motor output. Acute neglect symptoms (e.g., immediate symptoms after the stroke that are possibly reversible) can be distinguished from chronic neglect symptoms (e.g., permanent symptoms that last for years after the stroke happened). While neglect of the visual field contralateral to a brain lesion can happen after both right- and left-hemispheric brain lesions, neglect most often occurs after a right-hemispheric lesion to the posterior parietal and/or superior and middle temporal cortex (see Fig. 7.3).[9–12]

Stone et al. investigated neglect in 171 patients with acute unilateral stroke—69 patients with right-hemispheric stroke and 102 patients with left-hemispheric stroke.[13] They found that 82% of patients with right-hemispheric lesions and 65% of patients with left-hemispheric lesions showed neglect. Moreover, patients with right-hemispheric lesions

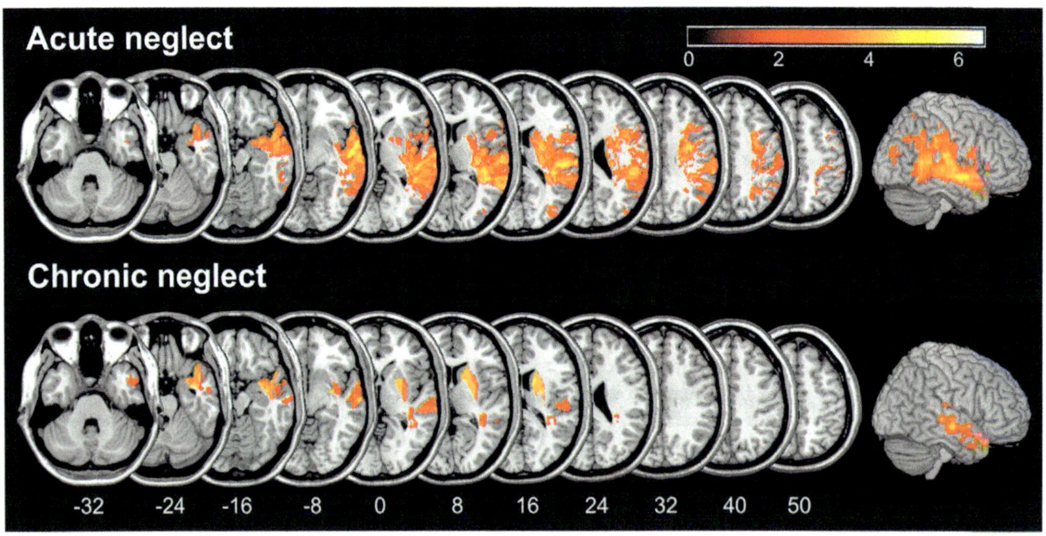

FIGURE 7.3 Lesion behavior mapping in acute and chronic neglect in patients with right-hemispheric brain damage. Injury in the highlighted regions predicts acute (upper panel) and chronic (lower panel) neglect.[9,10] *Reproduced with permission from the author and the publisher from Karnath H-O, Rennig J, Johannsen L, Rorden C. The anatomy underlying acute versus chronic spatial neglect: a longitudinal study.* Brain. *2011;134:903–912; Karnath H-O, Rorden C. The anatomy of spatial neglect.* Neuropsychologia. *2012;50:1010–1017.*

typically show more severe neglect symptoms and take longer to recover—if they recover at all—than patients with left-hemispheric lesions.[14] Moreover, an activation likelihood estimation (ALE) meta-analysis of 20 neuroimaging studies on lesion sites in patients who experienced unilateral spatial neglect revealed nine significant clusters associated with neglect—all of them in the right hemisphere and mostly in the parietal and temporal lobe.[15] Comparably, a similar ALE meta-analysis on lesion-symptom mapping studies in neglect reported 15 significant clusters, all of them in the right hemisphere.[16] Thus neglect is more common and more severe after right-hemispheric lesions than after left-hemispheric lesions.[8] This asymmetry has been interpreted to reflect a fundamental asymmetry in visuospatial attention.[17,18]

In the late 1970s, Kenneth M. Heilman from the University of Florida, United States, suggested that the left hemisphere attended only to contralateral stimuli, while the right hemisphere attended to both contralateral and ipsilateral stimuli.[19–21] To test this model, Sandra Weintraub and Marek-Marsel Mesulam from Harvard Medical School, United States, asked neglect patients to perform a shape cancellation task.[22] They were handed a sheet of paper with different shapes—for example, suns, stars, and crosses. Each quadrant of the sheet contained 90 different shapes, 15 of which were target shapes. Patients had to circle all the target shapes, and Weintraub and Mesulam determined how many target shapes were omitted on the left side of the midline of the sheet and how many were omitted on the right side of the midline. They found that patients with right-hemispheric lesions omitted more targets on the left side of the paper (17.12 out of 30) than patients with left-hemispheric lesions (1.25) or healthy controls (0.56). For targets on the right side of the paper, patients with left-hemispheric lesions made more omissions (2.38) than healthy controls (0.33). However, this difference was not significant. Surprisingly, patients with right-hemispheric lesions were significantly worse than the two other groups (8.00). Based on these findings, Weintraub and Mesulam concluded that neglect is much more severe after right-hemispheric lesions. Importantly, patients with right-hemispheric lesions also showed signs of ipsilateral neglect, supporting Heilman's idea that the right hemisphere is dominant for visuospatial attention, in the sense that the left hemisphere only contains neural units attending to the right hemispace, while the right hemisphere contains neural units attending to the left hemispace and also to the right hemispace, albeit to a somewhat lesser extent. Today, the general idea that a deficit of spatial attention underlies neglect is still pretty much agreed upon, but has since been extended. In 2011, Maurizio Corbetta and Gordon L. Shulman wrote an influential review article on neglect in which they concluded that neglect should not be understood as the consequence of damage to specific circumscribed brain regions but as a dysfunction in the interaction of two brain networks that control attention.[2] Corbetta and Shulman pointed out that the spatial attention functions that are impaired in neglect patients were controlled by a bilateral dorsal frontoparietal network, including occipital but also superior parietal and superior frontal areas.[23] This dorsal attention network, however, typically is not damaged in neglect patients. Instead, the structural damage observed in neglect patients is located in areas involved in a second attention network, the ventral attention network. The ventral attention network, located in the inferior frontal and inferior parietal lobe, controls nonspatial attention functions such as arousal, inhibition, reorienting, and detection.[24–27] In contrast to the bilateral dorsal attention network, it is strongly right-lateralized. Corbetta and Shulman argue that the two networks interact strongly. For example, it has been shown that arousal affects spatial attention, with lower arousal leading to a rightward shift in attention. Along these lines, damage to the lateralized ventral attention network impairs the function of the bilateral dorsal attention network, causing typical neglect symptoms.

How can we assess whether a patient suffers from neglect? First, there are several telltale signs in everyday behavior. For example, patients who suffer from a right-hemispheric stroke often look to the right side. Even if addressed directly in a conversation, they often do not look at the other person but rather somewhere to the right side of the other person. Persons or items in the left visual field are ignored, and the patient's attention has to be directly brought to them, so they interact with them. Also, patients often forget about items, such as keys, when these are stored in left-sided pockets. Moreover, it has been reported that patients forget to comb their hair or shave their faces on the left side. Despite the absence of motor symptoms, patients also often bump into obstacles if they are in the left visual field. In addition to observing patient behavior, several standardized tests have been developed to reliably diagnose neglect:

- Cancellation task: Patients are presented with a sheet of paper with a large number of symbols or letters on it. One specific symbol or letter is the target that patients have to mark or circle. While healthy subjects usually find equal numbers of targets on both sides of the sheet, neglect patients typically omit a larger number of targets on the left side (see Fig. 7.4).[28–30] Today, this task is sometimes also conducted with touchscreen devices.[31,32]

FIGURE 7.4 Typical results by a neglect patient in a cancellation task. While most targets (the letter A) on the right side of the paper sheet are correctly encircled, there is a large number of omissions on the left side of the paper.

- Clock drawing test: Patients are asked to draw a clock into an empty circle. While healthy subjects are more or less able to accurately space the numbers 1–12 in the circle, the drawings of neglect patients typically contain a vast amount of empty space on the left side of the clock, while the numbers and hands are mostly placed on the right side (see Fig. 7.5).[33]
- Copying task: Patients are shown a drawing of an object (e.g., a house) and are asked to copy that drawing. While healthy subjects are more or less able to copy the full drawing, neglect patients typically only copy the right side while omitting the left (see Fig. 7.6).[34]
- Writing assessment: Patients are asked to write down a sentence. Even if they are forced to start in the left visual field (e.g., by placing the tip of the pen there), they typically continue to only write on the right side of the paper after the first line break (see Fig. 7.7).
- Extinction task: When two objects are shown in the left and right visual field at the same time, neglect patients often report only seeing the object presented in the right visual field.[35]
- Line bisection: Patients are asked to bisect a horizontal line exactly in the middle.[36] Typically, they place the middle somewhere close to the middle of the portion of the line that is in the right visual field.

The line bisection task and pseudoneglect

As mentioned above, one of the oldest and simplest ways to test whether a patient suffers from neglect is the so-called line bisection task. Interestingly, this task has also been used extensively in healthy subjects to test for hemispheric asymmetries in visuospatial attention. Before you continue reading, you can use Box 7.1 to test your own spatial attention using a short version of this test.

When healthy participants are asked to mark the center of a horizontal line on the line bisection task, they show the opposite pattern compared to neglect patients with right-hemispheric lesions: They reliably show a small leftward bias, indicating that the perceived center of the line is shifted to the left of the veridical center (see Fig. 7.9).

FIGURE 7.5 Typical results of a neglect patient in the clock drawing test.

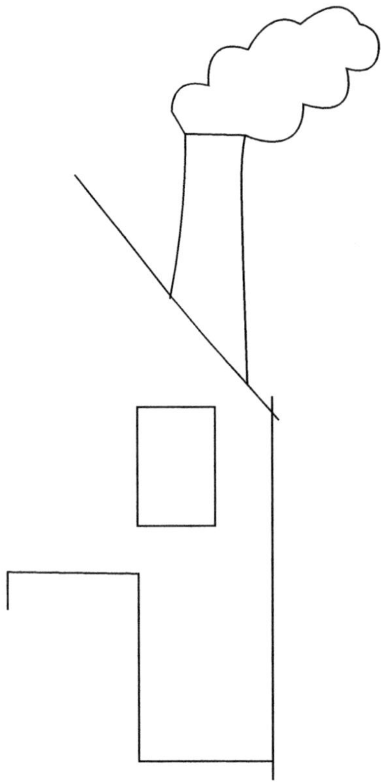

FIGURE 7.6 Typical results of a neglect patient in the copying task.

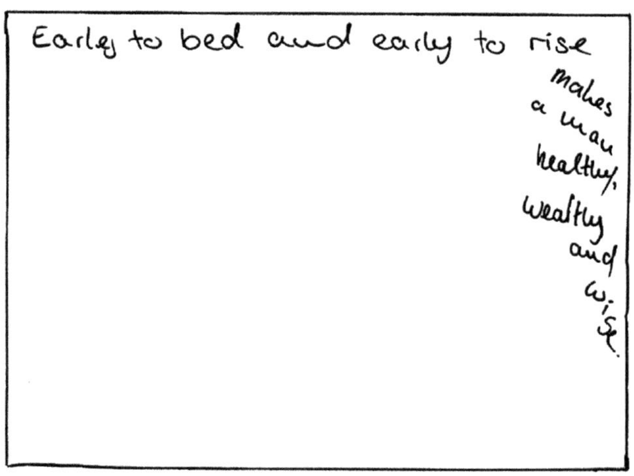

FIGURE 7.7 Typical results of a neglect patient when writing. Even if the patient is forced to start in the left visual field (e.g., by placing the tip of the pen there), they typically continue to only write on the right side of the paper after the first line break.

Today, this leftward bisection bias in healthy individuals is commonly referred to as pseudoneglect, since this bias is in the opposite direction of what has been found for neglect patients. Pseudoneglect is an attentional bias toward the left hemispace.[37] It is

BOX 7.1

Test your spatial attention.

With the lines in Fig. 7.8 you can test your spatial attention. First, place the book on a table in front of you. The page with Fig. 7.8 should be exactly centered in front of you. Now take a pencil in your right hand and quickly mark the exact middle of the first five lines. After that, take the pencil in your left hand and quickly mark the exact middle of the last five lines.

Now use a ruler to determine the exact center of each line and write down how many millimeters you deviated from the exact center of each line. Write down deviations to the left with a minus and to the right with a plus. Now sum up these 10 numbers and calculate your average bias. Do you have a positive or a negative value? While neglect patients typically show a strong rightward bias (a high positive value), healthy individuals typically show a small bias to the left (a small negative value). Almost no one can mark the center of all lines 100% correctly.

FIGURE 7.8 Follow the instructions in Box 7.1 to use these lines to test your spatial attention.

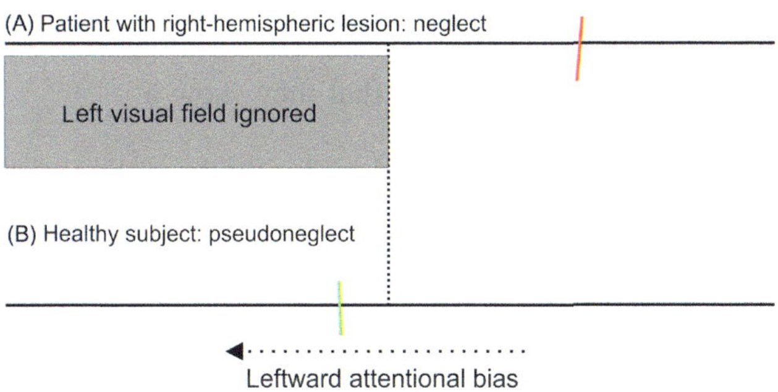

FIGURE 7.9 (A) Typical result by a neglect patient in the line bisection task. Neglect patients show a pronounced bias to the right. (B) Typical results by a healthy individual in the line bisection task. Healthy individuals show so-called pseudoneglect, that is, a slight bias to the left.

thought to result from right-hemispheric dominance for spatial attention, which leads to systematic overrepresentation of the left as compared to the right visual half-field.[38,39] This overrepresentation of the left visual field then leads to the misjudgment that the center of the line is marked somewhat to the left of the true center of the line.

The term pseudoneglect was first used in a 1980 paper by Dawn Bowers and Kenneth M. Heilman from the University of Florida, United States.[40] Interestingly, Bowers and

Heilman did not use the paper-and-pencil line bisection task in this classic study but a tactile version of the line bisection task. The stimuli in their study were 40 different sticks of balsa wood that were mounted on a piece of cardboard. Subjects were blindfolded and asked to determine the midpoint of each balsa wood stick with their index finger. Bowers and Heilman found that when the stimuli were placed exactly in the midline of the subjects, they showed a leftward bias irrespective of whether they used the left or the right index finger to determine the midline.

Subsequently, pseudoneglect was also reported by many studies using a variety of tasks, including the classic paper-and-pencil line bisection task. In 2000, George Jewell and Mark E. McCourt from North Dakota State University in Fargo, United States, performed a meta-analysis of 73 studies that assessed visual and nonvisual line bisection in healthy individuals.[38] Overall, 2,191 subjects were included in the analysis. Jewell and McCourt found that across all studies, healthy subjects showed a significant leftward bisection error with the effect size of Cohen's d between 0.37 and 0.44. An effect size is a quantitative measure of the strength of an effect that is commonly used in meta-analyses. One advantage of effect sizes is that they are independent of the *p*-value. As meta-analyses typically include thousands of individual participants, even very small effects can reach significance, making it hard to interpret the functional significance of significant effects. Cohen's d, the effect size that Jewell and McCourt used in their meta-analysis, is one of many different effect size measures. Cohen's d can be calculated by dividing the difference between the means of two groups by the pooled standard deviation.[41,42] Cohen defined a small effect size for Cohen's d as 0.2, a medium effect size as 0.3, and a large effect size as 0.5.[41] Thus the pseudoneglect findings of Jewell and McCourt had a medium to large effect size, indicating that pseudoneglect is a stable and highly replicable phenomenon. Nevertheless, Jewell and McCourt identified several variables that influence pseudoneglect. These included:

- The hand used to perform the bisection: Participants show stronger pseudoneglect when using the left hand.
- Handedness: Right-handers show stronger pseudoneglect than left-handers.
- Sex: Men show stronger pseudoneglect than women.
- Age: Older subjects show weaker pseudoneglect to the extent that they show a rightward bisection bias.
- Testing method: Visual line bisection, tactile line bisection, and midsagittal-pointing tasks all elicit pseudoneglect, but kinesthetic tasks—for example, bisection of an aluminum rod using a cursor within a channel running the length of the rod—elicit a rightward bias.
- Direction of motor scanning with eye or hand: Left-to-right scan pattern leads to strong pseudoneglect, but right-to-left scan pattern causes rightward bias.

Why do these factors impact pseudoneglect? For hand use, the explanation is easy. Motor control of the hands is crossed. The left motor cortex controls the right hand and vice versa. Thus the left hand is directly controlled by the right hemisphere, which is dominant for spatial attention and is thought to cause pseudoneglect. In contrast, the connection between the right-hemispheric brain areas involved in spatial attention and the right hand is more remote and involves interhemispheric transfer over the corpus callosum. Therefore the absence of the corpus callosum should affect performance in the line bisection task. This hypothesis was tested by Hausmann et al. who tested four patients who had undergone partial or complete split-brain surgery with the line bisection task.[43] They found that two patients showed pseudoneglect just as healthy controls did. In these patients, only the anterior part of the corpus callosum had been cut. Thus the anterior corpus callosum does not seem to affect asymmetries in visuospatial attention. In contrast, one patient who had undergone complete split-brain surgery showed the strong rightward bias typical for neglect patients. In contrast, the fourth patient who had only the posterior part of the corpus callosum cut showed stronger pseudoneglect than healthy controls. Based on these results, Hausmann et al. concluded that the posterior corpus callosum is essential for performance in the line bisection task and that its absence leads to a more consistent bias than in healthy controls. The direction of this bias, however, is not fixed but depends on which hemisphere takes control over visuospatial attention. Further evidence supporting a role of the corpus callosum for line bisection was provided in a study with children with spina bifida.[44] Spina bifida is a birth defect that results in nervous system malformations.[45] Most notable is an incomplete closing of the backbone around the spinal cord, but additionally, dysmorphologies of the corpus callosum as well as of the posterior cortex and the midbrain have been reported. Dennis et al. found that children with spina bifida showed stronger pseudoneglect than healthy controls, mirroring the results by Hausmann et al. for the split-brain patient with resection of the posterior corpus callosum.[44]

Concerning the impact of sex on line bisection, the meta-analysis by Jewell and McCourt revealed slightly stronger lateralization of visuospatial attention in men than in women, which is also in line with another study investigating sex differences in hemispheric asymmetries in visuospatial attention (see Chapter 12 for a detailed discussion of sex differences in hemispheric asymmetries).[46] For handedness, the meta-analysis by Jewell and McCourt revealed stronger pseudoneglect in right- as compared to left-handers. Jewell and McCourt argued that this effect could likely be explained by the fact that left-handers show atypical hemispheric asymmetries more often than right-handers (see Chapter 6 for a detailed discussion on handedness). Thus the prevalence of individuals with atypical left-hemispheric dominance for spatial attention might be higher among left-handers than among right-handers, reducing the leftward line bisection bias. For age, it was found that older subjects do not show pseudoneglect but rather a more neglect-like result pattern. This finding is consistent with the so-called right hemi-aging model (see Chapter 11 for details). This model assumes that the right hemisphere is more vulnerable to the effects of aging than the left in the sense that both right and left hemisphere functions decline with age, but left hemisphere functions are less strongly impaired.[47–50] Thus Jewell and McCourt suggested that the shift from pseudoneglect to the neglect-like pattern in older subjects reflected an age-related decrease in the abilities of the right hemisphere to control visuospatial attention.[38]

In addition to the variables that Jewell and McCourt identified, more recent studies have reported several additional factors that influence pseudoneglect:

1. Active training: Pseudoneglect could be reduced by training participants in a tennis video game that required attention to both visual fields.[51] This shows that active training can reduce attentional biases.

2. Cultural factors: The writing direction of the language of participants influences the degree of pseudoneglect.[52] Right-to-left readers show a rightward shift of pseudoneglect compared to left-to-right readers.[53]
3. Viewing distance: Pseudoneglect is stronger in near than in far space.[54]
4. Stress: Increased stress has been linked to an increase in pseudoneglect,[55] potentially through an activation of right-hemispheric networks due to negative emotions caused by stress.
5. Sensory modality: A study found that participants demonstrated pseudoneglect in visual, auditory, and tactile tasks, but the performance measures across sensory modalities showed only small correlations with each other.[56] This suggests that pseudoneglect depends to some extent on the sensory modality.

Pseudoneglect in other tasks

Pseudoneglect has not only been observed in the line bisection task but also in several other tasks. These include the following:

1. The grayscales task: In this task, participants view two grayscales stimuli simultaneously on a video screen (see Fig. 7.10).[58,59]
   These two stimuli are left-right mirror reversals of each other. One of the two stimuli is slightly darker than the other, and participants are asked to judge which one is darker. Typically, participants show a leftward bias in this task, in that they select the stimulus that is darker on the left side rather than the stimulus that is darker overall [57].
2. Brightness, numerosity, and size judgments: Tasks with left-right reversed stimulus pairs, like those used in the grayscales task, have also been used for the judgment of brightness, numerosity, and size. Comparable to what has been found in the grayscales task, participants showed a leftward bias, more often selecting the stimulus with the relevant feature on the left side.[60]
3. Mental bisection of number intervals: Mentally, numerical information is typically represented in spatial relationships, with smaller numbers located on the left and larger numbers located on the right.[61] When mentally determining the midpoint between two numbers without calculating (e.g., participants are asked to quickly name the midpoint between 19 and 83), participants showed a bias to the left, resembling pseudoneglect.[62–67]

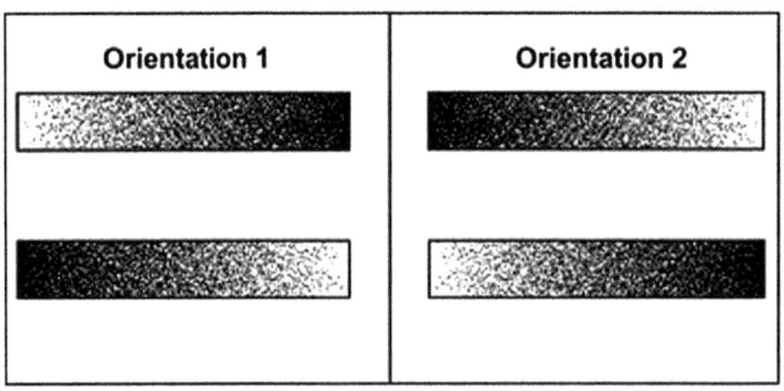

FIGURE 7.10 Examples of stimuli used in the grayscales task.[57] *Reproduced with permission from the author and the publisher from Loftus AM, Vijayakumar N, Nicholls MER. Prism adaptation overcomes pseudoneglect for the grayscales task.* Cortex. *2009;45:537–543.*

4. Mental bisection of alphabet lines: Just like numerical information, the mental representation of the alphabet is organized spatially with letters at the beginning of the alphabet located on the left and letters at the end of the alphabet located on the right side. When mentally analyzing letter triplets like "C-H-P" and asked to estimate whether there was a greater letter length on the left or the right side of the letter in the middle, participants showed a leftward bias and overestimated the letter length on the left.[68,69]
5. Auditory pseudoneglect: While most authors have investigated pseudoneglect in the visual or haptic modalities, there is also evidence for auditory pseudoneglect.[56] To test auditory pseudoneglect, participants were seated in a soundproof chamber with 13 loudspeakers, six to the left of the participant, six to the right of the participant, and one in the middle. A sound was played from two different speakers, and participants had to point at the middle between these two speakers. Participants showed a leftward deviation from the midpoint of the auditory midline, resembling pseudoneglect.
6. Online pseudoneglect tasks: While the line bisection task is a paper and pencil test that needs to be conducted in person, online measures of pseudoneglect have been developed. For example, horizontal lines can be shown on a computer screen and participants have to click on the middle of each line with a mouse cursor.[70]
7. Virtual reality pseudoneglect task: Similar to online pseudoneglect tasks, virtual reality tasks have been developed, such as a VR line bisection task.[71]

Visuospatial asymmetries in nonhuman animals

Evidence for a left-sided visuospatial bias indicating right-hemispheric control of visuospatial attention has not only been found in humans but also in other animals. Bettina Diekamp from the Biopsychology Department at Ruhr-University Bochum, Germany, and coworkers tested 2-week-old chicks (*Gallus domesticus*) and adult pigeons (*Columba livia*) with a paradigm that resembled the cancellation task used by Weintraub and Mesulam in their study with neglect patients. The animals were placed in front of an area in which pieces of grain were uniformly spread (see Fig. 7.11A) and were allowed to peck the grains.[72] While the birds could freely move their heads, their bodies were restrained centrally in front of the search area. To determine whether the birds showed a visuospatial bias, the area was subdivided into evenly spaced vertical sectors (in chicks: 17 sectors; in pigeons: nine sectors). Then, all pecks within each of these vertical sectors were counted in chicks. Since pigeons do not peck as much as chicks, a different strategy was used to analyze the data in this species: Here, the sectors were scored based on the order in which the pecks occurred, with the first peck given the highest score. The results (see Fig. 7.11B) were clear: Both chicks and pigeons pecked significantly more often in left-sided sectors. Thus like humans, these birds showed a leftward visuospatial bias and thus right-hemispheric control of visuospatial attention.

This finding has also been replicated in another study in chicks[73] as well as in an aging study in pigeons.[74] Moreover, 9-day-old chicks have been reported to show pseudoneglect in a version of the line bisection task in which they were trained to peck at the central bead of a line that was made of beads to get access to food.[75] The implication seems to be straightforward: It is conceivable that the asymmetries that result in neglect in human

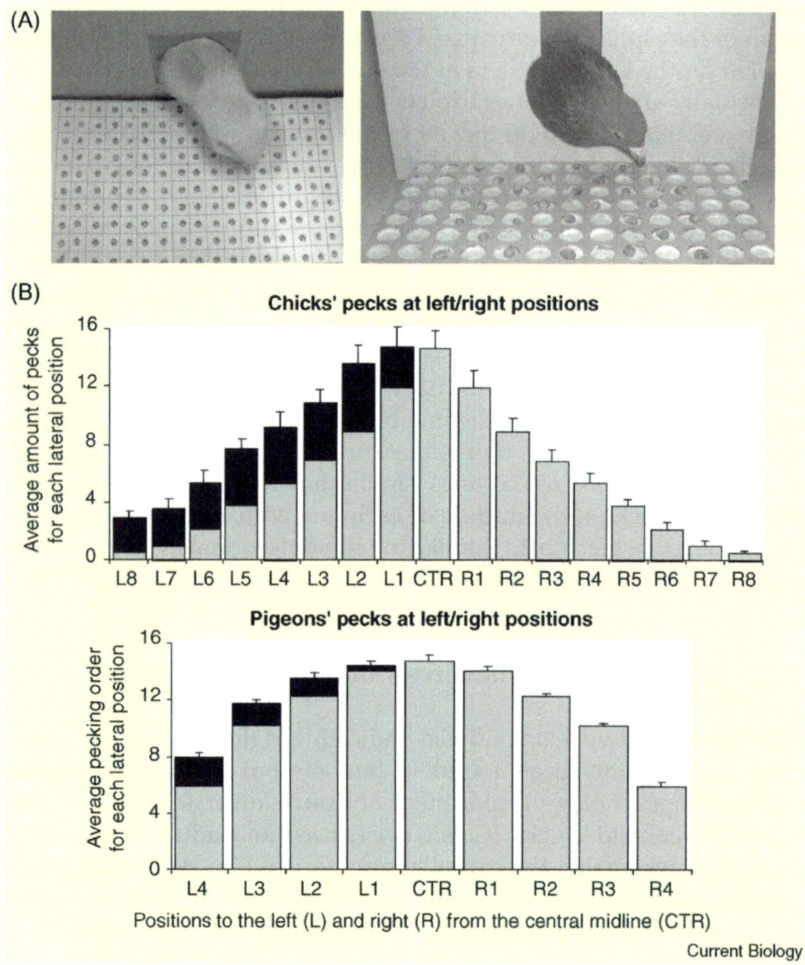

FIGURE 7.11 (A) The experimental design used by Diekamp et al.[72] Chicks and adult pigeons were restrained centrally in front of an area with evenly distributed grains and allowed to move their heads to peck for these grains. (B) The results of the study by Diekamp et al.[72] For chicks, all pecks within each vertical sector were counted. For pigeons, the sectors were scored based on the order in which the pecks occurred, with the first peck given the highest score. The birds pecked significantly more often (chicks) and earlier (pigeons) in left-sided sectors (L) than in the equivalent right-sided sectors (R). The number for each sector indicates the distance from the central midline (CTR). The black sections of the bar plots highlight the number of left-over-right pecks for chicks and earlier pecks within left sectors compared to corresponding right sectors for pigeons. *Reproduced with permission from the authors and the publisher from Diekamp B, Regolin L, Güntürkün O, Vallortigara G. A left-sided visuospatial bias in birds.* Curr Biol. *2005;15: R372–R373.*

patients have a neural fundament that is shared by a large number of animals, like birds. Since the last common ancestor of birds and mammals dates back about 300 million years,[76] this would imply that we are dealing with a truly ancient lateralized brain system (also see Chapter 2).

The neuroanatomy of visuospatial attention

As outlined in the previous sections, there is strong evidence for the dominance of the right hemisphere for visuospatial attention in both humans and nonhuman animals. However, which brain areas and networks are relevant for visuospatial attention? One of the first studies that investigated this question using modern neuroimaging techniques was conducted by José F. Pardo from the Washington University Medical Center in St. Louis, United States, and coworkers.[77] Pardo et al. used positron emission tomography (PET) to identify changes in regional blood flow in the brain during simple sustained attention tasks in 23 subjects. Subjects were tested with a somatosensory and a visual task. In the somatosensory sustained attention task, they were instructed to attend to either their left or their right big toe. A filament was used to touch the toes manually with a frequency between 3 and 5 Hz. However, sometimes this somatosensory stimulation was paused for 1–3 s, and subjects were instructed to count and report the number of these pauses during each scanning period. Therefore they had to permanently focus their attention on their big toes. Analysis of the PET scans revealed that focusing attention on the left big toe led to brain activation in the right prefrontal and right superior parietal cortex. The same brain areas were also active when subjects focused their attention on their right big toe, with an additional smaller activation cluster in the left superior parietal cortex. In the visual task, subjects were asked to fixate on a fixation cross that was presented on a video screen and to report whenever there was a luminance change. Thus they had to constantly focus their attention on the fixation cross to detect any luminance changes. Comparably to the somatosensory task, the visual task revealed strong activation in the right prefrontal cortex and right superior parietal cortex. Based on these findings, Pardo et al. concluded that the right prefrontal cortex and right superior parietal cortex form a neural network that is involved in sustained attention to sensory stimulation.

Pardo et al. used a central visual stimulus, but what happens when participants have to actively shift their attention toward the right or left visual half-field? This question was investigated in another early neuroimaging study on visuospatial attention that was conducted by Maurizio Corbetta and coworkers in 1993.[78] Also using PET, Corbetta et al. measured brain activation in 24 healthy right-handers while they performed a shifting attention task, in which they had to shift their attention to peripheral locations within the left or right visual half-field. They found that activation in the superior parietal and superior frontal cortex increased when participants shifted their attention to peripheral locations compared to when they maintained their attention at the center of gaze. Interestingly, there was a pronounced difference in superior parietal cortex activation between right and left visual field trials. Left visual field stimulus detection led to strong activation in the contralateral right superior parietal cortex and much weaker activation in the ipsilateral left superior parietal cortex. The magnitude of the right-hemispheric activation was about twice as large as that of the left-hemispheric activation. For right visual field stimulus detection, there was also activation in both the contralateral and the ipsilateral superior parietal cortex, but here activation was weaker and there was no strong leftward asymmetry. However, the right superior parietal cortex showed stronger activation in this condition than the left superior parietal cortex for left visual field detection. These findings suggest that the activation asymmetry in the superior parietal cortex might be one of the factors influencing hemispheric asymmetries in behavioral tests of visuospatial attention. Corbetta et al. concluded that this hemispheric asymmetry could potentially explain why neglect typically is observed more often and also more severe after right-hemispheric lesions compared to after left-hemispheric lesions.

Taken together, these two classic studies demonstrate a clear rightward activation asymmetry in frontoparietal networks during attention tasks. However, which brain areas are active when participants are tested with the line bisection task in the fMRI scanner? Since participants lie down in the scanner, it is not possible to let them perform the line bisection task while scanning their brains. Therefore researchers use variations of the line bisection task in the scanner in which participants are not actually bisecting a line on a piece of paper. Typically, the so-called landmark test is used: Participants are shown lines that are already bisected and they have to judge whether the shown lines were correctly bisected at the midline or not.[79–81] For example, in 2000, Fink et al. tested 12 male right-handers with the landmark task while recording brain activity with fMRI.[82] Just like the early PET experiment by Pardo and Corbetta, the findings Fink et al. obtained suggested an important role of the right parietal cortex for visuospatial attention. They found activation in the right superior and inferior parietal lobe but also in early visual processing areas bilaterally as well as in the cerebellar vermis and the left cerebellar hemisphere. A later study by Fink et al. also reported bilateral activation in the prefrontal cortex during the landmark task.[83] In 2009, Çiçek et al. tested participants with both the landmark test and a line bisection test, in which the subjects had to move a cursor to bisect a line on a screen.[84] They found that both tasks led to activations in the right intraparietal sulcus and lateral peristriate cortex.

Similarly, Céline Cavézian and coworkers tested participants with both a bisection and a landmark test version of the grayscales task.[85] For the bisection task, participants were outfitted with a gaze tracker system and their eye movements were recorded. They were asked to bisect the grayscale stimuli with their eye movements by fixating on the section of the stimulus they thought to be its center. For the landmark test, the same stimuli were used, but transection marks were added. Participants were falsely told that none of the transection marks correctly showed the true center of the stimulus. They had to indicate whether a transection mark was to the left or to the right of the true center of the stimulus. For the bisection test, Cavézian et al. found activation in a bilateral cortical network including the parietal and frontal lobes but also occipital areas and the cerebellum (see Fig. 7.12).

For the landmark test, Cavézian et al. found activations in parietal and frontal but also occipital and cerebellar areas, for which there was a rightward activation asymmetry (see Fig. 7.13).

The strongest rightward asymmetry was observed in activations of the superior parietal lobule. Cavézian et al. concluded that the more pronounced bilateral activity in the bisection task was due to a different cognitive strategy in this task compared to the landmark task. In the bisection task, participants often directly fixated on what they thought was the veridical center of the stimulus and after that did not deviate much from their initial fixation. In the landmark test, participants explored the stimuli more fully, potentially activating more right-lateralized networks for visual attention. Similar results were also found in a more recent fMRI study on perceptual pseudoneglect.[86] Here, a right-hemispheric network including occipital and parietal brain areas, as well as the frontal eye fields and the right ventral cortex, were identified using partial least square imaging analysis.

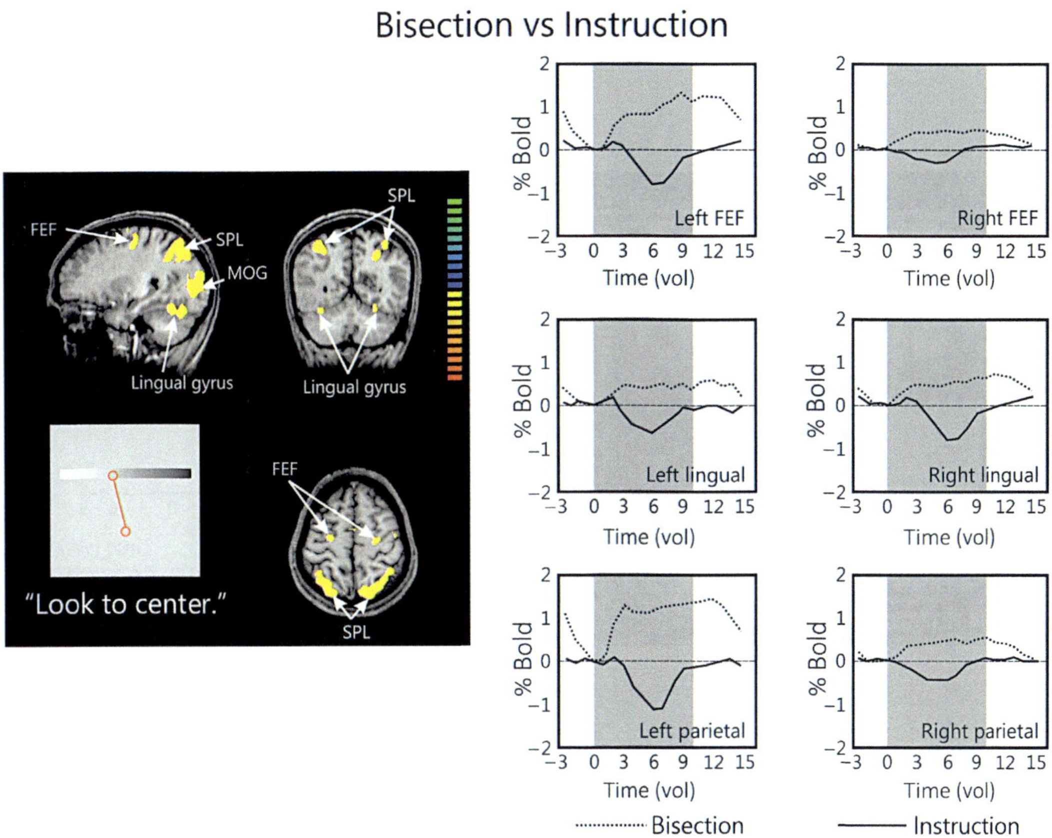

FIGURE 7.12 Results of the ocular grayscales bisection task by Cavézian et al.[85] Left panel: Brain activations observed during the bisection task contrasted with the instruction period. FEF = frontal eye field; SPL = superior parietal lobule; MOG = middle occipital gyrus. Right panel: Average percentage of event-related signal changes in the BOLD answer for the regions shown in the left panel. Brain activity during the instruction period is shown as solid lines, and brain activity during the bisection task is shown as dotted lines. Gray shading represents the stimulus period.[85] *Reproduced with permission from the publisher from Cavézian C, Valadao D, Hurwitz M, Saoud M, Danckert J. Finding centre: ocular and fMRI investigations of bisection and landmark task performance.* Brain Res. *2012;1437:89–103.*

The brain areas relevant for visuospatial attention have not only been investigated using fMRI but also with neurostimulation techniques such as transcranial direct current stimulation (tDCS). In 2012, Loftus and Nicholls used tDCS to stimulate the left or right posterior parietal cortex of participants who performed the grayscales task.[87] A tDCS setup consists of a device that delivers the current and two electrodes, one cathode (negatively charged electrode) and one anode (positively charged electrode). When the tDCS device is switched on, electric current flows from the positive anode through the brain tissue between the two electrodes to the negative cathode. When the anode is placed over the brain area of interest, tDCS stimulation increases the cortical excitability of that brain area. In contrast, if the cathode is placed over the brain area of interest, the cortical excitability of that brain area is reduced. Moreover, most experimental designs contain a sham

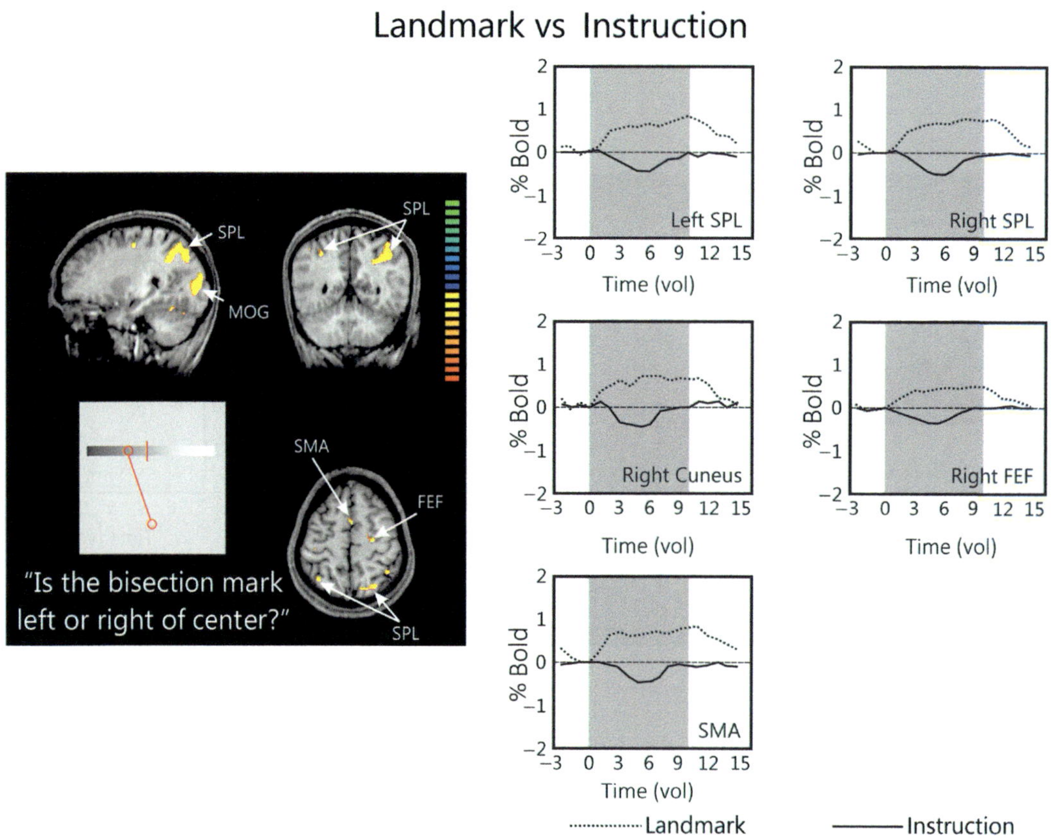

FIGURE 7.13 Results of the landmark task by Cavézian et al.[85] Left panel: Brain activations observed during the landmark task contrasted to the instruction period. FEF = frontal eye field; SPL = superior parietal lobule; MOG = middle occipital gyrus; SMA = supplementary motor area. Right panel: Average percentage of event-related signal changes in the BOLD answer for the regions shown in the left panel. Brain activity during the instruction period is shown as solid lines, and brain activity during the landmark task is shown as dotted lines. Gray shading represents the stimulus period. *Reproduced with permission from the publisher from Cavézian C, Valadao D, Hurwitz M, Saoud M, Danckert J. Finding centre: ocular and fMRI investigations of bisection and landmark task performance.* Brain Res. *2012;1437:89–103.*

condition in which no stimulation is applied to rule out placebo effects. Loftus and Nicholls found that anodal tDCS stimulation of the left posterior parietal cortex decreased pseudoneglect. Thus increased cortical excitability in this brain area led to a reduction of functional hemispheric asymmetries for visuospatial attention. In contrast, cathodal or sham stimulation of the left posterior parietal cortex or any form of stimulation over the right posterior parietal cortex did not affect pseudoneglect at all. Loftus and Nicholls concluded that these results supported the idea of an activation–orientation model of pseudoneglect. This model assumes that pseudoneglect is caused by overactivation of right-hemispheric attention networks.[88] By stimulating the left parietal cortex, this overactivation was balanced out, thereby reducing pseudoneglect.

Lateralization in white matter networks for visuospatial attention

There is not much empirical evidence for strong anatomical asymmetries in the fronto-parietal gray matter areas related to asymmetries in visuospatial attention. While Jäncke et al. reported an asymmetry in the brain area called the planum parietale, an inferior parietal area that shows strong rightward asymmetry in right-handed men and left-handed women,[89] the relevance of this asymmetry for behavioral asymmetries remains unclear. There is, however, convincing evidence that structural asymmetries in frontoparietal white matter networks might play a role in functional asymmetries in visuospatial attention. Evidence for a larger parietofrontal attention network in the right than in the left hemisphere was provided in a 2011 diffusion tensor imaging study by Michel Thiebaut de Schotten from King's College, London, United Kingdom.[90] The main white matter tract connecting the frontal and parietal lobes of the human brain is the superior longitudinal fasciculus (SLF) (see Fig. 7.14).

The SLF can be subdivided into the following:

- The dorsal superior SLF I
- The middle SLF II
- The ventral SLF III

Thiebaut de Schotten et al. measured the volume of the three SLF branches in the left and the right hemispheres of 20 right-handed subjects. Moreover, subjects were tested with the line bisection task and the classic Posner paradigm.[91] In this task, subjects had to detect targets in the left or right visual field as quickly as possible. By comparing detection times for left and right visual field trials, asymmetries in attention orientation could be quantified. Thiebaut de Schotten et al. found that the SLF I was symmetric, the SLF II showed a trend toward rightward asymmetry, and the SLF III showed clear rightward asymmetry (Fig. 7.15A). Importantly, anatomical asymmetries in the SLF were related to hemispheric asymmetries in the two behavioral tasks, implicating a relevance of anatomical asymmetries in the SLF for functional hemispheric asymmetries in visuospatial attention. As expected, subjects showed pseudoneglect in the line bisection task. Interestingly, the extent of pseudoneglect was correlated with anatomical asymmetries in the SLF II, with larger SLF II volumes in the right hemisphere associated with stronger pseudoneglect (Fig. 7.15B). Moreover, subjects with larger SLF II volumes in the right hemisphere also showed faster detection times of targets in the left visual field in the Posner task (Fig. 7.15C). Pseudoneglect and detection times in the Posner paradigm were also correlated, with stronger pseudoneglect related to faster detection times for left visual field targets (Fig. 7.15D).

Taken together, anatomical asymmetries of the SLF II can predict behavioral lateralization in visuospatial attention tasks, implicating it as one of the factors causing rightward dominance for visuospatial attention. This network approach to the neuroanatomical basis of visuospatial attention fits well with the previously mentioned Corbetta and Shulman model that also assumes that interacting networks, not specific areas, drive the right-hemispheric dominance for visuospatial attention.

More recently, asymmetries of the SLF have also been related to visuomotor processing. For example, Budisavljevic et al. reported that larger right SLF II and SLF III were related to faster visuomotor processing.[92]

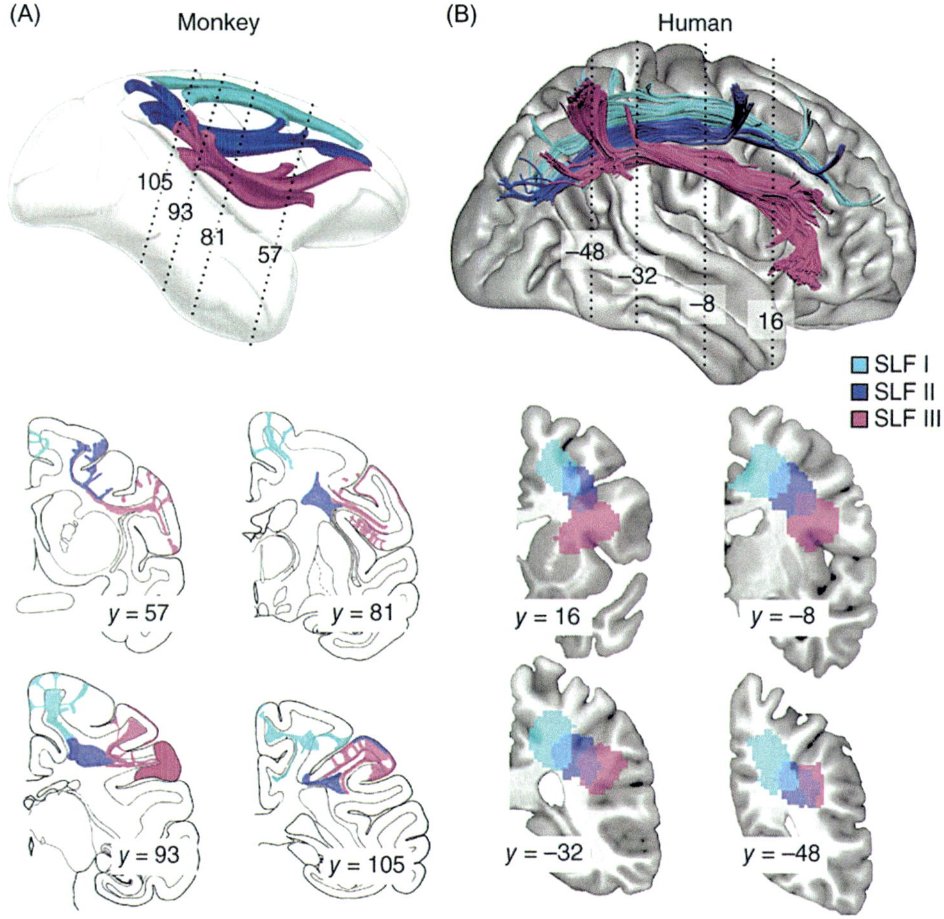

FIGURE 7.14 The three branches of the superior longitudinal fasciculus (SLF I, II, and III) in monkeys (A) and humans (B).[90] *Reproduced with permission from the author and the publisher from Thiebaut de Schotten M, Dell'Acqua F, Forkel SJ, et al. A lateralized brain network for visuospatial attention.* Nat Neurosci. *2011;14:1245–1246.*

Hemispheric asymmetries in spatial abilities

Visuospatial attention as measured with the line bisection task and similar paradigms is not the only form of spatial processing for which a dominance of the right hemisphere has been proposed. In 2003, Jennifer J. Vogel from the University of Central Florida in Orlando, United States, and coworkers conducted a meta-analysis to systematically integrate early literature on hemispheric asymmetries in spatial processing across different types of tasks.[93] Vogel et al. distinguished between three different forms of spatial tasks:

1. Spatial visualization tasks: This family of tasks included paradigms in which subjects were asked to perform mental rotation of objects. These paradigms were typically based on the classic mental rotation task.[94–97] In this task, subjects view an array of visually

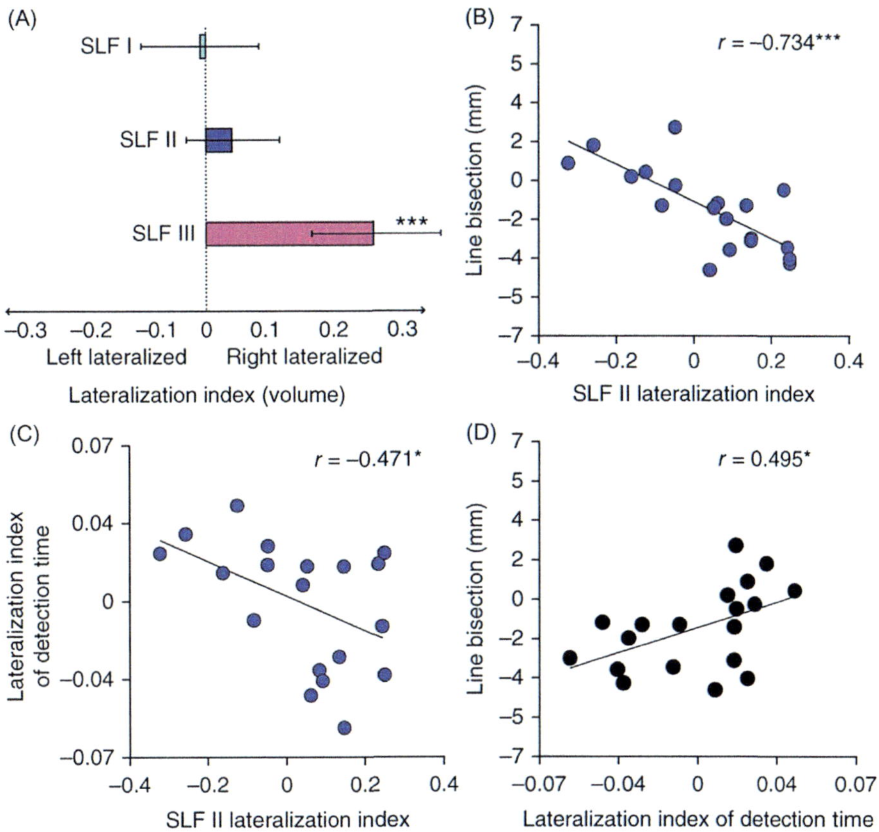

FIGURE 7.15 Results of the study by Thiebaut de Schotten et al.[90] (A) Anatomical asymmetries of the three SLF branches. Error bars show 95% confidence intervals. (B) Correlation between the anatomical asymmetry of the SLF II and pseudoneglect measured with the line bisection task. (C) Correlation between anatomical asymmetry of the SLF II and lateralization of the detection time in the Posner paradigm. (D) Correlation between pseudoneglect measured with the line bisection task and detection time in the Posner paradigm. Note that * indicates $P < 0.05$ and *** indicates $P < 0.001$. *Reproduced with permission from the author and the publisher from Thiebaut de Schotten M, Dell'Acqua F, Forkel SJ, et al. A lateralized brain network for visuospatial attention.* Nat Neurosci. *2011;14:1245–1246.*

complex block figures. At the beginning of each task, there is a target stimulus, followed by four test stimuli. Two of the test stimuli show the same block figure as the target stimulus but rotated around its vertical axis to varying degrees. The other two test stimuli show different block figures. Participants have to indicate which two test stimuli show the same block figure as the target stimulus as fast as possible. To do this, they must mentally rotate the test stimuli until they have the same spatial configuration as the target stimulus. To behaviorally assess asymmetries in this task, authors often use the divided visual field technique. For example, Jones and Anuza showed their participants visual stimuli or their mirror images in the left or right visual field and asked them to judge whether the presented stimulus had its standard form or was a mirror image.[98]

2. Spatial orientation tasks: This family of tasks included paradigms in which subjects were asked to react to symbols arranged in specific visual patterns. For example, Kimura presented a single dot in the left or the right visual field and then asked subjects to identify the location of the previously shown dot on a spatial map of all possible dot locations.[99]
3. Manual manipulation tasks: This family of tasks included paradigms in which subjects were asked to manually react to haptic stimuli in specific spatial configurations. For example, Ernest used a plywood box into which the participant's hands were inserted through holes.[100] Inside the box, the patient was handed cardboard stimuli in different shapes. Participants had to first explore a target stimulus and afterward indicate whether a second cardboard stimulus that was handed to them after the target stimulus was identical to the target stimulus or not. By comparing right-hand and left-hand performance, hemispheric asymmetries could be assessed.

Vogel et al. found an overall significant rightward asymmetry for spatial tasks. This rightward asymmetry was, however, modulated by several moderator variables. For example, when the three task types were analyzed separately, spatial orientation tasks and manual manipulation tasks yielded strong rightward asymmetry, while no asymmetry was observed for spatial visualization tasks. This is in line with a meta-analysis study of neuroimaging research on the mental rotation task that also reported largely bilateral activations.[101] In addition, a systematic review of mental rotation studies by Corballis revealed some evidence for right-hemispheric dominance for mental rotation but also some counter-evidence suggesting leftward dominance.[96]

Moreover, Vogel et al. reported that while right-handers showed strong rightward asymmetries for spatial abilities, this effect did not reach significance for left-handers. Also, the sex of participants affected the results. When male and female subjects were analyzed together, there was a significant rightward asymmetry. However, when the sexes were analyzed separately, the asymmetry only reached significance for male but not for female subjects. Vogel et al. suggested that this pattern might be explained by female subjects relying more often on verbal strategies in spatial tasks than male subjects[102]—for example, by internally naming complex shapes or verbally recapitulating strategies to solve a task. Since the left hemisphere controls most aspects of language,[103] this sex difference could lead to stronger activation of the left hemisphere during spatial tasks in female compared to male subjects. The notion of sex differences for hemispheric asymmetries in spatial processing was subsequently supported by an fMRI study on mental rotation by Hattemer et al.[104] The authors presented their participants with pairs of three-dimensional transparent cubes. The cubes were shown from different perspectives, and participants had to indicate from which perspective of the original image the rotated image was taken. Hattemer et al. found that this mental rotation task led to activations in a large bilateral network that included occipital, parietal, and frontal areas. Interestingly, there was no significant rightward lateralization of brain activity for the whole sample, but men showed stronger right-hemispheric asymmetry than women.

Remarkably, the evidence for hemispheric asymmetries of visuospatial tasks in nonhuman primates is also quite inconsistent. While one early study reported left-hemispheric dominance for visuospatial tasks in monkeys,[105] Oleksiak recently conducted a systematic review of published studies on hemispheric asymmetries in spatial functioning in nonhuman primates.[106] He concluded that there was no clear evidence for hemispheric asymmetries for spatial processing in monkeys, and that lateralization of spatial cognition in humans is a rather new feature from an evolutionary perspective. This idea is, however, contradicted by the finding that birds do show functional hemispheric asymmetries for visuospatial attention. This contradiction highlights the importance of investigating hemispheric asymmetries in spatial abilities across a wide range of species, as we currently do not have a solid understanding of the evolutionary bases of this form of lateralization.

Hemispheric asymmetries in spatial navigation

The visuospatial tasks described in the previous section typically involve participants mentally rotating or reacting to visual or haptic objects. A further form of spatial abilities that is highly relevant in real life and has been investigated regarding hemispheric asymmetries is spatial navigation, for example, the ability to navigate successfully toward a destination.[107] This can be investigated in humans using virtual reality fMRI paradigms,[107] but also using the Morris Water Maze or other forms of mazes in rodents.[108] For human participants, a recent ALE meta-analysis integrated the findings of 47 fMRI studies on spatial navigation.[107] In general, participants had to navigate two types of spaces in the included studies that differed in spatial scale. Vista spaces are environments in which the participant can see the whole space from a single viewpoint such as a room. An environmental space (such as a city) is generally larger than a vista space. In such an environment, participants must integrate spatial information from different viewpoints.

Eleven significant activation clusters were observed for environmental spaces and two for vista spaces. Interestingly, two significant activation clusters were shared for both conditions across studies, both located in the right hemisphere. These clusters included the following brain regions in the right hemisphere:

1. The parahippocampal place area: This brain area is important for the representation of the local spatial scene.
2. The retrosplenial complex: This brain area is important for locating the viewed scene within a larger spatial context.
3. The occipital place area: This brain area responds to possible choices when navigating the environment.

These findings support the idea that the core networks underlying spatial navigation in both environmental and vista spaces are in the right hemisphere. For the more complex navigation in environmental spaces, additional bilateral networks are involved in navigation. A comparative study assessing Morris maze performance in rats similarly concluded that spatial navigation performance in rodents is lateralized to the right.[108]

Spatial navigation through environments is also affected by pseudoneglect. In a study of lateralization of everyday collisions, 276 participants had to walk through a narrow doorway and the researchers recorded collisions with the doorframe.[109] There were significantly more collisions on the right side. This reflects an attentional bias to the left side comparable to pseudoneglect in the line bisection test, as more collisions happened on the less-attended right side. Therefore a right-hemispheric dominance for spatial attention can be observed during navigation through obstacles that could lead to collisions.

Sample questions

- Define and describe spatial hemineglect. Which tests can be used to assess whether a patient suffers from this condition?
- Name and describe six neuropsychological tests that can be used to determine whether a patient suffers from neglect.
- What is pseudoneglect, and how can it be tested experimentally?
- What is the grayscales task, and how can it be used to detect pseudoneglect?
- What are the typical results for neglect patients and neurologically healthy individuals in the line bisection task? Why are these results observed?
- Do animals show right-hemispheric dominance for visuospatial attention? Describe an experiment that answers this question.
- What is known about the role of structural asymmetries in white matter tracts for visuospatial attention? Describe one key experiment/finding.
- What is the mental rotation test?
- What is known about hemispheric asymmetries during spatial navigation?

References

1. Bisiach E, Luzzatti C. Unilateral neglect of representational space. *Cortex*. 1978;14(1):129–133. Available from: https://doi.org/10.1016/s0010-9452(78)80016-1.
2. Corbetta M, Shulman GL. Spatial neglect and attention networks. *Annu Rev Neurosci*. 2011;34:569–599. Available from: https://doi.org/10.1146/annurev-neuro-061010-113731.
3. Li K, Malhotra PA. Spatial neglect. *Pract Neurol*. 2015;15(5):333–339. Available from: https://doi.org/10.1136/practneurol-2015-001115.
4. Emanuele B, Santini B, Talacchi A, Gerosa M, Savazzi S. Pre- and post-operative assessment of visuo-spatial functions in right hemisphere tumour patients: a pilot study. *J Neurooncol*. 2012;108(2):261–267. Available from: https://doi.org/10.1007/s11060-012-0820-9.
5. Ishiai S, Koyama Y, Seki K, et al. Unilateral spatial neglect in AD: significance of line bisection performance. *Neurology*. 2000;55(3):364–370. Available from: https://doi.org/10.1212/WNL.55.3.364.
6. Venneri A, Pentore R, Cotticelli B, Della Sala S. Unilateral spatial neglect in the late stage of Alzheimer's Disease. *Cortex*. 1998;34(5):743–752. Available from: https://doi.org/10.1016/s0010-9452(08)70777-9.
7. Chen P, Ward I, Khan U, Liu Y, Hreha K. Spatial neglect hinders success of inpatient rehabilitation in individuals with traumatic brain injury: a retrospective study. *Neurorehabil Neural Repair*. 2016;30(5):451–460. Available from: https://doi.org/10.1177/1545968315604397.
8. Kerkhoff G. Spatial hemineglect in humans. *Prog Neurobiol*. 2001;63(1):1–27. Available from: https://doi.org/10.1016/s0301-0082(00)00028-9.
9. Karnath HO, Rennig J, Johannsen L, Rorden C. The anatomy underlying acute versus chronic spatial neglect: a longitudinal study. *Brain*. 2011;134(Pt 3):903–912. Available from: https://doi.org/10.1093/brain/awq355.
10. Karnath HO, Rorden C. The anatomy of spatial neglect. *Neuropsychologia*. 2012;50(6):1010–1017. Available from: https://doi.org/10.1016/j.neuropsychologia.2011.06.027.
11. Vallar G, Perani D. The anatomy of unilateral neglect after right-hemisphere stroke lesions. A clinical/CT-scan correlation study in man. *Neuropsychologia*. 1986;24(5):609–622. Available from: https://doi.org/10.1016/0028-3932(86)90001-1.

12. Karnath HO, Ferber S, Himmelbach M. Spatial awareness is a function of the temporal not the posterior parietal lobe. *Nature*. 2001;411(6840):950–953. Available from: https://doi.org/10.1038/35082075.
13. Stone SP, Halligan PW, Greenwood RJ. The incidence of neglect phenomena and related disorders in patients with an acute right or left hemisphere stroke. *Age Ageing*. 1993;22(1):46–52. Available from: https://doi.org/10.1093/ageing/22.1.46.
14. Stone SP, Wilson B, Wroot A, et al. The assessment of visuo-spatial neglect after acute stroke. *J Neurol Neurosurg Psychiatry*. 1991;54(4):345–350. Available from: https://doi.org/10.1136/jnnp.54.4.345.
15. Molenberghs P, Sale MV, Mattingley JB. Is there a critical lesion site for unilateral spatial neglect? A meta-analysis using activation likelihood estimation. *Front Hum Neurosci*. 2012;6:78. Available from: https://doi.org/10.3389/fnhum.2012.00078.
16. Chechlacz M, Rotshtein P, Humphreys GW. Neuroanatomical dissections of unilateral visual neglect symptoms: ALE meta-analysis of lesion-symptom mapping. *Front Hum Neurosci*. 2012;6:230. Available from: https://doi.org/10.3389/fnhum.2012.00230.
17. Vallar G. Spatial hemineglect in humans. *Trends Cogn Sci*. 1998;2(3):87–97. Available from: https://doi.org/10.1016/s1364-6613(98)01145-0.
18. Bartolomeo P, Seidel Malkinson T. Hemispheric lateralization of attention processes in the human brain. *Curr Opin Psychol*. 2019;29:90–96. Available from: https://doi.org/10.1016/j.copsyc.2018.12.023.
19. Heilman KM, Valenstein E. Mechanisms underlying hemispatial neglect. *Ann Neurol*. 1979;5(2):166–170. Available from: https://doi.org/10.1002/ana.410050210.
20. Heilman KM, van den Abell T. Right hemispheric dominance for mediating cerebral activation. *Neuropsychologia*. 1979;17(3–4):315–321. Available from: https://doi.org/10.1016/0028-3932(79)90077-0.
21. Heilman KM, van den Abell T. Right hemisphere dominance for attention: the mechanism underlying hemispheric asymmetries of inattention (neglect). *Neurology*. 1980;30(3):327–330. Available from: https://doi.org/10.1212/WNL.30.3.327.
22. Weintraub S, Mesulam MM. Right cerebral dominance in spatial attention. Further evidence based on ipsilateral neglect. *Arch Neurol*. 1987;44(6):621–625. Available from: https://doi.org/10.1001/archneur.1987.00520180043014.
23. Kastner S, Pinsk MA, de Weerd P, Desimone R, Ungerleider LG. Increased activity in human visual cortex during directed attention in the absence of visual stimulation. *Neuron*. 1999;22(4):751–761. Available from: https://doi.org/10.1016/s0896-6273(00)80734-5.
24. Stevens MC, Calhoun VD, Kiehl KA. Hemispheric differences in hemodynamics elicited by auditory oddball stimuli. *Neuroimage*. 2005;26(3):782–792. Available from: https://doi.org/10.1016/j.neuroimage.2005.02.044.
25. Aron AR, Robbins TW, Poldrack RA. Inhibition and the right inferior frontal cortex. *Trends Cogn Sci*. 2004;8(4):170–177. Available from: https://doi.org/10.1016/j.tics.2004.02.010.
26. Aron AR, Robbins TW, Poldrack RA. Inhibition and the right inferior frontal cortex: one decade on. *Trends Cogn Sci*. 2014;18(4):177–185. Available from: https://doi.org/10.1016/j.tics.2013.12.003.
27. Dosenbach NUF, Visscher KM, Palmer ED, et al. A core system for the implementation of task sets. *Neuron*. 2006;50(5):799–812. Available from: https://doi.org/10.1016/j.neuron.2006.04.031.
28. Small M, Cowey A, Ellis S. How lateralised is visuospatial neglect? *Neuropsychologia*. 1994;32(4):449–464. Available from: https://doi.org/10.1016/0028-3932(94)90090-6.
29. Mark VW, Monson N. Two-dimensional cancellation neglect a review and suggested method of analysis. *Cortex*. 1997;33(3):553–562. Available from: https://doi.org/10.1016/s0010-9452(08)70236-3.
30. Prasad S, Berkowitz AL. Modified target cancellation in hemispatial neglect. *Pract Neurol*. 2014;14(4):277. Available from: https://doi.org/10.1136/practneurol-2013-000799.
31. Keller I, Volkening K, Garbacenkaite R. Stimulus-dependent modulation of visual neglect in a touch-screen cancellation task. *Neuropsychology*. 2015;29(3):417–420. Available from: https://doi.org/10.1037/neu0000146.
32. Hopfner S, Kesselring S, Cazzoli D, et al. Neglect and motion stimuli–insights from a touchscreen-based cancellation task. *PLoS One*. 2015;10(7):e0132025. Available from: https://doi.org/10.1371/journal.pone.0132025.
33. Chen P, Goedert KM. Clock drawing in spatial neglect: a comprehensive analysis of clock perimeter, placement, and accuracy. *J Neuropsychol*. 2012;6(2):270–289. Available from: https://doi.org/10.1111/j.1748-6653.2012.02028.x.
34. Agrell BM, Dehlin OI, Dahlgren CJ. Neglect in elderly stroke patients: a comparison of five tests. *Psychiatry Clin Neurosci*. 1997;51(5):295–300. Available from: https://doi.org/10.1111/j.1440-1819.1997.tb03201.x.

35. Vossel S, Eschenbeck P, Weiss PH, et al. Visual extinction in relation to visuospatial neglect after right-hemispheric stroke: quantitative assessment and statistical lesion-symptom mapping. *J Neurol Neurosurg Psychiatry*. 2011;82(8):862–868. Available from: https://doi.org/10.1136/jnnp.2010.224261.
36. Schenkenberg T, Bradford DC, Ajax ET. Line bisection and unilateral visual neglect in patients with neurologic impairment. *Neurology*. 1980;30(5):509–517. Available from: https://doi.org/10.1212/WNL.30.5.509.
37. Ludwig K, Schenk T. Definition: pseudoneglect. *Cortex*. 2022;155:390–391. Available from: https://doi.org/10.1016/j.cortex.2022.07.007.
38. Jewell G, McCourt ME. Pseudoneglect: a review and meta-analysis of performance factors in line bisection tasks. *Neuropsychologia*. 2000;38(1):93–110. Available from: https://doi.org/10.1016/s0028-3932(99)00045-7.
39. Zago L, Petit L, Jobard G, et al. Pseudoneglect in line bisection judgement is associated with a modulation of right hemispheric spatial attention dominance in right-handers. *Neuropsychologia*. 2017;94:75–83. Available from: https://doi.org/10.1016/j.neuropsychologia.2016.11.024.
40. Bowers D, Heilman KM. Pseudoneglect: effects of hemispace on a tactile line bisection task. *Neuropsychologia*. 1980;18(4-5):491–498. Available from: https://doi.org/10.1016/0028-3932(80)90151-7.
41. Cohen J. A power primer. *Psychol Bull*. 1992;112(1):155–159. Available from: https://doi.org/10.1037/0033-2909.112.1.155.
42. Cohen J. *Statistical Power Analysis for the Behavioral Sciences*. Elsevier Science; 2013.
43. Hausmann M, Corballis MC, Farbi M. Line bisection in the split brain. *Neuropsychology*. 2003;17(4):602–609. Available from: https://doi.org/10.1037/0894-4105.17.4.602.
44. Dennis M, Edelstein K, Frederick J, et al. Peripersonal spatial attention in children with spina bifida: associations between horizontal and vertical line bisection and congenital malformations of the corpus callosum, midbrain, and posterior cortex. *Neuropsychologia*. 2005;43(14):2000–2010. Available from: https://doi.org/10.1016/j.neuropsychologia.2004.10.014.
45. Mohd-Zin SW, Marwan AI, Abou Chaar MK, Ahmad-Annuar A, Abdul-Aziz NM. Spina Bifida: pathogenesis, mechanisms, and genes in mice and humans. *Scientifica*. 2017;2017:5364827. Available from: https://doi.org/10.1155/2017/5364827.
46. McCourt ME, Mark VW, Radonovich KJ, Willison SK, Freeman P. The effects of gender, menstrual phase and practice on the perceived location of the midsagittal plane. *Neuropsychologia*. 1997;35(5):717–724. Available from: https://doi.org/10.1016/s0028-3932(96)00115-7.
47. Dolcos F, Rice HJ, Cabeza R. Hemispheric asymmetry and aging: right hemisphere decline or asymmetry reduction. *Neurosci Biobehavioral Rev*. 2002;26(7):819–825. Available from: https://doi.org/10.1016/s0149-7634(02)00068-4.
48. Salthouse TA. Differential age-related influences on memory for verbal-symbolic information and visual-spatial information? *J Gerontol B Psychol Sci Soc Sci*. 1995;50(4):P193–P201. Available from: https://doi.org/10.1093/geronb/50B.4.P193.
49. Hatta T, Iwahara A, Hatta T, et al. Developmental trajectories of verbal and visuospatial abilities in healthy older adults: comparison of the hemisphere asymmetry reduction in older adults model and the right hemi-ageing model. *Laterality*. 2015;20(1):69–81. Available from: https://doi.org/10.1080/1357650X.2014.917656.
50. Baltes PB, Lindenberger U. Emergence of a powerful connection between sensory and cognitive functions across the adult life span: a new window to the study of cognitive aging? *Psychol Aging*. 1997;12(1):12–21. Available from: https://doi.org/10.1037/0882-7974.12.1.12.
51. Gambino G, Pia L, Ferraro G, et al. Reducing visuospatial pseudoneglect in healthy subjects by active video gaming. *Brain Sci*. 2023;13(6). Available from: https://doi.org/10.3390/brainsci13060877.
52. Muayqil TA, Alhaidari GM, Alkuwaiz LA, et al. Effect of culturally mediated right-favoritism on the direction of pseudoneglect on line bisection tasks. *Front Psychol*. 2021;12:756492. Available from: https://doi.org/10.3389/fpsyg.2021.756492.
53. Chokron S, Imbert M. Influence of reading habits on line bisection. *Brain Res Cogn Brain Res*. 1993;1(4):219–222. Available from: https://doi.org/10.1016/0926-6410(93)90005-p.
54. Hartmann M, Sommer NR, Diana L, Müri RM, Eberhard-Moscicka AK. Further to the right: viewing distance modulates attentional asymmetries ('pseudoneglect') during visual exploration. *Brain Cogn*. 2019;129:40–48. Available from: https://doi.org/10.1016/j.bandc.2018.11.008.
55. Somma F, Bartolomeo P, Vallone F, et al. Further to the left: stress-induced increase of spatial pseudoneglect during the COVID-19 lockdown. *Front Psychol*. 2021;12:573846. Available from: https://doi.org/10.3389/fpsyg.2021.573846.

56. Eardley AF, Darling S, Dumper P, Browne D, van Velzen J. Related but different: examining pseudoneglect in audition, touch and vision. *Brain Cogn.* 2017;113:164–171. Available from: https://doi.org/10.1016/j.bandc.2017.01.011.
57. Loftus AM, Vijayakumar N, Nicholls MER. Prism adaptation overcomes pseudoneglect for the greyscales task. *Cortex.* 2009;45(4):537–543. Available from: https://doi.org/10.1016/j.cortex.2007.12.011.
58. Mattingley JB, Bradshaw JL, Nettleton NC, Bradshaw JA. Can task specific perceptual bias be distinguished from unilateral neglect? *Neuropsychologia.* 1994;32(7):805–817. Available from: https://doi.org/10.1016/0028-3932(94)90019-1.
59. Giglhuber K, Maurer S, Zimmer C, Meyer B, Krieg SM. Mapping visuospatial attention: the greyscales task in combination with repetitive navigated transcranial magnetic stimulation. *BMC Neurosci.* 2018;19(1):40. Available from: https://doi.org/10.1186/s12868-018-0440-1.
60. Nicholls MER, Bradshaw JL, Mattingley JB. Free-viewing perceptual asymmetries for the judgement of brightness, numerosity and size. *Neuropsychologia.* 1999;37(3):307–314. Available from: https://doi.org/10.1016/s0028-3932(98)00074-8.
61. Thomas NA, Rose W, Nicholls MER. The influence of distractors and numerical direction on mental number line bisection. *Laterality.* 2017;22(1):31–48. Available from: https://doi.org/10.1080/1357650X.2015.1108329.
62. Rotondaro F, Merola S, Aiello M, Pinto M, Doricchi F. Dissociation between line bisection and mental-number-line bisection in healthy adults. *Neuropsychologia.* 2015;75:565–576. Available from: https://doi.org/10.1016/j.neuropsychologia.2015.07.016.
63. Cattaneo Z, Fantino M, Silvanto J, Tinti C, Vecchi T. Blind individuals show pseudoneglect in bisecting numerical intervals. *Atten Percept Psychophys.* 2011;73(4):1021–1028. Available from: https://doi.org/10.3758/s13414-011-0094-x.
64. Brugger P, Schubiger M, Loetscher T. Leftward bias in number space is modulated by magical ideation. *Cognitive and Behavioral Neurology.* 2010;23(2):119–123. Available from: https://doi.org/10.1097/WNN.0b013e3181d74901.
65. Loftus AM, Nicholls MER, Mattingley JB, Chapman HL, Bradshaw JL. Pseudoneglect for the bisection of mental number lines. *Q J Exp Psychol (Hove).* 2009;62(5):925–945. Available from: https://doi.org/10.1080/17470210802305318.
66. Longo MR, Lourenco SF. Spatial attention and the mental number line: evidence for characteristic biases and compression. *Neuropsychologia.* 2007;45(7):1400–1407. Available from: https://doi.org/10.1016/j.neuropsychologia.2006.11.002.
67. Longo MR, Lourenco SF. Bisecting the mental number line in near and far space. *Brain Cogn.* 2010;72(3):362–367. Available from: https://doi.org/10.1016/j.bandc.2009.10.016.
68. Nicholls MER, Kamer A, Loftus AM. Pseudoneglect for mental alphabet lines is affected by prismatic adaptation. *Exp Brain Res.* 2008;191(1):109–115. Available from: https://doi.org/10.1007/s00221-008-1502-x.
69. Nicholls MER, Loftus AM. Pseudoneglect and neglect for mental alphabet lines. *Brain Res.* 2007;1152:130–138. Available from: https://doi.org/10.1016/j.brainres.2007.03.036.
70. Mitchell AG, Kandt PO, McIntosh RD. On line bisection: Validity and reliability of online measures of pseudoneglect. *Laterality.* 2022;27(4):443–466. Available from: https://doi.org/10.1080/1357650X.2022.2109657.
71. Kállai J, Páll T, Topa K, Zsidó AN. Physically real and virtual reality exposed line bisection response patterns: visuospatial attention allocation in virtual reality. *Front Psychol.* 2023;14:1176379. Available from: https://doi.org/10.3389/fpsyg.2023.1176379.
72. Diekamp B, Regolin L, Güntürkün O, Vallortigara G. A left-sided visuospatial bias in birds. *Current Biol.* 2005;15(10):R372–R373. Available from: https://doi.org/10.1016/j.cub.2005.05.017.
73. Chiandetti C. Pseudoneglect and embryonic light stimulation in the avian brain. *Behav Neurosci.* 2011;125(5):775–782. Available from: https://doi.org/10.1037/a0024721.
74. Wilzeck C, Kelly DM. *Avian visual pseudoneglect: the effect of age and sex on visuospatial side biases. Behavioral Lateralization in Vertebrates.* Berlin, Heidelberg: Springer; 2013:55–70.
75. Regolin L. The case of the line-bisection: when both humans and chickens wander left. *Cortex.* 2006;42(1):101–103. Available from: https://doi.org/10.1016/s0010-9452(08)70330-7.
76. Güntürkün O, Stacho M, Ströckens F. The brains of reptiles and birds. In: Kaas J, ed. *Evolution of Nervous Systems.* 2nd ed. Academic Press; 2017:171–221.
77. Pardo JV, Fox PT, Raichle ME. Localization of a human system for sustained attention by positron emission tomography. *Nature.* 1991;349(6304):61–64. Available from: https://doi.org/10.1038/349061a0.

78. Corbetta M, Miezin FM, Shulman GL, Petersen SE. A PET study of visuospatial attention. *J Neurosci*. 1993; 13(3):1202–1226. Available from: https://doi.org/10.1523/JNEUROSCI.13-03-01202.1993.
79. Milner AD, Harvey M. Distortion of size perception in visuospatial neglect. *Current Biology*. 1995;5(1):85–89. Available from: https://doi.org/10.1016/s0960-9822(95)00020-0.
80. Milner AD, Harvey M, Roberts RC, Forster SV. Line bisection errors in visual neglect: misguided action or size distortion? *Neuropsychologia*. 1993;31(1):39–49. Available from: https://doi.org/10.1016/0028-3932(93)90079-f.
81. Bisiach E, Ricci R, Lualdi M, Colombo MR. Perceptual and response bias in unilateral neglect: two modified versions of the milner landmark task. *Brain Cogn*. 1998;37(3):369–386. Available from: https://doi.org/10.1006/brcg.1998.1003.
82. Fink GR, Marshall JC, Shah NJ, et al. Line bisection judgments implicate right parietal cortex and cerebellum as assessed by fMRI. *Neurology*. 2000;54(6):1324–1331. Available from: https://doi.org/10.1212/WNL.54.6.1324.
83. Fink GR, Marshall JC, Weiss PH, Zilles K. The neural basis of vertical and horizontal line bisection judgments: an fMRI study of normal volunteers. *Neuroimage*. 2001;14(1 Pt 2):S59–S67. Available from: https://doi.org/10.1006/nimg.2001.0819.
84. Çiçek M, Deouell LY, Knight RT. Brain activity during landmark and line bisection tasks. *Front Hum Neurosci*. 2009;3:7. Available from: https://doi.org/10.3389/neuro.09.007.2009.
85. Cavézian C, Valadao D, Hurwitz M, Saoud M, Danckert J. Finding centre: ocular and fMRI investigations of bisection and landmark task performance. *Brain Res*. 2012;1437:89–103. Available from: https://doi.org/10.1016/j.brainres.2011.12.002.
86. Chen J, Lee ACH, O'Neil EB, Abdul-Nabi M, Niemeier M. Mapping the anatomy of perceptual pseudoneglect. A multivariate approach. *Neuroimage*. 2020;207:116402. Available from: https://doi.org/10.1016/j.neuroimage.2019.116402.
87. Loftus AM, Nicholls MER. Testing the activation-orientation account of spatial attentional asymmetries using transcranial direct current stimulation. *Neuropsychologia*. 2012;50(11):2573–2576. Available from: https://doi.org/10.1016/j.neuropsychologia.2012.07.003.
88. Kinsbourne M. The cerebral basis of lateral asymmetries in attention. *Acta Psychol (Amst)*. 1970;33:193–201. Available from: https://doi.org/10.1016/0001-6918(70)90132-0.
89. Jäncke L, Schlaug G, Huang Y, Steinmetz H. Asymmetry of the planum parietale. *Neuroreport*. 1994; 5(9):1161–1163. Available from: https://doi.org/10.1097/00001756-199405000-00035.
90. de Schotten MT, Dell'Acqua F, Forkel S, et al. A lateralized brain network for visuo-spatial attention. *Nat Prec*. 2011;1. Available from: https://doi.org/10.1038/npre.2011.5549.1.
91. Posner MI. Orienting of attention. *Q J Exp Psychol*. 1980;32(1):3–25. Available from: https://doi.org/10.1080/00335558008248231.
92. Budisavljevic S, Dell'Acqua F, Zanatto D, et al. Asymmetry and structure of the fronto-parietal networks underlie visuomotor processing in humans. *Cereb Cortex*. 2017;27(2):1532–1544. Available from: https://doi.org/10.1093/cercor/bhv348.
93. Vogel JJ, Bowers CA, Vogel DS. Cerebral lateralization of spatial abilities: a meta-analysis. *Brain Cogn*. 2003;52(2):197–204. Available from: https://doi.org/10.1016/s0278-2626(03)00056-3.
94. Peters M, Laeng B, Latham K, Jackson M, Zaiyouna R, Richardson C. A redrawn Vandenberg and Kuse mental rotations test: different versions and factors that affect performance. *Brain Cogn*. 1995;28(1):39–58. Available from: https://doi.org/10.1006/brcg.1995.1032.
95. Searle JA, Hamm JP. Mental rotation: an examination of assumptions. *Wiley Interdiscip Rev Cogn Sci*. 2017; 8(6):e1443. Available from: https://doi.org/10.1002/wcs.1443.
96. Corballis MC. Mental rotation and the right hemisphere. *Brain Lang*. 1997;57(1):100–121. Available from: https://doi.org/10.1006/brln.1997.1835.
97. Vandenberg SG, Kuse AR. Mental rotations, a group test of three-dimensional spatial visualization. *Percept Mot Skills*. 1978;47(2):599–604. Available from: https://doi.org/10.2466/pms.1978.47.2.599.
98. Jones B, Anuza T. Effects of sex, handedness, stimulus and visual field on "mental rotation". *Cortex*. 1982; 18(4):501–514. Available from: https://doi.org/10.1016/s0010-9452(82)80049-x.
99. Kimura D. Spatial localization in left and right visual fields. *Can J Psychol. Revue Canad Psychol*. 1969; 23(6):445–458. Available from: https://doi.org/10.1037/h0082830.

100. Ernest CH. Spatial ability and lateralization in the haptic modality. *Brain Cogn*. 1998;36(1):1–20. Available from: https://doi.org/10.1006/brcg.1997.0959.
101. Zacks JM. Neuroimaging studies of mental rotation: a meta-analysis and review. *J Cogn Neurosci*. 2008; 20(1):1–19. Available from: https://doi.org/10.1162/jocn.2008.20013.
102. McGlone J, Kertesz A. Sex differences in cerebral processing of visuospatial tasks. *Cortex*. 1973;9(3):313–320. Available from: https://doi.org/10.1016/S0010-9452(73)80009-7.
103. Ocklenburg S, Beste C, Arning L, Peterburs J, Güntürkün O. The ontogenesis of language lateralization and its relation to handedness. *Neurosci Biobehav Rev*. 2014;43:191–198. Available from: https://doi.org/10.1016/j.neubiorev.2014.04.008.
104. Hattemer K, Plate A, Heverhagen JT, et al. Determination of hemispheric dominance with mental rotation using functional transcranial Doppler sonography and FMRI. *J Neuroimaging*. 2011;21(1):16–23. Available from: https://doi.org/10.1111/j.1552-6569.2009.00402.x.
105. Jason GW, Cowey A, Weiskrantz L. Hemispheric asymmetry for a visuo-spatial task in monkeys. *Neuropsychologia*. 1984;22(6):777–784. Available from: https://doi.org/10.1016/0028-3932(84)90102-7.
106. Oleksiak A, Postma A, van der Ham IJM, Klink PC, van Wezel RJA. A review of lateralization of spatial functioning in nonhuman primates. *Brain Res Rev*. 2011;67(1-2):56–72. Available from: https://doi.org/10.1016/j.brainresrev.2010.11.002.
107. Li J, Zhang R, Liu S, et al. Human spatial navigation: Neural representations of spatial scales and reference frames obtained from an ALE meta-analysis. *Neuroimage*. 2021;238:118264. Available from: https://doi.org/10.1016/j.neuroimage.2021.118264.
108. Cowell PE, Waters NS, Denenberg VH. The effects of early environment on the development of functional laterality in Morris maze performance. *Laterality*. 1997;2(3-4):221–232. Available from: https://doi.org/10.1080/713754274.
109. Nicholls MER, Loftus A, Mayer K, Mattingley JB. Things that go bump in the right: the effect of unimanual activity on rightward collisions. *Neuropsychologia*. 2007;45(5):1122–1126. Available from: https://doi.org/10.1016/j.neuropsychologia.2006.07.015.

CHAPTER

8

Recognizing yourself and others—the role of the right hemisphere for face and self-perception

OUTLINE

The Lateralized Brain

DOI: https://doi.org/10.1016/B978-0-323-99737-9.00013-6

Introductory short story

Dr. Bodamer was puzzled. He reviewed the patient's file containing the results from previous medical examinations again, highlighting the important parts with his pen. The patient, Lieutenant HA, had been born in 1908 and was ambidextrous. On October 22, 1944, he had been wounded during an attack on one of the command posts of the German forces in Eastern Prussia. All other German officers in the command post had died immediately, but HA had been lucky. He had survived, even though a shell splinter had severely injured the right side of his head. After he had awoken, the shell splinter was removed from his head in a field hospital, but the patient continued to complain about visual impairments and paralysis of his left arm and leg. On November 25th, he had been transferred to a neurological hospital after a lack of improvement of his left hemiparesis. To this point, everything about this patient sounded like a typical hemiparetic symptom complex, following damage to the right hemisphere, but Dr. Bodamer had the feeling something was special about HA. He had noticed that the patient would only react to him or the nurses when they started speaking as if he could not correctly identify them just by looking at them. When asked about this, the patient said that he was unable to identify people just by their faces. Only when they started speaking or when he could spot characteristic features such as the height or noticeable glasses would he finally recognize them. Dr. Bodamer suspected that the patient showed a form of face blindness. He had shared this thought with the head nurse, but she had remained skeptical. HA would greet her every day even when spotting her from far away, so she was convinced that the patient could recognize her face. Dr. Bodamer decided to conduct a critical test to see whether he or the head nurse was right. He asked the head nurse and another younger nurse to come to his office and stand in front of the patient. Both nurses were known to the patient and due to a 20-year age gap between them easily identifiable. Dr. Bodamer had asked the two women to take off all items that could identify them, such as their glasses, and to wear identical uniforms. Moreover, they had to stand completely still and not say a word. During the test, HA stood in front of the two nurses, staring at their faces. He was completely unable to identify them. On the following day, Dr. Bodamer conducted an even less difficult face recognition test with HA. This time, he had asked the patient's wife to come to the hospital. They had been married for more than 7 years, so if HA was unable to recognize his wife's face, Dr. Bodamer would have proven once and for all a specific inability to recognize faces in the patient. When HA's wife arrived, Dr. Bodamer first asked her to go into the nurses' changing room and change into a standard nurse uniform. Subsequently, he asked her to line up with several real nurses, all of whom had a similar stature and height as HA's wife. The women were instructed to not move at all and to be completely silent. Dr. Bodamer then led HA into the room and asked him to identify his wife. When asked beforehand whether he would be able to identify his wife in such a line-up, HA had been confident that he could, but his behavior contradicted this confidence. Again and again, he walked from one woman to the next, staring at their faces, unable to say which one he was married to. Only after more than five minutes did he suddenly correctly identify his wife. When asked how he did identify his wife, he answered that he sensed something vaguely familiar about the look in her eyes. Thus even though the patient was able to identify his wife, he did not do it because of her face but because of recognizing a specific individual feature. Dr. Bodamer now was sure: The patient's right-hemispheric brain damage had led to a specific inability to recognize faces. Over the next years, Dr. Bodamer would collect data on similar cases and in 1947 finally publish a paper in which he coined the term for the disorder he discovered that is still used today: Prosopagnosia.

Inspired by Bodamer (1947).[1]

Where in the brain is the self?

What is the self? And where in the brain is it located? These questions have fascinated scientists for centuries.[2,3] In his 2009 book *From Axons to Identity: Neurological Explorations of the Nature of the Self*, Todd Feinberg, one of the leading experts in neurological research on the self, defined the self as follows:

The self is a unity of consciousness in perception and action that persists in time.[4]

Thus the self is experienced as a spatial unity of self and body that is distinct from other individuals and constant across time and space. Judging from the literature, the question of whether one hemisphere dominates self-awareness in general is difficult to answer. The left hemisphere is dominant for language,[5] a cognitive function that is essential to verbally reflect on oneself. In addition, in the context of split-brain research (see Chapter 3), it has been suggested that the left hemisphere acts as an interpreter—generating a unitary sense of consciousness even in a disconnected brain.[6] While these findings would argue for left-hemispheric dominance for self-awareness, there is also empirical evidence supporting an essential role of the right hemisphere in self-awareness. The right hemisphere is dominant for the recognition of faces of other people[7] and also self-face recognition,[8,9] and distinguishing one's own face from other faces is an essential part of self-awareness. Moreover, the right hemisphere is also dominant for several aspects of emotional processing[10] that are important to reflect on one's own emotional status. Additionally, it is dominant for spatial processing,[11] a cognitive function that might be essential to feel localized in one's own body.

In general, it is very unlikely that a highly complex concept such as the self can be localized in one single brain area or even one hemisphere. What can be done, however, is to identify different aspects of the self and to determine which brain areas are related to these specific functions. Moreover, dominance of one hemisphere for specific aspects of the self can be assessed. Three main aspects of the self have been identified, which we will discuss in the following sections. These are as follows:

1. Visual self-awareness: The ability to recognize one's own face or body in a mirror or in pictures.[12]
2. Sense of body ownership: Knowledge of which body parts belong to one's own body.[13]
3. Self-concept: Personal experiences and memories, ideals, or moral concepts.[14]

Visual self-awareness

Visual self-awareness refers to the ability to recognize one's own face or body in a mirror or in pictures. While all humans (except for very young children and some very rare neurological cases) possess this ability, it has long been a matter of debate whether nonhuman animals also show visual self-awareness. In January 1970, Gordon G. Gallup Jr. from the Psychology Department at Tulane University, New Orleans, United States, published a paper in *Science* that finally settled this debate.[15] In this paper, he showed experimentally that chimpanzees (*Pan troglodytes*) showed visual self-awareness. In doing so, he also established the most commonly used empirical test to determine whether a species shows visual self-awareness or not: The mirror test. Nonhuman animals, when confronted with their own reflection in a mirror, typically act as if they were interacting with another animal, that is, by showing social or antagonistic behavior. However, it has been suggested that at least some species could learn to recognize themselves just like humans do, after prolonged experience with mirrors. Due to their evolutionary closeness to humans, primates and especially great apes were the most likely candidates to show visual self-recognition. Thus, Gallup studied visual self-recognition in two male and two female preadolescent chimpanzees with little to no experience with mirrors. At the beginning of the experiment, mirrors were placed in front of the animals' cages and the chimpanzees' behavior was observed. After 10 days of exposure to the mirrors, the animals were anesthetized and removed from their cages. While the animals were unconscious, their upper eyebrow ridge was marked with clearly visible red dye. The animals were then brought back into their cages from which the mirrors had been removed and interaction with the marked area of their faces was observed (see Fig. 8.1).

Then, the mirrors were reintroduced. Interestingly, the animals' interaction with the marked area of their faces increased dramatically (between 400% and 1000%) once the mirrors were reintroduced. For example, one animal inspected the area with its finger and then sniffed at the finger for olfactory cues about the dye. Thus the animals were able to conclude that the dye was on their own faces by looking at their reflections in the mirror. Gallup concluded that these findings confirmed visual self-awareness in great apes.

Gallup initially proposed that only humans and great apes showed visual self-awareness in the mirror test. Indeed, the finding that humans, chimpanzees, and orangutans show self-recognition in the mirror test has been replicated many times by different research groups.[16] In addition, intense efforts have been made to determine whether other species outside the *Primates* order are also capable of recognizing themselves in the mirror. It has been reported that Asian elephants (*Elephas maximus*),[17,18] bottlenose dolphins (*Tursiops truncatus*),[19] killer whales (*Orcinus orca*),[20] and magpies (*Pica pica*)[21] might show self-recognition in the mirror test. In contrast, many species have failed the mirror test,[16] including several monkey species such as stump-tailed macaques (*Macaca arctoides*),[15] rhesus monkeys (*Macaca mulatta*),[15] and

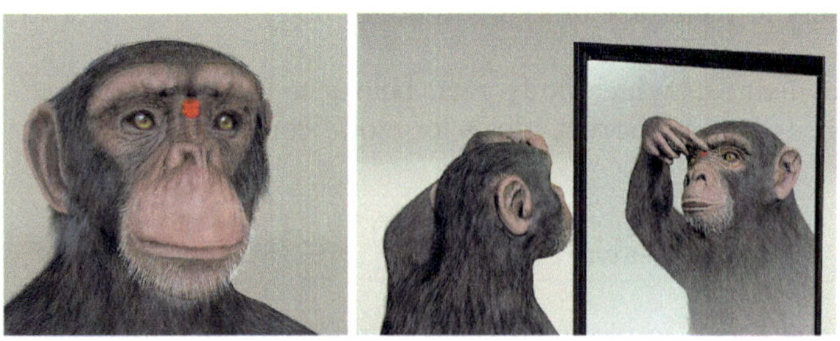

FIGURE 8.1 The Mirror test for self-recognition developed by Gallup.[15] Left panel: The upper eyebrow ridge of a chimpanzee is marked with dye. Right panel: The animal is placed in front of a mirror and shows increased manipulation behavior of the marked spot, indicating visual self-recognition.

cynomolgus monkeys (*Macaca fascicularis*)[15] as well as giant pandas (*Ailuropoda melanoleuca*),[22] sea lions (*Zalophus californianus*),[20] and jackdaws (*Corvus monedula*)[23]—to name just a few.

In humans, the mirror test was first used by Beulah Amsterdam from the University of California, United States, in a 1972 study on mirror self-recognition in children.[24] In this study, mothers were asked to put a spot of rouge on the side of their child's nose before the child's behavior in front of a mirror was observed. None of the children below the age of 18 months showed behavior that would be indicative of self-recognition in the mirror (e.g., manipulating the spot of rouge in their faces). At 18–20 months of age, 42% of children showed evidence of self-recognition, and at 20–24 months of age, 63% of children showed evidence of self-recognition. In line with this, the general understanding today is that the ability to recognize oneself in the mirror arises at around 2 years of age[25] and is subsequently normally retained for the rest of life. There are, however, some rare neurological cases in which self-recognition is lost without also displaying the symptom complex of prosopagnosia (see section on disorders of body ownership below). In chimpanzees, only animals that grew up in groups—but not individuals that grew up in isolation—display mirror self-recognition starting at an age of about 4 years.[26] Also, mirror recognition is sometimes lost in senescence.

So, what about hemispheric asymmetries in visual self-awareness? There is empirical evidence supporting the idea that the right hemisphere is dominant for self-recognition in humans. The first evidence for this view came from a 2001 study by Julian Paul Keenan and his colleagues from Harvard Medical School in Massachusetts, United States.[27] These authors used the classic Wada test (see Fig. 4.1) to assess how the hemispheres differ in processing one's own face. Five epileptic patients who underwent the Wada test before surgery to treat epilepsy were presented with stimuli that were generated by morphing their own faces with the faces of famous people—for example, Marilyn Monroe (see Fig. 8.2).

When the left hemisphere was anesthetized, 100% of patients (5 out of 5) indicated that the morphed face showed their own face. However, when the right hemisphere was anesthetized, 80% of patients (4 out of 5) indicated that the face of the famous person had been shown. Based on these results, Keenan and coworkers concluded that the right hemisphere is dominant for recognizing one's own face. These results obtained in a clinical sample with the Wada test have also been confirmed by functional magnetic resonance imaging (fMRI). Keenan and colleagues contrasted viewing one's own face with viewing the face of a famous person—for example, Bill Clinton—and found an activation asymmetry of the right prefrontal cortex for the own face.[28] In 2023, Istvan Molnar-Szakacs and Lucina Q. Uddin reviewed electroencephalography (EEG), transcranial magnetic stimulation, positron emission tomography (PET), and fMRI studies on self-face recognition and also concluded that the right hemisphere is dominant for self-face recognition.[8] Interestingly, they identified a specific neuronal network for self-recognition that is distinct from recognition of the faces of other people (see sections on face perception later in this chapter). This network involved mainly the right inferior frontoparietal cortex, but also bilateral regions of the occipitotemporal cortex. It has been suggested that the right inferior frontoparietal network is crucial for generating the feeling that a face is the face of the person looking at it.[29] A right-hemispheric dominance for self-face processing was also confirmed by a recent ALE meta-analysis.[9]

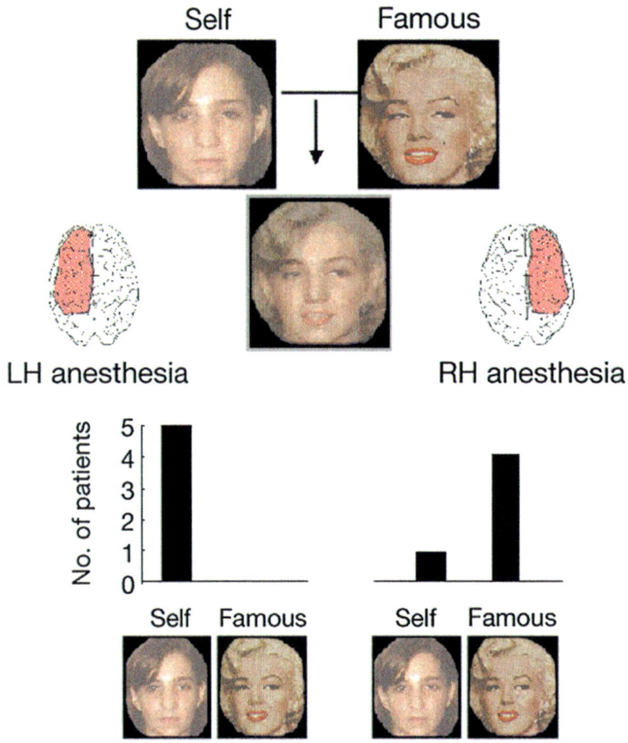

FIGURE 8.2 The results of the study by Keenan and colleagues suggest dominance for the right hemisphere for visual self-recognition.[27] Upper panel: Stimuli were generated by morphing each patient's face with the face of a famous person—shown here is Marilyn Monroe. Lower panel: Under left hemisphere anesthesia, five out of five patients indicated that the morphed face showed their own face. Under right hemisphere anesthesia, four out of five patients indicated that the face of the famous person had been presented. *Figure reproduced with permission from Keenan JP, Nelson A, O'Connor M, Pascual-Leone A. Self-recognition and the right hemisphere.* Nature. *2001;409:305.*

Sense of body ownership

Feeling localized in one's own body and not somewhere in the outside world is a major aspect of the self. The questions "What belongs to me?" and "Where are the parts of my body?" seem trivial to healthy humans, but the sense of body ownership (e.g., the knowledge of which body parts belong to one's own body[13]) can be disturbed in several neurological disorders[30,31] (see below). One of the major experimental paradigms to assess the sense of body ownership is the so-called rubber hand illusion.[32–34] In this paradigm, first published by Matthew Botvinick and Jonathan Cohen from the University of Pittsburgh, United States, in 1998,[35] participants sit in front of a desk on which a rubber hand is placed in an anatomically plausible position. The participant's real hand is hidden behind an occluder—for example, a wooden panel or a curtain—so that only the rubber hand is seen by the participant (see Fig. 8.3).

Typically, participants wear some sort of gown so that they cannot see that the rubber hand is not attached to their body. An experimenter sitting in front of the participant then starts to stroke both the rubber hand and the real hand with two paintbrushes in perfect synchrony. After a brief period of this type of stimulation, most participants report the illusionary perception of touch sensation arising from the rubber hand, the so-called rubber hand illusion. When the rubber hand and the real hand are stroked asynchronously, participants do not experience the rubber hand illusion.

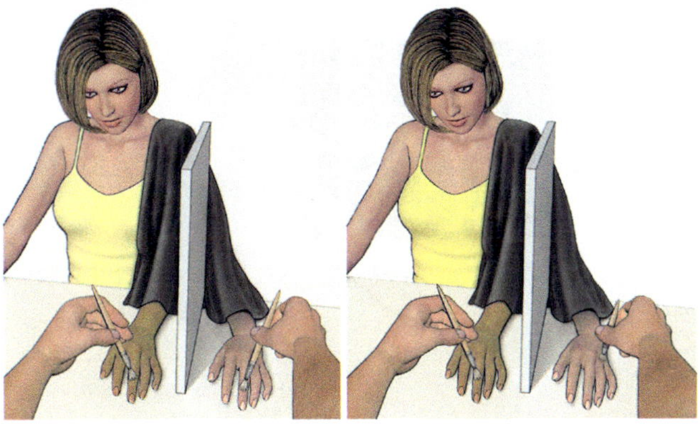

FIGURE 8.3 The rubber hand illusion. Left panel: The experimental condition. In this condition, an experimenter strokes the rubber hand and the participant's real hand with a paintbrush in perfect synchrony. The participant can only see the rubber hand, while the real hand is hidden behind an occluder. Right panel: The control condition. In this condition, the experimenter strokes the rubber hand and the real hand asynchronously.

How can the experience of the rubber hand illusion be measured? Three different techniques have been proposed:

1. Questionnaire: In the original study by Botvinick and Cohen[35], the vividness of the illusion was measured using a nine-item questionnaire. Participants had to indicate to which extent they agreed with statements such as "I felt as if the rubber hand was my hand." This measure, however, might not be optimal given that subjective self-reports could be biased by participants who wanted to respond to task demands even if they did not experience the illusion. Therefore two more objective, physiological measurement techniques for the rubber hand illusion have been developed (see below).
2. Proprioceptive drift: The rubber hand illusion leads to a drift of the perceived location of the real hand toward the location of the rubber hand. Using a special setup for the rubber hand illusion in which both rubber and real hand are located underneath the desk and only the rubber hand is visible due to a mirror, the magnitude of this proprioceptive drift can be determined. To this end, participants are asked to indicate the perceived position of their unseen real hand underneath the desk using a ruler that is mounted on top of the desk.[36,37]
3. Skin conductance response to threatening the rubber hand: Another way to determine the vividness of the rubber hand illusion is to measure skin conductance responses to watching the rubber hand being harmed (see Fig. 8.4). In this technique, the rubber hand illusion is elicited for a set amount of time—for example, 3 minutes. After the end of this period, the experimenter surprisingly "harms" the rubber hand—for example, by beating it with a hammer or stabbing it with a syringe or knife. Participants wear electrodes to measure the skin conductance response, which is modulated by sweating. Since increased arousal—for example, in response to a threat—is associated with increased sweating, the skin conductance response to watching the rubber hand being harmed is seen as autonomic nervous system arousal in anticipation of pain. In the control condition that entails asynchronous stroking of the rubber hand and real hand, there is no increased sweating in anticipation of pain. Since autonomic nervous system responses cannot be deliberately faked by the participant, this technique is considered a more objective way to measure illusion vividness than self-reports.

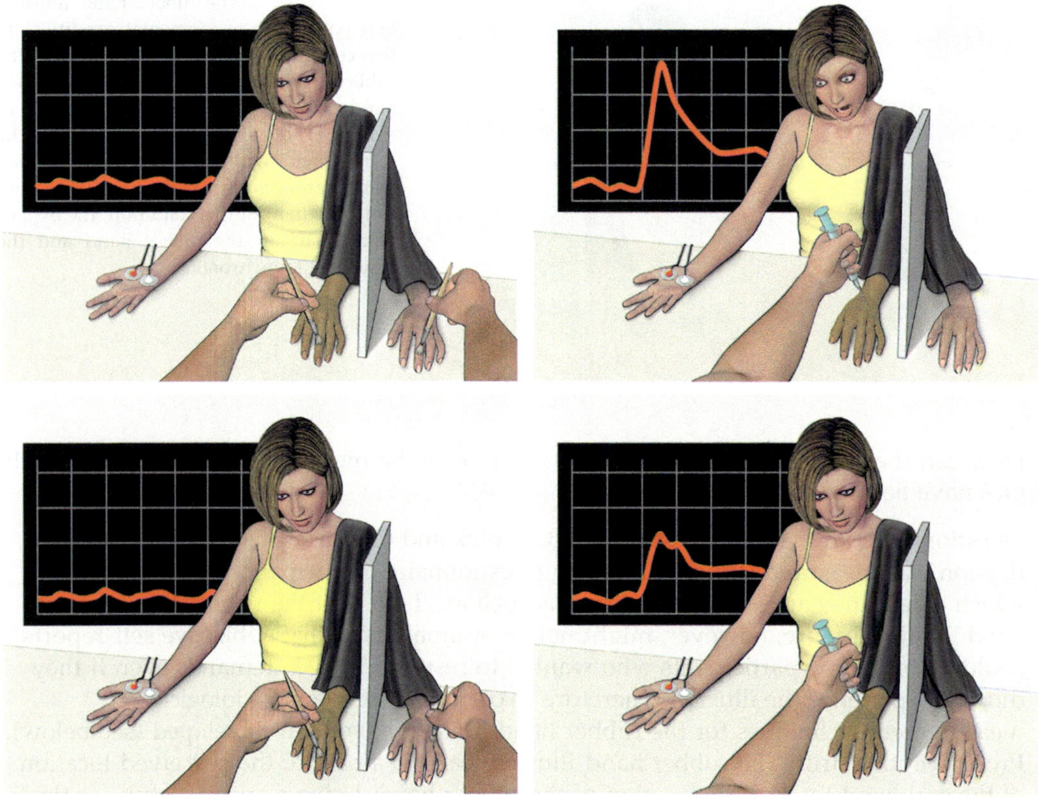

FIGURE 8.4 Using skin conductance responses to watching the rubber hand being threatened with a syringe to measure the vividness of the illusion. Upper panel: The experimental condition. When the rubber hand and real hand are stroked in synchrony, the participant shows a strong increase in the skin conductance response when the rubber hand is harmed. Lower panel: When the rubber hand and real hand are stroked asynchronously, the participant only shows a weak increase in the skin conductance response when the rubber hand is harmed.

Phenomena such as the rubber hand illusion indicate that the sense of body ownership is not constant. Integration of information from different senses permanently updates cognitive representations of the body. Neuroimaging studies show that the neural representation of the rubber hand resembles that of a real hand during the rubber illusion. The feeling of ownership of the rubber hand is reflected by neural activity in the premotor region,[38] and threatening the rubber hand elicits a neural anxiety response that is comparable to that elicited by threatening a real hand.[39]

What do we know about hemispheric asymmetries in the rubber hand illusion? A study that elicited the rubber hand illusion on the left and right side in left- and right-handers revealed that handedness did not affect the vividness of the illusion.[40] In both left- and right-handers, a stronger skin conductance response to threatening the rubber hand was observed when the illusion was elicited on the left hand (see Fig. 8.5).

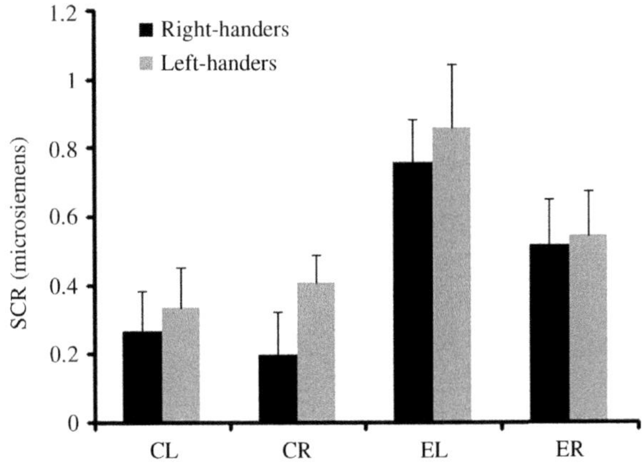

FIGURE 8.5 Results of the experiment by Ocklenburg and colleagues investigating laterality in the rubber hand illusion.[40] Skin conductance responses (in microsiemens) are shown for right-handers (black bars) and left-handers (gray bars) for the control (CL: left hand; CR: right hand) and the experimental condition (EL: left hand; ER: right hand). *Figure reproduced with permission from Ocklenburg S, Rüther N, Peterburs J, Pinnow M, Güntürkün O. Laterality in the rubber hand illusion.* Laterality. *2011;16:174–187.*

This asymmetry suggests right-hemispheric dominance for sense of body ownership. This asymmetry might be related to the right-hemispheric dominance for spatial processing,[11] as this cognitive function might be essential to feel localized in one's body.

The role of the right hemisphere for delusions of misidentification

Another line of research providing evidence for a role of the right hemisphere in the sense of body ownership comes from patient studies on so-called delusions of misidentification. Delusion of misidentification is an umbrella term for a large number of delusions of the self, of other people, places, or objects.[41] These include, for example:

- Capgras syndrome: Patients are sure that a familiar person has been replaced by an identical-looking imposter.[42,43]
- Anosognosia: Patients are unaware of a neurologic deficit (e.g., hemiplegia) and hence insist that they can still normally move paralyzed limbs. When confronted with evidence of paralysis, they may produce elaborate excuses and confabulations.[44,45]
- Asomatognosia: Patients deny ownership of their limbs, often claiming they belong to someone else.[46,47]
- Cotard syndrome: Patients strongly believe that they are dead.[48,49]

Typically, patients with these disorders show right-hemispheric lesions. Todd E. Feinberg and Julian Paul Keenan analyzed 29 previously published cases of delusions of misidentifications and found that in the frontal, temporal, and parietal lobe of the brain right-hemispheric lesions were significantly more common than left-hemispheric lesions[2] (see Fig. 8.6).

The most pronounced differences were observed in the frontal lobe, for which 96.6% of the cases showed right-sided lesions compared to only 48.3% for the left frontal lobe. Similarly, 31% of cases showed right temporal damage, but only 6.9% of cases showed left temporal damage. In the parietal lobe, 44.8% of cases showed right-hemispheric damage

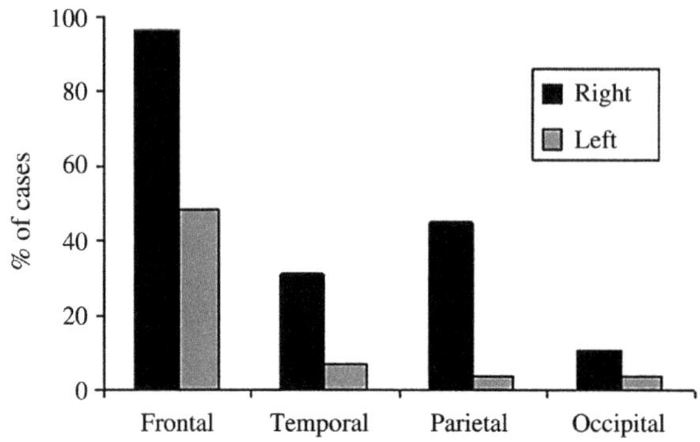

FIGURE 8.6 Results of the analysis of lesion location in patients with delusions of misidentification by Feinberg and Keenan.[2] Data show that right-hemispheric damage is more prevalent in these patients than left-hemispheric damage. *Figure reproduced with permission from Feinberg TE, Keenan JP. Where in the brain is the self?* Conscious Cogn. *2005;14:661–678.*

and only 3.4% showed left-hemispheric damage. Only for the occipital lobe did the comparison fail to reach significance—10.3% of cases showed right-hemispheric damage, and only 3.4% showed left-hemispheric damage.

Capgras syndrome

Patients with the Capgras syndrome are sure that a person close to them—for example, their significant other or a close friend—has been substituted by an identical imposter or that at least two versions of the same individual exist. It was named after Joseph Capgras (1873–1950), a French psychiatrist who described the first case, Madame M, who had the delusion that her husband had been exchanged for a double.[50] Capgras syndrome has repeatedly been reported in late-stage Alzheimer's dementia[51–53] but can also occur without dementia. Without dementia, Capgras syndrome typically develops after lesions in the right hemisphere.[41,54]

Todd E. Feinberg and Raymond M. Shapiro reported the case of a 77-year-old woman (SM) with mirror-related Capgras syndrome.[55] The patient had right-hemispheric temporoparietal atrophy with right-sided enlargement of the lateral ventricle. The patient had been deaf since childhood due to an infection with scarlet fever at the age of five and communicated by using sign language. In December 1985, she suddenly started having visual hallucinations and experienced increased forgetfulness. In June 1986, her son one day found her sitting in front of a mirror and using sign language to communicate with her mirror image. When her son asked her what she was doing, she responded that she was talking to the "other SM" who also happened to be deaf and only visible in mirrors. This "other SM" had attended the same school as SM, but they had not known each other at that time. The "other SM" also had a son who had the same name as hers and also looked like him. In general, both SMs were pretty much alike, but the "other SM" talked too much and did not use sign language as well as SM. SM recognized the reflections of objects or other persons in the mirror as normal reflections. It was only her own reflection that she mistook for a real person. SM had an average IQ of 97. All other cognitive values were also average. However, she had reduced visuospatial skills and borderline values in face recognition but no prosopagnosia (see below). Interestingly, Capgras patients do not show differential affective skin conductance responses to known faces.[56] Face recognition is mostly mediated by the right hemisphere. This finding suggests that Capgras syndrome might be related to dysfunctions in right-hemispheric neuronal networks related to face recognition.

Asomatognosia

Asomatognosia describes the inability to perceive parts of one's own body as belonging to oneself. For example, patients might be unable to perceive their own hand or arm as part of their body. This disorder typically occurs after damage to the right hemisphere, especially in the parietal region. Jean Paysant and colleagues from the University of Nancy, France, investigated 16 patients with unilateral right-hemispheric stroke and found that 14 out of the 16 patients showed manifestations of asomatognosia.[57] However, these symptoms mostly disappeared after 2 months. In a similar study, Daniel Antoniello and Reena Gottesman from the Albert Einstein College of Medicine in New York, United States, assessed asomatognosia in 36 patients with hyperacute right middle cerebral artery infarct.[58] They found that 61% of patients experienced asomatognosia at the onset but only 15% after 1 week, concluding that limb misidentification is common after right hemisphere infarcts.

Somatoparaphrenia is a disorder that is similar to asomatognosia. While asomatognosia refers to the unawareness of ownership of a limb, patients with somatoparaphrenia additionally show more severe delusional misidentifications and confabulations.[59] While patients with asomatognosia typically present with temporoparietal and/or medial frontal lesions, patients with somatoparaphrenia often show additional orbitofrontal lesions.[59] Interestingly, more than 90% of reported cases of somatoparaphrenia had delusions of disownership of left-sided body parts, which were associated with extensive lesions in right-hemispheric brain regions.[60]

Anosognosia

Anosognosia is a delusion in which the patient who suffers from a disability or illness is unable to understand that they are disabled or ill.[44] Anosognosia is commonly observed in neglect patients (see Chapter 7). When asked why they are in a hospital or why they cannot move one side of their body, patients typically resort to confabulation. As is the case for Capgras syndrome, anosognosia has often been reported in patients with Alzheimer's dementia[61–63] but can also occur in patients without neurodegenerative diseases. In this group, it typically occurs after right-hemispheric parietal or fronto-temporoparietal lesions.[64] Patients deny severe stroke-related motor symptoms such as left-sided hemiplegia—paralysis of the left half of the body.[65] A special form of visual anosognosia is Anton–Babinski syndrome. Patients with this syndrome are cortically blind and show anosognosia with vivid confabulations.[66]

Cotard syndrome

Cotard syndrome—named after the French neurologist Jules Cotard (1840–89)—is a rare syndrome in which patients are convinced that they are dead or deny the existence of their own body.[48,67] Yoshiyuki Nishio and Etsuro Mori from Tohoku University in Japan reported a case of delusions of death in a patient with a right-hemispheric infarct (see Fig. 8.7).[68]

The patient, a 69-year-old right-handed retired company owner, had an infarct affecting right frontal, temporal, and parietal regions as well as the thalamus and showed left hemiparesis and left neglect. After being hospitalized for 2 weeks, he became delusional and thought that Kim Jong-il (leader of North Korea at that time) was staying on the same floor as him and that his therapist was the grandson of Puyi (China's last Emperor). Two weeks later he developed feelings of unreality and asked his wife whether he was alive or dead—"I guess I am dead. I'd like to ask for your opinion." Gradually, he became increasingly sure that he was dead—"My death certificate is registered. You walk with a dead man." However, he was not depressed or anxious. When he was asked whether a dead man could speak, he understood the logical problem but could not change it. Two months after his infarct, he became less convinced of his death: "I feel I am dead. But now I am talking with you in this world. I am unable to realize that I am alive." The delusion disappeared entirely 4 months later. But 1 year after the stroke, he still believed in the memories that he had formed during his delusional state. He said, "Now I am alive. But I was once dead at that time," and "I saw Kim Jong-il in the hospital where I stayed."

Right-hemispheric stroke: insights into the sense of body ownership

While the delusions of misidentification described in the previous section provide highly interesting insights into the role of the right hemisphere for body ownership, the number of patients is often small and their symptoms and lesions are highly individual. That makes it difficult to draw overarching conclusions about which brain areas in the right hemisphere

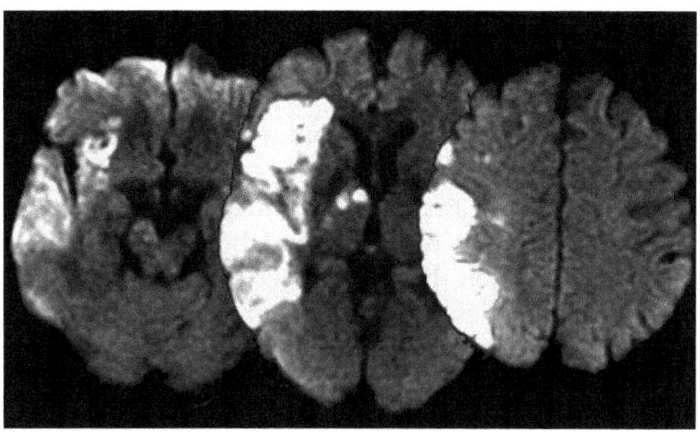

FIGURE 8.7 Brain MRI scans of the case reported by Nishio and Mori.[68] The right side of the brain is shown on the left side of the scan and vice versa. The white areas show an infarct in the right frontal, temporal, and parietal lobes as well as in the thalamus. *Figure reproduced with permission from Nishio Y, Mori E. Delusions of death in a patient with right hemisphere infarction.* Cogn Behav Neurol. *2012;25:216–223.*

constitute the core network for the sense of body ownership. Therefore research on body ownership has also been conducted in a much more common group of patients: individuals with a stroke in the right hemisphere. For example, in a recent study, lesion analysis was performed on neuroimaging data of 49 patients who had experienced a stroke in the right hemisphere.[69] 23 of these patients experienced limb disownership, and 26 did not experience limb disownership. The results showed that it was not only lesions in gray matter regions that mattered for the sense of body ownership but also specific white matter connections. The authors found that disturbed sensation of limb ownership was associated with damage to several gray matter regions in the right hemisphere, specifically pars opercularis of the inferior frontal gyrus, the precentral gyrus, the postcentral gyrus, and the supramarginal gyrus. Moreover, disturbed sensation of limb ownership was associated with damage to several white matter tracts in the right hemisphere, that form an extended fronto-insular-parietal network. These findings highlight the importance of taking right hemisphere white matter networks into account when assessing the sense of body ownership. Functionally, the authors suggested that the identified areas and networks imply that brain networks for sensory processing and multisensory integration, but also networks involved in salience monitoring and belief updating are relevant for the sense of body ownership.

The integrity of the bodily self

Integration of multisensory signals from the entire body is critical for bodily self-consciousness or self-identification and depends on right-hemispheric temporoparietal networks. Lukas Heydrich from the Ecole Polytechnique Fédérale de Lausanne in Switzerland and colleagues reported the case of an epileptic patient with a lesion in the right posterior intraparietal sulcus.[70] This patient experienced delusional body perceptions during seizures. His seizures would start with the experience of increased pressure on the left side of the body. Subsequently, he would have the feeling that a stranger invaded the left side of his body. While this feeling lasted, the patient experienced the sensation that the left half of his head, the upper part of his left trunk, his left arm, and his left leg no longer belonged to him. He thought that these parts were disconnected from the rest of his body and that his body was divided into two parts (see Fig. 8.8). During these episodes, he localized the self as only within the right side of his body.

Taken together, evidence from various delusions of misidentifications and related neurological disorders suggests that the right hemisphere is dominant for integrating feelings of body ownership. This comprises feelings of bodily integrity and even the feeling that we are alive.

Thinking about oneself: the self-concept and the left hemisphere

The last of the three aspects of the self that have been investigated concerning hemispheric asymmetries is the self-concept. The self-concept includes knowledge about personality traits, autobiographic memory, and thoughts such as: "How did I react in the past

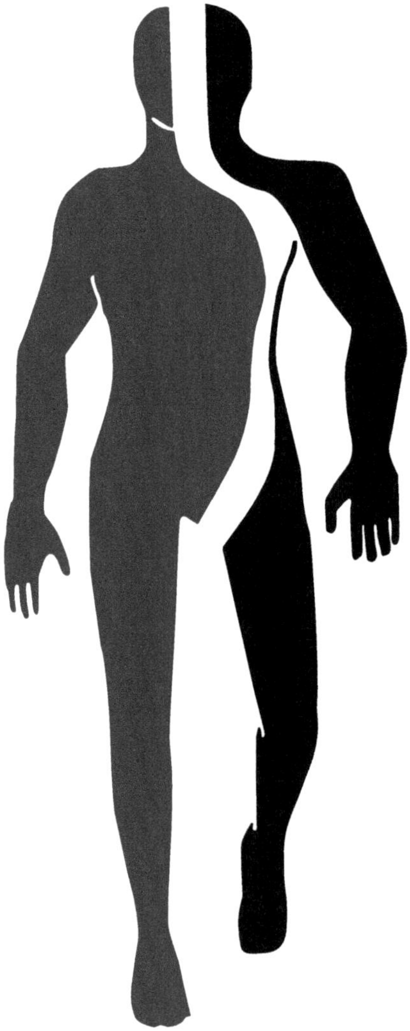

FIGURE 8.8 The patient described by Heydrich[70] reported the feeling that the left half of his head, the upper part of his left trunk, his left arm, and his left leg no longer belonged to him (in black). He described the self as being located on the right side of his body (in gray). *Figure reproduced with permission from Heydrich L, Dieguez S, Grunwald T, Seeck M, Blanke O. Illusory own body perceptions: case reports and relevance for bodily self-consciousness.* Conscious Cogn. *2010;19:702–710.*

when confronted with specific situations? How do I see myself in relation to others?" Interestingly, this part of the self seems to be largely mediated by the left hemisphere. For example, the left hemisphere is more strongly activated when a participant hears their own name compared to hearing the names of other people.[71] Moreover, it is also more active when a participant is thinking about adjectives that may characterize their personality[72] or when thinking about statements that may have a relation to autobiographic memory.[73] Thus this verbal and declarative aspect of the self is linked to left-hemispheric asymmetries for language processes. The verbal superiority of the left hemisphere results in higher activity levels when speaking to oneself about oneself.

Summary: the self and the right hemisphere

Taken together, the question "Where in the brain is the self?" cannot be answered by identifying a single brain area. Instead, findings about the neuroscience of the self can be summed up as follows:

- Self-awareness depends on different cognitive components including the bodily and autobiographic self. Behavioral and neuroimaging studies indicate that the sense of self is widely distributed throughout the brain. Accordingly, left- as well as right-hemispheric neuronal networks differentially contribute to its generation. There is no specific isolated "ego-center" in the brain.
- Visual self-awareness and sense of body ownership seem to be mainly controlled by right-hemispheric networks. This asymmetry might be linked to rightward asymmetries in visuospatial processes.
- In contrast, the self-concept is mainly controlled by left-hemispheric networks as self-declarative verbal processes are involved that require left-hemispheric networks.
- Interhemispheric communication and coordination seem to be necessary for an integrated feeling of one's actions and body parts. This supports the intimate interrelation between intra- and interhemispheric processing for appropriate cognition.

Facial asymmetries

Given right-hemispheric dominance for self-recognition and sense of body ownership, it seems intuitive to suspect that recognition of the faces of other people also shows rightward hemispheric asymmetries. Indeed, hemispheric asymmetries have been thoroughly investigated both regarding the anatomy of faces as well as regarding the cognitive correlates of processing the faces of other people. As outlined in detail in Chapter 1, the dawn of research on facial asymmetries was on April 8, 1820,[74] when the Venus of Milo (see Fig. 8.9) was rediscovered in the Ottoman Empire.

The critique that Henke[75] wrote about the statue led Hasse to point out that facial asymmetries are normal and made him conduct the first empirical study on facial asymmetries in humans.[76] Subsequently, Hallervorden developed the mirror technique to determine facial asymmetries that is still used to this day.[77] He asked his subjects to mimic the facial expressions associated with different emotions and then photographed them. For each photograph, he would print two versions: One that showed the original and one that was mirror-reversed. He would then cut both pictures exactly at the midline of the face and create two new composite pictures by merging the two pictures depicting the left half of the face and the two pictures depicting the right half of the face. If a face shows no asymmetry, both of the two composite pictures should look identical to the original picture. This, however, is rarely the case (see Fig. 8.10).

In 1933, Werner Wolff used the technique developed by Hallervorden in a series of psychoanalytic experiments.[78] Today, we know that facial asymmetries are a quite common

FIGURE 8.9 The Venus of Milo. *Figure reproduced from https://commons.wikimedia.org/w/index.php?curid = 54858451. Created by Livioandronico2013 and used under a CC BY-SA 4.0 license.*

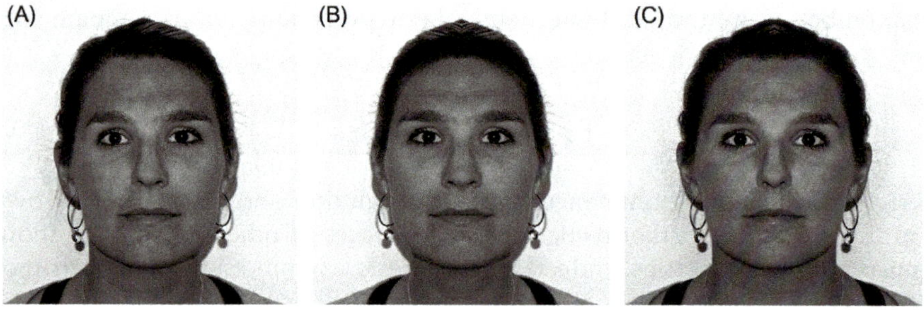

FIGURE 8.10 Frontal photograph of a female face. (A) Original picture. (B) Left-left composite picture. (C) Right-right composite picture. Note that the original picture and the mirror-reversed images differ from each other.

feature in the general population.[79] For example, Tammy R. Severt and William R. Proffit from the University of North Carolina, United States, surveyed facial asymmetries in 1,460 patients who had appointments in the Dentofacial Clinic at the University of North Carolina.[80] They found that 34% of these patients showed facial asymmetry, with 5% of the affected individuals showing asymmetries in the upper face only, 36% showing asymmetries in the midface, for example, asymmetries of the nose, and 74% showing chin asymmetries. Thus asymmetries of the lower face seem to be more common than of the upper or midface.

Interestingly, facial asymmetries are not limited to the anatomy of the face itself but are also evident in its function. In a classic study from 1978, Harold A. Sackeim, Ruben C. Gur, and Marcel C. Saucy from the University of Pennsylvania, United States, cut pictures of emotional faces in half and created left-side-only and right-side-only composite pictures. Participants were asked to indicate how intensely the faces in the pictures expressed emotions. Results revealed that left-side-only composite pictures were rated as expressing emotions more intensely than right-side-only composite pictures, indicating dominance of the right hemisphere for the expression of emotions in the face. The finding that humans show stronger emotions on the left side of the face has also been confirmed by several subsequent studies, for example, those showing that people perceive the left hemiface as being more expressive than the right hemiface.[81] Facial asymmetries have also been reported for speech production. Most people open the right side of their mouth further than the left side, which is thought to reflect left-hemispheric language dominance.[82–84]

Hemispheric asymmetries in behavioral markers of face perception

While asymmetries in facial anatomy and functional characteristics of faces certainly are interesting, the majority of work on hemispheric asymmetries in this domain has focused on the perception and recognition of the faces of other individuals. Much of the early behavioral work on hemispheric asymmetries in face perception was carried out using the divided visual field technique described in Box 3.3.[85] To quickly recapitulate: In paradigms using this technique, the participant is asked to fixate on a cross that is presented in the center of a screen. Then, a picture is presented very briefly on either the left or the right side of the fixation cross, and the participant has to respond to this stimulus. Since the nasal portions of the visual input are processed by the hemisphere contralateral to the stimulus and the distal portions of the visual input are processed by the hemisphere ipsilateral to the stimulus, everything on the left side of the screen is projected to the right hemisphere and vice versa (see Fig. 3.11). Thus stimuli presented in this way are initially only processed by one hemisphere, and by comparing performance after stimulus presentation in the left- and right visual field, hemispheric dominance can be determined. In studies on face perception, normal, upright pictures of faces are typically contrasted against altered pictures of faces. For example, Hausmann and Güntürkün compared unaltered normal faces to faces in which some facial features had been translocated (see Fig. 8.11).[86]

Participants were required to indicate as quickly and as accurately as possible whether they saw original or altered faces. Other authors have used upright versus inverted faces[87,88] or intact versus fractured faces.[89] Another paradigm involved naming famous faces as fast as possible.[90] Typically, all of these paradigms reveal robust right-hemispheric superiority, with participants responding faster and more accurately to face stimuli presented in the left visual field.[86–93] This superiority can even be found for inverted faces or "exploded" face stimuli that have been fractured into many different parts.[89]

fMRI studies on face perception: finding the fusiform face area

What are the neural correlates of face perception and are they asymmetric? The quest to identify the brain areas responsible for face recognition in healthy individuals started in the early 1990s when advanced neuroimaging techniques became available. In 1992, Justine Sergent from the Montreal Neurological Institute at McGill University in Montreal,

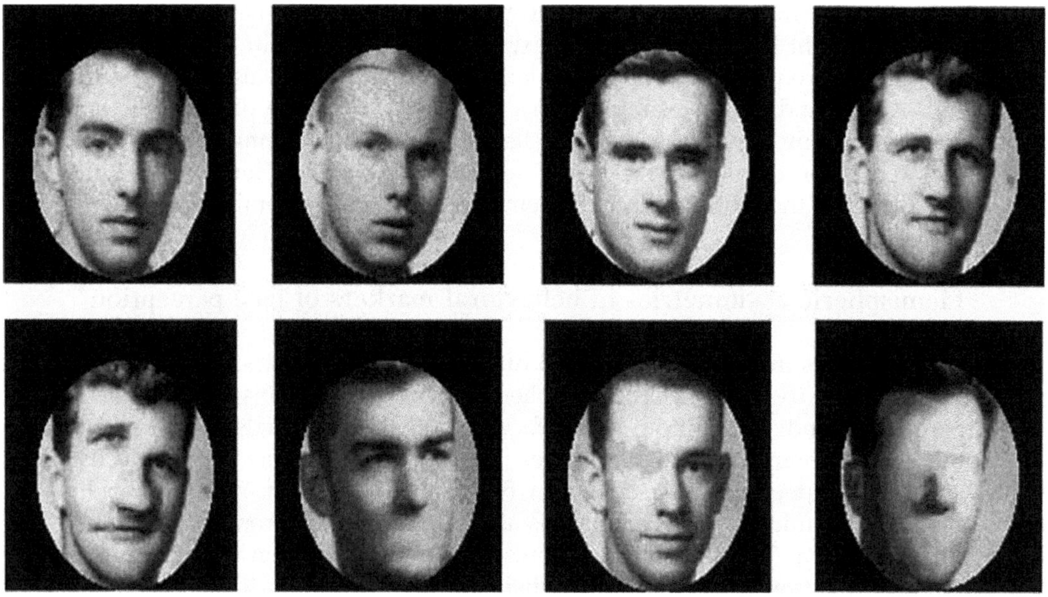

FIGURE 8.11 Examples of stimuli used in the divided visual field face discrimination task by Hausmann and Güntürkün.[86] Upper panel: Unchanged faces. Lower panel: Altered faces with translocated facial features. *Figure reproduced with permission from Hausmann M, Güntürkün O. Sex differences in functional cerebral asymmetries in a repeated measures design.* Brain Cogn. *1999;41:263–275.*

Canada, and colleagues published a PET study on face recognition in the scientific journal *Brain*.[94] In this study, Sergent determined regional cerebral blood flow changes in seven male right-handers during six different conditions:

1. Fixation on a fixation point in the center of the screen.
2. Looking at unfamiliar faces without performing any task.
3. Looking at horizontal and vertical sine wave gratings and indicating their orientation.
4. Looking at unfamiliar faces and indicating their gender.
5. Looking at famous faces and indicating whether the depicted individuals were actors or not.
6. Looking at objects and indicating whether the objects were natural or man-made.

For the face-gender categorization task, Sergent found that the right ventral extrastriate cortex played a critical role in face perception. The extrastriate cortex is the part of the occipital cortex that is adjacent to the primary visual cortex which is sometimes called striate cortex due to its striped appearance in microscopic images. Interestingly, the left extrastriate cortex was not activated significantly. For the famous face identification task, Sergent found activations in the right lingual gyrus and bilaterally in the fusiform gyrus. The strongest activations were observed in the right parahippocampal gyrus and the right anterior temporal lobe. Sergent concluded that while the left hemisphere participated in face recognition, the contribution of the right hemisphere to this cognitive function is both necessary and sufficient. Within the right hemisphere, particularly ventromedial areas were crucial for face recognition.

Is it possible to identify the brain areas relevant to face recognition with even greater accuracy? In 1997, Nancy Kanwisher from Harvard University, United States, used a then-new technique called fMRI (see Box 5.1) to find an answer to this question.[95,96] Compared to PET, which was the technique that Sergent had used, fMRI had several advantages: Participants were not exposed to potentially harmful radiation, the images were much sharper, and the technique had a much better temporal resolution, allowing recording of an image every few seconds. PET, on the other hand, had a low temporal resolution—between 10 seconds and several minutes, depending on the specific PET technique used—which made it impossible to assess cognitive processing in the brain in real time, given that most cognitive processes such as face recognition occur much faster than that.

Kanwisher asked 15 participants to passively view freshman ID photographs or pictures of objects such as spoons or cars. She then determined individually for each subject which brain areas showed activity related to face perception. In 12 out of 15 participants, a significant activation difference was found in the fusiform gyrus. When looking only at the 10 right-handed subjects, Kanwisher found that half of them showed a significant activation difference between looking at faces and objects only in the right fusiform gyrus. While the five left-handers also showed a significant activation difference in the left fusiform gyrus, the right-hemispheric area in this group was decidedly larger (1 cm^3) than the left-hemispheric area (0.5 cm^3). Kanwisher named this face-sensitive area the "fusiform face area" (FFA)—a term that still is used today[97] (see Fig. 8.12).

Kanwisher then proceeded to conduct several control experiments to prove that the FFA was indeed selectively involved in face processing. She could show that the FFA showed significantly stronger activation when:

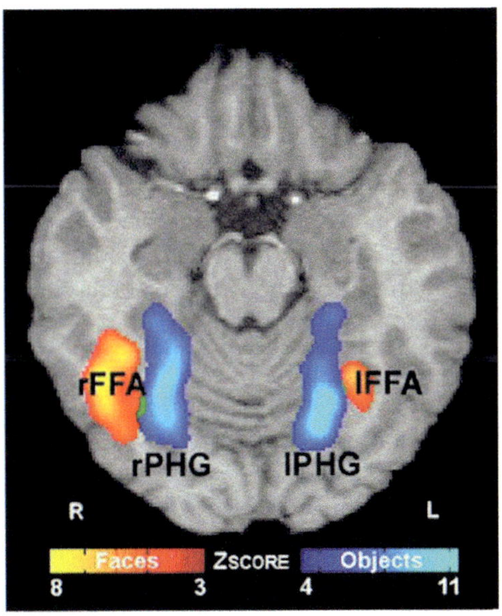

FIGURE 8.12 The right and left fusiform face area (FFA) as well as the left and right parahippocampal gyrus, an object-sensitive area. Note that the right FFA is bigger than the left FFA.[97] *Figure reproduced with permission from Caldara R, Seghier ML, Rossion B, Lazeyras F, Michel C, Hauert C-A. The fusiform face area is tuned for curvilinear patterns with more high-contrasted elements in the upper part.* Neuroimage. *2006;31:313–319.*

1. Participants passively viewed intact faces compared to scrambled faces.
2. Participants passively viewed pictures of faces compared to pictures of houses.
3. Participants passively viewed pictures of faces compared to pictures of hands.
4. Participants performed a consecutive matching task with pictures of faces compared to pictures of hands.

These control procedures allowed Kanwisher to reject other possible explanations for the function of the FFA (such as the general processing of living things) and to demonstrate that it is selectively involved in face perception.

Following Kanwisher's initial research, thousands of studies have used fMRI to study face perception. In general, a robust right-sided activation asymmetry is found in these studies. For example, Gjurgjica Badzakova-Trajkov from the University of Auckland, New Zealand, and colleagues investigated cerebral activation asymmetries during face processing and other tasks in monozygotic twins using fMRI.[98] Overall, 21 pairs of twins concordant for handedness and 21 pairs of twins discordant for handedness were asked to watch short video clips of faces with emotional expressions or video clips of moving objects such as a blender. Using the temporal lobe as the region of interest for their analysis, Badzakova-Trajkov observed robust right-sided activation asymmetry during face perception. In 92.3% of twin pairs discordant for handedness, both individuals showed rightward activation asymmetry when watching faces. In only 7.7% of twin pairs discordant for handedness, one individual showed right-sided activation dominance, while the other showed left-sided activation. Similar numbers were observed in the twin pairs concordant for handedness. Here, in 83.3% of the twin pairs, both individuals showed rightward activation asymmetry when watching faces. In only 16.6% of the twin pairs, one individual showed right-sided activation dominance, while the other showed left-sided activation. In both groups, there was no twin pair in which both individuals showed leftward activation asymmetry. Similar findings supporting rightward activation asymmetry in the FFA were also reported by several other authors,[99–105] but some authors also highlighted that the rightward asymmetry for face perception is weaker and more variable than the leftward asymmetry for speech processing.[106] Part of this heterogeneity may be explained by handedness since it has been shown that the rightward activation asymmetry during face processing is weaker in left-handers compared to right-handers.[107] Moreover, at least two studies found a reversed left-hemispheric FFA activation asymmetry in left-handers.[108,109] Importantly, the magnitude of right-hemispheric activation asymmetry in the FFA has been shown to strongly correlate with left visual field superiority in a divided visual field paradigm with facial stimuli, supporting the idea that it might be the cause of behavioral asymmetries in face recognition.[110]

Electrical brain stimulation studies on face perception

A special role of the right fusiform gyrus for face recognition was also confirmed by a study applying electrical brain stimulation in patients with intracranial electrodes implanted in the ventral temporal cortex, a larger brain area containing the fusiform gyrus. Such electrodes are sometimes implanted in patients with intractable epilepsy to localize the source of their seizures. Vinitha Rangarajan from Stanford University, United States, and colleagues investigated face perception in five patients with electrodes implanted in the right ventral temporal cortex and five patients implanted with electrodes in the left ventral temporal cortex.[111] First, face-sensitive electrode sites were determined by recording the signal obtained from the electrode while the patients viewed pictures of faces. Face-sensitive sites were found in both the right and left fusiform gyrus. In the second step, face-sensitive electrode sites were stimulated while the patients viewed pictures of faces and other stimuli. Patients were asked to report their conscious experience of the face stimuli (e.g., whether they experienced any perceptual effects or changes in the stimuli). Here, a striking hemispheric asymmetry was observed: Only stimulation of electrode sites in the right fusiform gyrus led to changes in the conscious perception of faces, while stimulation of left-sided electrodes only affected the visual experience of nonface stimuli. Therefore Rangarajan and colleagues concluded that this clear functional dissociation between the left and the right fusiform gyrus indicated dominance of the right fusiform gyrus for conscious perception of faces. Interestingly, in a follow-up study, Rangarajan reported a reversed pattern in two patients with atypical language lateralization and left-handedness, concluding that the functional architecture of the ventral temporal cortex might be linked to handedness and language lateralization.[112] This idea assumed that when the right hemisphere is language-dominant, the left might in turn be dominant for face perception at least in some individuals.

Electrophysiological asymmetries during face perception

Further evidence for the crucial role of the right hemisphere for face recognition also comes from EEG studies (see Box 5.2). In the context of EEG research on face perception, the most widely investigated event-related potential component is the so-called N170, a negative component with a peak of about 130–170 ms after stimulus presentation.[113–120] Typically, it is centered over occipitotemporal brain areas with a source in the fusiform gyrus and surrounding areas.[121] The N170 is thought to reflect the early structural encoding of faces[122] and is more pronounced for faces compared to objects[123] and also for face-like make-up stimuli compared to animal-like make-up stimuli.[124] Interestingly, the N170 has repeatedly been shown to be more negative over the right compared to the left hemisphere,[115,119,125–127] showing that strong rightward lateralization already exists during the early stages of face recognition (see Fig. 8.13).

Prosopagnosia

One way to further investigate this notion is to investigate patients who suffer from neurological disorders affecting face perception. Patients experiencing difficulties in recognizing faces were first described in the 19th century, for example, by the famous French neurologist Jean-Martin Charcot (1825–93) in 1883.[128] Today, the inability to recognize faces despite otherwise unimpaired visual system function is called prosopagnosia (from the Greek word *"prosopon"* meaning face). This term was coined by the German

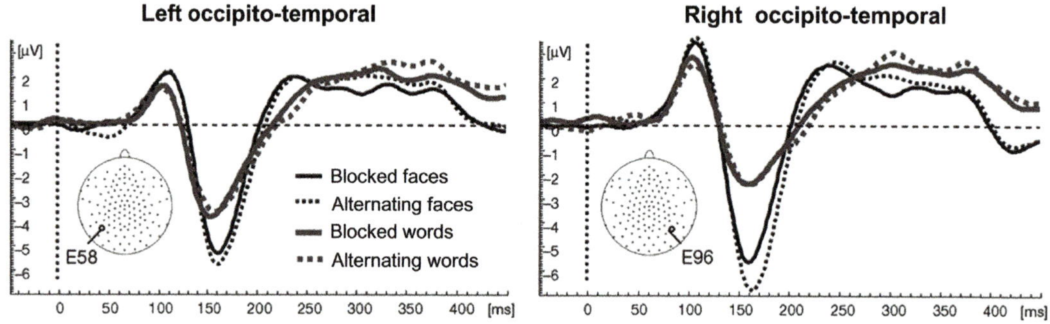

FIGURE 8.13 Rightward lateralization of the N170 for facial stimuli. Typically, the N170 is more negative for faces (black lines) than for words (gray lines). Moreover, the N170 for faces is more negative at right-hemispheric occipitotemporal electrodes as compared to left-hemispheric occipitotemporal electrodes.[126] *Figure published as open access and reproduced from Maurer U, Rossion B, McCandliss BD. Category specificity in early perception: face and word n170 responses differ in both lateralization and habituation properties.* Front Hum Neurosci. *2008;2:18.*

neurologist Joachim Bodamer (1910–85) in a 1947 paper published in the German journal *Archiv für Psychiatrie und Nervenkrankheiten.*[1] In this article, Bodamer described three cases of prosopagnosia, which he had observed in his clinical practice (see introductory short story), all of which showed severe difficulties in recognizing faces, even of familiar people such as their significant others. Despite this impairment, prosopagnosia patients are not necessarily unable to identify anyone. Indeed, they typically use other information such as the voice or height or typical accessories such as glasses or jewelry to recognize people.[128] While there is presently no way of healing prosopagnosia, these strategies can be a tremendous help for patients to function in everyday life.

What is the cause of prosopagnosia? While many patients with prosopagnosia presented with bilateral damage in the temporal and occipital lobe, it has been argued that lesions in the posterior part of the right hemisphere—for example, in and around the right FFA—are critical for the development of the disorder.[109] For example, in 1970, Robert K. Yin from the Massachusetts Institute of Technology in Cambridge, United States, tested 37 patients who had suffered penetrating head wounds with brain tissue loss during the Korean War (1950–53).[129] These patients were grouped into five different categories:

1. Patients with bilateral brain tissue loss.
2. Patients with unilateral left frontal brain tissue loss.
3. Patients with unilateral right frontal brain tissue loss.
4. Patients with unilateral left posterior brain tissue loss.
5. Patients with unilateral right posterior brain tissue loss.

Patients were asked to look at a series of 40 pictures of faces—studio portraits of adult males—and houses. Then they were asked to look at a second set of pictures that consisted of 50% pictures that had been shown in the previous set and 50% new pictures. Patients were asked to indicate whether they recognized each picture or not. Subsequently, the same procedure was repeated, but this time all the pictures were shown upside-down. Results showed that patients with unilateral right posterior brain tissue loss made more errors in recognizing upright pictures of faces than healthy control subjects or the three other patient groups with unilateral brain damage. The number of errors they made was almost identical to the number of errors made by patients with bilateral brain damage. Based on these findings, Yin concluded that patients with lesions in the right posterior cortex had a specific deficit in recognizing faces. To date, many cases of prosopagnosia with unilateral right-sided lesions have been reported, further supporting the idea that damage to the right hemisphere may critically affect face perception, while the left hemisphere plays only a minor role in this function.[109,130–138] Taken together, data on prosopagnosia are in line with behavioral and neuroimaging data in healthy participants, suggesting a pivotal role of the right hemisphere in face recognition.

Lateralization of human body recognition

A research field that is strongly connected to the investigation of hemispheric asymmetries in face processing is the investigation of hemispheric asymmetries in body processing. This is largely because the relevant neuronal structures are located closely to each other and also partly overlap. This comes as no surprise as we rarely perceive a face without a body or a body without a face in real-life interactions. Just like the FFA for faces, there is a specific brain area in the fusiform gyrus that shows specific activations when pictures of bodies are shown: the fusiform body area (FBA).[139–141] The FBA is adjacent to the FFA and often overlaps with it.[141] An fMRI study using a simple one-back task with different kinds of pictures including faces, bodies, chairs, and scenes found that both left-handers and right-handers show a rightward activation asymmetry in the brain for processing pictures of human bodies.[107] However, the rightward activation asymmetry was smaller in left-handers than in right-handers. Fig. 8.14 shows the rightward activation asymmetry for body recognition for both groups.

Interestingly, a right-hemispheric dominance for body perception has been observed not only in the visual domain but also for olfactory stimuli. A recent ALE meta-analysis on neuroimaging studies on the processing of human body odors also found a right-hemispheric dominance, with a significant convergence of activation results in the right inferior frontal gyrus.[142]

Sample questions

- Please describe anosognosia, asomatognosia, and Capgras syndrome. Briefly outline the deficits of the respective patients. Which hemisphere is mostly affected in each syndrome?
- What are the respective contributions of the left hemisphere and right hemisphere to the feeling of self?
- What is the Wada test, how does it work, and what can we learn from it about the feeling of self-identity?

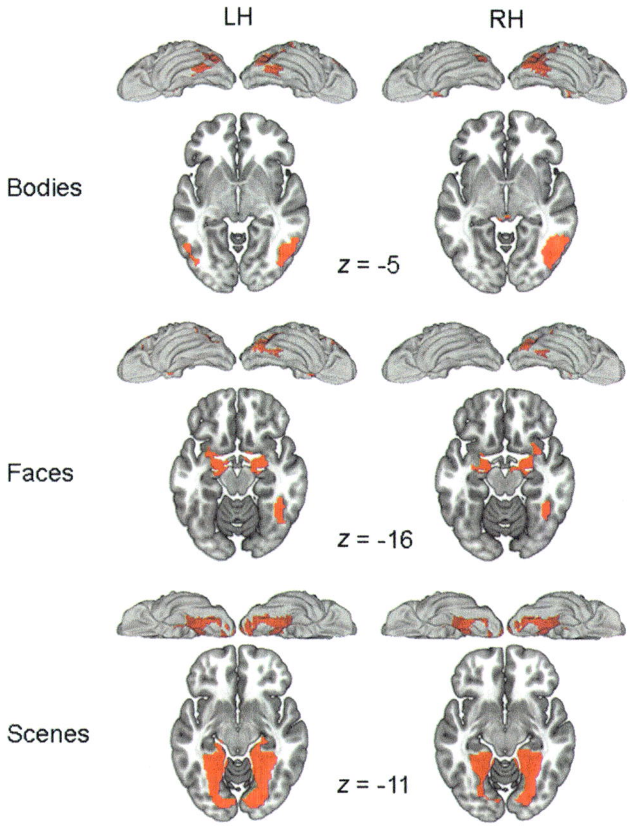

FIGURE 8.14 Group activation maps for perception of bodies, faces, and scenes for left-handers (LH) and right-handers from the study by Johnstone et al. (2021).[107] *Figure published open access under the terms of the Creative Commons CC BY license and reproduced from Johnstone LT, Karlsson EM, Carey DP. Left-Handers Are Less Lateralized Than Right-Handers for Both Left and Right Hemispheric Functions.* Cereb Cortex. *2021;31:3780–3787.*

- What is the rubber hand illusion? Describe three different techniques to assess the vividness of the illusion. What are the brain mechanisms underlying the illusion, and what can we learn about body ownership from this illusion?
- Are there hemispheric asymmetries for face recognition? How have they been investigated empirically?
- Please describe the neuronal correlates of face recognition. Which brain area plays a central role in face recognition?
- What is prosopagnosia?
- What is known about hemispheric asymmetries in body perception?

References

1 Bodamer J. Die Prosop-Agnosie. *Arch f Psychiatr u Z Neur.* 1947;179:6–54.
2. Feinberg TE, Keenan JP. Where in the brain is the self? *Conscious Cognit.* 2005;14(4):661–678. Available from: https://doi.org/10.1016/j.concog.2005.01.002.
3. Dary Z, Lopez C. Understanding the neural bases of bodily self-consciousness: recent achievements and main challenges. *Front Integr Neurosci.* 2023;17:1145924. Available from: https://doi.org/10.3389/fnint.2023.1145924.
4. Feinberg TE. *From Axons to Identity: Neurological Explorations of the Nature of the Self.* W. W. Norton; 2009.

5. Corballis MC. What's left in language? Beyond the classical model. *Ann NY Acad Sci.* 2015;1359(1):14–29. Available from: https://doi.org/10.1111/nyas.12761.
6. Gazzaniga MS. Brain and conscious experience. *Adv Neurol.* 1998;77:181–192. Available from: https://europepmc.org/article/med/9709824.
7. Gainotti G. Laterality effects in normal subjects' recognition of familiar faces, voices and names. Perceptual and representational components. *Neuropsychologia.* 2013;51(7):1151–1160. Available from: https://doi.org/10.1016/j.neuropsychologia.2013.03.009.
8. Molnar-Szakacs I, Uddin LQ. Laterality and hemispheric specialization of self-face recognition. *Neuropsychologia.* 2023;186:108586. Available from: https://doi.org/10.1016/j.neuropsychologia.2023.108586.
9. Hu C, Di X, Eickhoff SB, et al. Distinct and common aspects of physical and psychological self-representation in the brain: A meta-analysis of self-bias in facial and self-referential judgments. *Neurosci Biobehav Rev.* 2016;61:197–207. Available from: https://doi.org/10.1016/j.neubiorev.2015.12.003.
10. Davidson RJ. Anterior cerebral asymmetry and the nature of emotion. *Brain Cognition.* 1992;20(1):125–151. Available from: https://doi.org/10.1016/0278-2626(92)90065-t.
11. Hellige JB. Hemispheric asymmetry for visual information processing. *Acta Neurobiol Exp.* 1996;56:485–497.
12. Suddendorf T, Butler DL. The nature of visual self-recognition. *Trends Cognit Sci.* 2013;17(3):121–127. Available from: https://doi.org/10.1016/j.tics.2013.01.004.
13. Butler AA, Héroux ME, Gandevia SC. Body ownership and a new proprioceptive role for muscle spindles. *Acta Physiol.* 2017;220(1):19–27. Available from: https://doi.org/10.1111/apha.12792.
14. Babic MJ, Morgan PJ, Plotnikoff RC, Lonsdale C, White RL, Lubans DR. Physical activity and physical self-concept in youth: systematic review and meta-analysis. *Sports Med.* 2014;44(11):1589–1601. Available from: https://doi.org/10.1007/s40279-014-0229-z.
15. Gallup GG. Chimpanzees: self-recognition. *Science.* 1970;167(3914):86–87. Available from: https://doi.org/10.1126/science.167.3914.86.
16. Anderson JR, Gallup GG. Mirror self-recognition: a review and critique of attempts to promote and engineer self-recognition in primates. *Primates.* 2015;56(4):317–326. Available from: https://doi.org/10.1007/s10329-015-0488-9.
17. Plotnik JM, de Waal FBM, Moore D, Reiss D. Self-recognition in the Asian elephant and future directions for cognitive research with elephants in zoological settings. *Zoo Biol.* 2010;29(2):179–191. Available from: https://doi.org/10.1002/zoo.20257.
18. Plotnik JM, de Waal FBM, Reiss D. Self-recognition in an Asian elephant. *Proc Natl Acad Sci U S A.* 2006;103(45):17053–17057. Available from: https://doi.org/10.1073/pnas.0608062103.
19. Reiss D, Marino L. Mirror self-recognition in the bottlenose dolphin: a case of cognitive convergence. *Proc Natl Acad Sci U S A.* 2001;98(10):5937–5942. Available from: https://doi.org/10.1073/pnas.101086398.
20. Delfour F, Marten K. Mirror image processing in three marine mammal species: killer whales (Orcinus orca), false killer whales (Pseudorca crassidens) and California sea lions (Zalophus californianus). *Behavioural Process.* 2001;53(3):181–190. Available from: https://doi.org/10.1016/S0376-6357(01)00134-6.
21. Prior H, Schwarz A, Güntürkün O. Mirror-induced behavior in the magpie (Pica pica): evidence of self-recognition. *PLOS Biol.* 2008;6(8):e202. Available from: https://doi.org/10.1371/journal.pbio.0060202.
22. Ma X, Jin Y, Luo B, Zhang G, Wei R, Liu D. Giant pandas failed to show mirror self-recognition. *Anim Cogn.* 2015;18(3):713–721. Available from: https://doi.org/10.1007/s10071-015-0838-4.
23. Soler M, Pérez-Contreras T, Peralta-Sánchez JM. Mirror-mark tests performed on jackdaws reveal potential methodological problems in the use of stickers in avian mark-test studies. *PLOS ONE.* 2014;9(1):e86193. Available from: https://doi.org/10.1371/journal.pone.0086193.
24. Amsterdam B. Mirror self-image reactions before age two. *Dev Psychobiol.* 1972;5(4):297–305. Available from: https://doi.org/10.1002/dev.420050403.
25. Anderson JR. The development of self-recognition: a review. *Dev Psychobiol.* 1984;17(1):35–49. Available from: https://doi.org/10.1002/dev.420170104.
26. Lin AC, Bard KA, Anderson JR. Development of self-recognition in chimpanzees (Pan troglodytes). *J Comp Psychol.* 1992;106(2):120–127. Available from: https://doi.org/10.1037/0735-7036.106.2.120.
27. Keenan JP, Nelson A, O'Connor M, Pascual-Leone A. Self-recognition and the right hemisphere. *Nature.* 2001;409(6818):305. Available from: https://doi.org/10.1038/35053167.
28. Keenan JP, Wheeler MA, Gallup GG, Pascual-Leone A. Self-recognition and the right prefrontal cortex. *Trends Cognit Sci.* 2000;4(9):338–344. Available from: https://doi.org/10.1016/s1364-6613(00)01521-7.

29. Morita T, Saito DN, Ban M, et al. Self-Face recognition begins to share active region in right inferior parietal lobule with proprioceptive illusion during adolescence. *Cereb Cortex*. 2018;28(4):1532–1548. Available from: https://doi.org/10.1093/cercor/bhy027.
30. Feinberg TE. Neuropathologies of the self: clinical and anatomical features. *Conscious Cogn*. 2011;20(1):75–81. Available from: https://doi.org/10.1016/j.concog.2010.09.017.
31. Carota A, Bogousslavsky J. Minor Hemisphere Major Syndromes. *Front Neurol Neurosci*. 2018;41:1–13. Available from: https://doi.org/10.1159/000475690.
32. Castro F, Lenggenhager B, Zeller D, Pellegrino G, D'Alonzo M, Di Pino G. From rubber hands to neuroprosthetics: neural correlates of embodiment. *Neurosci Biobehav Rev*. 2023;153:105351. Available from: https://doi.org/10.1016/j.neubiorev.2023.105351.
33. Golaszewski S, Frey V, Thomschewski A, et al. Neural mechanisms underlying the Rubber Hand Illusion: a systematic review of related neurophysiological studies. *Brain Behav*. 2021;11(8):e02124. Available from: https://doi.org/10.1002/brb3.2124.
34. Riemer M, Trojan J, Beauchamp M, Fuchs X. The rubber hand universe: on the impact of methodological differences in the rubber hand illusion. *Neurosci Biobehav Rev*. 2019;104:268–280. Available from: https://doi.org/10.1016/j.neubiorev.2019.07.008.
35. Botvinick M, Cohen J. Rubber hands 'feel' touch that eyes see. *Nature*. 1998;391(6669):756. Available from: https://doi.org/10.1038/35784.
36. Costantini M, Haggard P. The rubber hand illusion: sensitivity and reference frame for body ownership. *Conscious Cognition*. 2007;16(2):229–240. Available from: https://doi.org/10.1016/j.concog.2007.01.001.
37. Tsakiris M, Haggard P. The rubber hand illusion revisited: visuotactile integration and self-attribution. *J Exp Psychol Hum Percept Perform*. 2005;31(1):80–91. Available from: https://doi.org/10.1037/0096-1523.31.1.80.
38. Ehrsson HH, Spence C, Passingham RE. That's my hand! Activity in premotor cortex reflects feeling of ownership of a limb. *Science*. 2004;305(5685):875–877. Available from: https://doi.org/10.1126/science.1097011.
39. Ehrsson HH, Wiech K, Weiskopf N, Dolan RJ, Passingham RE. Threatening a rubber hand that you feel is yours elicits a cortical anxiety response. *Proc Natl Acad Sci U S A*. 2007;104(23):9828–9833. Available from: https://doi.org/10.1073/pnas.0610011104.
40. Ocklenburg S, Rüther N, Peterburs J, Pinnow M, Güntürkün O. Laterality in the rubber hand illusion. *Laterality*. 2011;16(2):174–187. Available from: https://doi.org/10.1080/13576500903483515.
41. Devinsky O. Delusional misidentifications and duplications: right brain lesions, left brain delusions. *Neurology*. 2009;72(1):80–87. Available from: https://doi.org/10.1212/01.wnl.0000338625.47892.74.
42. Devinsky O. Behavioral neurology. The neurology of Capgras syndrome. *Rev Neurol Dis*. 2008;5(2):97–100.
43. Barrelle A, Luauté J-P. Capgras syndrome and other delusional misidentification syndromes. *Front Neurol Neurosci*. 2018;42:35–43. Available from: https://doi.org/10.1159/000475680.
44. Vuilleumier P. Anosognosia: the neurology of beliefs and uncertainties. *Cortex*. 2004;40(1):9–17. Available from: https://doi.org/10.1016/s0010-9452(08)70918-3.
45. Gainotti G. History of anosognosia. *Front Neurol Neurosci*. 2019;44:75–82. Available from: https://doi.org/10.1159/000494954.
46. Orjuela-Rojas JM, Ramírez-Bermúdez J, Martínez-Juárez IE, Kerik NE, Diaz Meneses I, Pérez-Gay FJ. Visual hallucinations of autobiographic memory and asomatognosia: a case of epilepsy due to brain cysticercosis. *Neurocase*. 2015;21(5):635–641. Available from: https://doi.org/10.1080/13554794.2014.962548.
47. Jenkinson PM, Moro V, Fotopoulou A. Definition: asomatognosia. *Cortex*. 2018;101:300–301. Available from: https://doi.org/10.1016/j.cortex.2018.02.001.
48. Solimine S, Chan S, Morihara SK. Cotard syndrome: "I'm dead, so why do i need to eat?.". *Prim Care Companion CNS Disord*. 2016;18(2). Available from: https://doi.org/10.4088/PCC0.15l01862.
49. Robertson C, Dunn T. "A ghost doesn't need insulin," Cotard's delusion leading to diabetic ketoacidosis and a body-mass index of 15: a case presentation. *BMC Psychiatry*. 2023;23(1):551. Available from: https://doi.org/10.1186/s12888-023-05039-6.
50. Ramachandran VS. Consciousness and body image: lessons from phantom limbs, Capgras syndrome and pain asymbolia. *Philos Trans R Soc Lond B Biol Sci*. 1998;353(1377):1851–1859. Available from: https://doi.org/10.1098/rstb.1998.0337.
51. Kaufman KR, Newman NB, Dawood A. Capgras delusion with violent behavior in Alzheimer dementia: case analysis with literature review. *Ann Clin Psychiatry*. 2014;26(3):187–191.

52. Jedidi H, Daury N, Capa R, et al. Brain metabolic dysfunction in capgras delusion during Alzheimer's disease: a positron emission tomography study. *Am J Alzheimers Dis Other Demen.* 2015;30(7):699–706. Available from: https://doi.org/10.1177/1533317513495105.
53. Fischer C, Keeler A, Fornazzari L, Ringer L, Hansen T, Schweizer TA. A rare variant of Capgras syndrome in Alzheimer's disease. *Can j neurol sci.* 2009;36(4):509–511. Available from: https://doi.org/10.1017/s0317167100007903.
54. Feinberg TE. Neuropathologies of the self and the right hemisphere: a window into productive personal pathologies. *Front Hum Neurosci.* 2013;7:472. Available from: https://doi.org/10.3389/fnhum.2013.00472.
55. Feinberg TE, Shapiro RM. Misidentification-reduplication and the right hemisphere. *Neuropsych Neuropsychol Behav Neurol.* 1989;2(1):39–48.
56. Brighetti G, Bonifacci P, Borlimi R, Ottaviani C. "Far from the heart far from the eye": evidence from the Capgras delusion. *Cognit Neuropsych.* 2007;12(3):189–197. Available from: https://doi.org/10.1080/13546800600892183.
57. Paysant J, Beis JM, Le Chapelain L, André JM. Mirror asomatognosia in right lesions stroke victims. *Neuropsychologia.* 2004;42(7):920–925. Available from: https://doi.org/10.1016/j.neuropsychologia.2003.12.001.
58. Antoniello D, Gottesman R. Limb misidentification: a clinical-anatomical prospective study. *J Neuropsych Clin Neurosci.* 2017;29(3):284–288. Available from: https://doi.org/10.1176/appi.neuropsych.16090169.
59. Feinberg TE, Venneri A, Simone AM, Fan Y, Northoff G. The neuroanatomy of asomatognosia and somatoparaphrenia. *J Neurol Neurosurg Psychiatry.* 2010;81(3):276–281. Available from: https://doi.org/10.1136/jnnp.2009.188946.
60. Vallar G, Ronchi R. Somatoparaphrenia: a body delusion. A review of the neuropsychological literature. *Exp Brain Res.* 2009;192(3):533–551. Available from: https://doi.org/10.1007/s00221-008-1562-y.
61. Wilson RS, Sytsma J, Barnes LL, Boyle PA. Anosognosia in dementia. *Curr Neurol Neurosci Rep.* 2016;16(9)77. Available from: https://doi.org/10.1007/s11910-016-0684-z.
62. Sunderaraman P, Cosentino S. Integrating the constructs of anosognosia and metacognition: a review of recent findings in dementia. *Curr Neurol Neurosci Rep.* 2017;17(3)27. Available from: https://doi.org/10.1007/s11910-017-0734-1.
63. Starkstein SE. Anosognosia in Alzheimer's disease: diagnosis, frequency, mechanism and clinical correlates. *Cortex.* 2014;61:64–73. Available from: https://doi.org/10.1016/j.cortex.2014.07.019.
64. Vallar G, Ronchi R. Anosognosia for motor and sensory deficits after unilateral brain damage: a review. *Restor Neurol Neurosci.* 2006;24(4-6):247–257.
65. Moro V, Pernigo S, Zapparoli P, Cordioli Z, Aglioti SM. Phenomenology and neural correlates of implicit and emergent motor awareness in patients with anosognosia for hemiplegia. *Behavioural Brain Res.* 2011;225 (1):259–269. Available from: https://doi.org/10.1016/j.bbr.2011.07.010.
66. Chen J-J, Chang H-F, Hsu Y-C, Chen D-L. Anton-Babinski syndrome in an old patient: a case report and literature review. *Psychogeriatrics.* 2015;15(1):58–61. Available from: https://doi.org/10.1111/psyg.12064.
67. Grover S, Aneja J, Mahajan S, Varma S. Cotard's syndrome: two case reports and a brief review of literature. *J Neurosci Rural Pract.* 2014;5(Suppl 1):S59–62. Available from: https://doi.org/10.4103/0976-3147.145206.
68. Nishio Y, Mori E. Delusions of death in a patient with right hemisphere infarction. *Cogn Behav Neurol.* 2012;25 (4):216–223. Available from: https://doi.org/10.1097/WNN0.0b013e31827504c7.
69. Moro V, Pacella V, Scandola M, et al. A fronto-insular-parietal network for the sense of body ownership. *Cereb Cortex.* 2023;33(3):512–522. Available from: https://doi.org/10.1093/cercor/bhac081.
70. Heydrich L, Dieguez S, Grunwald T, Seeck M, Blanke O. Illusory own body perceptions: case reports and relevance for bodily self-consciousness. *Conscious Cogn.* 2010;19(3):702–710. Available from: https://doi.org/10.1016/j.concog.2010.04.010.
71. Carmody DP, Lewis M. Brain activation when hearing one's own and others' names. *Brain Res.* 2006;1116 (1):153–158. Available from: https://doi.org/10.1016/j.brainres.2006.07.121.
72. Macrae CN, Moran JM, Heatherton TF, Banfield JF, Kelley WM. Medial prefrontal activity predicts memory for self. *Cereb Cortex.* 2004;14(6):647–654. Available from: https://doi.org/10.1093/cercor/bhh025.
73. Maguire EA, Mummery CJ. Differential modulation of a common memory retrieval network revealed by positron emission tomography. *Hippocampus.* 1999;9(1):54–61. https://doi.org/10.1002/(SICI)1098-1063(1999)9:1 < 54:AID-HIPO6 > 3.0.CO;2-O.
74. Güntürkün O. The Venus of Milo and the dawn of facial asymmetry research. *Brain Cognition.* 1991;16 (2):147–150. Available from: https://doi.org/10.1016/0278-2626(91)90003-q.

75. Henke W. Glossen zur Venus von Melos. *Z für bildende Kunst*. 1886;21:194–199.
76. Hasse C. *Über Gesichtsasymmetrien. Archiv für Anatomie und Entwickelungsgeschichte*. Veit & Co; 1887:119–125.
77. Hallervorden J. Eine neue methode experimenteller physiognomik. *Psychiatrisch-Neurologische Wochenschr*. 1902;4:309–311.
78. Wolff W. The experimental study of forms of expression. *Character Personality*. 1933;2(2):168–176.
79. Thiesen G, Gribel BF, Freitas MPM. Facial asymmetry: a current review. *Dental Press J Orthod*. 2015;20 (6):110–125. Available from: https://doi.org/10.1590/2177-6709.20.6.110-125.sar.
80. Severt TR, Proffit WR. The prevalence of facial asymmetry in the dentofacial deformities population at the University of North Carolina. *Int J Adult Orthodon Orthognath Surg*. 1997;12(3):171–176.
81. Lindell A. Lateralization of the expression of facial emotion in humans. *Prog Brain Res*. 2018;238:249–270. Available from: https://doi.org/10.1016/bs.pbr.2018.06.005.
82. Hausmann M, Behrendt-Körbitz S, Kautz H, Lamm C, Radelt F, Güntürkün O. Sex differences in oral asymmetries during wordrepetition. *Neuropsychologia*. 1998;36(12):1397–1402. Available from: https://doi.org/10.1016/s0028-3932(98)00027-x.
83. Wolf ME, Goodale MA. Oral asymmetries during verbal and non-verbal movements of the mouth. *Neuropsychologia*. 1987;25(2):375–396. Available from: https://doi.org/10.1016/0028-3932(87)90026-1.
84. Graves R, Goodglass H, Landis T. Mouth asymmetry during spontaneous speech. *Neuropsychologia*. 1982;20 (4):371–381. Available from: https://doi.org/10.1016/0028-3932(82)90037-9.
85. Bourne VJ. The divided visual field paradigm: methodological considerations. *Laterality*. 2006;11(4):373–393. Available from: https://doi.org/10.1080/13576500600633982.
86. Hausmann M, Güntürkün O. Sex differences in functional cerebral asymmetries in a repeated measures design. *Brain Cognition*. 1999;41(3):263–275. Available from: https://doi.org/10.1006/brcg.1999.1126.
87. Ellis HD, Shepherd JW. Recognition of upright and inverted faces presented in the left and right visual fields. *Cortex*. 1975;11(1):3–7. Available from: https://doi.org/10.1016/s0010-9452(75)80014-1.
88. Leehey S, Carey S, Diamond R, Cahn A. Upright and inverted faces: the right hemisphere knows the difference. *Cortex*. 1978;14(3):411–419. Available from: https://doi.org/10.1016/s0010-9452(78)80067-7.
89. Prete G., Marzoli D., Tommasi L. Upright or inverted, entire or exploded: right-hemispheric superiority in face recognition withstands multiple spatial manipulations. PeerJ. 2015; 3: e1456. doi:10.7717/peerj.1456
90. Levine SC, Koch-Weser MP. Right hemisphere superiority in the recognition of famous faces. *Brain Cognition*. 1982;1(1):10–22. Available from: https://doi.org/10.1016/0278-2626(82)90003-3.
91. Cormier P, Jackson T. Effects of instruction on a divided visual-field task of face processing. *Percept Mot Skills*. 1995;80(3 Pt 1):923–930. Available from: https://doi.org/10.2466/pms.1995.80.3.923.
92. Hausmann M, Güntürkün O. Steroid fluctuations modify functional cerebral asymmetries: the hypothesis of progesterone-mediated interhemispheric decoupling. *Neuropsychologia*. 2000;38(10):1362–1374. Available from: https://doi.org/10.1016/S0028-3932(00)00045-2.
93. Hausmann M, Becker C, Gather U, Güntürkün O. Functional cerebral asymmetries during the menstrual cycle: a cross-sectional and longitudinal analysis. *Neuropsychologia*. 2002;40(7):808–816. Available from: https://doi.org/10.1016/s0028-3932(01)00179-8.
94. Sergent J, Ohta S, MacDonald B. Functional neuroanatomy of face and object processing. A positron emission tomography study. *Brain*. 1992;115(Pt 1(1)):15–36. Available from: https://doi.org/10.1093/brain/115.1.15.
95. Kanwisher N. The quest for the FFA and where it led. *J Neurosci*. 2017;37(5):1056–1061. Available from: https://doi.org/10.1523/JNEUROSCI.1706-16.2016.
96. Kanwisher N, McDermott J, Chun MM. The fusiform face area: a module in human extrastriate cortex specialized for face perception. *J Neurosci*. 1997;17(11):4302–4311. Available from: https://doi.org/10.1523/JNEUROSCI.17-11-04302.1997.
97. Caldara R, Seghier ML, Rossion B, Lazeyras F, Michel C, Hauert C-A. The fusiform face area is tuned for curvilinear patterns with more high-contrasted elements in the upper part. *Neuroimage*. 2006;31(1):313–319. Available from: https://doi.org/10.1016/j.neuroimage.2005.12.011.
98. Badzakova-Trajkov G, Häberling IS, Corballis MC. Cerebral asymmetries in monozygotic twins: an fMRI study. *Neuropsychologia*. 2010;48(10):3086–3093. Available from: https://doi.org/10.1016/j.neuropsychologia.2010.06.020.
99. Goffaux V, Duecker F, Hausfeld L, Schiltz C, Goebel R. Horizontal tuning for faces originates in high-level Fusiform Face Area. *Neuropsychologia*. 2016;81:1–11. Available from: https://doi.org/10.1016/j.neuropsychologia.2015.12.004.

100. Rossion B, Caldara R, Seghier M, Schuller A-M, Lazeyras F, Mayer E. A network of occipito-temporal face-sensitive areas besides the right middle fusiform gyrus is necessary for normal face processing. *Brain.* 2003;126(Pt 11):2381–2395. Available from: https://doi.org/10.1093/brain/awg241.
101. Badzakova-Trajkov G, Häberling IS, Roberts RP, Corballis MC. Cerebral asymmetries: complementary and independent processes. *PLOS ONE.* 2010;5(3):e9682. Available from: https://doi.org/10.1371/journal.pone.0009682.
102. Harris A, Aguirre GK. Neural tuning for face wholes and parts in human fusiform gyrus revealed by FMRI adaptation. *J Neurophysiol.* 2010;104(1):336–345. Available from: https://doi.org/10.1152/jn.00626.2009.
103. Pinel P, Lalanne C, Bourgeron T, et al. Genetic and Environmental Influences on the Visual Word Form and Fusiform Face Areas. *Cereb Cortex.* 2015;25(9):2478–2493. Available from: https://doi.org/10.1093/cercor/bhu048.
104. Zhen Z, Yang Z, Huang L, et al. Quantifying interindividual variability and asymmetry of face-selective regions: a probabilistic functional atlas. *Neuroimage.* 2015;113:13–25. Available from: https://doi.org/10.1016/j.neuroimage.2015.03.010.
105. Rossion B, Lochy A. Is human face recognition lateralized to the right hemisphere due to neural competition with left-lateralized visual word recognition? A critical review. *Brain Struct Funct.* 2022;227(2):599–629. Available from: https://doi.org/10.1007/s00429-021-02370-0.
106. Thome I, García Alanis JC, Volk J, Vogelbacher C, Steinsträter O, Jansen A. Let's face it: the lateralization of the face perception network as measured with fMRI is not clearly right dominant. *Neuroimage.* 2022;263:119587. Available from: https://doi.org/10.1016/j.neuroimage.2022.119587.
107. Johnstone LT, Karlsson EM, Carey DP. Left-handers are less lateralized than right-handers for both left and right hemispheric functions. *Cereb Cortex.* 2021;31(8):3780–3787. Available from: https://doi.org/10.1093/cercor/bhab048.
108. Willems RM, Peelen MV, Hagoort P. Cerebral lateralization of face-selective and body-selective visual areas depends on handedness. *Cereb Cortex.* 2010;20(7):1719–1725. Available from: https://doi.org/10.1093/cercor/bhp234.
109. Bukowski H, Dricot L, Hanseeuw B, Rossion B. Cerebral lateralization of face-sensitive areas in left-handers: only the FFA does not get it right. *Cortex.* 2013;49(9):2583–2589. Available from: https://doi.org/10.1016/j.cortex.2013.05.002.
110. Yovel G, Tambini A, Brandman T. The asymmetry of the fusiform face area is a stable individual characteristic that underlies the left-visual-field superiority for faces. *Neuropsychologia.* 2008;46(13):3061–3068. Available from: https://doi.org/10.1016/j.neuropsychologia.2008.06.017.
111. Rangarajan V, Hermes D, Foster BL, et al. Electrical stimulation of the left and right human fusiform gyrus causes different effects in conscious face perception. *J Neurosci.* 2014;34(38):12828–12836. Available from: https://doi.org/10.1523/JNEUROSCI.0527-14.2014.
112. Rangarajan V, Parvizi J. Functional asymmetry between the left and right human fusiform gyrus explored through electrical brain stimulation. *Neuropsychologia.* 2016;83:29–36. Available from: https://doi.org/10.1016/j.neuropsychologia.2015.08.003.
113. Bieniek MM, Frei LS, Rousselet GA. Early ERPs to faces: aging, luminance, and individual differences. *Front Psychol.* 2013;4:268. Available from: https://doi.org/10.3389/fpsyg.2013.00268.
114. Eimer M. Effects of face inversion on the structural encoding and recognition of faces. *Cognit Brain Res.* 2000;10(1-2):145–158. Available from: https://doi.org/10.1016/S0926-6410(00)00038-0.
115. Rossion B, Joyce CA, Cottrell GW, Tarr MJ. Early lateralization and orientation tuning for face, word, and object processing in the visual cortex. *Neuroimage.* 2003;20(3):1609–1624. Available from: https://doi.org/10.1016/j.neuroimage.2003.07.010.
116. Rossion B, Gauthier I. How does the brain process upright and inverted faces? *Behav Cognit Neurosci Rev.* 2002;1(1):63–75. Available from: https://doi.org/10.1177/1534582302001001004.
117. Ocklenburg S, Ness V, Güntürkün O, Suchan B, Beste C. Response inhibition is modulated by functional cerebral asymmetries for facial expression perception. *Front Psychol.* 2013;4:879. Available from: https://doi.org/10.3389/fpsyg.2013.00879.
118. Itier RJ, Latinus M, Taylor MJ. Face, eye and object early processing: what is the face specificity? *Neuroimage.* 2006;29(2):667–676. Available from: https://doi.org/10.1016/j.neuroimage.2005.07.041.
119. Bentin S, Allison T, Puce A, Perez E, McCarthy G. Electrophysiological studies of face perception in humans. *J Cogn Neurosci.* 1996;8(6):551–565. Available from: https://doi.org/10.1162/jocn.1996.8.6.551.

120. Itier RJ, Taylor MJ. Inversion and contrast polarity reversal affect both encoding and recognition processes of unfamiliar faces: a repetition study using ERPs. *Neuroimage.* 2002;15(2):353–372. Available from: https://doi.org/10.1006/nimg.2001.0982.
121. Corrigan NM, Richards T, Webb SJ, et al. An investigation of the relationship between fMRI and ERP source localized measurements of brain activity during face processing. *Brain Topogr.* 2009;22(2):83–96. Available from: https://doi.org/10.1007/s10548-009-0086-5.
122. Herrmann CS, Knight RT. Mechanisms of human attention: event-related potentials and oscillations. *Neurosci Biobehav Rev.* 2001;25(6):465–476. Available from: https://doi.org/10.1016/s0149-7634(01)00027-6.
123. Yovel G. Neural and cognitive face-selective markers: an integrative review. *Neuropsychologia.* 2016;83:5–13. Available from: https://doi.org/10.1016/j.neuropsychologia.2015.09.026.
124. Luo S, Luo W, He W, Chen X, Luo Y. P1 and N170 components distinguish human-like and animal-like makeup stimuli. *NeuroReport.* 2013;24(9):482–486. Available from: https://doi.org/10.1097/WNR0.0b013e328361cf08.
125. Mercure E, Dick F, Halit H, Kaufman J, Johnson MH. Differential lateralization for words and faces: category or psychophysics? *J Cogn Neurosci.* 2008;20(11):2070–2087. Available from: https://doi.org/10.1162/jocn.2008.20137.
126. Maurer U, Rossion B, McCandliss BD. Category specificity in early perception: face and word n170 responses differ in both lateralization and habituation properties. *Front Hum Neurosci.* 2008;2:18. Available from: https://doi.org/10.3389/neuro.09.018.2008.
127. Balconi M, Lucchiari C. Event-related potentials related to normal and morphed emotional faces. *J Psychol.* 2005;139(2):176–192. Available from: https://doi.org/10.3200/JRLP0.139.2.176-192.
128. Hecaen H, Angelergues R. Agnosia for faces (prosopagnosia). *Arch Neurol.* 1962;7(2):92–100. Available from: https://doi.org/10.1001/archneur.1962.04210020014002.
129. Yin RK. Face recognition by brain-injured patients: a dissociable ability? *Neuropsychologia.* 1970;8(4):395–402. Available from: https://doi.org/10.1016/0028-3932(70)90036-9.
130. Barton JJS. Structure and function in acquired prosopagnosia: lessons from a series of 10 patients with brain damage. *J Neuropsychol.* 2008;2(1):197–225. Available from: https://doi.org/10.1348/174866407X214172.
131. Busigny T, Graf M, Mayer E, Rossion B. Acquired prosopagnosia as a face-specific disorder: ruling out the general visual similarity account. *Neuropsychologia.* 2010;48(7):2051–2067. Available from: https://doi.org/10.1016/j.neuropsychologia.2010.03.026.
132. de Renzi E. Prosopagnosia in two patients with CT scan evidence of damage confined to the right hemisphere. *Neuropsychologia.* 1986;24(3):385–389. Available from: https://doi.org/10.1016/0028-3932(86)90023-0.
133. de Renzi E, Faglioni P, Grossi D, Nichelli P. Apperceptive and Associative Forms of Prosopagnosia. *Cortex.* 1991;27(2):213–221. Available from: https://doi.org/10.1016/s0010-9452(13)80125-6.
134. Landis T, Regard M, Bliestle A, Kleihues P. Prosopagnosia and agnosia for noncanonical views. An autopsied case. *Brain.* 1988;111(6):1287–1297. Available from: https://doi.org/10.1093/brain/111.6.1287.
135. Wada Y, Yamamoto T. Selective impairment of facial recognition due to a haematoma restricted to the right fusiform and lateral occipital region. *J Neurol Neurosurg Psychiatry.* 2001;71(2):254–257. Available from: https://doi.org/10.1136/jnnp.71.2.254.
136. Sergent J, Signoret JL. Varieties of functional deficits in prosopagnosia. *Cereb Cortex.* 1992;2(5):375–388. Available from: https://doi.org/10.1093/cercor/2.5.375.
137. Susilo T, Wright V, Tree JJ, Duchaine B. Acquired prosopagnosia without word recognition deficits. *Cogn Neuropsychol.* 2015;32(6):321–339. Available from: https://doi.org/10.1080/02643294.2015.1081882.
138. Schroeger A, Kaufmann JM, Zäske R, Kovács G, Klos T, Schweinberger SR. Atypical prosopagnosia following right hemispheric stroke: a 23-year follow-up study with M.T. *Cogn Neuropsychol.* 2022;39(3-4):196–207. Available from: https://doi.org/10.1080/02643294.2022.2119838.
139. Downing PE, Peelen MV. Body selectivity in occipitotemporal cortex: Causal evidence. *Neuropsychologia.* 2016;83:138–148. Available from: https://doi.org/10.1016/j.neuropsychologia.2015.05.033.
140. Peelen MV, Downing PE. Selectivity for the human body in the fusiform gyrus. *J Neurophysiol.* 2005;93(1):603–608. Available from: https://doi.org/10.1152/jn.00513.2004.
141. Schwarzlose RF, Baker CI, Kanwisher N. Separate face and body selectivity on the fusiform gyrus. *J Neurosci.* 2005;25(47):11055–11059. Available from: https://doi.org/10.1523/JNEUROSCI.2621-05.2005.
142. Dal Bò E, Gentili C, Cecchetto C. Human chemosignals and brain activity: a preliminary meta-analysis of the processing of human body odors. *Chem Senses.* 2020;45(9):855–864. Available from: https://doi.org/10.1093/chemse/bjaa067.

CHAPTER

9

Hemispheric asymmetries in emotion processing and social behaviors

OUTLINE

Introductory short story

In the late 1980s, she was working as a teacher in a small town in Germany. Her life was completely uneventful, and that was OK for her. She was a bit of a loner, rarely met other people, and never used to hang out with friends. That was the way she was, and she

DOI: https://doi.org/10.1016/B978-0-323-99737-9.00004-5

never felt the need to change anything about it. It could have been like that forever, but one day something strange happened. She felt a surge of something that had no name. Like a wave, it hit her out of nowhere. Suddenly, everything felt dangerous—like imminent death but without any obvious reasons. Usually, it happened only for a short moment, sometimes just for a few seconds, but it was very strange. These waves started to come again and again over the course of the next couple of months. Mostly, it felt like some abstract form of danger. Rarely, it felt like sudden, bright sunshine and pure lust. She had no idea what it was, and she had no name for it, but she was worried. Suddenly, it occurred to her that these waves were pure emotions: Mostly anxious ones but sometimes also moments of pure lust. By understanding that she was occasionally visited by emotions she also realized that she never seemed to have experienced them before. Worried about that, she started a long journey through hospitals and psychotherapy units. She was dispatched to psychiatry, then to neurology, and ended up in front of cognitive behavioral therapists and all kinds of other psychotherapists. None of them were able to come up with a proper diagnosis, and nobody could explain her surges. Eventually, one day she stood in front of the information desk of the university clinic in Bochum and started to explain what she was looking for. By chance, the head of the neuropsychology ward was standing close to her. Her hoarse voice alerted him. He scrutinized her: The skin was wrinkly and dry, and even from a distance, he could see the beadings around her eyelids. He came close and started hearing what she was complaining about. Finally, he introduced himself and asked if he could invite her for some diagnostics and also run a brain scan. She accepted. Some days later she sat in front of his desk and heard him say that she was suffering from Urbach–Wiethe disease, a rare recessive genetic disorder that, during puberty, slowly destroys a part of her brain called the amygdala. The neuropsychologist explained the role of the amygdala in emotions, its asymmetrical functions in emotion regulation, and what all this could mean for her strange condition. Then he took out the pictures of her brain scans and showed them to her. She had never seen anything like that before. The neuropsychologist moved with his fingers to a spot at the lower part of the image and explained that this was the location of the amygdala. In disbelief, she stared at the tip of his finger: It pointed to an empty hole in her brain.

Inspired by a story from *Biologische Psychologie*.[1]

Emotions—some basics

One of the prime examples of functional lateralization in the human brain is emotion processing.[2] Before we discuss emotional lateralization, let us quickly recapitulate some basic facts about emotions. What is an emotion? Emotions represent valanced, discrete, and consistent responses to internal or external stimuli or situations which have a particular significance for the organism. They have a vital role in facilitating adaptive behavior and optimizing decision-making concerning events in the environment.[3] Emotions are brief and consist of a coordinated set of responses, which may include verbal, physiological, behavioral, and neuronal mechanisms. They can involve multiple types of appraisal processes that access the significance of a stimulus to current goals. Thus, emotions are

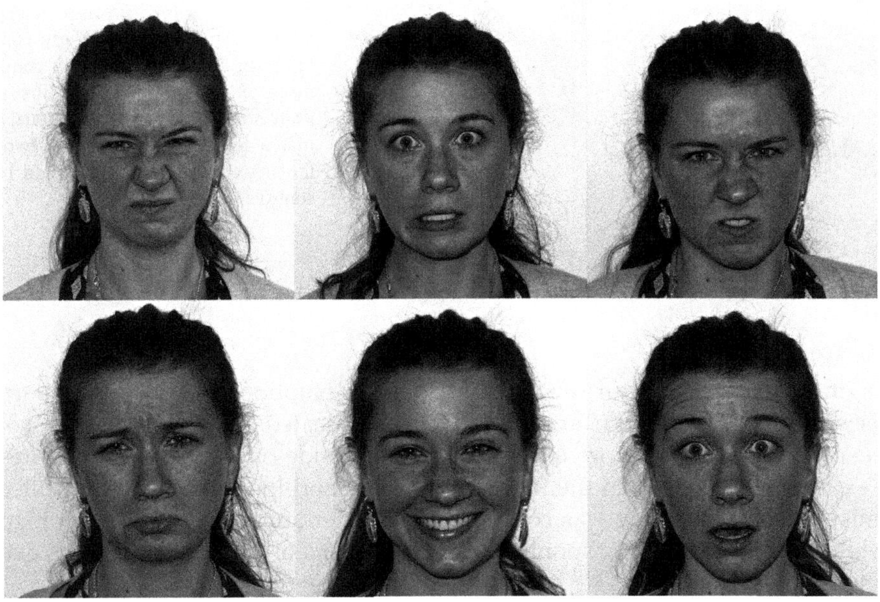

FIGURE 9.1 Facial expressions for the six basic emotions: Upper panel: disgust, fear, and anger; lower panel: sadness, happiness, and surprise.

dependent on the present situation, previous experiences, and/or personality. Emotions consist of three main components:

1. A physiological reaction to a stimulus (e.g., increased heartbeat and sweating)
2. A behavioral response (e.g., running toward a loved person)
3. A feeling (e.g., the subjective experience of happiness)

Six basic emotions that are evident in distinct facial expressions (see Fig. 9.1) have been identified by researchers:[4]

1. Disgust
2. Fear
3. Anger
4. Sadness
5. Happiness
6. Surprise

Left-right differences in facial emotion expression

Facial movements such as smiling, frowning, scowling, or sneering represent one of the primary nonverbal means to cosnsciously or unconsciously communicate internal emotional states to other individuals. As mentioned in Chapter 8, emotional expression in the face seems to be lateralized. When asked to indicate how intensely left-side-only and

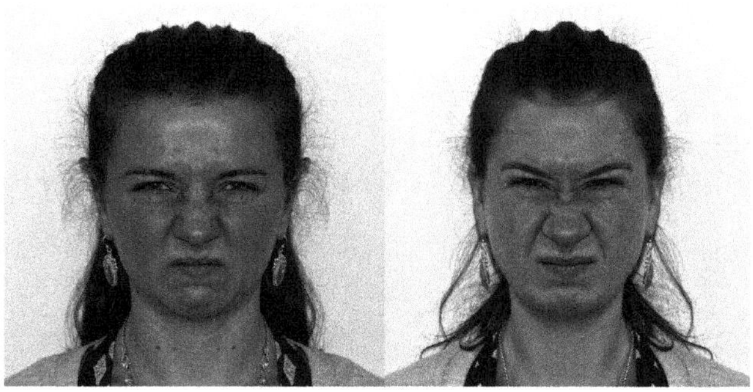

FIGURE 9.2 Facial emotion expression is lateralized and more strongly controlled by the right hemisphere. Thus a composite face made of two right hemifaces (first panel) typically shows less intense emotional expression than a composite face made of two left hemifaces (second panel) as the right hemisphere controls the left hemiface.

right-side-only composite pictures of faces expressed emotions (see Fig. 9.2), participants in the classic study by Sackeim and colleagues [5] indicated that left-side-only composite pictures expressed emotions more intensely than right-side-only composite pictures.

Thus, expressing emotions in the face is lateralized and more strongly controlled by the right hemisphere. This finding was confirmed by a later study in which participants were filmed while making posed and spontaneous expressions of emotion[6], and was even replicated in nonhuman primates.[3,7–10]

Interestingly, the fact that the left hemiface is emotionally more expressive directly influences how we perceive paintings or pictures. In 2002, Michael Nicholls and coworkers systematically investigated whether asymmetries in facial emotional expression affected how participants perceived pictures that featured one side of the face more prominently than the other.[11] They photographed the faces of six different models in three different poses (see Fig. 9.3A):

1. Midline condition: The model's body and head faced directly toward the camera
2. Left-pose condition: The model's body midline faced directly toward the camera, but the head was turned 15 degrees to the left
3. Right-pose condition: The model's body midline faced directly toward the camera, but the head was turned 15 degrees to the right

Subsequently, these pictures were shown to 384 university students, who were asked to evaluate the emotional expressivity of the pictures by answering a short questionnaire. Results showed that midline and left hemiface pictures were rated as significantly more emotionally expressive than right hemiface portraits (see Fig. 9.3B). This effect was also replicated when portraits were mirror-reversed (see Fig. 9.3B), proving that it was not caused by potential perceptual biases. Thus even a relatively slight leftward posing bias can influence perceived emotional expressivity for pictures.

Interestingly, asymmetries in perceived emotional expression seem to affect how painters ask their models to pose for a painting. In 1973, McManus and Humphries analyzed posing bias in more than 1400 Western European portraits painted between the 16th and the 20th century.[12] Only paintings that showed parts of the body and the face of a single person were included in the study. Overall, there was a significant 60% bias to show more of the left cheek than of the right cheek in all paintings. Interestingly, this left-cheek bias was more pronounced in women (68%) than in men (56%). Following the study by McManus and Humphries, several authors investigated the left-cheek bias in different

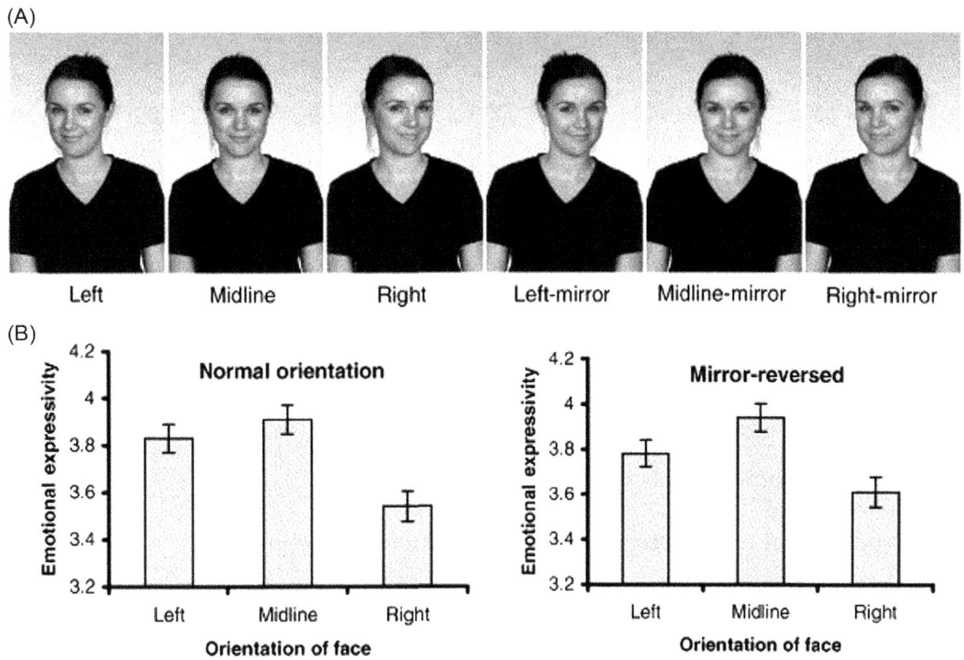

FIGURE 9.3 Stimuli and results from a study on posing and emotional expressivity by Nicholls and coworkers.[11] (A) Examples of the stimuli used in the study. For each model, left, right, and midline pose photographs were created (left side). All three pictures were also presented in a mirror-reversed version (right side). (B) Bar charts showing the results of the study for normal (left side) and mirror-reversed stimuli (right side). For both conditions, mean emotional expressivity is shown for the left, right, and midline poses. Error bars show standard errors. *Reprinted with permission of the author and the publisher from Nicholls MER, Wolfgang BJ, Clode D, et al. The effect of left and right poses on the expression of facial emotion.* Neuropsychologia. *2002;40:1s662–1665.*

contexts.[13,14] For example, the left-cheek bias was replicated in portraits in school yearbooks,[15] religious artworks depicting Jesus,[16] celebrity chef cookbook covers,[17] and even in smartphone selfie pictures [18–20] and in social media ads.[21] Interestingly, it has been shown that on social media, left-cheek poses gain more "likes" than right-cheek poses.[22]

It has been shown that individuals who want to be perceived as nonemotional, for example in the context of their jobs, show a reversal of the typical left-cheek bias. Owen Churches from the University of South Australia in Adelaide and coworkers investigated posing bias in 5,829 homepage pictures of academics from different fields.[23] They found that both male and female academics from the fields of chemistry, mathematics, and engineering showed their right cheek more often than their left. Such an effect was not observed for English academics who showed the typical leftward bias.

Hemispheric asymmetries in emotion processing

Hemispheric asymmetries have been found not only for emotional expressivity in the face but also for emotion processing in the brain.[24,25] Three major models have been

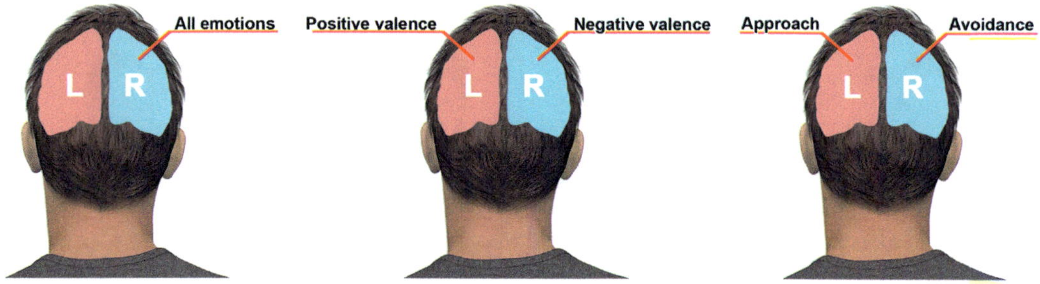

FIGURE 9.4 The three major models of emotional lateralization. First panel: The right-hemisphere model; second panel: the valence model; third panel: the frontal cortical asymmetry model of motivational direction.

suggested for emotional lateralization (see Fig. 9.4), which will be outlined in the next sections. These models include:

1. The right-hemisphere model
2. The valence model
3. The frontal cortical asymmetry model of motivational direction

The right-hemisphere model

The oldest model of emotional lateralization is the right-hemisphere model (see Fig. 9.4A). This theory assumes that the right hemisphere shows general dominance for processing emotional stimuli, independent of their specific valence.[26–33] The first support of this idea was provided in an 1879 paper by the British neurologist John Hughlings Jackson (1835–1911),[34] in which he investigated how lesions in the brain affected patients' ability to speak. In accordance with what Broca had proposed 30 years earlier (see Chapter 1), Hughlings Jackson noticed that speechless patients typically had lesions affecting the left side of the brain. He further noticed that even with extensive damage to the left hemisphere, most patients were still able to produce short utterances like "pittymy" or "committymy." How was it possible that these patients were still able to speak at all, given that they were otherwise speechless? Hughlings Jackson assumed that this capability was caused by strong, suddenly induced emotion triggered by bad circumstances the patients were in and the resulting fear that even worse might come. The finding that severely speech-impaired patients with left-hemispheric lesions had somewhat preserved emotional language implied that the right hemisphere likely played a major role in emotions. After this initial finding by Hughlings Jackson, empirical support for the idea that the right hemisphere shows a general dominance for the processing of emotional stimuli has been provided by several different lines of research in healthy individuals and patients.

The right-hemisphere model: evidence from patient studies

Following the patient study by Hughlings Jackson, several other authors published patient studies in support of the idea that the right hemisphere plays a major role in emotion processing.[27] For the auditory modality, Kenneth M. Heilman from the University of Florida, United States, investigated comprehension of affective speech in six patients with lesions in the left temporal and parietal lobe and six patients with lesions in the right temporal and parietal lobe.[35] The patients had to listen to 32 tape-recorded sentences and to either indicate the semantic content of the sentence or the emotional mood of the person who spoke the sentence by pointing toward drawings of emotional faces. All patients showed 100% correct responses for semantic content, but there was a significant difference between right- and left-lesioned patients for emotional mood identification. Patients with right-hemispheric lesions showed substantially lower accuracy (26.06% correct responses) than patients with left-hemispheric lesions (63.56% correct responses). Heilman concluded that patients with right-hemispheric lesions exhibited deficits in the comprehension of affective speech, a condition he termed auditory affective agnosia.

Similar findings have also been reported for the visual modality. In 1980, Steven T. DeKosky from the University of Florida, United States, and colleagues tested discrimination and comprehension of emotional faces and pictures in nine patients with right-hemispheric lesions, nine patients with left-hemispheric lesions, and nine neurological control patients without unilateral lesions.[36]

The patients were tested on six different tests:

1. Prosopagnosia test: Patients viewed two pictures of faces and had to indicate whether they showed the same individual or two different individuals.
2. Name emotional face test: Patients viewed pictures of emotional faces and had to identify the emotions the faces were expressing.
3. Choose emotional face test: Patients were told a specific emotion and had to choose a picture showing this emotion from several pictures showing emotional faces.
4. Same–different emotional face test: Patients viewed two pictures of the same person at a time and had to indicate whether they showed the same emotion or different emotions.
5. Name emotional scene test: Patients viewed drawings of emotional scenes (see Fig. 9.5), and the experimenter pointed toward one person shown in this scene. Patients had to indicate the emotional status of this person in the drawing.

FIGURE 9.5 Example of one of the emotional scenes used in the study by DeKosky et al.[36] *Reprinted with permission from the publisher from DeKosky ST, Heilman KM, Bowers D, Valenstein E. Recognition and discrimination of emotional faces and pictures. Brain Lang. 1980;9:206–214.*

6. Choose emotional scene test: Patients viewed four drawings of emotional scenes and had to select one that showed a specific emotion.

The results were clear: Patients with right-hemispheric lesions performed significantly worse than controls in all six tests and also worse than patients with left-hemispheric lesions in most of them (see Fig. 9.6).

Based on these observations, DeKosky and colleagues concluded that the right hemisphere is responsible for processing emotional faces and scenes.

More evidence for a major role of the right hemisphere in emotion processing was provided in a neuroimaging study on blindsight. Blindsight is a phenomenon that sometimes develops in neurological patients with bilateral damage to the primary visual cortex. While these patients have no conscious experience of seeing their environment, they nevertheless can react properly to moving targets or identify an object's orientation. For example, Pegna et al. reported a case of a patient with a lesion affecting both visual cortices who, despite cortical blindness, was able to correctly identify whether a face showed a happy or angry expression.[37] To investigate which cortical structure allowed this patient to differentiate happy from angry faces, Pegna et al. showed the patient pictures of angry, happy, fearful, or neutral faces while an fMRI scan was performed. When contrasting the activation for the three types of emotional faces with neutral faces, Pegna et al. found a significant effect for only one brain area: The right amygdala, an almond-shaped structure deep in the temporal lobe that plays an important part in emotion processing and emotional memory in the brain[38] (see Fig. 9.7).

What happens when the amygdala does not function anymore can be observed in patients with Urbach–Wiethe disease. Urbach–Wiethe disease is a rare genetic disorder

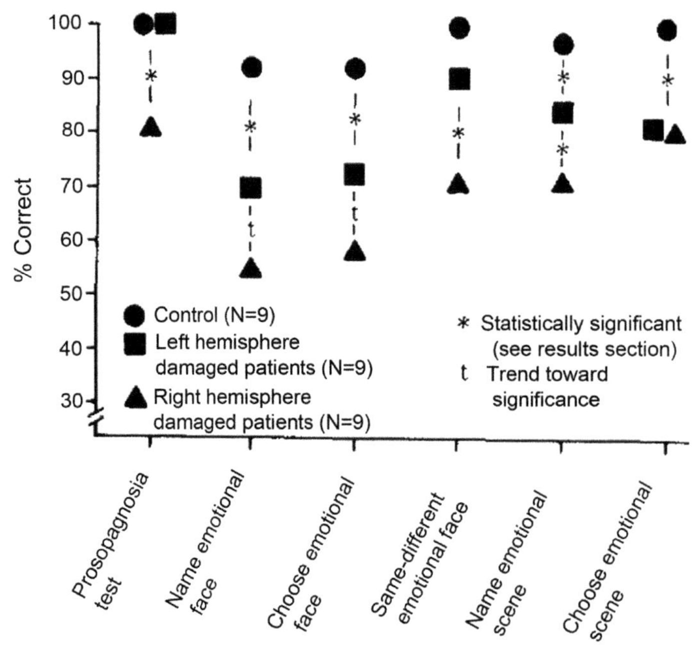

FIGURE 9.6 Results from the patient study on emotion recognition by DeKosky et al.[36] *Reprinted with permission from the publisher from DeKosky ST, Heilman KM, Bowers D, Valenstein E. Recognition and discrimination of emotional faces and pictures. Brain Lang. 1980;9:206–214.*

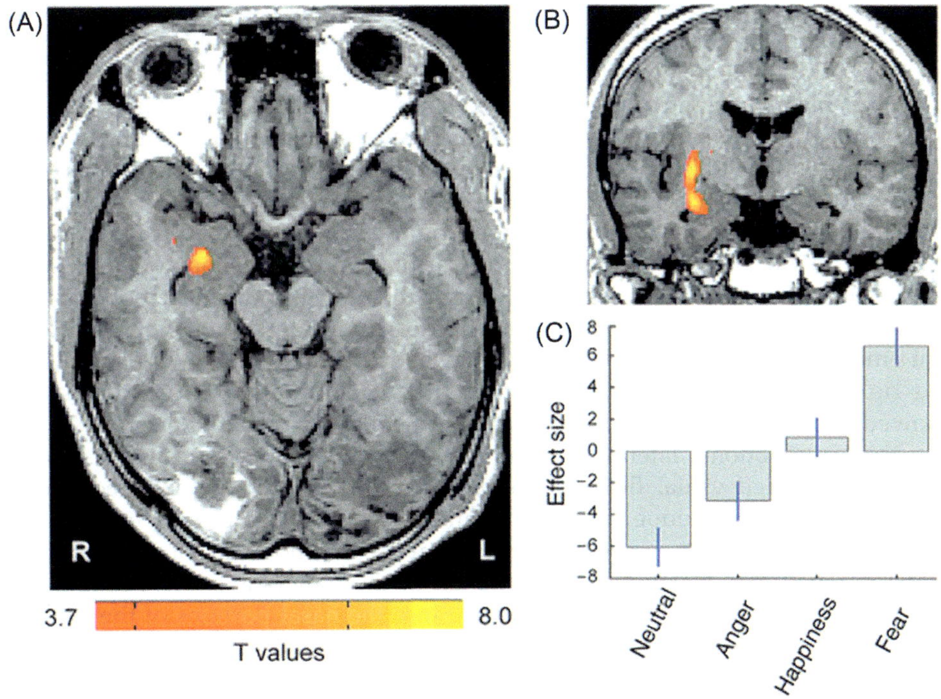

FIGURE 9.7 Results from a case study on emotion recognition in a blindsighted patient by Pegna et al.[37] (A) and (B) The patient showed significant activation of the right amygdala when viewing emotional faces compared to neutral faces. (C) Parameter estimates for the different emotional faces all differed from neutral faces. *Reprinted with permission from the publisher from Pegna AJ, Khateb A, Lazeyras F, Seghier ML. Discriminating emotional faces without primary visual cortices involves the right amygdala. Nat Neurosci. 2005;8:24–25.*

that has both dermatological and neurological symptoms. On the one hand, it leads to a thickening of the skin and mucous membranes, resulting in a hoarse voice, dry and wrinkly skin, and beadings around the eyelids. On the other hand, Urbach–Wiethe disease in many cases leads to calcifications of the blood vessels supplying the amygdala.[39,40] These calcifications lead to a hardening of the blood vessels and finally destroy them. This causes lesions in the amygdala, as it did in the patient in the introductory short story. While the consequences vary from patient to patient, bilateral amygdala lesions caused by Urbach–Wiethe disease typically lead to severe impairments in emotion recognition, and some patients report a complete inability to experience fear and other emotions.[41]

The right-hemisphere model: evidence from behavioral studies in healthy individuals

Evidence for a major role of the right hemisphere in emotion processing in healthy individuals comes from research with the emotional dichotic listening task and with divided visual field paradigms with emotional faces. To test emotional lateralization in the auditory modality, Amiram Carmon and Israel Nachshon from the Hadassah University Hospital in Jerusalem developed a modified version of the dichotic listening task.[42] Instead of words or syllables as in the classic verbal dichotic listening task,[43] pairs of nonverbal human emotional sounds—crying, shrieking, and laughing of a child, woman, and man—were used as stimuli. Subjects had to react by pointing toward the picture showing the sound they heard best on a board of nine pictures showing a pictorial depiction of all nine possible sounds. In contrast to the strong right-ear advantage typically observed for verbal stimuli in the dichotic listening task, Carmon and Nachshon observed a left-ear advantage, indicating the dominance of the right hemisphere during emotional dichotic listening. This effect was later replicated by Bryden and MacRae for words spoken with different emotional intonations.[44] When subjects had to attend to the words, they showed a strong right-ear advantage. However, when they had to attend to emotional prosody, they showed a left-ear advantage, indicating right-hemispheric dominance.

To investigate emotional lateralization in the visual modality, most authors have used variants of the divided visual field technique[45] (see Box 3.3). To quickly recapitulate: In paradigms using this technique, the participant is asked to fixate on a fixation cross in the center of a screen. Then, a picture—for example, an emotional face or an emotional scene—is presented very briefly on either the left or the right side relative to fixation, and the participant has to respond to this stimulus. Since the nasal portions of the visual input are processed by the hemisphere contralateral to the stimulus and the distal portions of the visual input are processed by the hemisphere ipsilateral to the stimulus, everything on the left side of the screen is projected to the right hemisphere and vice versa (see Fig. 3.11). Thus, stimuli presented this way are at first only processed by one hemisphere, and by comparing performance for stimuli presented in the left and the right visual field, hemispheric dominance for the respective stimulus type can be determined.

In 1979, Robert G. Ley and M.P. Bryden from the University of Waterloo in Canada assessed whether healthy participants showed lateralization of facial emotion recognition.[46] To this end, they used a divided visual field paradigm with cartoon drawings of faces showing five different emotional expressions. Participants had to indicate whether the emotional expression in a laterally presented face matched that of a subsequently presented face in the center of the screen. Participants showed better performance for stimuli presented in the left visual field, leading Ley and Bryden to conclude that the right hemisphere is dominant for processing emotional stimuli. Similar results were found by Esther Strauss and Morris Moscovitch from the University of Toronto, Canada, who used photographs of faces as stimuli.[47] Participants were presented with pairs of faces in the left or right visual field. The two faces could either show the same person or two different persons. Moreover, each face could show one of three possible emotional expressions: Happy, sad, or surprised. Thus, four categories of face pairs were shown to the participants (see Fig. 9.8):

1. SFSE: Same face, same expression
2. DFSE: Different face, same expression
3. SFDE: Same face, different expression
4. DFDE: Different face, different expression

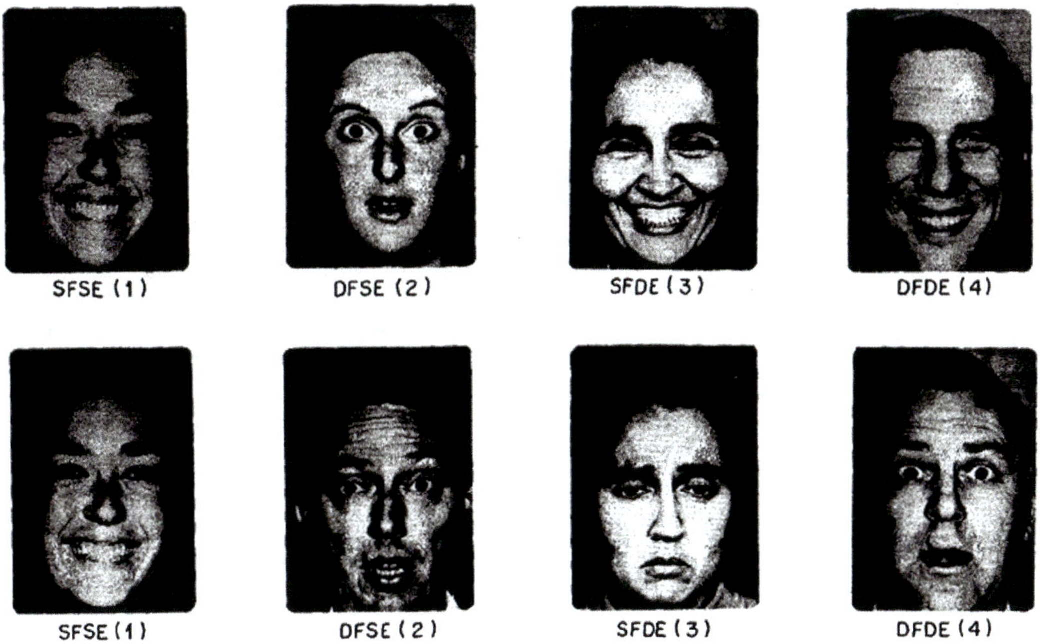

FIGURE 9.8 Examples of the stimuli used in the divided visual field study by Strauss and Moscovitch.[47] The four possible combinations of two faces are shown: SFSE: Same face, same expression; DFSE: Different face, same expression; SFDE: Same face, different expression; DFDE: Different face, different expression. *Reprinted with permission of the publisher from Strauss E, Moscovitch M. Perception of facial expressions. Brain Lang. 1981;13:308–332.*

There were two different versions of the task. In the first, participants had to indicate by pressing a button whether the emotional expression of the face had changed, irrespective of the identity of the face. In the second, they had to indicate whether the identity of the face had changed, irrespective of emotional expression. Results revealed that for both facial identity and emotion expression perception, participants responded faster when face pairs were presented in the left visual field. Therefore the right hemisphere seems to have an advantage in controlling visual analysis of emotional expressions.

Right hemisphere superiority for recognition of emotional faces was subsequently replicated by several authors.[30–32,48] Taken together, these data strongly argue for a major role of the right hemisphere in recognizing and producing emotions. However, several findings contradict the right-hemisphere model (see next section).

The valence model

Despite empirical support for the right-hemisphere model, there are some findings this model struggles to explain. First, modern neuroimaging studies have shown strong activations in the left hemisphere during emotional tasks. For example, Daan Baas from the University of Utrecht in the Netherlands and coworkers conducted a meta-analysis of 54 fMRI and PET studies that investigated amygdala activations during emotion processing.[49] They found that 80.39% of the studies included in their meta-analysis reported activation in the left amygdala, but only 55.56% reported activation in the right amygdala. This data strongly contradicts the idea that the right hemisphere alone processes emotion.

Second, it was reported as early as in the 1930s that the two hemispheres might be differentially involved in different types of emotions. For example, in his book *The Organism: A Holistic Approach to Biology Derived from Pathological Data in Man*, first published in 1939, German-American neurologist Kurt Goldstein (1878–1965) described that patients with left- and right-frontal lesions showed striking differences in their emotionality.[50] Goldstein found that what he called "catastrophic reaction"—for example, severe depression and agitation—was generally observed more often in aphasic patients with lesions in the left frontal cortex. In contrast, patients with right frontal damage more often show euphoric reactions—for example, intensely positive emotionality. Thus, lesions in the left and right frontal lobes differentially affected the patients' emotionality. These observations were also in line with the results of a Wada test experiment from the 1960s. In 1964, Terzian described asymmetries in emotional responses during Wada testing.[51] Terzian reported that in most right-handed patients, anesthesia of the left hemisphere led to catastrophic reactions similar to what had been described by Goldstein for aphasic patients with left-frontal lesions. Terzian described this as follows:

> *The patient, especially when spoken to, despairs and expresses a sense of guilt, of nothingness, of indignity, and worries about his own future or that of his relatives.* **From Terzian.**[51]

In contrast, anesthesia of the right hemisphere typically produced the opposite emotional reaction in Terzian's patients. Here, patients showed a euphoric reaction that in some patients bordered on maniacal behavior. Thus, both Goldstein's and Terzian's studies suggested that the left hemisphere might be dominant for positive emotions, while the right might be dominant for negative emotions.

However, it took until the 1980s for this idea to enter the scientific mainstream. At that time, Richard J. Davidson from the University of Wisconsin in Madison developed what is now commonly referred to as the valence model of emotional lateralization (see Fig. 9.4B). In contrast to the right-hemisphere model, the valence model assumes that only negative or withdrawal-related emotions are controlled by the right hemisphere, while positive or approach-related emotions are controlled by the left hemisphere. Specifically, the left prefrontal cortex is thought to mediate positive emotions, while the right prefrontal cortex is assumed to control negative affect (NA).[52–54]

Empirical evidence for the valence model was provided in 1981 with a divided visual field paradigm study by Patricia Reuter-Lorenz and Richard J. Davidson.[55] In this study, 28 young adults were shown pairs of facial stimuli, one in the right visual field and the other in the left. One of the stimuli showed a neutral facial expression, while the other one either showed a positively valenced facial expression or a negatively valenced facial expression. Subjects had to indicate as quickly as possible in which visual field the emotional face was shown. For happy faces, subjects were faster for right than for left visual field presentation, indicating dominance of the left hemisphere. The reversed pattern was observed for sad faces. Here, subjects were faster for left than for right visual field presentation, indicating right-hemispheric dominance. Reuter-Lorenz and Davidson concluded that the right-hemisphere model was wrong and that the two hemispheres differentially contributed to emotion processing, with the left hemisphere being dominant for positive emotions and the right hemisphere being dominant for negative emotions. Results similar to those reported by Reuter-Lorenz and Davidson were also found in divided visual field studies with emotional words.[56,57] Interestingly, a differential valence-based asymmetry in processing emotional faces has been observed under natural, free-viewing conditions. Jansari et al. presented their subjects with pairs of faces, among which one had a neutral expression and the other showed a very faint emotional expression.[58] Results revealed a clear valence-based dissociation. Subjects were significantly better when distinguishing a face with a positive emotional expression from a neutral face when it was presented on the right side of the neutral face. In contrast, they were significantly better at distinguishing a face with a negative emotional expression from a neutral face when presented on the left side of the neutral face. Thus right-hemispheric specialization for processing negative emotions and left-hemispheric specialization for processing positive emotions can be explained in this paradigm.

The major body of empirical evidence supporting the valence hypothesis comes from studies with electroencephalography (EEG). As explained in Box 5.2, the raw EEG signal (see Fig. 5.11) is composed of signals from millions of neurons firing at different frequencies. By using Fourier transformation, different frequency bands can be identified within the raw EEG signal that are related to sleeping/waking cycles and cognitive activity. Typically, the following frequency bands can be obtained from the raw EEG signal:

1. Delta-band (0.1–4 Hz): This frequency band is dominant in deep sleep.
2. Theta-band (4–8 Hz): This frequency band is commonly observed during the change from sleeping to waking.
3. Alpha-band (8–12 Hz): This frequency band is commonly observed in a relaxed state and when eyes are closed.
4. Beta-band (14–25 Hz): This frequency band is dominant in a state of mental activity.
5. Gamma-band (>30 Hz): This frequency band is observed during focused concentration on a task.

For research on emotional lateralization, in particular, alpha-band EEG activity has played a vital role.[59–61] Alpha-band activity is typically observed when subjects are relaxed and do not engage in any demanding cognitive tasks. Once they are actively engaging in a task, alpha-band activity tends to decrease. Due to this inverse relation of alpha-band activity and cognitive activity, the power of the alpha-band (i.e., how much electrical energy is produced during a given time period in this frequency band) is thought to reflect the absence of cognitive activity. By comparing alpha-power between electrode sites on the left and right side of the head, hemispheric asymmetries in cognitive activity can be assessed. For example, if alpha-power in the right hemisphere is higher than in the left during a task, it is assumed that the participant shows greater activation of the left hemisphere.

EEG alpha-band asymmetries, particularly in the frontal lobes, have repeatedly been linked to emotional processing. For example, Geoffrey L. Ahern and Gary L. Schwartz from Yale University recorded EEG data from 33 female Yale undergraduates while they answered 60 different questions that were designed to evoke emotions.[62] The questions were either neutral ("What is the primary distinction between logical and rational?"), positive ("Picture and describe the last situation in which you laughed."), or negative ("Picture and describe a person who scares you."). In accordance with the valence hypothesis, Ahern and Schwartz found that questions aimed at inducing happiness led to more alpha-band activity in the right frontal lobe, implying greater cognitive activity in the left hemisphere. In contrast, questions aimed at inducing fear led to less alpha-band activity in the right frontal lobe, implying greater cognitive activity in the right hemisphere. Similar results were also obtained in a study by Davidson et al., in which EEG activity to emotionally arousing movies was recorded.[63] After a neutral film clip, participants watched two positive film clips (a puppy playing with flowers and monkeys playing in the zoo) and two negative film clips (a leg amputation and a third-degree burn victim). EEG data revealed that participants showed greater cortical activation—as indexed by less alpha-band power—in the frontal and anterior temporal regions of the right hemisphere when watching negative movie clips. In contrast, positive movie clips elicited greater cortical activity in anterior temporal regions of the left hemisphere.

Comparable results have also been reported by Nancy Aaron Jones and Nathan A. Fox from the University of Maryland in a 1991 EEG study.[64] These authors asked 308 undergraduates to fill out a personality questionnaire. From this sample, 12 subjects who rated themselves high in positive affectivity and 11 subjects who rated themselves high in negative affectivity were selected. Subjects watched nine different video clips—one neutral video clip and two video clips each that were aimed at inducing happiness, sadness, disgust, and anger—while EEG was recorded. In accordance with the valence model, results showed right-hemispheric specialization—decreased alpha power—for processing negative movie clips and left-hemispheric specialization for processing positive movie clips.

Interestingly, a link between frontal EEG alpha-band asymmetries and emotional processing has also been found when individuals with left- or right-hemispheric asymmetries in resting state EEG were tested regarding their emotionality. Andrew J. Tomarken from Vanderbilt University and coworkers recorded EEG in 90 right-handed women without giving them any task at all.[65] Afterward, participants filled out the Positive and Negative Affect Schedule (PANAS),[66] a widely used mood questionnaire that consists of 10 positive words—for example, proud or excited—and 10 negative words—for example, ashamed or afraid. Subjects were asked to rate how much these words accurately described the way they felt at the moment of testing on five-point scales ranging from "not at all" to "extremely." From these answers, two values can be calculated, one that measures positive affect (PA) and another that measures NA. Out of the 90 initially tested subjects, Tomarken et al. selected two extreme groups: One with relatively greater rightward EEG alpha-band asymmetry (left hemisphere activation) and one with relatively leftward EEG alpha-band asymmetry (right hemisphere activation). When comparing the PANAS results of these two groups, Tomarken et al. found an interaction that was in line with the valence hypothesis. The group that showed greater activation of the left hemisphere experienced increased PA and decreased NA compared to the group that showed greater activation of the right hemisphere. Thus frontal alpha-band asymmetry is related to the general affective style. Similar results were also reported by Wheeler et al., who found that subjects with leftward activation asymmetry showed more PA in responses to positive movie clips, while subjects with rightward activation asymmetries showed more NA to negative movie clips.[67]

In addition to EEG studies, support for the valence hypothesis has been provided by studies using other neuroscientific techniques. For example, in an early fMRI study from 1998, Canli et al. showed positively and negatively valenced pictures (e.g., happy couples or puppies vs. angry people or spiders) to 14 female participants. Pictures were matched for the level of emotional arousal they induced in participants.[68] Overall brain activity to positive pictures was lateralized to the left, while overall brain activity in response to negative pictures was lateralized to the right, directly supporting the valence hypothesis.

More support for the valence model came from a study applying electrical stimulation to the amygdala. Laura Lanteaume from the CNRS at the University of Provence and coworkers tested the effects of electrical stimulation in the left and right amygdala in eight epileptic patients who had been implanted with intracerebral electrodes for presurgical evaluation.[69] When asked to report which emotions they felt during amygdala stimulation, patients reported an increase of negative emotions, in particular fear and sadness, after stimulation of the right amygdala. In contrast, stimulation of the left amygdala induced happiness, a positive emotional state. Lanteaume et al. concluded that their findings strongly supported the notion of functional lateralization in the amygdala in line with the valence hypothesis.

Interestingly, evidence for the valence model also comes from research on emotional lateralization in nonhuman animals. Lisette M.C. Leliveld from the Leibniz Institute for Farm Animal Biology in Dummerstorf, Germany, systematically analyzed published studies on emotional lateralization in a diverse range of vertebrate species.[70] She and her coworkers found that except for fish, all vertebrate classes showed a similar valence-dependent pattern of lateralization. While the left hemisphere was dominant for processing positive emotions (such as those elicited by a food reward), the right hemisphere was dominant for negative emotions (such as fear and aggression).

Taken together, there is strong empirical evidence that both hemispheres play a role in emotion processing and that the left hemisphere might be specialized for positive emotions, while the right hemisphere might be specialized for negative emotions. Along these lines, the valence model might provide a better explanation for emotional lateralization than the right-hemisphere model. However, another theoretical account has been proposed more recently that might also explain data patterns without including emotional valence.

The frontal cortical asymmetry model of motivational direction

While research from the 1980s highlights the importance of emotional valence for hemispheric asymmetries, more recently it has been suggested that motivational rather than emotional valence might better explain the observed data patterns. Jeffrey Allan Grey (1934–2004), a British psychologist, suggested that two general motivational systems might underlie behavior in humans and other organisms:[71,72]

1. The behavioral approach system (BAS) is believed to regulate appetitive motives in which the goal is to move toward something desired.

2. The behavioral inhibition system (BIS) is thought to regulate aversive motives in which the goal is to move away from something unpleasant.

Individual differences in the sensitivity of these systems can be assessed using the BIS/BAS scales developed by Carver and White.[73] In this 24-item questionnaire, subjects read different statements regarding their appetitive motives (e.g., being excited to win a contest) and their aversion to negative experiences (e.g., feeling worried about having performed poorly at an important event). Subjects then rate each of these statements on four-point scales ranging from "very true for me" to "very false for me." Based on their answers, BIS and BAS scores are calculated for individual subjects. Sutton and Davidson asked 46 undergraduates to fill out the BIS/BAS scales and recorded resting state EEG.[54] They found that subjects with greater left prefrontal activation—less right-hemispheric EEG alpha-band activity—reported higher levels of behavioral approach tendencies. In contrast, subjects with greater right prefrontal activation—less right-hemispheric EEG alpha-band activity—reported higher levels of behavioral avoidance tendencies.

Along these lines, it appears plausible that early research on the valence model has confounded an emotional state—positive versus negative—with a motivational direction—approach versus avoidance. Indeed, this is what Eddie Harmon-Jones from the University of Wisconsin in Madison suggested in his frontal cortical asymmetry model of motivational direction[74–77] (see Fig. 9.4C). He argued that in previous studies, motivational direction and emotional valence were often confounded, given that only positive emotions were associated with approach motivation, while only negative emotions were associated with avoidance motivation. Thus, it was impossible to judge from these findings whether hemispheric asymmetries were determined by emotional valence, motivational direction, or a combination of both factors.[76]

The critical test to distinguish between the valence model and the frontal cortical asymmetry model of motivational direction is to test the lateralization of anger since anger is a negatively valenced emotion that typically evokes behavioral tendencies of approach.[78,79] While the valence model would suggest right-hemispheric dominance for anger, the frontal cortical asymmetry model of motivational direction would suggest left-hemispheric dominance. Data collected by Harmon-Jones and Sigelman suggested that anger is predominantly associated with the left hemisphere.[74] The authors investigated the relationship between state-induced anger and prefrontal asymmetries. In their study, 42 undergraduates were told that they were participating in a study on personality, psychophysiology, and perception. Moreover, they were told that in a second room, another participant was tested by another experimenter and that the two test sessions would be linked. Participants then filled out a baseline affect scale and baseline EEG was recorded. Subsequently, they were asked to write a one-page essay on a topic that was of personal importance to them. After that, the essay was collected by the experimenter, and the participant was told that the other participant would now rate the essay. However, in reality, there was no other participant, and participants received evaluation letters that had been prepared by the experimenters to be either neutral or so negative that they were insulting—for example, "I can't believe an educated person would think like this" or similar remarks. Then, EEG was recorded again, and the participants' anger was assessed. As intended, anger was induced in the group of participants who had received the insulting evaluation. Moreover, this group showed reduced left-hemispheric alpha oscillations in the EEG data, indicating increased cognitive activity in the left frontal cortex. Thus, anger seems to be associated with relative left prefrontal activity, demonstrating that asymmetrical prefrontal activity might reflect motivational direction rather than emotional valence.

This idea was corroborated in a slow repetitive transcranial magnetic stimulation (rTMS) study by d'Alfonso.[80] rTMS is a technique in which a coil (see Fig. 9.9) that generates a magnetic field is placed close to the head of the participant. The magnetic field induces small electric currents in the brain areas underneath the coil which can induce or suppress neuronal activity.

rTMS can be used to selectively inhibit the left or right prefrontal cortex:

1. rTMS applied to the right prefrontal cortex decreases its activation and causes the left prefrontal cortex to become more active.
2. rTMS applied to the left prefrontal cortex causes activation of the right prefrontal cortex.

Using an emotional Stroop task, d'Alfonso et al. [80] found that rTMS applied to the right prefrontal cortex caused participants to selectively attend to angry faces. In contrast, rTMS applied to the left prefrontal cortex caused participants to attend less to angry faces. Thus, an increase in left prefrontal activity led participants to attentionally approach angry faces just like in an aggressive confrontation. In contrast, an increase in right prefrontal activity led participants to attentionally avoid angry faces just like in fear-based avoidance.

Further evidence supporting the frontal cortical asymmetry model of motivational direction was provided in a 2012 study by Hortensius et al.[81] These authors predicted that anger associated with greater relative left frontal cortical activation would be more likely to result in aggression. To test this hypothesis, they combined transcranial direct current stimulation (tDCS) over the frontal cortex with interpersonal provocation. tDCS, like rTMS, is a neurostimulation technique.[82–85] In contrast to rTMS, no magnetic field is induced. Instead, low currents are directly applied to brain regions by an electrode placed on the head. A typical tDCS setup (see Figs. 9.10 and 9.11) consists of a device that delivers a current and two electrodes, one cathode—a negatively charged electrode—and one anode—a positively charged electrode.

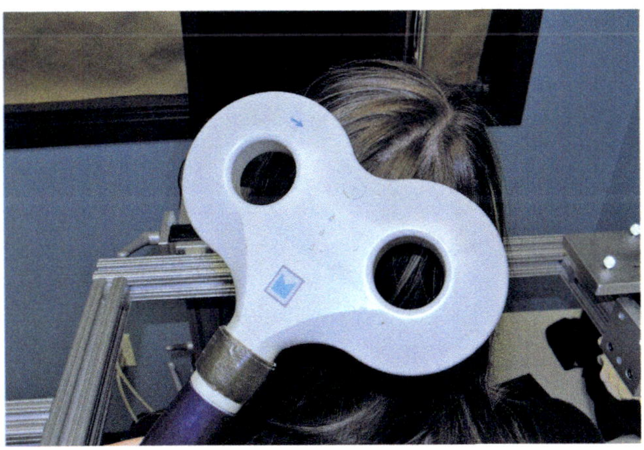

FIGURE 9.9 A typical figure-of-eight coil used in rTMS research.

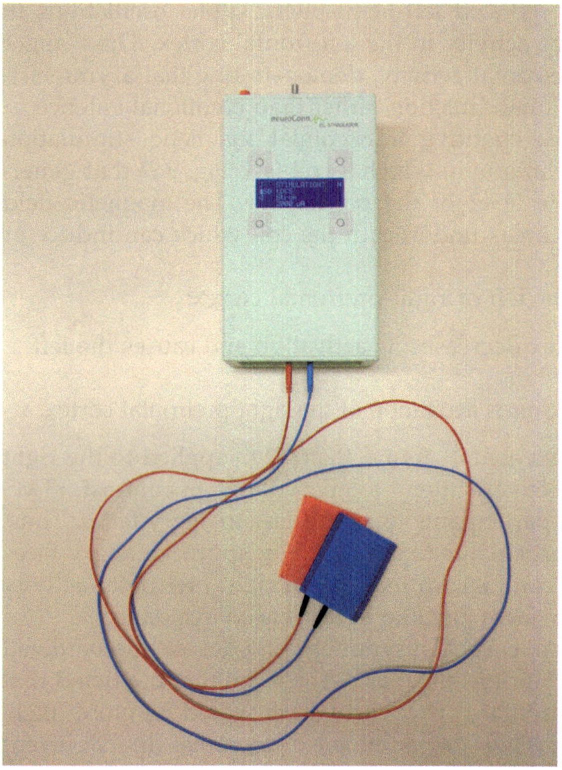

FIGURE 9.10 A typical tDCS setup consisting of a device that delivers the current and two electrodes, the cathode (negatively charged electrode) and the anode (positively charged electrode). *Picture courtesy Christian Bellebaum and Laura Bechtold.*

FIGURE 9.11 A participant wearing a typical tDCS setup during an experiment. *Picture courtesy Christian Bellebaum and Laura Bechtold.*

When the tDCS device is switched on, electric current flows from the positive anode through the brain tissue between the two electrodes to the negative cathode. When the anode is placed over the brain area of interest, tDCS stimulation increases the cortical excitability of that brain area. In contrast, if the cathode is placed over the brain area of interest, the cortical excitability of that brain area is reduced.

In the study by Hortensius et al., three groups of participants were first asked to write an essay on a controversial issue just like in the study by d'Alfonso.[80] Then they were stimulated with tDCS for 15 min:

1. Anodal to left and cathodal to right frontal cortex: This stimulation increased left frontal activity.
2. Anodal to right and cathodal to left frontal cortex: This stimulation increased right frontal activity.
3. Sham: In this condition, participants were only stimulated for 30 s, and then the tDCS unit was switched off. Since stimulation was so short, no effect on the brain should have occurred.

After tDCS stimulation ended, participants received insulting feedback on their previously written essays by the alleged other participant to induce anger. Subsequently, an aggression paradigm[86] was conducted with the participants. They were told that they had to play a competitive reaction time game against the other participant who had given them insulting feedback on their essays. They had to press the left or right shift key on a keyboard as fast as possible whenever a green plus sign was shown on the left or right side of a screen. They were told that whenever they had won a trial, they were able to give the opponent a loud noise blast, but when they had lost a trial, they would receive a noise blast themselves. However, the outcome of the game was not linked to actual performance, and each participant won and lost on the same number of trials. On won trials, the participants could determine the length of the noise blast which served as a measure of their aggression toward the alleged opponent. For each participant, an aggression score was calculated. Individuals who had received tDCS to increase relative left frontal cortical activity behaved more aggressively if the insulting feedback made them angry (see Fig. 9.12).

No relation between anger and aggression was observed in the group in which relative right frontal cortical activity was increased or in the group exposed to the sham condition. These results concur with the frontal cortical asymmetry model of motivational direction, as left frontal activity is associated with anger-induced aggression.

Support for the notion of left-hemispheric control of approach motivation and right-hemispheric control of avoidance motivation has been found not only in humans but also in nonhuman animal species. Animals seem to show two complementary side biases: When avoiding predators, they flee to the left side, implying greater right-hemispheric control of escape behavior. In contrast, when approaching prey, animals tend to have a right-sided bias, implying greater left-hemispheric control of hunting behavior. Right-hemispheric control of escape behavior has, for example, been reported in several fishes such as the Australian lungfish (*Neoceratodus forsteri*)[87] or the goldbelly topminnow (*Girardinus falcatus*),[88] different frog and toad species,[89–91] chicken (*Gallus gallus domesticus*),[92] and the stripe-faced dunnart (*Sminthopsis macroura*).[93] Left-hemispheric control in attacking prey or

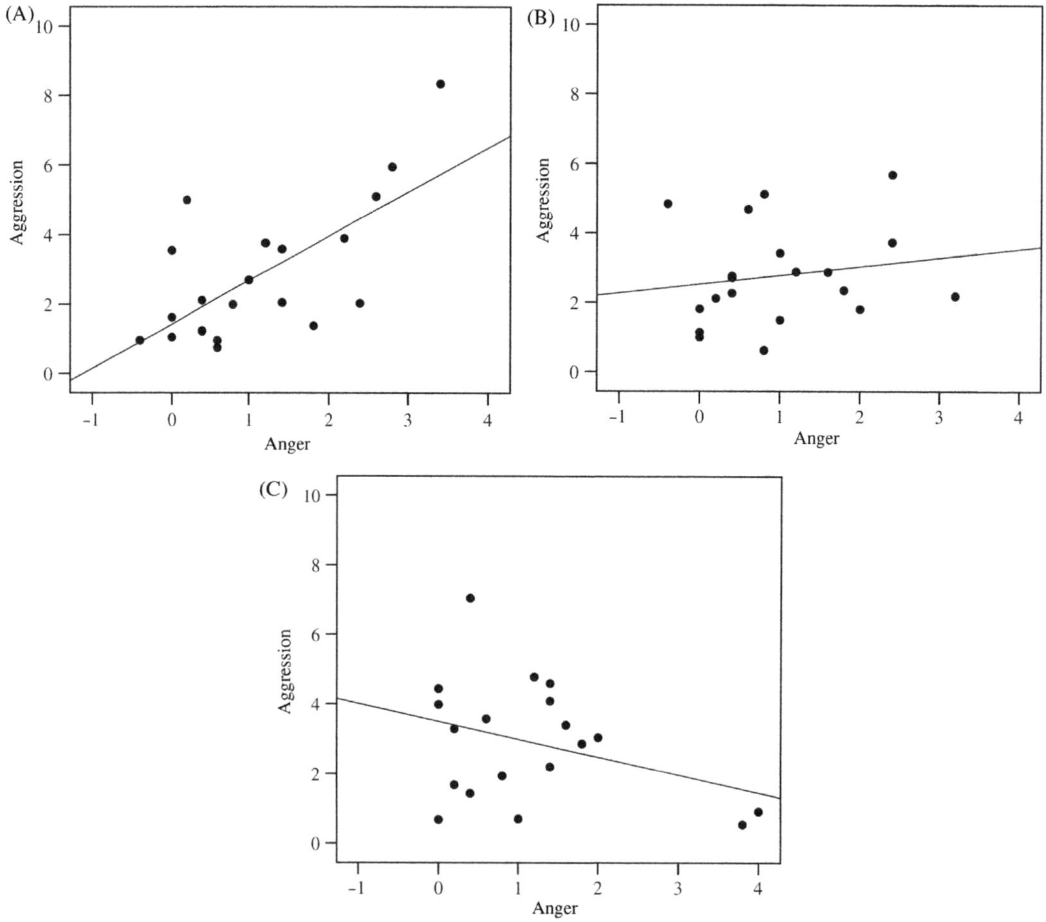

FIGURE 9.12 Results of the study by Hortensius et al.[81] (A) Participants who had received tDCS to increase cortical activity in the left prefrontal cortex showed a significant influence of anger on aggression, with more anger leading to more aggression. (B) Participants who had received tDCS to increase cortical activity in the right prefrontal cortex did not show a significant influence of anger on aggression. (C) Participants who had received sham tDCS also did not show a significant influence of anger on aggression. *Reprinted with permission from the publisher from Hortensius R, Schutter DJ, Harmon-Jones E. When anger leads to aggression: induction of relative left frontal cortical activity with transcranial direct current stimulation increases the anger-aggression relationship. Soc Cogn Affect Neurosci. 2012;7:342–347.*

in feeding behavior has also been found in several species—for example, in the European common cuttlefish (*Sepia officinalis*),[94] the Australian lungfish (*Neoceratodus forsteri*),[87] the common wall lizard (*Podarcis muralis*),[95] the ornate dragon lizard (*Ctenophorus ornatus*),[96] several species of toads,[97] and chickens (*Gallus gallus domesticus*).[92] These comparative findings provide a strong argument that an ancient asymmetry of approach and withdrawal might also underlie human emotional lateralization.

Integrating the models

Taken together, the empirical evidence outlined in the last sections indicates that emotions are lateralized. Theoretical positions on emotional lateralization have slowly moved from an overall right hemisphere approach to a left-right difference for PA and NA and later on to left hemisphere dominance for approach and right hemisphere superiority for avoidance. So, which of the three models for emotional lateralization is the best? While this might sound like a worthwhile question, recent research efforts have moved from pitting these hypotheses against each other toward a more integrative approach. Since all of them are supported by at least some empirical evidence, scientists are focusing more on identifying specific situations or brain networks for which right-hemispheric emotional processing or valence/approach-avoidance emotional processing is relevant. For example, in 2007, William D.S. Killgore and Deborah A. Yurgelun-Todd from Harvard Medical School published an article in the scientific journal *Social Cognitive and Affective Neuroscience*, asking the intriguing question: "The right-hemisphere and valence hypotheses: could they both be right (and sometimes left)?" [98] In their fMRI study, 12 right-handed female participants performed two facial affect perception tasks in the MRI scanner, one in which chimeric happy faces (e.g., left half happy and right half neutral) or chimeric sad faces (e.g., left half sad and right half neutral) were presented. In accordance with the right-hemisphere model, Killgore, and Yurgelun-Todd found an activation of the posterior right hemisphere during nonconscious perception of emotional faces that was independent of stimulus valence. However, they also reported evidence for valence-specific activations in anterior brain regions that were consistent with the valence model. Based on these findings, Killgore and Yurgelun-Todd proposed that the different models of emotional lateralization do not really oppose each other but that instead two interrelated but differentially lateralized brain networks underlie emotional face processing. On the one hand, there seems to be a cortical network in the posterior right hemisphere that is specialized for general emotion perception independent of specific valence. On the other hand, there also seems to be a cortical network in the posterior left hemisphere that is linked to valence-specific activation of the orbitofrontal cortex and ventral striatum. Thus the different theories on emotional lateralization seem to reflect different aspects of the complex neuronal network involved in emotion processing.

In line with the ideas put forward by Killgore and Yurgelun-Todd, other researchers have suggested that the right-hemisphere model and the valence and approach/avoidance model might reflect different aspects of emotional processing. While the right-hemisphere model seems to apply mainly to emotion perception and reflects activation of the right posterior cortex, the valence model seems to reflect the emotional experience and has been linked to the asymmetries in the dorsolateral prefrontal cortex (dlPFC).[98–102]

Moreover, all three theories have been criticized for viewing emotion processing as too much of an isolated process by not taking into account cognition–emotion interactions—for example, how prefrontal executive functions modulate emotion processing.[103] One important factor in this regard seems to be whether the specific task used to test emotional lateralization is mainly bottom-up driven—that is, relying on sensory-driven processes without involvement of higher cognitive processes—or involves a top-down processing component—that is, prefrontally mediated higher cognitive processes such as decision-making or development of expectancies. For example, it has been suggested that valence-specific processing occurs primarily when tasks involve expectancies, a typical top-down component that engages frontal processing networks.[99,100] Further elaborating on the relation of emotional lateralization and frontal top-down processing, Gina M. Grimshaw from Victoria University of Wellington in New Zealand and David Carmel from the University of Edinburgh in the United Kingdom developed the so-called asymmetric inhibition model of hemispheric differences in emotional processing.[103] This model highlights the role of inhibitory executive control processes for hemispheric asymmetries in emotion processing. It predicts that valence-based effects are most likely to be observed when tasks involve the use of prefrontal top-down control mechanisms, as the dlPFC is thought to be a key structure for emotion–cognition interaction that behaves in a valence-specific way. The dlPFC is located in the middle frontal gyrus and is one of the key areas for executive functions such as planning, inhibition, and cognitive control.[104–108] Additionally, several authors could show that it also plays a key role in emotion regulation.[109–112] Based on these findings, Grimshaw and Carmel suggested that emotional lateralization in the prefrontal cortex may not reflect asymmetries in emotional experience but in the top-down control of emotional information. While the right dlPFC is dominant for inhibition or control of positive emotion, the left dlPFC is specialized for inhibition or control of negative information.

Emotional lateralization in the age of systems neuroscience: moving toward models incorporating multiple distinct networks

Both the valence model and the right-hemisphere model were developed before neuroimaging techniques to monitor human brain functions in-vivo such as fMRI became widely available to researchers. Large-scale neuroimaging meta-analyses of studies on emotion processing generally indicate that none of the two models can perfectly explain the observed brain activation patterns.[113] Brain activation is highly dependent not only on the specific emotion, but also on sensory modalities (emotions elicited by watching emotional faces, hearing emotional voices, or hearing emotional music) and the empirical paradigms employed in emotion studies.[114–116]

These and other findings have recently been reviewed by Nicola Palomero-Gallagher and Katrin Amunts from the Research Centre Jülich in Germany.[2] In their review article, they highlight that emotions are not processed by a single neural network, but by multiple neural networks (e.g., the amygdala-centered network linking sensory experience to emotion and the hippocampus-centered network linking emotion to memory and cognition) that coordinate themselves with each other and may show different levels of lateralization. Moreover, these networks show differential activation during diverse aspects of emotion processing such as generation, perception, and regulation of positive and negative emotions. Based on these findings, Palomero-Gallagher and Amunts concluded that in order for emotion laterality research to advance, the field needs to move away from static models that assume overall hemispheric specialization for emotion processing (e.g., the left hemisphere is always dominant for processing of all positive emotions in all situations).

Instead, more complex dynamic models are needed, which include differential lateralization patterns for multiple interrelated networks.[2]

A similar reasoning can also be found in a recent in-depth analysis of hemispheric asymmetries in the amygdala, one of the core structures in emotion processing, especially for fear.[113] While the amygdala shows a clear rightward volume asymmetry in structure[117] (see Chapter 4), functional asymmetries in amygdala activation do not follow a clear left-right dominance across all empirical paradigms. Fig. 9.13 shows an integration of the results of different neuroimaging studies and neuroimaging meta-analyses on hemispheric asymmetries in amygdala activation.[113]

In general, three major factors affect the direction of amygdala activation asymmetries in a neuroimaging task:[113]

1. Temporal dynamics: In experimental tasks with sustained (conscious) processing of emotional stimuli, there is a leftward amygdala activation asymmetry, while very short or unconscious processing of emotional stimuli leads to a rightward amygdala activation asymmetry.[118]
2. Emotional valence: Animal research on fear processing and pain research suggests a strong rightward functional asymmetry of the amygdala for processing negative emotions. Lesion studies in humans similarly suggest that there is a dominance of the

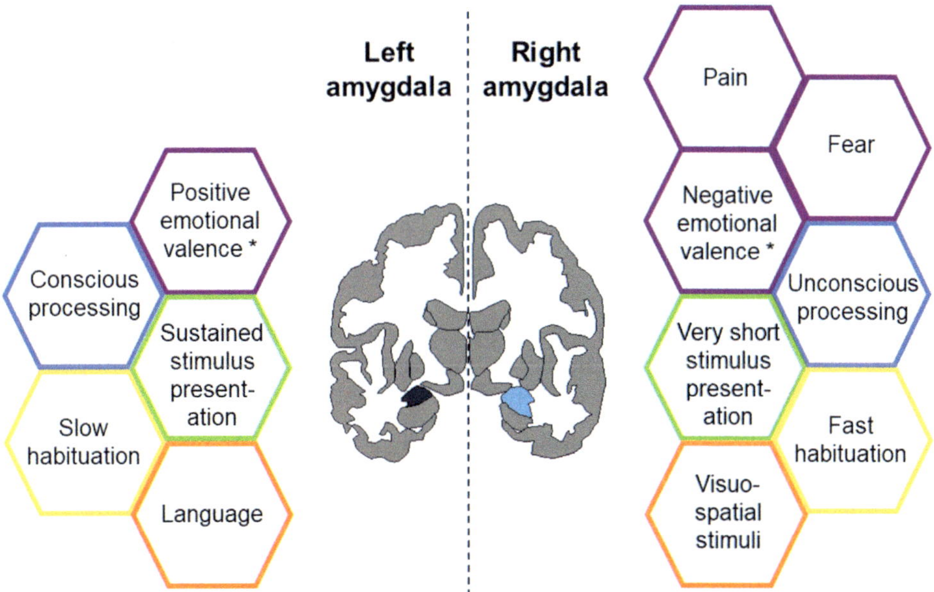

FIGURE 9.13 Factors that affect functional hemispheric asymmetries in amygdala activation.[113] Colors code functions that are associated with each other (e.g., in orange: verbal stimuli lead to left amygdala activation and visuospatial stimuli to right amygdala activation). *: empirical evidence in animal lesion studies points in this direction, but there are also studies showing contradicting results in human participants. *Reprinted with permission of the authors and the publisher from Ocklenburg S, Peterburs J, Mundorf A. Hemispheric asymmetries in the amygdala: A comparative primer. Prog Neurobiol. 2022;214:102283.*

right amygdala for processing negative emotions and of the left amygdala for positive emotions. However, neuroimaging meta-analyses in human participants do not show this effect,[49] which may be due to moderating effects of the other two factors and needs further investigation. Another possible explanation could be that human volunteers know that they are in an experiment and no harm will happen to them, so their fear levels are low. In contrast, animals in fear studies experience real harm (e.g., from electroshocks) and are not aware that they are in an experiment. Therefore their fear level is much higher, which may also be a relevant factor for rightward amygdala activation. This would also be the line with the findings that in human participants, an immersive fear paradigm in virtual reality elicited stronger right-hemispheric asymmetries in electrophysiological activity than looking at fear-inducing pictures on a computer monitor.[119]

3. Perceptual properties: Functional hemispheric asymmetries for the stimuli used in an experiment are crucial for amygdala asymmetries (e.g., dominance of the left amygdala for processing linguistic stimuli and a dominance of the right amygdala for processing spatial stimuli).

This so-called "TEP" (temporal dynamics, emotional valence, perceptual properties) model assumes that amygdala asymmetries are determined by the interactions of the neural networks relevant to a specific task. For example, if participants have to react to negative words that are shown for several seconds, a rightward asymmetry for negative stimuli, a leftward asymmetry for processing words, and a leftward asymmetry for processing stimuli with sustained presentation would interact. Thus overall, a weak leftward asymmetry of amygdala activation would be observed, even though the stimulus material is negative, and both the right hemisphere theory and the valence theory would predict dominance of the right amygdala.

The lateralized social brain

A field of laterality research that is strongly connected to emotional lateralization, but not completely overlapping, is hemispheric asymmetry in social-interactive situations.[120–125] Different combinations of social interaction and emotional valence can occur, for example:

1. Social situation with a positive emotional valence: Two individuals who just got married kiss each other happily.
2. Social situation with a neutral emotional valence: Two individuals who barely know each other greet each other in the elevator at work.
3. Social situation with a negative emotional valence: Two soccer fans that support rival teams insult each other.
4. Nonsocial situation with a positive emotional valence: Somebody is sitting alone in front of their computer and feels a sense of accomplishment because they just handed in their thesis.
5. Nonsocial situation with a neutral emotional valence: Somebody is alone in their room and reads the instruction manual for their new vacuum cleaner.

6. Nonsocial situation with a negative emotional valence: Somebody is crying alone in their room because they feel lonely.

Thus, many but not all social situations contain an emotional element, and many but not all situations in which people experience emotion contain a social element. Therefore it makes sense to distinguish between laterality research on emotion processing and social laterality, even though they frequently overlap. This idea is also supported by a recent large-scale ALE meta-analysis [126] of studies using different social tasks that assessed four social constructs within NIMH's Research Domain Criteria (RDoC) framework. These included:

1. Affiliation and attachment
2. Social communication
3. Perception and understanding of self
4. Perception and understanding of others

Neuroimaging meta-analysis of 864 fMRI contrasts revealed convergence in several brain areas that have also been linked to emotion processing such as the amygdala, the prefrontal cortex, the cingulate cortex, the insula, and the precuneus. Moreover, there were effects for the fusiform gyrus, which is linked to face processing, the temporoparietal junction, a brain area linked to attention and social cognition,[127] and the thalamus. This suggests that the brain networks underlying social and emotional processing overlap but are not identical.

In addition, social laterality research in general focuses on the interaction between two individuals, while emotion research is often more focused on one individual and not a dyad. Finally, there is a line of strong comparative research on social laterality in species for which the assessment of emotional experiences is empirically challenging since it is difficult to assess the feeling component of emotions, such as social and nonsocial bees.[128–130]

In a widely cited review article from 2009, Alfredo Brancucci from the University of Chieti in Italy and his coworkers gave an overview of visual, auditory, and chemical asymmetries in the social brain.[120] These included:

1. Auditory perception of voices: While there is a clear leftward asymmetry for processing the linguistic content of speech, there is a rightward asymmetry for processing socially relevant nonlinguistic features of listening to voices such as pitch timbre.
2. Visual perception of faces and gazes: There is a rightward asymmetry for the processing of both faces and gazes, which are highly relevant in social situations.
3. Visual perception of biological motion: In general, socially relevant biological motion is primarily processed by right-hemispheric networks, but processing of biological motion that signifies linguistic meaning (such as sign language gestures) also depends on left-hemispheric language networks (see Chapter 5).
4. Olfactory perception and processing of pheromones: Independent of the side of nostril stimulation, the right hemisphere is dominant for the recognition and evaluation of olfactory stimuli. For the processing of human pheromones, there is a strong rightward dominance in male individuals, which is less pronounced in female individuals.

Based on these findings, Brancucci and coworkers concluded that for all three sensory modalities discussed in the review article, the right hemisphere is dominant for social perception.[120] They further concluded that social laterality may be the complex net result of other forms of laterality (e.g., for face perception) that are largely processed by right-hemispheric networks. Whether or not there is truly an independent lateralization for social stimuli or whether the findings described in the review article and other studies completely rely on other forms of laterality relevant to social contexts is an interesting question that is still not fully answered.

Asymmetries in social touch

One very active field of research in the context of social laterality is hemispheric asymmetries in social touch.[131] Social touch, in general, refers to touch used as a form of nonverbal communication between two or more individuals, that is typically used to communicate positive emotions like love or gratitude without words, but can also be used to communicate negative emotions such as envy or anger.[132] In laterality research, three forms of social touch have been investigated in particular (see Fig. 9.14):

1. Cradling in parent–child interaction
2. Romantic kissing between two adults
3. Hugging between two adults

There are also many other forms of social touch such as patting someone on the shoulder to cheer them up, giving someone a handshake, and giving someone a high-five that could be investigated in terms of laterality. However, these forms of social touch have not been within the focus of laterality research yet.

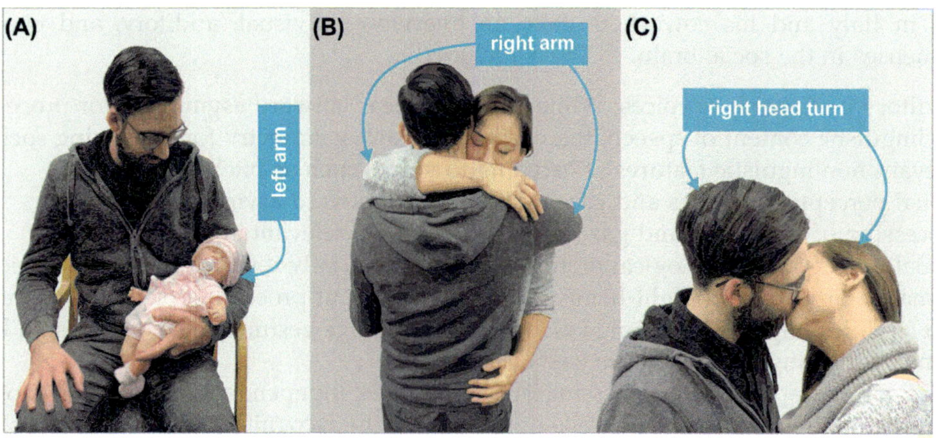

FIGURE 9.14 Typical behavioral asymmetries in human social touch.[131] (A) An adult cradling an infant with his left arm. (B) Two adults hugging with their right arms leading the hug. (C) Two adults kissing with their heads turned to the right side. *Reprinted with permission of the authors and the publisher from Ocklenburg S, Packheiser J, Schmitz J, Rook N, Güntürkün O, Peterburs J, Grimshaw GM. Hugs and kisses - The role of motor preferences and emotional lateralization for hemispheric asymmetries in human social touch. Neurosci Biobehav Rev. 2018;95:353–360.*

On the empirical level, cradling is probably the most widely investigated form of social touch in parental relationships.[123–125,131] There is a long research tradition in investigating cradling laterality, with the earliest widely acclaimed publication being a 1960 paper by Lee Salk from the City Hospital at Elmhurst, New York City.[133] Salk found that both left-handed and right-handed mothers had a strong bias to cradle their babies with their left hand (left-handed mothers: 78.1% leftward cradling, 21.9% rightward cradling; right-handed mothers: 83.1% leftward cradling, and 16.9% rightward cradling). This cradling bias was investigated by many further studies, and in 2019, laterality researcher Julian Packheiser from Ruhr University Bochum in Germany and coworkers published a set of three meta-analyses based on data from 40 different cradling studies to investigate the effects of handedness and sex on lateral preferences during cradling.[134] Overall, 9790 cases of leftward cradling (65.66%) and 5119 cases of rightward cradling (34.34%) were reported in the 40 studies included in the meta-analyses. The meta-analysis showed a significant effect with a leftward cradling preference having a prevalence between 66% and 72%. Therefore, the leftward cradling bias first reported by Salk is a robust and replicable phenomenon across different studies.

Packheiser and coworkers also investigated whether handedness and sex affected the leftward cradling bias. They found that female individuals had a significantly stronger leftward cradling bias compared to male individuals. Moreover, left-handers showed a less pronounced leftward cradling bias (61%) compared to right-handers (74%). This indicates that handedness may contribute to some extent to the leftward cradling bias. For example, it may be beneficial for right-handers to use their nondominant hand for cradling, as it may leave the more skilled dominant hand able to perform other tasks.[135] However, this theory cannot explain why most left-handers use their dominant left hand and not their nondominant right hand. This suggests that cradling bias is not fully explained by motor asymmetries, but that there is a clear right-hemispheric dominance for cradling that is largely unaffected by handedness. This may reflect the right-hemispheric dominance for social processes described by Brancucci.[120] Interestingly, there is evidence that humans are not the only species that show a leftward cradling bias. A comparative study found that both chimpanzees (74% leftward) and gorillas (75% leftward) showed a significant leftward cradling bias, while Orang-Utans and gibbons did not show a significant side preference in cradling.[136] Furthermore, a comprehensive review focusing on lateralized social behavior between mothers and offspring across different mammalian species reported a robust left-sided preference for the mother-offspring position, even in species without arms.[122] The authors performed a random-effects meta-analysis of the data from different mammalian species (e.g., walrus, horse, reindeer, muskox, beluga whale, and orca) on the active choice of the spatial position of an infant in relation to its mother. The overall effect showed a significant leftward bias, with 530 out of 835 choices across species being leftward (63.47%). This is particularly interesting as a leftward bias was even observed in species without hands (e.g., beluga whale), which suggests an ancient evolutionary base for the human leftward cradling bias and further supports the idea that it is not driven by handedness alone. The authors of the comparative study argued that their findings support the idea that the origin of the leftward behavioral asymmetry reflects a right-hemispheric dominance for social perceptual processing.

In addition to cradling, another form of social touch that has been widely investigated in laterality research is kissing.[131] For romantic kissing, a study conducted at international airports revealed a significant rightward bias, with 64.5% of couples kissing to the right and 35.5% of couples kissing to the left.[137] This rightward bias for romantic kissing was also replicated by several subsequent studies,[138–141] indicating that it is a robust phenomenon. Interestingly, nonromantic forms of kissing do not show such a strong rightward asymmetry and are possibly strongly influenced by cultural factors.[142,143] In line with previous studies, an analysis comparing romantic kissing between two parents with parental kissing on the forehead from parent to child found a rightward bias for romantic kissing.[143] For parental kissing, however, there was a leftward bias, and the author suggested that this may have been the result of a learning effect as the leftward cradling bias may have a strong effect on the direction of parental kissing. Further support for the idea that social learning effects may affect laterality for nonromantic kissing comes from a French study in which laterality for cheek kissing was assessed in 10 cities in France.[142] In general, individuals within one city showed significant population-level asymmetries, but the direction of the asymmetry was dependent on the city. This shows that social laterality may, to some extent, be determined by social pressures.

A form of social touch that sometimes but not always cooccurs with kissing is hugging, but fewer studies have been published on hugging laterality than on kissing laterality.[131] Comparable to kissing, there is a rightward hugging bias.[144–146] This bias is affected by hemispheric asymmetries for emotion processing, with less pronounced rightward asymmetry in emotional compared to neutral situations.[145] This indicates a stronger involvement of right-hemispheric networks in emotional hugs compared to nonemotional hugs (e.g., greeting someone). Moreover, handedness also influences hugging laterality to some extent, as right-handers had a more rightward hugging bias than left-handers. As only 9.6% of the variance in hugging laterality was explained by handedness, the study indicates that hugging laterality cannot be fully explained by handedness. In general, these findings support the abovementioned idea by Brancucci and coworkers that social laterality is often influenced by different other forms of laterality.[120s]

Comparative research on social asymmetries in insects

The idea that social asymmetries are not unique to the human brain but have evolutionary old roots is supported not only by the findings on positional biases in infant–mother positioning[122] but also by several studies in invertebrates that link social behavior and hemispheric asymmetries.[128–130] For example, it has been shown that the honeybee (a social species), but not the mason bee (a solitary species), shows olfactory lateralization when recalling an olfactory memory.[147] Moreover, annual social bumble bees show olfactory asymmetries,[148] and social ants show population-level asymmetries during social antenna contacts.[149] The most direct evidence for social asymmetries in insects comes from a study on antenna use in social situations in honeybees.[150] In this study, it was shown that honeybees show a strong preference to use their right antenna in social situations. This shows that social asymmetries emerge even in comparably primitive nervous systems.

Sample questions

1. What is known about asymmetries in facial emotion expression, and how can they be assessed experimentally?
2. Name and describe the three major models of emotional lateralization. Which model is best suited to explain the empirical data? Please support your answer by describing one relevant empirical study in detail.
3. Describe the right-hemisphere model of emotional lateralization. Which empirical findings support this model?
4. Describe the valence model of emotional lateralization. Which empirical findings support this model?
5. Describe the frontal cortical asymmetry model of motivational direction. Which empirical findings support this model?
6. Describe the asymmetric inhibition model of hemispheric differences in emotional processing.
7. Describe the main findings on hemispheric asymmetries of social touch. Which three forms of social touch are often investigated in laterality research and what are the major findings about their laterality?

References

1. Güntürkün O. *Biologische Psychologie*. 2., aktualisierte Auflage. Hogrefe; 2019.
2. Palomero-Gallagher N, Amunsts K. A short review on emotion processing: a lateralized network of neuronal networks. *Brain Struct Funct*. 2022;227(2):673–684. Available from: https://doi.org/10.1007/s00429-021-02331-7.
3. Lindell AK. Continuities in emotion lateralization in human and non-human primates. *Front Hum Neurosci*. 2013;7:464. Available from: https://doi.org/10.3389/fnhum.2013.00464.
4. Sauter DA, Eisner F, Ekman P, Scott SK. Cross-cultural recognition of basic emotions through nonverbal emotional vocalizations. *PNAS*. 2010;107(6):2408–2412. Available from: https://doi.org/10.1073/pnas.0908239106.
5. Sackeim HA, Gur RC. Asymmetry in facial expression. *Science*. 1980;209(4458):834–836. Available from: https://doi.org/10.1126/science.7403854.
6. Borod JC, Koff E, White B. Facial asymmetry in posed and spontaneous expressions of emotion. *Brain Cognit*. 1983;2(2):165–175. Available from: https://doi.org/10.1016/0278-2626(83)90006-4.
7. Hauser MD, Akre K. Asymmetries in the timing of facial and vocal expressions by rhesus monkeys: implications for hemispheric specialization. *Anim Behav*. 2001;61(2):391–400. Available from: https://doi.org/10.1006/anbe.2000.1588.
8. Fernández-Carriba S, Loeches A, Morcillo A, Hopkins WD. Functional asymmetry of emotions in primates: new findings in chimpanzees. *Brain Res Bull*. 2002;57(3–4):561–564. Available from: https://doi.org/10.1016/s0361-9230(01)00685-2.
9. Fernández-Carriba S, Loeches Á, Morcillo A, Hopkins WD. Asymmetry in facial expression of emotions by chimpanzees. *Neuropsychologia*. 2002;40(9):1523–1533. Available from: https://doi.org/10.1016/s0028-3932(02)00028-3.
10. Wallez C, Vauclair J. First evidence of population-level oro-facial asymmetries during the production of distress calls by macaque (Macaca mulatta) and baboon (Papio anubis) infants. *Behavioural Brain Res*. 2012;234(1):69–75. Available from: https://doi.org/10.1016/j.bbr.2012.06.004.
11. Nicholls ME, Wolfgang BJ, Clode D, Lindell AK. The effect of left and right poses on the expression of facial emotion. *Neuropsychologia*. 2002;40(10):1662–1665. Available from: https://doi.org/10.1016/s0028-3932(02)00024-6.
12. McManus C, Humphrey NK. Turning the left cheek. *Nature*. 1973;243(5405):271–272. Available from: https://doi.org/10.1038/243271a0.

13. Lindell AK. The silent social/emotional signals in left and right cheek poses: a literature review. *Laterality*. 2013;18(5):612–624. Available from: https://doi.org/10.1080/1357650X.2012.737330.
14. Conesa J, Brunold-Conesa C, Miron M. Incidence of the half-left profile pose in single-subject portraits. *Percept Mot Skills*. 1995;81(3 Pt 1):920–922. Available from: https://doi.org/10.2466/pms.1995.81.3.920.
15. Labar M. Turning the left cheek examined using modern photography. *Nature*. 1973;245(5424):338. Available from: https://doi.org/10.1038/245338a0.
16. Duerksen KN, Friedrich TE, Elias LJ. Did Buddha turn the other cheek too? A comparison of posing biases between Jesus and Buddha. *Laterality*. 2016;21(4–6):633–642. Available from: https://doi.org/10.1080/1357650X.2015.1087554.
17. Lindell AK. Celebrity chefs put their left cheek forward: cover image orientation in celebrity cookbooks. *Laterality*. 2017;22(5):515–520. Available from: https://doi.org/10.1080/1357650X.2016.1223090.
18. Lindell AK. Consistently showing your best side? Intra-individual consistency in #selfie pose orientation. *Front Psychol*. 2017;8:246. Available from: https://doi.org/10.3389/fpsyg.2017.00246.
19. Bruno N, Bertamini M, Protti F. Selfie and the city: a world-wide, large, and ecologically valid database reveals a two-pronged side bias in naïve self-portraits. *PLOS ONE*. 2015;10(4):e0124999. Available from: https://doi.org/10.1371/journal.pone.0124999.
20. Bruno N, Bode C, Bertamini M. Composition in portraits: selfies and wefies reveal similar biases in untrained modern youths and ancient masters. *Laterality*. 2017;22(3):279–293. Available from: https://doi.org/10.1080/1357650X.2016.1185108.
21. Messina J, Lindell AK. Instagram advertisements favor the left cheek. *Cyberpsychol Behav Soc Netw*. 2020;23(11):768–772. Available from: https://doi.org/10.1089/cyber.2020.0073.
22. Lindell AK. Left cheek poses garner more likes: the effect of pose orientation on Instagram engagement. *Laterality*. 2019;24(5):600–613. Available from: https://doi.org/10.1080/1357650X.2018.1556278.
23. Churches O, Callahan R, Michalski D, et al. How academics face the world: a study of 5829 homepage pictures. *PLOS ONE*. 2012;7(7):e38940. Available from: https://doi.org/10.1371/journal.pone.0038940.
24. Demaree HA, Everhart DE, Youngstrom EA, Harrison DW. Brain lateralization of emotional processing: historical roots and a future incorporating "dominance.". *Behav Cognit Neurosci Rev*. 2005;4(1):3–20. Available from: https://doi.org/10.1177/1534582305276837.
25. Harmon-Jones E, Gable PA, Peterson CK. The role of asymmetric frontal cortical activity in emotion-related phenomena: a review and update. *Biol Psychol*. 2010;84(3):451–462. Available from: https://doi.org/10.1016/j.biopsycho.2009.08.010.
26. Borod JC. Interhemispheric and intrahemispheric control of emotion: a focus on unilateral brain damage. *J Consulting Clin Psychol*. 1992;60(3):339–348. Available from: https://doi.org/10.1037/0022-006x.60.3.339.
27. Borod JC, Bloom RL, Brickman AM, Nakhutina L, Curko EA. Emotional processing deficits in individuals with unilateral brain damage. *Appl Neuropsychology*. 2002;9(1):23–36. Available from: https://doi.org/10.1207/S15324826AN0901_4.
28. Borod JC, Kent J, Koff E, Martin C, Alpert M. Facial asymmetry while posing positive and negative emotions: support for the right hemisphere hypothesis. *Neuropsychologia*. 1988;26(5):759–764. Available from: https://doi.org/10.1016/0028-3932(88)90013-9.
29. Christman SD, Hackworth MD. Equivalent perceptual asymmetries for free viewing of positive and negative emotional expressions in chimeric faces. *Neuropsychologia*. 1993;31(6):621–624. Available from: https://doi.org/10.1016/0028-3932(93)90056-6.
30. Hugdahl K, Iversen PM, Johnsen BH. Laterality for facial expressions: does the sex of the subject interact with the sex of the stimulus face. *Cortex*. 1993;29(2):325–331. Available from: https://doi.org/10.1016/s0010-9452(13)80185-2.
31. McLaren J, Bryson SE. Hemispheric asymmetries in the perception of emotional and neutral faces. *Cortex*. 1987;23(4):645–654. Available from: https://doi.org/10.1016/s0010-9452(87)80054-0.
32. Moreno CR, Borod JC, Welkowitz J, Alpert M. Lateralization for the expression and perception of facial emotion as a function of age. *Neuropsychologia*. 1990;28(2):199–209. Available from: https://doi.org/10.1016/0028-3932(90)90101-s.
33. Tamietto M, Latini Corazzini L, des Gelder B, Geminiani G. Functional asymmetry and interhemispheric cooperation in the perception of emotions from facial expressions. *Exp Brain Res*. 2006;171(3):389–404. Available from: https://doi.org/10.1007/s00221-005-0279-4.

34. Hughlings Jackson J. On affections of speech from disease of the brain. *Brain*. 1879;1(3):304–330. Available from: https://doi.org/10.1093/brain/1.3.304.
35. Heilman KM, Scholes R, Watson RT. Auditory affective agnosia. *Disturb Compr Affect speech J Neurol Neurosurg Psychiatry*. 1975;38(1):69–72. Available from: https://doi.org/10.1136/jnnp.38.1.69.
36. DeKosky ST, Heilman KM, Bowers D, Valenstein E. Recognition and discrimination of emotional faces and pictures. *Brain Lang*. 1980;9(2):206–214. Available from: https://doi.org/10.1016/0093-934x(80)90141-8.
37. Pegna AJ, Khateb A, Lazeyras F, Seghier ML. Discriminating emotional faces without primary visual cortices involves the right amygdala. *Nat Neurosci*. 2005;8(1):24–25. Available from: https://doi.org/10.1038/nn1364.
38. Pessoa L. A Network model of the emotional brain. *Trends Cognit Sci*. 2017;21(5):357–371. Available from: https://doi.org/10.1016/j.tics.2017.03.002.
39. Parida JR, Misra DP, Agarwal V. Urbach-Wiethe syndrome. Case Reports; 2015:bcr2015212443. Available from: https://doi.org/10.1136/bcr-2015-212443.
40. Jansen S, Quante G, Luers JC, Beutner D. Urbach-Wiethe disease in a young woman: a case report. *ENT: Ea, Nose Throat J*. 2016;95(1).
41. Adolphs R, Tranel D, Damasio H, Damasio A. Impaired recognition of emotion in facial expressions following bilateral damage to the human amygdala. *Nature*. 1994;372(6507):669–672. Available from: https://doi.org/10.1038/372669a0.
42. Carmon A, Nachshon I. Ear asymmetry in perception of emotional non-verbal stimuli. *Acta Psychologica*. 1973;37(6):351–357. Available from: https://doi.org/10.1016/0001-6918(73)90002-4.
43. Hugdahl K, Westerhausen R, Alho K, Medvedev S, Laine M, Hämäläinen H. Attention and cognitive control: unfolding the dichotic listening story. *Scand J Psychol*. 2009;50(1):11–22. Available from: https://doi.org/10.1111/j.1467-9450.2008.00676.x.
44. Bryden MP, MacRae L. Dichotic laterality effects obtained with emotional words. *Neuropsychiatry Neuropsychol Behav Neurol*. 1988;1(3):171–176.
45. Bourne VJ. The divided visual field paradigm: methodological considerations. *Laterality*. 2006;11(4):373–393. Available from: https://doi.org/10.1080/13576500600633982.
46. Ley RG, Bryden MP. Hemispheric differences in processing emotions and faces. *Brain Lang*. 1979;7(1):127–138. Available from: https://doi.org/10.1016/0093-934x(79)90010-5.
47. Strauss E, Moscovitch M. Perception of facial expressions. *Brain Lang*. 1981;13(2):308–332. Available from: https://doi.org/10.1016/0093-934x(81)90098-5.
48. Alves NT, Aznar-Casanova JA, Fukusima SS. Patterns of brain asymmetry in the perception of positive and negative facial expressions. *Laterality*. 2009;14(3):256–272. Available from: https://doi.org/10.1080/13576500802362927.
49. Baas D, Aleman A, Kahn RS. Lateralization of amygdala activation: a systematic review of functional neuroimaging studies. *Brain Res Brain Res Rev*. 2004;45(2):96–103. Available from: https://doi.org/10.1016/j.brainresrev.2004.02.004.
50. Goldstein K. *The Organism: A Holistic Approach to Biology Derived from Pathological Data in Man*. American Book Company; 1939.
51. Terzian H. Behavioural and EEG effects of intracarotid sodium amytal injection. *Acta Neurochirurgica*. 1964;12(2):230–239. Available from: https://doi.org/10.1007/BF01402095.
52. Davidson RJ. Cerebral asymmetry, emotion, and affective style. In: Davidson RJ, Hugdahl K, eds. *Brain Asymmetry*. MIT Press; 1995:361–387.
53. Silberman EK, Weingartner H. Hemispheric lateralization of functions related to emotion. *Brain Cognit*. 1986;5(3):322–353. Available from: https://doi.org/10.1016/0278-2626(86)90035-7.
54. Sutton SK, Davidson RJ, Donzella B, Irwin W, Dottl DA. Manipulating affective state using extended picture presentations. *Psychophysiology*. 1997;34(2):217–226. Available from: https://doi.org/10.1111/j.1469-8986.1997.tb02135.x.
55. Reuter-Lorenz P, Davidson RJ. Differential contributions of the two cerebral hemispheres to the perception of happy and sad faces. *Neuropsychologia*. 1981;19(4):609–613. Available from: https://doi.org/10.1016/0028-3932(81)90030-0.
56. Jończyk R. Hemispheric asymmetry of emotion words in a non-native mind: a divided visual field study. *Laterality*. 2015;20(3):326–347. Available from: https://doi.org/10.1080/1357650X.2014.966108.
57. Holtgraves T, Felton A. Hemispheric asymmetry in the processing of negative and positive words: a divided field study. *Cogn Emot*. 2011;25(4):691–699. Available from: https://doi.org/10.1080/02699931.2010.493758.

58. Jansari A, Tranel D, Adolphs R. A valence-specific lateral bias for discriminating emotional facial expressions in free field. *Cogn Emot*. 2000;14(3):341–353. Available from: https://doi.org/10.1080/026999300378860.
59. Allen JJB, Coan JA, Nazarian M. Issues and assumptions on the road from raw signals to metrics of frontal EEG asymmetry in emotion. *Biol Psychol*. 2004;67(1-2):183–218. Available from: https://doi.org/10.1016/j.biopsycho.2004.03.007.
60. Coan JA, Allen JJB. Frontal EEG asymmetry as a moderator and mediator of emotion. *Biol Psychol*. 2004;67(1-2):7–49. Available from: https://doi.org/10.1016/j.biopsycho.2004.03.002.
61. Hagemann D. Individual differences in anterior EEG asymmetry: methodological problems and solutions. *Biol Psychol*. 2004;67(1-2):157–182. Available from: https://doi.org/10.1016/j.biopsycho.2004.03.006.
62. Ahern GL, Schwartz GE. Differential lateralization for positive and negative emotion in the human brain: EEG spectral analysis. *Neuropsychologia*. 1985;23(6):745–755. Available from: https://doi.org/10.1016/0028-3932(85)90081-8.
63. Davidson RJ, Ekman P, Saron CD, Senulis JA, Friesen WV. Approach-withdrawal and cerebral asymmetry: emotional expression and brain physiology: I. *J Personality Soc Psychol*. 1990;58(2):330–341. Available from: https://doi.org/10.1037/0022-3514.58.2.330.
64. Jones NA, Fox NA. Electroencephalogram asymmetry during emotionally evocative films and its relation to positive and negative affectivity. *Brain Cognition*. 1992;20(2):280–299. Available from: https://doi.org/10.1016/0278-2626(92)90021-d.
65. Tomarken AJ, Davidson RJ, Wheeler RE, Doss RC. Individual differences in anterior brain asymmetry and fundamental dimensions of emotion. *J Personality Soc Psychol*. 1992;62(4):676–687. Available from: https://doi.org/10.1037/0022-3514.62.4.676.
66. Watson D, Clark LA, Tellegen A. Development and validation of brief measures of positive and negative affect: the PANAS scales. *J Personality Soc Psychol*. 1988;54(6):1063–1070. Available from: https://doi.org/10.1037/0022-3514.54.6.1063.
67. Wheeler RE, Davidson RJ, Tomarken AJ. Frontal brain asymmetry and emotional reactivity: a biological substrate of affective style. *Psychophysiology*. 1993;30(1):82–89. Available from: https://doi.org/10.1111/j.1469-8986.1993.tb03207.x.
68. Canli T, Desmond JE, Zhao Z, Glover G, Gabrieli JDE. Hemispheric asymmetry for emotional stimuli detected with fMRI. *NeuroReport*. 1998;9(14):3233.
69. Lanteaume L, Khalfa S, Régis J, Marquis P, Chauvel P, Bartolomei F. Emotion induction after direct intracerebral stimulations of human amygdala. *Cereb Cortex*. 2007;17(6):1307–1313. Available from: https://doi.org/10.1093/cercor/bhl041.
70. Leliveld LM, Langbein J, Puppe B. The emergence of emotional lateralization: Evidence in non-human vertebrates and implications for farm animals. *Appl Anim Behav Sci*. 2013;145(1-2):1–14. Available from: https://doi.org/10.1016/j.applanim.2013.02.002.
71. Gray JA, MacNaughton N. *The Neuropsychology of Anxiety: An Enquiry into the Functions of the Septo-Hippocampal System*. Oxford Univ. Press; 2000.
72. Rutherford HJV, Lindell AK. Thriving and Surviving: Approach and Avoidance Motivation and Lateralization. *Emot Rev*. 2011;3(3):333–343. Available from: https://doi.org/10.1177/1754073911402392.
73. Carver CS, White TL. Behavioral inhibition, behavioral activation, and affective responses to impending reward and punishment: the BIS/BAS Scales. *J Personality Soc Psychol*. 1994;67(2):319–333. Available from: https://doi.org/10.1037/0022-3514.67.2.319.
74. Harmon-Jones E, Sigelman J. State anger and prefrontal brain activity: evidence that insult-related relative left-prefrontal activation is associated with experienced anger and aggression. *J Personality Soc Psychol*. 2001;80(5):797–803. Available from: https://doi.org/10.1037/0022-3514.80.5.797.
75. Schutter DJLG, Harmon-Jones E. The corpus callosum: a commissural road to anger and aggression. *Neurosci Biobehav Rev*. 2013;37(10 Pt 2):2481–2488. Available from: https://doi.org/10.1016/j.neubiorev.2013.07.013.
76. Harmon-Jones E. Clarifying the emotive functions of asymmetrical frontal cortical activity, Early Career Award*Psychophysiology*. 2003;40(6):838–848. Available from: https://doi.org/10.1111/1469-8986.00121.
77. Schutter DJLG, Weijer AD, de Meuwese JDI, Morgan B, van Honk J. Interrelations between motivational stance, cortical excitability, and the frontal electroencephalogram asymmetry of emotion: a transcranial magnetic stimulation study. *Hum Brain Mapp*. 2008;29(5):574–580. Available from: https://doi.org/10.1002/hbm.20417.

78. Carver CS, Harmon-Jones E. Anger is an approach-related affect: evidence and implications. *Psychological Bull*. 2009;135(2):183–204. Available from: https://doi.org/10.1037/a0013965.
79. Harmon-Jones E, Allen JJB. Anger and frontal brain activity: EEG asymmetry consistent with approach motivation despite negative affective valence. *J Personality Soc Psychol*. 1998;74(5):1310–1316. Available from: https://doi.org/10.1037/0022-3514.74.5.1310.
80. d'Alfonso AA, van Honk J, Hermans E, Postma A, Haan EHde. Laterality effects in selective attention to threat after repetitive transcranial magnetic stimulation at the prefrontal cortex in female subjects. *Neurosci Lett*. 2000;280(3):195–198. Available from: https://doi.org/10.1016/s0304-3940(00)00781-3.
81. Hortensius R, Schutter DJLG, Harmon-Jones E. When anger leads to aggression: induction of relative left frontal cortical activity with transcranial direct current stimulation increases the anger-aggression relationship. *Soc Cogn Affect Neurosci*. 2012;7(3):342–347. Available from: https://doi.org/10.1093/scan/nsr012.
82. Maeda K, Yamaguchi T, Tatemoto T, Kondo K, Otaka Y, Tanaka S. Transcranial direct current stimulation does not affect lower extremity muscle strength training in healthy individuals: a triple-blind, sham-controlled study. *Front Neurosci*. 2017;11:179. Available from: https://doi.org/10.3389/fnins.2017.00179.
83. Lee HY, Choi MS, Chang DS, Cho C-S. Combined bifrontal transcranial direct current stimulation and tailor-made notched music training in chronic tinnitus. *J Audiol Otol*. 2017;21(1):22–27. Available from: https://doi.org/10.7874/jao.2017.21.1.22.
84. Buch ER, Santarnecchi E, Antal A, et al. Effects of tDCS on motor learning and memory formation: a consensus and critical position paper. *Clin Neurophysiol*. 2017;128(4):589–603. Available from: https://doi.org/10.1101/064204.
85. Lefaucheur J-P, Antal A, Ayache SS, et al. Evidence-based guidelines on the therapeutic use of transcranial direct current stimulation (tDCS). *Clin Neurophysiol*. 2017;128(1):56–92. Available from: https://doi.org/10.1016/j.clinph.2016.10.087.
86. Anderson CA, Lindsay JJ, Bushman BJ. Research in the psychological laboratory. *Curr Dir Psychol Sci*. 1999;8(1):3–9. Available from: https://doi.org/10.1111/1467-8721.00002.
87. Lippolis G, Joss JMP, Rogers LJ. Australian lungfish (Neoceratodus forsteri): a missing link in the evolution of complementary side biases for predator avoidance and prey capture. *BBE*. 2009;73(4):295–303. Available from: https://doi.org/10.1159/000230674.
88. Cantalupo C, Bisazza A, Vallortigara G. Lateralization of predator-evasion response in a teleost fish (*Girardinus falcatus*). *Neuropsychologia*. 1995;33(12):1637–1646. Available from: https://doi.org/10.1016/0028-3932(95)00043-7.
89. Lippolis G, Bisazza A, Rogers LJ, Vallortigara G. Lateralisation of predator avoidance responses in three species of toads. *Laterality*. 2002;7(2):163–183. Available from: https://doi.org/10.1080/13576500143000221.
90. Wassersug RJ, Naitoh T, Yamashita M. Turning bias in tadpoles. *J Herpetology*. 1999;33(4):543. Available from: https://doi.org/10.2307/1565570.
91. Yamashita M, Naitoh T, Wassersug RJ. Startle response and turning bias in microhyla tadpoles. *Zool Sci*. 2000;17(2):185–189. Available from: https://doi.org/10.2108/zsj.17.185.
92. Rogers LJ. Evolution of hemispheric specialization: advantages and disadvantages. *Brain Lang*. 2000;73(2):236–253. Available from: https://doi.org/10.1006/brln.2000.2305.
93. Lippolis G, Westman W, McAllan BM, Rogers LJ. Lateralisation of escape responses in the stripe-faced dunnart, Sminthopsis macroura (Dasyuridae: Marsupialia). *Laterality*. 2005;10(5):457–470. Available from: https://doi.org/10.1080/13576500442000210.
94. Schnell AK, Hanlon RT, Benkada A, Jozet-Alves C. Lateralization of eye use in cuttlefish: opposite direction for anti-predatory and predatory behaviors. *Front Physiol*. 2016;7:620. Available from: https://doi.org/10.3389/fphys.2016.00620.
95. Bonati B, Csermely D, Romani R. Lateralization in the predatory behaviour of the common wall lizard (Podarcis muralis). *Behavioural Process*. 2008;79(3):171–174. Available from: https://doi.org/10.1016/j.beproc.2008.07.007.
96. Robins A, Chen P, Beazley LD, Dunlop SA. Lateralized predatory responses in the ornate dragon lizard (Ctenophorus ornatus). *NeuroReport*. 2005;16(8):849. Available from: https://doi.org/10.1097/00001756-200505310-00014.
97. Vallortigara G, Rogers LJ, Bisazza A, Lippolis G, Robins A. Complementary right and left hemifield use for predatory and agonistic behaviour in toads. *NeuroReport*. 1998;9(14):3341. Available from: https://doi.org/10.1097/00001756-199810050-00035.

98. Killgore WDS, Yurgelun-Todd DA. The right-hemisphere and valence hypotheses: could they both be right (and sometimes left)? *Soc Cogn Affect Neurosci*. 2007;2(3):240–250. Available from: https://doi.org/10.1093/scan/nsm020.
99. Rodway P, Schepman A. Valence specific laterality effects in prosody: expectancy account and the effects of morphed prosody and stimulus lead. *Brain Cognition*. 2007;63(1):31–41. Available from: https://doi.org/10.1016/j.bandc.2006.07.008.
100. Schepman A, Rodway P, Geddes P. Valence-specific laterality effects in vocal emotion: interactions with stimulus type, blocking and sex. *Brain Cognit*. 2012;79(2):129–137. Available from: https://doi.org/10.1016/j.bandc.2012.03.001.
101. Heller W. Neuropsychological mechanisms of individual differences in emotion, personality, and arousal. *Neuropsychology*. 1993;7(4):476–489. Available from: https://doi.org/10.1037/0894-4105.7.4.476.
102. Heller W, Nitschke JB, Miller GA. Lateralization in emotion and emotional disorders. *Curr Dir Psychol Sci*. 1998;7(1):26–32. Available from: https://doi.org/10.1111/1467-8721.ep11521823.
103. Grimshaw GM, Carmel D. An asymmetric inhibition model of hemispheric differences in emotional processing. *Front Psychol*. 2014;5:489. Available from: https://doi.org/10.3389/fpsyg.2014.00489.
104. Friedman NP, Miyake A. Unity and diversity of executive functions: Individual differences as a window on cognitive structure. *Cortex*. 2017;86:186–204. Available from: https://doi.org/10.1016/j.cortex.2016.04.023.
105. Yuan P, Raz N. Prefrontal cortex and executive functions in healthy adults: a meta-analysis of structural neuroimaging studies. *Neurosci Biobehav Rev*. 2014;42:180–192. Available from: https://doi.org/10.1016/j.neubiorev.2014.02.005.
106. Chudasama Y. Animal models of prefrontal-executive function. *Behav Neurosci*. 2011;125(3):327–343. Available from: https://doi.org/10.1037/a0023766.
107. Kaller CP, Reisert M, Katzev M, et al. Predicting planning performance from structural connectivity between left and right mid-dorsolateral prefrontal cortex: moderating effects of age during postadolescence and midadulthood. *Cereb Cortex*. 2015;25(4):869–883. Available from: https://doi.org/10.1093/cercor/bht276.
108. Silton RL, Heller W, Towers DN, et al. The time course of activity in dorsolateral prefrontal cortex and anterior cingulate cortex during top-down attentional control. *NeuroImage*. 2010;50(3):1292–1302. Available from: https://doi.org/10.1016/j.neuroimage.2009.12.061.
109. Golkar A, Lonsdorf TB, Olsson A, et al. Distinct contributions of the dorsolateral prefrontal and orbitofrontal cortex during emotion regulation. *PLOS ONE*. 2012;7(11):e48107. Available from: https://doi.org/10.1371/journal.pone.0048107.
110. Rêgo GG, Lapenta OM, Marques LM, et al. Hemispheric dorsolateral prefrontal cortex lateralization in the regulation of empathy for pain. *Neurosci Lett*. 2015;594:12–16. Available from: https://doi.org/10.1016/j.neulet.2015.03.042.
111. Rudebeck PH, Saunders RC, Prescott AT, Chau LS, Murray EA. Prefrontal mechanisms of behavioral flexibility, emotion regulation and value updating. *Nat Neurosci*. 2013;16(8):1140–1145. Available from: https://doi.org/10.1038/nn.3440.
112. Zwanzger P, Steinberg C, Rehbein MA, et al. Inhibitory repetitive transcranial magnetic stimulation (rTMS) of the dorsolateral prefrontal cortex modulates early affective processing. *NeuroImage*. 2014;101:193–203. Available from: https://doi.org/10.1016/j.neuroimage.2014.07.003.
113. Ocklenburg S, Peterburs J, Mundorf A. Hemispheric asymmetries in the amygdala: a comparative primer. *Prog Neurobiol*. 2022;214:102283. Available from: https://doi.org/10.1016/j.pneurobio.2022.102283.
114. Koelsch S. A coordinate-based meta-analysis of music-evoked emotions. *NeuroImage*. 2020;223:117350. Available from: https://doi.org/10.1016/j.neuroimage.2020.117350.
115. Mauchand M, Zhang S. Disentangling emotional signals in the brain: an ALE meta-analysis of vocal affect perception. *Cogn Affect Behav Neurosci*. 2023;23(1):17–29. Available from: https://doi.org/10.3758/s13415-022-01030-y.
116. Vytal K, Hamann S. Neuroimaging support for discrete neural correlates of basic emotions: a voxel-based meta-analysis. *J Cogn Neurosci*. 2010;22(12):2864–2885. Available from: https://doi.org/10.1162/jocn.2009.21366.
117. Guadalupe T, Mathias SR, vanErp TGM, et al. Human subcortical brain asymmetries in 15,847 people worldwide reveal effects of age and sex. *Brain Imaging Behav*. 2017;11(5):1497–1514. Available from: https://doi.org/10.1007/s11682-016-9629-z.

118. Sergerie K, Chochol C, Armony JL. The role of the amygdala in emotional processing: a quantitative meta-analysis of functional neuroimaging studies. *Neurosci & Biobehav Rev*. 2008;32(4):811–830. Available from: https://doi.org/10.1016/j.neubiorev.2007.12.002.
119. El Basbasse Y, Packheiser J, Peterburs J, et al. Walk the plank! Using mobile electroencephalography to investigate emotional lateralization of immersive fear in virtual reality. *R Soc Open Sci*. 2023;10(5):221239. Available from: https://doi.org/10.1098/rsos.221239.
120. Brancucci A, Lucci G, Mazzatenta A, Tommasi L. Asymmetries of the human social brain in the visual, auditory and chemical modalities. *Philos Trans R Soc Lond B Biol Sci*. 2009;364(1519):895–914. Available from: https://doi.org/10.1098/rstb.2008.0279.
121. Forrester GS, Davis R, Mareschal D, Malatesta G, Todd BK. The left cradling bias: an evolutionary facilitator of social cognition. *Cortex*. 2019;118:116–131. Available from: https://doi.org/10.1016/j.cortex.2018.05.011.
122. Karenina K, Giljov A. Mother and offspring lateralized social behavior across mammalian species. *Prog Brain Res*. 2018;238:115–141. Available from: https://doi.org/10.1016/bs.pbr.2018.06.003.
123. Malatesta G, Marzoli D, Piccioni C, Tommasi L. The relationship between the left-cradling bias and attachment to parents and partnser. *Evol Psychol*. 2019;17(2). Available from: https://doi.org/10.1177/1474704919848117.
124. Malatesta G, Marzoli D, Prete G, Tommasi L. Human lateralization, maternal effects and neurodevelopmental disorders. *Front Behav Neurosci*. 2021;15:668520. Available from: https://doi.org/10.3389/fnbeh.2021.668520.
125. Malatesta G, Marzoli D, Tommasi L. Environmental and genetic determinants of sensorimotor asymmetries in mother-infant interaction. *Front Behav Neurosci*. 2022;16:1080141. Available from: https://doi.org/10.3389/fnbeh.2022.1080141.
126. Pintos Lobo R, Bottenhorn KL, Riedel MC, et al. Neural systems underlying RDoC social constructs: An activation likelihood estimation meta-analysis. *Neurosci Biobehav Rev*. 2023;144:104971. Available from: https://doi.org/10.1016/j.neubiorev.2022.104971.
127. Krall SC, Rottschy C, Oberwelland E, et al. The role of the right temporoparietal junction in attention and social interaction as revealed by ALE meta-analysis. *Brain Struct Funct*. 2015;220(2):587–604. Available from: https://doi.org/10.1007/s00429-014-0803-z.
128. Frasnelli E. Brain and behavioral lateralization in invertebrates. *Front Psychol*. 2013;4:939. Available from: https://doi.org/10.3389/fpsyg.2013.00939.
129. Frasnelli E, Haase A, Rigosi E, Anfora G, Rogers LJ, Vallortigara G. The bee as a model to investigate brain and behavioural asymmetries. *Insects*. 2014;5(1):120–138. Available from: https://doi.org/10.3390/insects5010120.
130. Frasnelli E, Vallortigara G, Rogers LJ. Left-right asymmetries of behaviour and nervous system in invertebrates. *Neurosci Biobehav Rev*. 2012;36(4):1273–1291. Available from: https://doi.org/10.1016/j.neubiorev.2012.02.006.
131. Ocklenburg S, Packheiser J, Schmitz J, et al. Hugs and kisses - the role of motor preferences and emotional lateralization for hemispheric asymmetries in human social touch. *Neurosci Biobehav Rev*. 2018;95:353–360. Available from: https://doi.org/10.1016/j.neubiorev.2018.10.007.
132. Saarinen A, Harjunen V, Jasinskaja-Lahti I, Jääskeläinen IP, Ravaja N. Social touch experience in different contexts: a review. *Neurosci Biobehav Rev*. 2021;131:360–372. Available from: https://doi.org/10.1016/j.neubiorev.2021.09.027.
133. Salk L. The effects of the normal heartbeat sound on the behaviour of the newborn infant: Implications for mental health. *World Ment Health*. 1960;12:168–175. Available from: https://cir.nii.ac.jp/crid/1573105974473334912.
134. Packheiser J, Schmitz J, Berretz G, Papadatou-Pastou M, Ocklenburg S. Handedness and sex effects on lateral biases in human cradling: three meta-analyses. *Neurosci Biobehav Rev*. 2019;104:30–42. Available from: https://doi.org/10.1016/j.neubiorev.2019.06.035.
135. van der Meer A, Husby A. Handedness as a major determinant of functional cradling bias. *Laterality*. 2006;11(3):263–276. Available from: https://doi.org/10.1080/13576500500513565.
136. Manning JT, Heaton R, Chamberlain AT. Left-side cradling: similarities and differences between apes and humans. *J Hum Evolution*. 1994;26(1):77–83. Available from: https://doi.org/10.1006/jhev.1994.1005.
137. Güntürkün O. Human behaviour: adult persistence of head-turning asymmetry. *Nature*. 2003;421(6924):711. Available from: https://doi.org/10.1038/421711a.
138. Barrett D, Greenwood JG, McCullagh JF. Kissing laterality and handedness. *Laterality*. 2006;11(6):573–579. Available from: https://doi.org/10.1080/13576500600886614.

139. Karim AKMR Proulx MJ, Sousa AAde, et al. The right way to kiss: directionality bias in head-turning during kissing. *Sci Rep*. 2017;7(1):5398. Available from: https://doi.org/10.1038/s41598-017-04942-9.
140. Ocklenburg S, Güntürkün O. Head-turning asymmetries during kissing and their association with lateral preference. *Laterality*. 2009;14(1):79–85. Available from: https://doi.org/10.1080/13576500802243689.
141. van der Kamp J, Canal-Bruland R. Kissing right? On the consistency of the head-turning bias in kissing. *Laterality*. 2011;16(3):257–267. Available from: https://doi.org/10.1080/13576500903530778.
142. Chapelain A, Pimbert P, Aube L, et al. Correction: can population-level laterality stem from social pressures? Evidence from cheek kissing in humans. *PLOS ONE*. 2016;11(1):e0148456. Available from: https://doi.org/10.1371/journal.pone.0148456.
143. Sedgewick JR, Elias LJ. Family matters: directionality of turning bias while kissing is modulated by context. *Laterality*. 2016;21(4-6):662–671. Available from: https://doi.org/10.1080/1357650X.2015.1136320.
144. Packheiser J, Schmitz J, Metzen D, et al. Asymmetries in social touch-motor and emotional biases on lateral preferences in embracing, cradling and kissing. *Laterality*. 2020;25(3):325–348. Available from: https://doi.org/10.1080/1357650X.2019.1690496.
145. Packheiser J, Rook N, Dursun Z, et al. Embracing your emotions: affective state impacts lateralisation of human embraces. *Psychol Res*. 2019;83(1):26–36. Available from: https://doi.org/10.1007/s00426-018-0985-8.
146. Turnbull OH, Stein L, Lucas MD. Lateral preferences in adult embracing: a test of the "hemispheric asymmetry" theory of infant cradling. *J Genet Psychol*. 1995;156(1):17–21. Available from: https://doi.org/10.1080/00221325.1995.9914802.
147. Anfora G, Frasnelli E, Maccagnani B, Rogers LJ, Vallortigara G. Behavioural and electrophysiological lateralization in a social (Apis mellifera) but not in a non-social (Osmia cornuta) species of bee. *Behav Brain Res*. 2010;206(2):236–239. Available from: https://doi.org/10.1016/j.bbr.2009.09.023.
148. Anfora G, Rigosi E, Frasnelli E, Ruga V, Trona F, Vallortigara G. Lateralization in the invertebrate brain: left-right asymmetry of olfaction in bumble bee, Bombus terrestris. *PLOS ONE*. 2011;6(4):e18903. Available from: https://doi.org/10.1371/journal.pone.0018903.
149. Frasnelli E, Iakovlev I, Reznikova Z. Asymmetry in antennal contacts during trophallaxis in ants. *Behav Brain Res*. 2012;232(1):7–12. Available from: https://doi.org/10.1016/j.bbr.2012.03.014.
150. Rogers LJ, Rigosi E, Frasnelli E, Vallortigara G. A right antenna for social behaviour in honeybees. *Sci Rep*. 2013;3:2045. Available from: https://doi.org/10.1038/srep02045.

CHAPTER

10

Ontogenesis of hemispheric asymmetries

OUTLINE

Introductory short story

Melbourne (Australia) in the late 1970s. A young ethologist and pharmacologist, Lesley Rogers, sits at her desk at Monash University and needs an explanation. In the 1970s, it had become increasingly clear that memory had a protein substrate. Consequently, a drug like cycloheximide that interferes with protein synthesis or any other substance that affects memory consolidation was expected to disturb learning and memory. Rogers had shown that injections of cycloheximide, or the transmitter glutamate, into the forebrain of young chicks could selectively slow learning.[1] However, this procedure could also be used for something quite different. In her next study, Rogers used 2-day-old chicks and injected cycloheximide into only one hemisphere, left or right. To her amazement, she found clear evidence for lateralization of function for visually guided discrimination and novelty

The Lateralized Brain
DOI: https://doi.org/10.1016/B978-0-323-99737-9.00020-3

detection. Rogers had discovered asymmetries of function in the avian brain![2] All colleagues to whom she told this major discovery were convinced that chicks should have a gene that programmed the asymmetry of the chick's brain. Rogers was not convinced. The geneticists had no evidence for their claim but could not think of alternatives. Rogers had never been keen on simplistic genetic explanations. In addition, she knew that alternatives existed. In 1978, Victor Denenberg had shown that an early experience, such as asymmetrical handling, affects the two brain hemispheres of rats.[3] In other words, the environment could play a key role in the ontogeny of asymmetry. However, Victor Denenberg could not convincingly demonstrate why handling had lateralized effects on rats nor were Rogers' chicks handled. Could her birds experience another external input that could shift their brains into asymmetries of function?

While pondering this question, she pulled out a classic textbook by Freeman and Vince[4] on the development of the avian embryo and started to browse through the appendix. The pictures on page 290 showed a late-stage embryo with its head turned toward its right side. The left eye was covered by the yolk sac and the chick's own body. She had seen this picture dozens of times before, but now it dawned on her that what she saw could be the key she was looking for. She started going backward in the book, page by page. The basic asymmetry persisted even when the little bird was too young to tuck its head on the right side. The embryo was floating on a huge yolk sac with its right eye looking toward the eggshell and its left eye covered by yolk. For her experiments, Rogers had always used chicken eggs that had been kept in a light incubator. Could it be that this light traversed the eggshell and reached predominantly the right eye of the developing embryo? If this were the case, the chick was exposed to asymmetric light stimulation that came in time to affect developing visual pathways. What a tantalizing possibility.

Quickly, she started two new batches of chicks, one incubated in darkness, the other in light. If light during ontogeny would indeed modify visual pathways by mainly activating right eye input, dark-incubated chicks should evince no behavioral asymmetry, while light-incubated birds would. Indeed, this was the outcome.[5] Lesley Rogers had not only discovered that visually driven behavioral asymmetries in birds result from a tight interaction of environmental—light input—and genetic—position of the developing embryo—factors, she also was the first scientist to discover a convincing mechanism that could explain why such factors can result in an asymmetrically organized brain. Her experiments showed that it was light exposure during the final stages of embryonic development, when the visual connections in the brain were developing and becoming functional, that generated the asymmetry of visual discrimination learning.

Breaking symmetry: how the embryonic left-right axis is determined

The vertebrate embryo starts as an egg cell and quickly grows into a largely symmetric ball of cells. It has no head or tail, no front or back, and no left and right. How it develops into the typical vertebrate body plan with an antero-posterior, dorsal-ventral, and left-right axis (see Fig. 10.1) has fascinated neurobiologists for decades.[6]

During embryogenesis, the anterior-posterior axis develops first, followed by the dorsal-ventral axis. With some delay, the left-right axis is added.[7] This process called

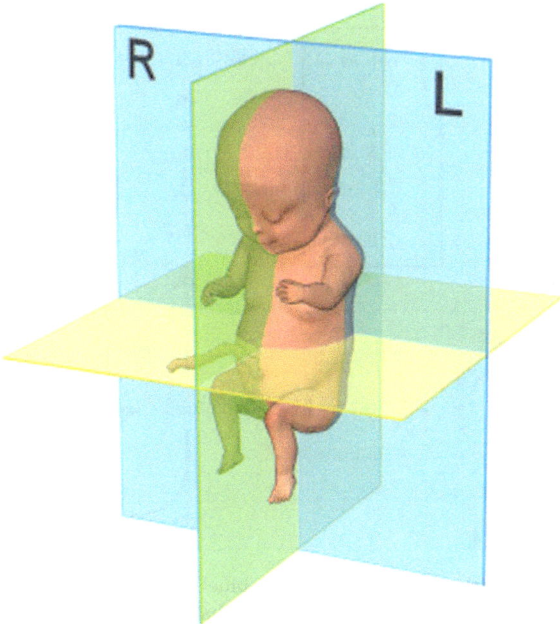

FIGURE 10.1 The three axes of the human body. The blue plane shows the left-right axis, the green plane shows the antero-posterior axis, and yellow plane the dorsoventral axis.

symmetry breaking has been investigated in many different species.[8–12] Breaking left-right symmetry in the embryo is controlled by the Nodal signaling cascade (see Fig. 10.2) and involves several separate steps, here described for the rodent embryo[13–15]:

1. At the posterior end of the developing embryo, there is a pit-shaped structure known as the ventral node (see Fig. 10.2A). Most, maybe even all, of our asymmetries depend in one way or another on this tiny little pit. It contains motile cilia, a type of cell organelle that is located on the cell surface in large numbers and can move in a coordinated fashion to regulate extracellular fluid flow (see Fig. 10.3 for scanning electron micrographs of node and cilia). The motile cilia in the pit move like seagrass, rhythmically moving back and forth by the currents of the waves above, but with one important difference. While seagrass is passive and just follows the waves, the motile cilia in the pit are actively moving and they are the ones making the waves. To be more specific, they cause a leftward flow of extracellular fluid, the so-called "nodal flow."
2. The unidirectional leftward flow of extracellular fluid causes a leftward movement of gene products and signaling molecules in the node. Immotile cilia, a second type of cilia that cannot move on their own and which are located in the periphery of the ventral node, mechanically sense the direction of fluid flow.[16] This generates a cue for asymmetric tissue patterning.[17]
3. This asymmetric signal is transferred to the lateral plate mesoderm, an embryonic structure that will later form the heart and other organs. Due to the leftward flow of extracellular fluid, Nodal, a signaling protein that is a member of the transforming growth factor β family of proteins, shows leftward expression asymmetries in the lateral plate mesoderm. In the left lateral plate mesoderm, Nodal signaling induces

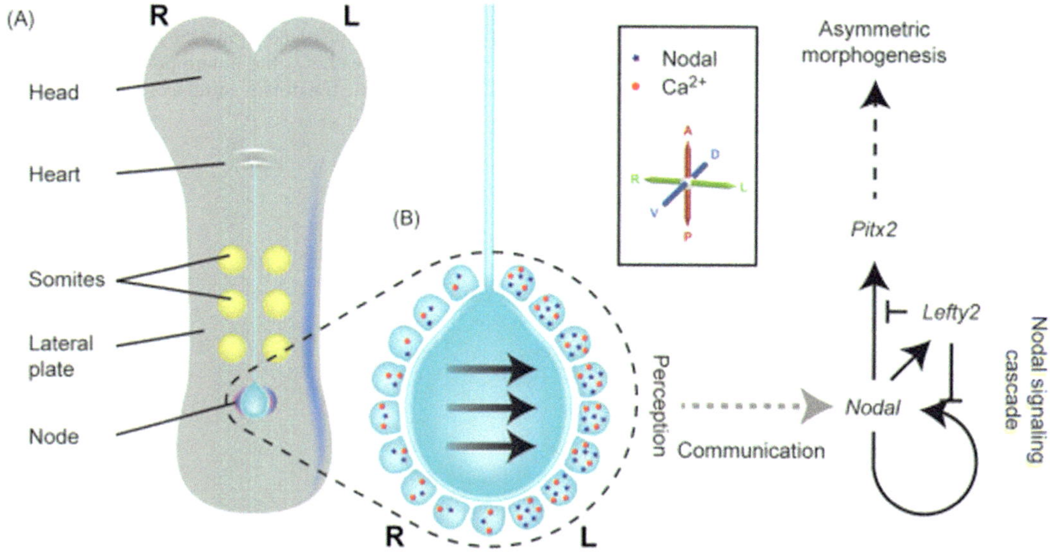

FIGURE 10.2 (A) A vertebrate embryo showing asymmetric *Nodal* expression (in blue) at the node and the left but not the right lateral plate. (B) The central pit of the node contains rotating cilia that induce a flow from the right to the left (symbolized by *black arrows*). This leads to asymmetric leftward expression of Nodal (*blue stars*) and leftward asymmetric Ca^{2+} signaling (*red dots*). This asymmetric gene expression at the node generates a leftward asymmetric signal that is transferred from the node to the lateral plate mesoderm. Here, the Nodal signaling pathway is activated, starting with asymmetric expression of *Nodal* that is regulated by *Lefty2*. Subsequently, leftward asymmetric *Pitx2* expression is induced, resulting in asymmetric morphogenesis.[13] *Republished under a Creative Commons Attribution License CC BY 2.0 from Norris DP. Cilia, calcium, and the basis of left-right asymmetry.* BMC Biol. 2012;10:102.

the expression of *Nodal* as well as of *Lefty2*, a gene that regulates *Nodal* expression (see Fig. 10.2B). The protein expressed by *Lefty2* reduces the diffusivity of Nodal, preventing it from reaching the right lateral plate mesoderm, thereby ensuring asymmetry.[13]

4. Subsequently, leftward asymmetric expression of *Pitx2* is induced, which produces the blueprint for our asymmetrically organized inner organs.[18]

Symmetry breaking occurs similarly, but not identically, across different animal species.[17] Comparable to the ventral node in mammals, other groups of animals show similar structures relevant for symmetry breaking such as Kupffer's vesicle in *Teleost* fish,[19] Hensen's node in birds,[20] or the gastrocoel roof plate in amphibians.[21] Together, these embryonic structures are called the L-R organizer (LRO).[22] Importantly, it has been shown that symmetry breaking in the LRO depends on the directional flow generated by motile cilia in humans, rodents, amphibians, and teleost fish. In contrast, symmetry breaking in birds and reptiles does not depend on cilial flow.[17] These animals do not have motile cilia in the LRO. Instead, specific cells outside the LRO generate leftward flow. This shows that different evolutionary routes can lead to the same result: breaking symmetry.

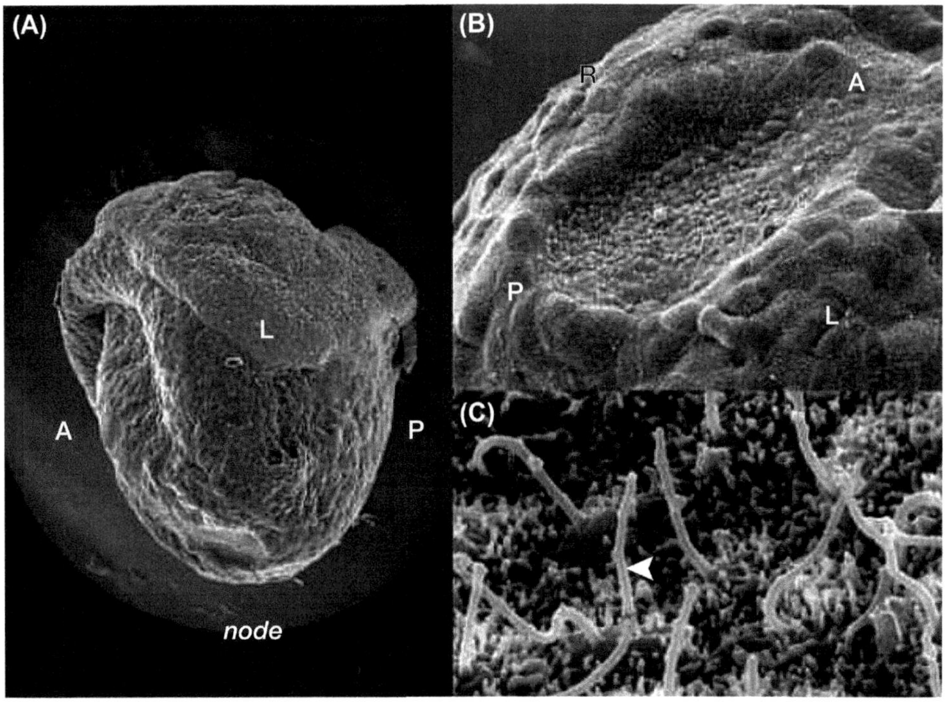

FIGURE 10.3 Scanning electron micrographs of the ventral node (A,B) and the cilia (C) in the brain of a mouse embryo. *A*, anterior; *P*, posterior; *L*, left; *R*, right.[14] *Figure reproduced with permission of the publisher from: Hamada H, Meno C, Watanabe D, Saijoh Y. Establishment of vertebrate left-right asymmetry.* Nat Rev Genet. *2002;3:103–113.*

From genes to lateralized behavior: what we can learn from zebrafish

Since the Nodal pathway is the most important signaling pathway for the establishment of bodily left-right asymmetries, it is logical to ask whether it not only affects the body but also the brain.[23,24] This can best be studied in a laboratory animal that breeds fast, shows an easily observable lateralized behavior, and offers diverse options to screen and manipulate genetic expression patterns. In this regard, the zebrafish is the ideal animal model.[25,26]

Most studies on the development of structural and functional asymmetries in zebrafish have focused on a brain structure called the epithalamus. The epithalamus constitutes the dorsal part of the diencephalon and contains the unpaired pineal complex and the bilateral habenular nuclei. The habenula and its associated fiber tracts form part of a conserved limbic conduction system that links the forebrain with both the ventral midbrain and a complex of serotonergic and dopaminergic midbrain nuclei.[27] As such, this system is involved in regulating all kinds of emotional behaviors related to pain, stress, and anxiety.[28] The past few years shed light on the function of this evolutionarily extremely old system by showing that the habenulae of monkeys are activated by stimuli that are associated with unpleasant events such as the absence of an expected reward or the occurrence of unpredictable punishment.[29] Interestingly, many animals freeze when something unpredictable and potentially dangerous happens.[30] According to Hikosaka,[31] the primary function of the habenula could be to suppress motor activity under such adverse conditions. Indeed, the lateral habenula can inhibit dopaminergic neurons which are key modulators of movement and motivation. This function of the habenula possibly dates back to very early evolutionary times when this pathway worked as the circadian system of ancient vertebrates. Indeed, the habenula is connected to the pineal gland, the so-called "third eye." In some fish, amphibians, and several reptile species, this gland is still photosensitive and signals day-night differences in luminance. It is conceivable that the habenula was once a brain system that could suppress motor activity based on day-night differences.[32] Only later did it change into a system that controls motivated behavior based on aversive events in a lateralized way.

As stated above, zebrafish are the ideal model species to inform us about the development of structural and functional asymmetries. In the following section, we will first review the general structural and connectional outline of the habenula system before discussing the ontogeny of its asymmetries.

The dorsal habenula of zebrafish can be divided into medial and lateral subnuclei.[33] Many connections within the epithalamic system are asymmetrically organized. For example, the right habenula receives privileged lateralized input from the anterior pallium and the olfactory bulb[34] (see Fig. 10.4). A similar asymmetry is also visible in the zebrafish pineal complex. This system is photosensitive and consists of the pineal and the parapineal organ. The pineal serves as a photosensitive clock that synchronizes physiological functions with the circadian cycle. The parapineal is situated left to the midline and projects only to the

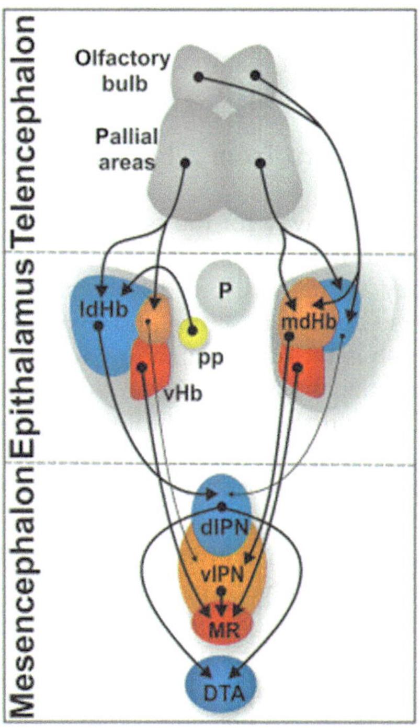

FIGURE 10.4 Schematic illustration of the zebrafish epithalamic system, its asymmetrical pallial input, and its lateralized projection to the mesencephalon.[35] The habenula of zebrafish is divided into a dorsal and a ventral component (vHb), with the dorsal habenula constituted by lateral (ldHb) and medial (mdHb) subnuclei of unequal sizes. Further components of the zebrafish epithalamus are the photosensitive pineal (P) and the asymmetrically organized parapineal organ (PP). ldHb and mdHb project to the dorsal (dIPN) and ventral (vIPN) nucleus interpeduncularis, respectively. Dorsal and ventral IPN project differentially to the dorsal tegmental area (DTA) and the serotonergic median raphe (MR), respectively.[35] *Reproduced with permission of the authors and the publisher from Güntürkün O, Ocklenburg S. Ontogenesis of lateralization.* Neuron. *2017;94:249–263.*

lateral subnucleus of the dorsal habenula on the left side.[36] Thus, the left and right habenula receive different input patterns and accordingly subserve different functions (Fig. 10.4).

These lateralized input patterns are also reflected in the physiological response patterns of habenula neurons. Left-sided neurons are mostly responsive to visual stimuli, while right habenula neurons are predominantly activated by olfactory cues.[37] If brain asymmetry is reversed, so is stimulus-selective asymmetry. Loss of asymmetry does not balance stimulus selectivity but produces a loss of responsibility for both visual and olfactory stimuli. Thus, functional properties of sensory processing[37] and fear-related behavior[38] in zebrafish depend on the presence of epithalamic connectional asymmetries.

About 24 h after fertilization, parapineal precursor cells of zebrafish embryos condense in the epithalamic midline and start to migrate toward the left side. There, they innervate the left habenula.[39] Without this unilateral migration, epithalamic asymmetry cannot be established.[40] Remember that both pineal and parapineal are photosensitive. Could light input during this critical period modify functional asymmetries throughout ontogeny? Budaev and Andrew demonstrated that the absence of light on day 1 after fertilization changes the preferred eye for predator inspection from left to right.[41] Dadda et al. utilized the fact that 10% of zebrafish larvae have a parapineal that aberrantly migrates to the right.[42] They discovered that animals with aberrant epithalamic asymmetries displayed different lateralized visual behaviors than animals with normal epithalamic asymmetry. This finding emphasizes that epithalamic asymmetry is directly linked to lateralized behavior.

The beauty of the zebrafish model system is the ease with which genetic factors can be analyzed. This is due to several reasons. On the one hand, zebrafish embryos are almost transparent, making it much easier to assess the development of organ systems than in most other vertebrates. Also, the phenotypic effects of many genetic mutations are well characterized. Furthermore, the genome of the zebrafish is fully sequenced. Techniques such as gene knockout—that is, a specific gene is made inoperative and the consequences of this on the phenotype are assessed—or gene overexpression—that is, expression of a specific gene is enhanced and the consequences of this on biological signaling cascades are assessed—can be administered reliably.[43] Genetic studies could show that the development of epithalamic asymmetries in zebrafish is regulated by at least four major genetic signaling cascades[44] (see Fig. 10.5):

1. The fibroblast growth factor (FGF) pathway: This is the very first genetic step for the onset of neuroanatomical asymmetries in zebrafish. FGF is a family of proteins that are involved in embryonic development, specifically in cell proliferation and differentiation. The FGF pathway induces the initial neuroanatomical symmetry break but is not relevant to the direction of the asymmetry. The FGF Fgf8 controls the initial migration of the parapineal from the midline to a lateralized position in the brain. Zebrafish *Fgf8* mutants do not develop epithalamic asymmetries.[45,46]
2. The Nodal pathway: In contrast to the FGF pathway, the Nodal pathway (see above) does not initiate the symmetry break in the parapineal but controls the direction of the asymmetry. If Nodal signaling is removed in *Fgf8* mutant zebrafish, local provision of Fgf8 leads to asymmetric migration of parapineal cells not to the left but toward the source of Fgf8 asymmetries.[46] There seems to be a division of labor: The FGF pathway induces asymmetry but without determining a direction. The Nodal pathway cannot induce the break of symmetry. However, once this first step is taken by the FGF pathway,

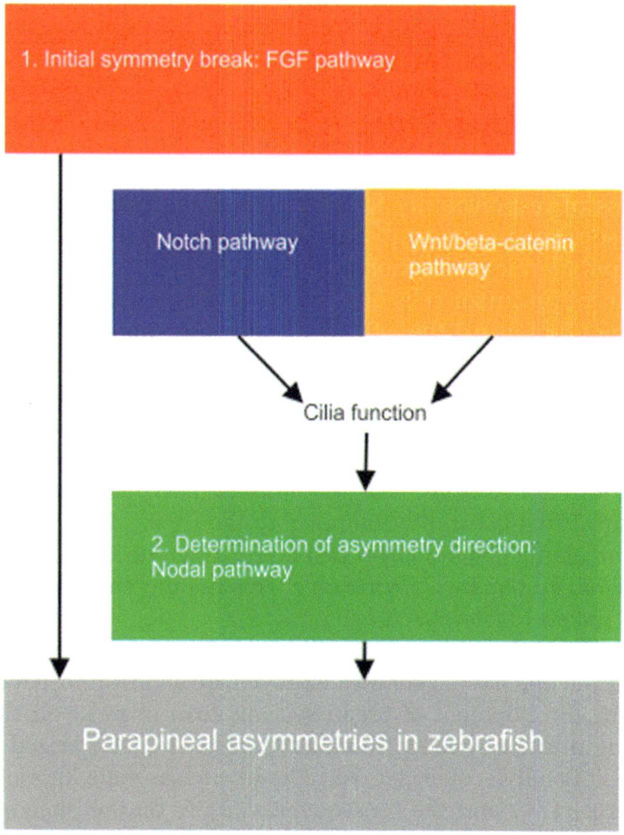

FIGURE 10.5 The four genetic signaling pathways involved in the formation of parapineal asymmetries in zebrafish.

the Nodal pathway skews the system to the left. Consequently, absence of Nodal pathway gene expression does not lead to neuroanatomical symmetry in the epithalamus. Epithalamic asymmetries are still established, but instead of the pronounced 90% leftward asymmetry found on the population level in wildtype fish, their direction is determined at random.[47,48] Taken together, FGF (asymmetry breaking) and Nodal (determining asymmetry direction) jointly pave the way for an asymmetrical zebrafish brain.

3. The Notch pathway: The Notch signaling pathway is a prerequisite for the proper functioning of the Nodal pathway. It plays a key role in development by managing interactions between adjacent cells, and it is involved in the control of cilia length.[49] Activity of the Notch pathway is necessary for Nodal expression and left-right asymmetries, and bilateral injection of Notch mRNA has been shown to induce bilateral expression of Nodal pathway genes which normally are only expressed in the left lateral plate mesoderm.[50,51]
4. The Wnt/beta-catenin pathway: Similar to the Notch pathway, the Wnt/beta-catenin pathway ("Wnt" stands for "wingless-related integration" site. The term "wingless" originates from the fact that one of the involved genes was first described in the *Drosophila Melanogaster* fly and has been shown to affect wing formation) is active in the lateral plate mesoderm before the first Nodal pathway genes are expressed, and it is

important for the proper functioning of the Nodal pathway, as it regulates ciliogenesis in zebrafish.[52] The pathway also later contributes to proper left-sided *Nodal* gene expression by repressing suppression of Nodal.[44]

After these genetic events have unfolded, the dorsal part of the habenula of young zebrafish is asymmetrically organized. This dorsal habenula consists of a lateral (ldHb) and a medial subnucleus (mdHb) that are grossly asymmetric concerning their sizes. While the left dorsal habenula has a large lateral and a small medial subnucleus, this pattern is inverted for the right dorsal habenula[39] (Fig. 10.4). Moreover, while the lateral subnucleus of the habenula targets the dorsal component of the nucleus interpeduncularis (dIPN) in the ventral mesencephalon, the medial habenula innervates the ventral component (vIPN). Due to the size asymmetry of lateral and medial habenular subnuclei, dIPN is mostly innervated from the left side, while vIPN receives input predominantly from the right habenula. This is a truly ingenious arrangement since left-right size asymmetries of the dorsal habenula are now converted to dorsoventral differences in the control of the interpeduncular nuclei.[53] As dorsal and ventral IPN project differentially to the dorsal tegmental area (DTA) and the serotonergic median raphe, respectively,[54] asymmetries of the habenula can modulate distinct downstream circuits with specific roles in emotional behavior.[40,55] Some of these findings can also shed new light on asymmetries in humans.

Human patients with primary ciliary dyskinesia, a disorder that impairs cilia function, have a 50% chance of situs inversus—a condition in which major visceral organs such as heart, stomach, and gut are mirror-reversed—possibly due to the absence of cilia movements and a subsequent lack of embryonal leftward flow of extracellular fluid at the node. Since these patients are mostly right-handed with left-hemispheric language dominance, it was concluded that Nodal pathway asymmetries cannot be the key factor for left-right differences of the human brain.[56] As shown by Barth et al.,[57] this is different in zebrafish. Situs inversus also exists in zebrafish. Another aspect of the beauty of this animal model is that scientists have developed a "frequent-situs-inversus" line of zebrafish, in which this condition occurs very often. Indeed, these animals show a coupling of visceral, habenulo-interpeduncular, and visual behavioral asymmetries. Thus, Nodal seems to orchestrate both visceral and some of the neural/behavioral asymmetries in zebrafish but not in humans. However, in zebrafish, two of the tested behavioral asymmetries do not reverse, suggesting that different parallel genetic routes influence functional lateralization during ontogeny.[57]

Specific manipulation of the lateralized epithalamic circuitry could provide further insight into behavioral lateralization. Inactivation of the ldHb induces freezing instead of flight when the animal is confronted with a fear stimulus.[58] Chou et al. went a step further by using agonistic encounters between dyads of male zebrafish as a behavioral model[55] (see Fig. 10.6). When male zebrafish are placed in one tank, they first start with display behavior and then launch biting attacks until one of the individuals surrenders. When a zebrafish had won the fight, Chou et al. observed higher activity levels in the ldHb→dIPN→DTA pathway (10.6A). In fish that had lost, there was a tendency for increased activation in vIPN. When the ldHb→dIPN pathway was experimentally silenced, the animals mostly lost fight encounters (10.6B). When the same manipulation was done with the mdHb→vIPN pathway, the animals turned into winners (10.6C). This differential effect was not due to fatigue or any other obvious behavioral difference between the two manipulations but

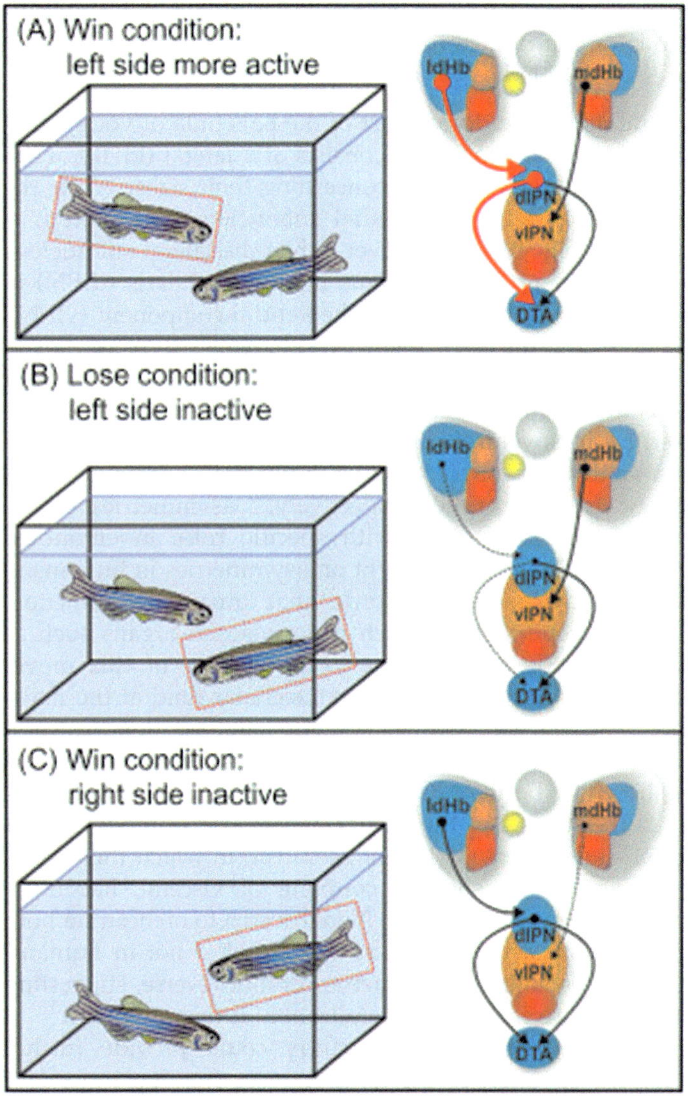

FIGURE 10.6 Habenular asymmetry modulates agonistic interactions in zebrafish.[55] (A) When a male zebrafish (the fish in the red frame) wins a fight against another male, its left ldHb → dlPN → DTA system is active. (B) Silencing this left pathway produces losers. (C) Silencing the right mdHb → vlPN results in a higher probability of winning a fight. *Based on Chou M-Y, Amo R, Kinoshita M, et al., Higashijima S-i, Okamoto H. Social conflict resolution regulated by two dorsal habenular subregions in zebrafish.* Science. *2016;352:87–90, with permission of AAAS. Zebrafish image: courtesy of Lizzy Griffiths.*

possibly resulted from differential resilience to the agonistic situation. Indeed, the DTA is known to play a key role in the binary switch between fight and flight.[59]

The reconstruction of these genetic, functional, and neuronal pathways in zebrafish, as well as their interactions with nongenetic factors, can shed light on the principles underlying ontogenetic mechanisms of brain asymmetries. In zebrafish, the involved anatomical system is an ancient pathway that connects the pallium via the diencephalon to the midbrain. The next model system that we will discuss is birds. Here, a minute light stimulation before hatching can ignite asymmetrical wiring of the visual system that also affects the telencephalon and creates left-right differences in cognition.

Avian lateralized visual cognition

As outlined in Chapter 2, the first discovery of reliable nonhuman asymmetries happened in the peripheral nervous system of the song pathways in chaffinches.[60] A few years later, Lesley Rogers demonstrated that 2-day-old chicks displayed profound asymmetry of visual function at the forebrain level.[2] For this, she used cycloheximide—a protein synthesis inhibitor that interferes with neural learning processes.[1] Injections of this drug into the left or right hemisphere showed that the learning speed in a visual discrimination task was significantly more affected by left-sided injections. In contrast, injections into the right hemisphere resulted in no performance impairment. Thus, the visual discrimination ability of chicks seems to be governed mainly by the left hemisphere.[2] Since the 1970s, subsequent studies with diverse avian species have revealed many further prominent examples of lateralization of visual processing. The left hemisphere excels in categorization, discrimination, and memorization of visual patterns,[35] as has been shown in chicks,[61] quails,[62] and pigeons.[63–65] Moreover, it excels in visually mediated compass information, as has been shown in chicks[66] and pigeons.[67–69] In contrast, the right hemisphere dominates visually guided interactions with emotionally charged stimuli (black-winged stilts,[70] quails,[71] and chicks[72]), attentional shifts (chicks and pigeons[73]), social interactions (chicks[74]), and relational analysis of pictorial and spatial information (chicks[75] and pigeons[65,76]).

These asymmetries also affect cognitive processes. A nice example is the mental number line where numbers are arranged along a continuum with smaller values on the left and larger ones on the right. Adult human subjects usually conceive a continuum of numbers from left to right, and it is still debated if this bias results from cultural experiences only or also has a biological basis.[77] The fact that newborn babies do the same makes a solely cultural interpretation very unlikely.[78] But what about nonhuman animals? Amazingly, both 3-day-old domestic chicks[79] as well as bees[80] choose the same left-to-right orientation. Results from chicks indicate that the tendency of the right hemisphere to allocate attention toward the left half-space possibly biases the starting point to the left, thereby resulting in a left-to-right direction of a sequence of ascending numerosity.[81]

The example with the mental number line nicely illustrates how diverse aspects of cognition can be affected by the asymmetry of the brain. This is also visible in categorization tasks with pigeons[65] (see Fig. 10.7). Here, birds were shown hundreds of photographs depicting everyday scenes with or without humans. Pigeons were rewarded for pecking at the pictures showing humans. Pecking on stimuli without humans was mildly punished. Soon the pigeons started to properly discriminate between the two photograph sets. Then the birds were confronted with novel pictures and had to apply their discrimination strategy to the new instances. As predicted, the pigeons spontaneously differentiated between the two new picture sets. Thus, the animals had acquired a general ability to categorize humans.

However, did the hemispheres solve this task by employing different strategies? To approach this question, the stimuli were modified in two different ways. In the first test, the stimuli were fractioned into increasingly smaller (Levels 1–6) squares that were arbitrarily scrambled (Fig. 10.8).

FIGURE 10.7 Pigeon working on a human/no-human categorization task. In the depicted trial, the bird is confronted with a novel photograph showing a birthday party with toddlers and is supposed to peck this picture to obtain a food reward from the food dispenser at the bottom.

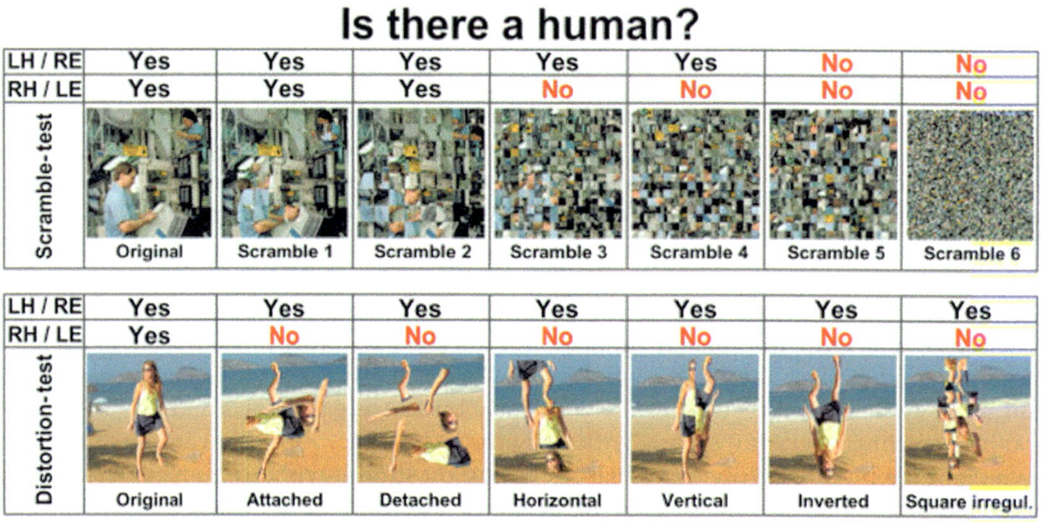

FIGURE 10.8 Hemispheric differences of visual categorization in pigeons as revealed by Yamazaki et al.[65] Pigeons had learned to discriminate pictures with and without humans. When being confronted with novel photographs, stimuli were altered either by increasingly scrambling them in six steps or by distorting the human image by six different procedures. The birds conducted this test either with the left hemisphere/right eye seeing (LH/RE) or with the right hemisphere/left eye (RH/LE). Scrambling affected LH/RE less, rendering it likely that the left hemisphere utilizes small pictorial features to solve the task. Every kind of distortion resulted in the RH/LE opting against the possibility that a human was depicted. Thus, the right hemisphere seems to rely on a configurational strategy to identify categorized objects.[35] *Reproduced with permission of the authors and the publisher from Güntürkün O, Ocklenburg S. Ontogenesis of lateralization.* Neuron. *2017;94:249–263.*

For Level 1, pictures were divided into 64 squares and only 8 were displaced. For Level 6, pictures were cut into 4096 squares that were all randomly rearranged. This procedure ensured that the birds had to use increasingly smaller features to decide between the presence or absence of humans. The pigeons worked on this task with eye caps that covered either the left or the right eye. Due to the complete crossover of the optic nerves, this ensures that only one hemisphere receives direct visual input. The result was clear-cut: Pigeons could tolerate increasing degrees of scrambling with the right eye (left hemisphere), while their performance quickly dropped down to chance level when categorizing even mildly scrambled pictures with the left eye (right hemisphere). Thus, the left hemisphere could discern diagnostic visual features in considerably smaller fragments than the right hemisphere. Since there were no acuity differences between the eyes, this difference had to result from differential specialization of the left and right visual systems.[82] In the second test, the pigeons were shown pictures in which the depicted humans were distorted in different ways such that the usual configuration of the human body was no longer given. In this test, performance of the left hemisphere was not affected, while the right hemisphere did not accept dismembered humans as positive stimuli. Thus, the right hemisphere used a configurational strategy to recognize humans. Taken together, visual asymmetry in birds reveals parallel and complementary functional specializations of the two hemispheres. But how does this lateralization develop? To understand this, we must travel back a century to the life and work of a scholar who pioneered behavioral epigenetics amidst the political turmoils in the Republic of China in the early 20th Century.[83]

How embryonic light asymmetry alters brain and behavior in birds

Already in 1932, Zing Yang Kuo had shown that birds take an asymmetrical position in the egg. The embryo is curled in such a way that the right eye is turned to the semitranslucent eggshell, while the left eye is covered by the embryo's body and yolk sac.[84] Due to this position, the right eye is stimulated by incident light, while the left is mostly light-deprived. During breeding, parental birds regularly turn their eggs and leave the nest for short periods of time.[4,85] Thus, the embryo's right eye is repeatedly stimulated by light. The lateralized position of the bird embryo is very likely mediated by the Nodal cascade since the processes underlying the asymmetrical positioning of the viscera are always accompanied by the embryonic torsion and a turn of the head to the right.[10,86] This occurs in all amniotes, including humans.[87]

The lateralized position of the avian embryo has an unexpected consequence: If chicken or pigeon eggs are incubated in darkness, the development of functional lateralization in object discrimination with dominance of the left hemisphere is prevented.[88,89] The same applies to other visually controlled lateralized behaviors such as aggression and copulation.[5] Embryonic[90] or posthatch[91] visual stimulation of the left eye can even reverse behavioral asymmetry in chicks and pigeons. Thus, normal rearing conditions correspond to right eye stimulation, resulting in left hemisphere superiority for visual object discrimination. This population bias is not genetically determined by factors within the visual system but by the lateralized environmental factor light that results from the genetically determined body position. Indeed, Lorenzi et al.[92] showed in a manganese-enhanced magnetic resonance imaging study that light stimulation before hatch induces asymmetrical brain activity in the thalamus and the forebrain Wulst of embryonic chicks.[92] The resulting asymmetry in visual object discrimination is mediated through activity differences between left- and right-retinal ganglion cells. Synaptic maturation of visual pathways is regulated by retinal activity and differential retinal gene expressions.[93] Transiently blocking right-eye retinal activity in pigeons reverses visual lateralization for the entire life.[67] Lateralized retinal activation asymmetrically regulates tectal neurons in the midbrain, which in turn seem to release brain-derived neurotrophic factor (BDNF) differentially on the left and the right side.[94,95] Possibly, BDNF is just one of many biochemical pathways that translate a transient embryonic visual stimulation asymmetry into structural left-right differences of the ascending visual system which ultimately determine lateralized visual object discrimination.

However, when talking about pigeons that categorized pictures into those that depicted and those that did not depict humans, we outlined that avian visual lateralization has a complementary organization, with the left and the right hemisphere specialized for processing different aspects of the image. As it turns out, the biased embryonic photic input only activates object discrimination asymmetries but does not affect right-hemispheric superiorities that mediate visual reactions to novelty,[96] social recognition,[97,98] or unilateral eye use during sleep.[99] Therefore, the neuronal effects of lateralized embryonic visual stimulation only mediate some of the left-hemispheric visually guided functions.

This is also true for a behavioral avian asymmetry that we discussed in Chapter 7. Here, we outlined that when pigeons and chicks are confronted with an area with lots of grains, they peck significantly more often at grains on the left side.[73] Such a visuospatial attentional bias is also evident in humans. Since the last common ancestor of birds and mammals lived about 324 million years ago, this is a truly ancient lateralized system.[100] In chicks, this left-sided attentional bias is affected by embryonic light stimulation.[101] Therefore, Sara Letzner and coworkers tested if pigeons without embryonic light stimulation cease to have an attentional bias to the left hemispace.[102] To this end, they analyzed the attentional direction of adult pigeons with and without embryonic light stimulation. As expected, light-exposed pigeons displayed a preference to first peck into the left hemifield. However, dark-incubated pigeons behaved unexpectedly: They preferred the right hemifield, thereby showing that pigeons without embryonic light experience hatch with a left hemispheric dominance for visuospatial attention. Light incubation thus does not induce a nonexistent asymmetry. The visuospatial asymmetry was already present with a right hemifield bias. What light incubation did was to shift visuospatial attentional asymmetry from the right toward the left hemifield. Such results demonstrate that embryonic light experience in pigeons profoundly alters cerebral asymmetries in the developing brain by inducing left-right patterns in some neural systems or reversing a preexisting bias in other pathways.

Let us close this part with a small gift to all those readers who love dinosaur stories. As most of them might know, birds are living dinosaurs.[103] Against this background, a paper by Xing et al.[104] reported a remarkable discovery.[104] The authors analyzed the exquisitely fossilized egg of a late Cretaceous (94–66 million years ago) oviraptorid theropod dinosaur embryo and revealed that the head of the embryo was positioned ventral to the body with the back curled in a posture very similar to modern bird embryos before hatch (Fig. 10.9). Unfortunately, the dinosaur specimen lacks the right forelimb and thus the

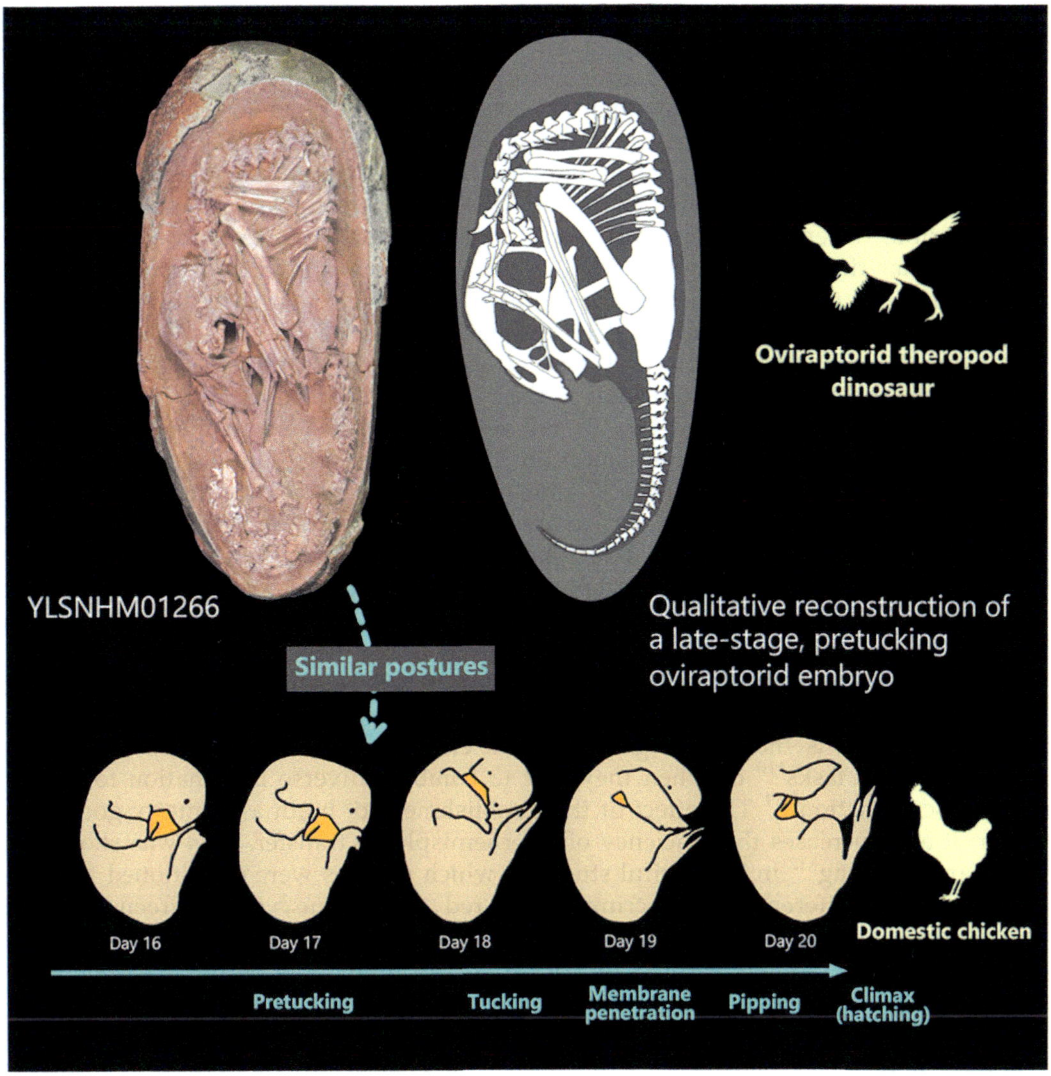

FIGURE 10.9 Fossil condition and reconstruction of an oviraptorid theropod dinosaur embryo from the late Cretaceous period that is exquisitely preserved in-ovo. Its curled position is like that of a chicken embryo. Since the embryonic position of chickens and pigeons results in biased prehatch light input and brain asymmetries, it is conceivable that similar conditions occurred in theropod dinosaurs.[104] *Reproduced with permission of the publisher from Xing, L., Niu, K., Ma, W., Zelenitsky, D. K., Yang, T. R., & Brusatte, S. L. (2021). An exquisitely preserved in-ovo theropod dinosaur embryo sheds light on avian-like prehatching postures.* iScience*;25:103516.*

exact position of the head relative to the missing limb cannot be finally decided. However, despite this limitation, the overall pattern opens the tantalizing possibility that theropod dinosaurs might also have had modern bird-like asymmetries of their visual and cognitive system that were ignited by biased light input before hatch.

In today's birds, there seem to be two different critical time points at which light can affect the onset of brain asymmetry in the avian embryo. Chiandetti et al. stimulated chicken eggs with light for 3 days after fertilization and subsequently kept them in the dark.[105] Usually, one would expect that this early light input has no effect since during this developmental period retinal photoreceptors are not yet functional.[106] However, to everyone's amazement, chicks hatched under these early light-stimulated conditions behaved in a detour task like birds that were light-stimulated toward the end of the incubation period—when photoreceptors were already functional. A subsequent study yielded similar findings and could show that this pattern differs from dark-incubated chicks.[107] How is this possible? First, it is important to note that early light-stimulated chicks are not in all aspects identical to late light-stimulated ones.[5,107,108] Hence, early light exposure has overlapping but also distinct effects from late light exposure. However, the most important question is this: How can light modify the brain of a 3-day-old embryo when photoreceptors are not yet functional? There are a couple of possibilities. It is known that light stimulation on day 3 of chicken embryogenesis results in some activity increase in the embryo,[4,61] suggesting the presence of some light-sensitive structures at this early stage. But where are they? One could be the pineal gland that was discussed earlier in the context of zebrafish asymmetry or the epiphysis-habenula route.[107] The others are deep photoreceptors embedded in the hypothalamus.[109] None of these systems are known to affect lateralized behavior in birds, but maybe it is worth doing the critical experiments.

In Chapter 2, we described that asymmetry pays. If this is true, should dark-incubated birds not show reduced performance compared to light-incubated, and thus lateralized, ones? Indeed, this is the case. Light-incubated chicks outperformed those hatched in darkness in dual tasks[110] or when they had to combine diverse information to master correct discrimination.[111] The effect of the establishment of brain asymmetry goes even further: It also increases the efficiency of interhemispheric transfer. This was shown by Manns and Römling[112] in a beautiful study, in which pigeons were conditioned to separately learn four different color discriminations, red > blue; blue > green; green > yellow; yellow > pink, leading them to prefer red over blue, blue over green, and so forth. The trick was that the animals learned the first two discriminations with their left hemisphere (seeing with their right eye) and the last two with their right hemisphere (seeing with their left eye). Subsequently, the pigeons performed these color discriminations with both eyes seeing—and thus with both hemispheres participating. From time to time, novel pairings like blue versus yellow were given. How would the pigeons decide? The correct decision had to be based on the transitive inference of the complete color sequence: red > blue > green > yellow > pink. To do this, however, the animals had to integrate the two different color discriminations learned with the left (red > blue; blue > green) with those learned with the right hemisphere (green > yellow; yellow > pink). Then, and only then, could the pigeons integrate all the information to conclude by transitive inference logic that blue was the correct color. Indeed, only light-incubated—and thus lateralized—pigeons could solve the task. This finding indicates that the efficiency of interhemispheric transfer depends on the proper establishment of brain asymmetries (Fig. 10.10). This finding will be of importance further below.

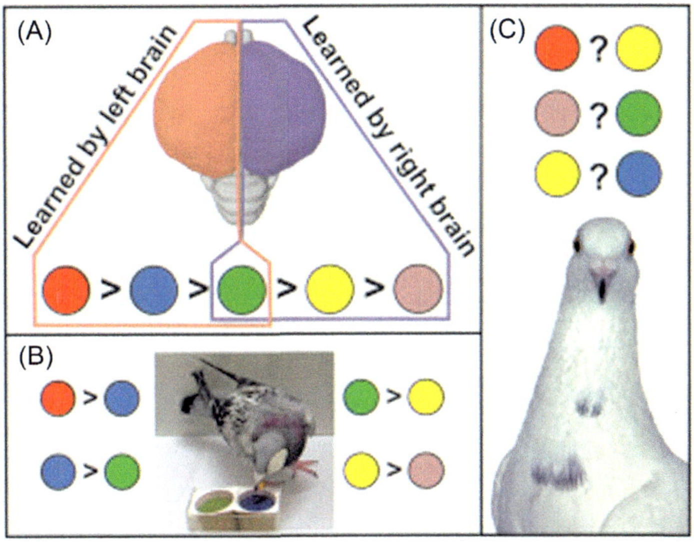

FIGURE 10.10 Test for interhemispheric transfer using the transitive inference task in pigeons. (A) Schematic depiction of the experimental rationale. The pigeons had to learn two different color discriminations with each hemisphere by means of eye caps. They had to choose red over blue with the left hemisphere (shown in red), yellow over pink with the right hemisphere (colored in blue), and so forth. Both hemispheres learned something about green; the left one to select blue over green, and the right chose green over yellow. (B) Depiction of the testing situation along with the pairs of colors that were trained with each hemisphere. The birds had to dig through grit of a certain color to find the covered food under pebbles of the correct color. (C) Finally, the birds were tested without eye caps (both hemispheres participate). Some of the test combinations are shown. They all required the ability to integrate information across hemispheres. The most critical pair was blue versus yellow since the reward history was equal for these colors.[112] *Based on Manns M, Römling J. The impact of asymmetrical light input on cerebral hemispheric specialization and interhemispheric cooperation.* Nat Commun. *2012;3:696, with permission from publisher.*

An asymmetrical anatomical system creates lateralized visual behavior

Visual information reaches the avian forebrain by two parallel ascending systems—the tectofugal and the thalamofugal pathways—which are homologous to the mammalian extrageniculocortical and geniculocortical systems, respectively. In pigeons, the tectofugal system is the most important pathway for visually guided behavior (Fig. 10.11). It runs from the retina via the optic tectum to the thalamic nucleus rotundus (Rt), which in turn projects to the entopallium in the telencephalon.[114] As outlined in detail below, the pigeon's tectofugal system shows multiple anatomical and physiological left-right differences that establish functional visual asymmetry. For example, the projection from the right midbrain tectum opticum (TO) to the left Rt is stronger than the projection from the left TO to the right Rt,[115] enabling integration of visual input from both eyes on the left but not on the right side.[116]

In contrast, visual asymmetry in chicken mostly depends on the thalamofugal system that runs from the retina to the contralateral nucleus geniculatus lateralis pars dorsalis (Gld) and from there bilaterally to a structure called "Wulst" in the telencephalon. In chicks, the

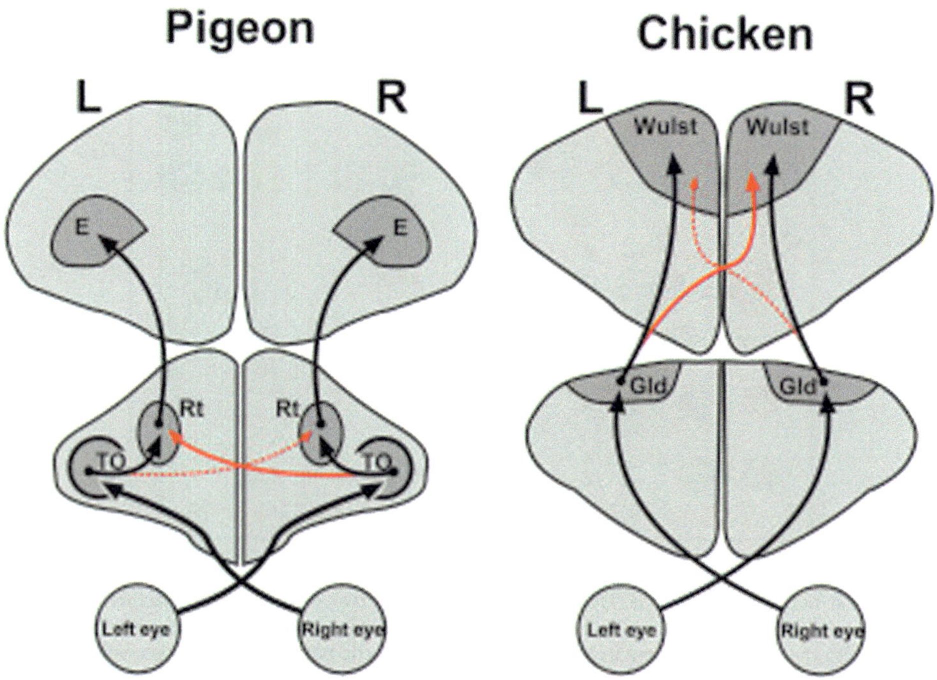

FIGURE 10.11 The tectofugal pathway of pigeons (left) and the thalamofugal pathway of juvenile chicken (right) are asymmetrically organized. In pigeons, the projection from right tectum opticum (TO) to left nucleus rotundus (Rt) is stronger than that from left TO to right Rt. In chicken, the contralateral projection from the left nucleus geniculatus lateralis pars dorsalis (Gld) to the right visual Wulst is stronger than that from the right Gld to the left visual Wulst. Contralateral projections within both pathways are depicted in red, with weak fiber systems marked by thin, dotted lines.[113] *E*, entopallium; *L*, left; *R*, right. *Based on Ströckens F, Freund N, Manns M, Ocklenburg S, Güntürkün O. Visual asymmetries and the ascending thalamofugal pathway in pigeons.* Brain Struct Funct. *2013;218:1197–1209, with permission from the publisher.*

contralateral projections from left Gld to the right Wulst are stronger than projections from the right Gld to the left visual Wulst.[117,118] This anatomical thalamofugal asymmetry only persists for about 3 weeks after hatch,[118] but as outlined above, results in a wide range of left-right differences of visually guided behavior.[119] Costalunga et al.[120] showed that incubation of chicken eggs under light or dark conditions had profound effects on Wulst neurons.[120] First, cells in both hemispheres were affected in different ways by embryonic light. Second, light exposure increased the light sensitivity of Wulst neurons, thereby promoting the maturation of the visual system. Third, multiple Wulst asymmetries were present in dark incubated animals, underlining the point that light incubation modifies asymmetries without constituting them. This finding is in line with the findings outlined above and shows that biased sensory stimulation during neural development alters and partly induces asymmetries by modifying both left- and right-hemispheric properties.[121]

All of these findings in chicken were confined to the thalamofugal system, and for a long time, it was thought that the tectofugal pathway in chicken is not asymmetrically organized.[122] Recently, Morandi-Raikova et al.[123] reported that dark-incubated chicks evince a higher density of possibly GABAergic parvalbumin-expressing neurons in the right entopallium.[123] Thus, also chicken may have a minor anatomical asymmetry in their tectofugal pathway.

Taken together, in pigeons and chicken, visual asymmetry looks deceptively similar when looking at ontogenetic mechanisms and behavioral effects.[35,124–126] However, the neural pathways that constitute these asymmetries differ (see Fig. 10.11). What is the possible reason? In a series of experiments, Ströckens et al.[113] painstakingly tested all conceivable explanations. In the end, only mundane species-difference remained as an explanation. This sobering finding yields an important take-home message: Two related species can be nearly identical in their lateralized behavior and in the ontogenetic conditions of the emergence of brain asymmetry. Yet, the fundamental brain systems subserving lateralized behavior can still be different. In other words, different neural pathways seem to be able to produce very similar behavior.

How commissural exchange modifies asymmetries

As outlined in Chapter 3, commissural exchange plays a key role for brain asymmetries. However, there is no consensus if this is achieved via inhibitory or excitatory commissural mechanisms.[128] A study in pigeons proposes a possibly smarter mechanism that could enable hemispheres to swiftly change between interhemispheric competition or cooperation.

The tectofugal visual pathway projects via a series of neural structures to the arcopallium, a motor area, from which descending fibers activate the birds' intended response. But the arcopallium also gives rise to the commissura anterior, which is the largest pathway for interhemispheric crosstalk in the avian forebrain.[129] To analyze the commissural mechanisms, Xiao and Güntürkün[127] recorded from left and right arcopallia while pigeons worked on a color discrimination task.[127] As expected, the birds responded faster when using their right eye (left hemisphere). Single cell recordings showed that left and right arcopallium neurons responded equally fast to the correct color, but strangely right arcopallial neurons needed more time to then activate the operant response. A closer analysis showed that most right hemispheric arcopallium neurons reached their maximal firing frequency after the left hemisphere had already executed the response. Thus, right-sided neurons consistently came too late to control the behavior of the animal (see Fig. 10.12).

To understand the reason for this disadvantage of the right arcopallium, the authors selectively anesthetized the left or the right arcopallium, while recording from the other side. By this, they wanted to see what arcopallium neurons do when freed from the commissural input of the competing hemisphere. Would the right arcopallium be faster when now no longer inhibited by the left hemisphere? This was not the case. Instead, anesthesia of the left arcopallium just drastically increased the variance of the timepoint of the maximal firing frequency of right arcopallial neurons. In other words, some right arcopallium neurons became faster, others slower. Thus, under normal conditions (without anesthesia), the dominant left hemisphere had tightly coordinated the cellular peak activities of the right arcopallium, thereby shifting them to a time point that was too late to decide the animal's response. The right hemisphere tried to do the same but was by far less successful in slowing down left arcopallial neurons.

Could these findings shed light on the possible role of commissural mechanisms for functional asymmetries? Commissural systems must enable cooperation between

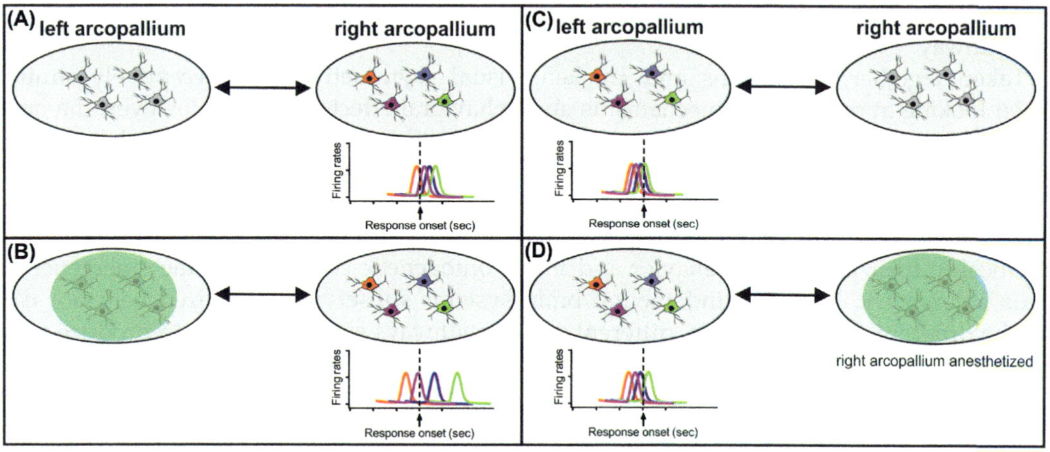

FIGURE 10.12 Schematic depiction of the asymmetries of temporal dynamics of arcopallial neurons that are modified by the contralateral hemisphere via the commissura anterior. (A) Under normal conditions of a color discrimination task, most right arcopallial neurons reach their peak activity after the onset of the animals' response. Thus, they come too late to control the animals' behavior, which is mostly governed by the left hemisphere. (B) If the dominant left arcopallium is temporarily anesthetized, peak times of right arcopallium neurons evince a higher variance, without much change in the latency of the population. (C) During color discrimination, left arcopallial neurons reach their peak activity before the onset of the animals' response and are thus able to execute the operant response. (D) If the right arcopallium is temporarily anesthetized, peak times of left arcopallium neurons are hardly affected. *Based on Xiao Q, Güntürkün O. Asymmetrical Commissural Control of the Subdominant Hemisphere in Pigeons. Cell Rep. 2018;25(5):1171–1180.e3. https://doi.org/10.1016/j.celrep.2018.10.011 but no direct reproduction from the paper.*

hemispheres and recruit bilateral hemispheric resources when jointly working on demanding tasks.[130] However, they should also be able to inhibit the other side under conditions of hemispheric competition.[131] Therefore, commissural exchange should be able to task dependently and flexibly switch back and forth between cooperation and competition. The results by Xiao & Güntürkün[127] offer a solution to this requirement. If hemispheres can modify the peak activity times of neurons of the other hemisphere, they can work in synchrony when cooperating or try to slow down the other hemisphere under conditions of competition. When competing, the hemisphere that is dominant for a certain task can win the race by being more successful in slowing down the subdominant side. Such a mechanism implies the ability to activate or dampen spike patterns of the other hemisphere, depending on task demands.

Anatomical and physiological asymmetries of visual pathways in pigeons

Let us now try to outline how structure and physiology of an asymmetrical neural system could generate lateralized behavior of object discrimination. Based on pigeon findings, there seem to be five main mechanisms (see Fig. 10.13):

1. Anatomical studies have demonstrated that the brainstem and thalamic components of the tectofugal system are characterized by diverse morphological asymmetries

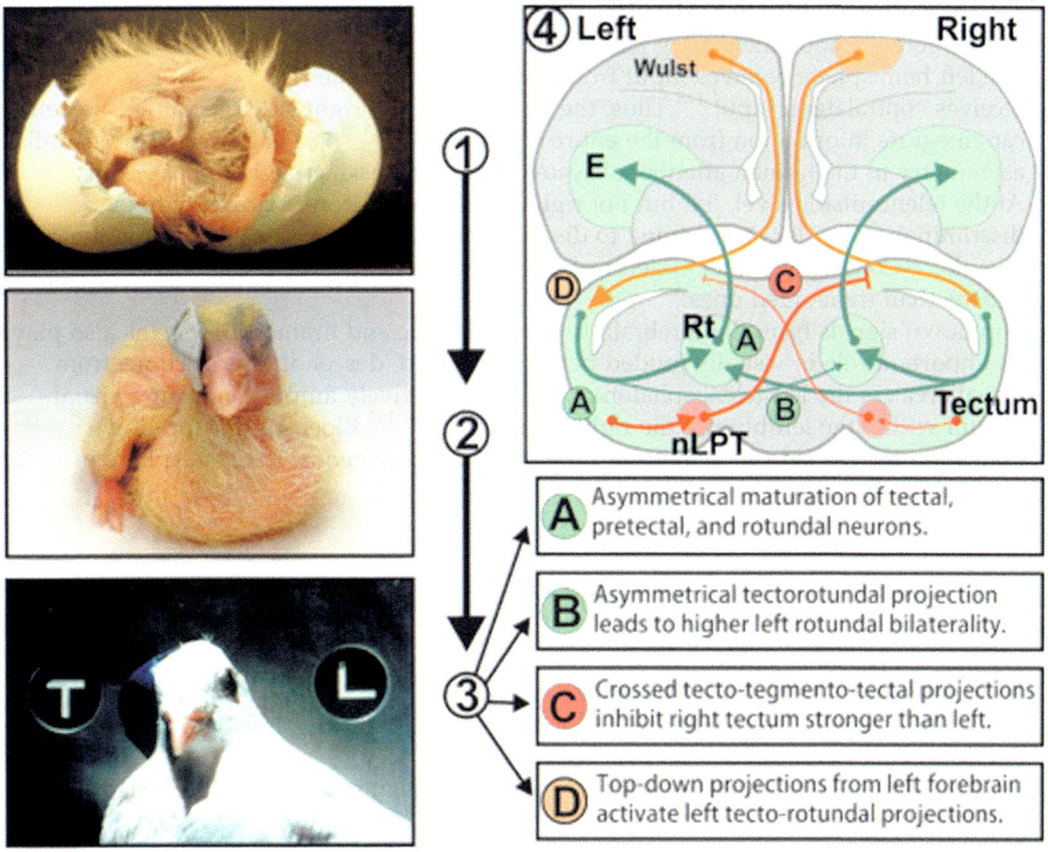

FIGURE 10.13 Ontogeny of visual asymmetry in pigeons. (1) Asymmetrical position of a pigeon hatchling. The animal is situated with its right eye toward the eggshell, while the left eye is covered by the body. As a result, the right eye is stimulated by incident light in the days before hatch, resulting in higher right retinal activity and lateralized release of BDNF in the developing visual system. (2) 2-day-old pigeon that undergoes right eye light deprivation. This procedure can reverse asymmetry in brain and behavior. (3) Adult pigeon with eye cap during a pattern discrimination task. Its behavioral asymmetry is related to four asymmetrical properties of the tectofugal system and associated pathways (A–D). (4) Schematic diagram of a frontal section through the pigeon brain depicting the tectofugal visual system (in green).[35] nLPT, the nucleus of the lateral pontomesencephalic tegmentum. *Reproduced with permission from the authors and the publisher from Güntürkün O, Ocklenburg S. Ontogenesis of lateralization.* Neuron *2017;94:249–263.*

that possibly modify the way visual object information is handled at hemispheric level.[91,95,132–134]

2. Commissural fibers between the tecta are asymmetrically organized. At the midbrain level, the left tectum inhibits its right counterpart to a greater extent than vice versa.[135,136] As a consequence, visual asymmetry drastically changes when the midbrain commissure is cut.[137] In addition, the left hemisphere controls via the commissura anterior the timing of right-sided arcopallial neurons. Thereby, the left hemisphere can delay the activity of right hemispheric neurons such that they come too late to

determine the animal's behavior. However, they could also synchronize with the other hemisphere to cooperate during joint tasks.[127]

3. The left hemisphere receives input from both eyes, while the right hemisphere mainly receives contralateral input.[115] Thus, the left- but not the right-sided tectofugal system can integrate information from the entire visual field.[116,138] We have described a similar asymmetry in the human attentional system when discussing neglect (Chapter 7).
4. At the telencephalic level, left but not right tectofugal neurons reach very high discrimination levels when having to discriminate between rewarded and nonrewarded stimuli.[139] Consequently, left-sided tectofugal lesions result in more severe visual deficits than right-sided ones.[140]
5. Top-down signals from the forebrain toward thalamic and brainstem systems also play an important role in visually guided behavior. Indeed, descending projections from visual areas in the left telencephalon of pigeons selectively amplify incoming visual stimuli within the left but not the right visual field.[141,142] Blocking this top-down system on the left side abolishes the superiority of the pigeons' left hemisphere in visual decision tasks.[142]

Sample questions

- What is the Nodal signaling cascade and how does it affect left-right axis differentiation in the developing vertebrate embryo?
- Name and describe the role of the four major genetic pathways involved in the development of parapineal asymmetries in zebrafish.
- Shortly outline the asymmetry of the zebrafish epithalamus and how it translates into lateralized behavior.
- Explain how visual asymmetries in birds emerge. Do not forget that there are two critical time periods.
- Describe the behavioral, visually driven asymmetries between the left and the right hemispheres in birds.
- How can commissures switch between a mode of cooperation and a mode of competition with the other hemisphere?

References

1. Rogers LJ, Drennen HD. Cycloheximide interacts with visual input to produce permanent slowing of visual learning in chickens. *Brain Res*. 1978;158(2):479–482. Available from: https://doi.org/10.1016/0006-8993(78)90694-7.
2. Rogers LJ, Anson JM. Lateralisation of function in the chicken fore-brain. *Pharmacology Biochem Behav*. 1979; 10(5):679–686. Available from: https://doi.org/10.1016/0091-3057(79)90320-4.
3. Denenberg VH, Garbanati J, Sherman DA, Yutzey DA, Kaplan R. Infantile stimulation induces brain lateralization in rats. *Science*. 1978;201(4361):1150–1152. Available from: https://doi.org/10.1126/science.684436.
4. Freeman BM, Vince MA. *Development of the Avian Embryo: A Behavioural and Physiological Study*. Springer Netherlands; 1974.
5. Rogers LJ. Light experience and asymmetry of brain function in chickens. *Nature*. 1982;297(5863):223–225. Available from: https://doi.org/10.1038/297223a0.

6. Schier AF, Shen MM. Nodal signalling in vertebrate development. *Nature*. 2000;403(6768):385–389. Available from: https://doi.org/10.1038/35000126.
7. Hackett BP. *Formation and Malformation of the Vertebrate Left-Right Axis*. Bentham Science Publishers; 2002.
8. Grimes DT. Making and breaking symmetry in development, growth and disease. *Development*. 2019;146(16). Available from: https://doi.org/10.1242/dev.170985.
9. Hamada H. Molecular and cellular basis of left-right asymmetry in vertebrates. *Proc Jpn Acad Ser B Phys Biol Sci*. 2020;96(7):273–296. Available from: https://doi.org/10.2183/pjab.96.021.
10. Levin M, Johnson RL, Stern CD, Kuehn M, Tabin C. A molecular pathway determining left-right asymmetry in chick embryogenesis. *Cell*. 1995;82(5):803–814. Available from: https://doi.org/10.1016/0092-8674(95)90477-8.
11. Little RB, Norris DP. Right, left and cilia: how asymmetry is established. *Semin Cell Dev Biol*. 2021;110:11–18. Available from: https://doi.org/10.1016/j.semcdb.2020.06.003.
12. Schweickert A, Ott T, Kurz S, et al. Vertebrate left-right asymmetry: what can nodal cascade gene expression patterns tell us? *J Cardiovasc Dev Dis*. 2017;5(1). Available from: https://doi.org/10.3390/jcdd5010001.
13. Norris DP. Cilia, calcium and the basis of left-right asymmetry. *BMC Biol*. 2012;10(1):102. Available from: https://doi.org/10.1186/1741-7007-10-102.
14. Hamada H, Meno C, Watanabe D, Saijoh Y. Establishment of vertebrate left-right asymmetry. *Nat Rev Genet*. 2002;3(2):103–113. Available from: https://doi.org/10.1038/nrg732.
15. Komatsu Y, Mishina Y. Establishment of left-right asymmetry in vertebrate development: the node in mouse embryos. *Cell Mol Life Sci*. 2013;70(24):4659–4666. Available from: https://doi.org/10.1007/s00018-013-1399-9.
16. Katoh TA, Omori T, Mizuno K, et al. Immotile cilia mechanically sense the direction of fluid flow for left-right determination. *Science*. 2023;379(6627):66–71. Available from: https://doi.org/10.1126/science.abq8148.
17. Hamada H, Tam P. Diversity of left-right symmetry breaking strategy in animals. *F1000Res*. 2020;9. Available from: https://doi.org/10.12688/f1000research.21670.1.
18. Fujinaga M. Development of sidedness of asymmetric body structures in vertebrates. *Int J Dev Biol*. 1997; 41(2):153–186.
19. Matsui T, Bessho Y. Left-right asymmetry in zebrafish. *Cell Mol Life Sci*. 2012;69(18):3069–3077. Available from: https://doi.org/10.1007/s00018-012-0985-6.
20. Katsu K, Tokumori D, Tatsumi N, Suzuki A, Yokouchi Y. BMP inhibition by DAN in Hensen's node is a critical step for the establishment of left-right asymmetry in the chick embryo. *Dev Biol*. 2012;363(1):15–26. Available from: https://doi.org/10.1016/j.ydbio.2011.12.015.
21. Walentek P, Schneider I, Schweickert A, Blum M. Wnt11b is involved in cilia-mediated symmetry breakage during Xenopus left-right development. *PLoS One*. 2013;8(9):e73646. Available from: https://doi.org/10.1371/journal.pone.0073646.
22. Szenker-Ravi E, Ott T, Khatoo M, et al. Discovery of a genetic module essential for assigning left-right asymmetry in humans and ancestral vertebrates. *Nat Genet*. 2022;54(1):62–72. Available from: https://doi.org/10.1038/s41588-021-00970-4.
23. Halpern ME, Liang JO, Gamse JT. Leaning to the left: laterality in the zebrafish forebrain. *Trends Neurosci*. 2003;26(6):308–313. Available from: https://doi.org/10.1016/S0166-2236(03)00129-2.
24. Liang JO, Etheridge A, Hantsoo L, et al. Asymmetric nodal signaling in the zebrafish diencephalon positions the pineal organ. *Development*. 2000;127(23):5101–5112. Available from: https://doi.org/10.1242/dev. 127.23.5101.
25. Facchin L, Burgess HA, Siddiqi M, Granato M, Halpern ME. Determining the function of zebrafish epithalamic asymmetry. *Philos Trans R Soc Lond B Biol Sci*. 2009;364(1519):1021–1032. Available from: https://doi.org/10.1098/rstb.2008.0234.
26. Jones WD, Mullins MC. Cell signaling pathways controlling an axis organizing center in the zebrafish. *Curr Top Dev Biol*. 2022;150:149–209. Available from: https://doi.org/10.1016/bs.ctdb.2022.03.005.
27. Herkenham M, Nauta WJ. Efferent connections of the habenular nuclei in the rat. *J Comp Neurol*. 1979; 187(1):19–47. Available from: https://doi.org/10.1002/cne.901870103.
28. Lecourtier L, Kelly PH. A conductor hidden in the orchestra? Role of the habenular complex in monoamine transmission and cognition. *Neurosci Biobehav Rev*. 2007;31(5):658–672. Available from: https://doi.org/10.1016/j.neubiorev.2007.01.004.
29. Matsumoto M, Hikosaka O. Lateral habenula as a source of negative reward signals in dopamine neurons. *Nature*. 2007;447(7148):1111–1115. Available from: https://doi.org/10.1038/nature05860.

30. Bracha HS. Freeze, flight, fight, fright, faint: adaptationist perspectives on the acute stress response spectrum. *CNS Spectr*. 2004;9(9). Available from: https://doi.org/10.1017/s1092852900001954.
31. Hikosaka O. The habenula: from stress evasion to value-based decision-making. *Nat Rev Neurosci*. 2010;11 (7):503–513. Available from: https://doi.org/10.1038/nrn2866.
32. Tavakoli-Nezhad M, Schwartz WJ. Hamsters running on time: is the lateral habenula a part of the clock? *Chronobiol Int*. 2006;23(1–2):217–224. Available from: https://doi.org/10.1080/07420520500521947.
33. Amo R, Aizawa H, Takahoko M, et al. Identification of the zebrafish ventral habenula as a homolog of the mammalian lateral habenula. *J Neurosci*. 2010;30(4):1566–1574. Available from: https://doi.org/10.1523/JNEUROSCI.3690-09.2010.
34. Miyasaka N, Morimoto K, Tsubokawa T, Higashijima S, Okamoto H, Yoshihara Y. From the olfactory bulb to higher brain centers: genetic visualization of secondary olfactory pathways in zebrafish. *J Neurosci*. 2009; 29(15):4756–4767. Available from: https://doi.org/10.1523/JNEUROSCI.0118-09.2009.
35. Güntürkün O, Ocklenburg S. Ontogenesis of lateralization. *Neuron*. 2017;94(2):249–263. Available from: https://doi.org/10.1016/j.neuron.2017.02.045.
36. Gamse JT, Kuan Y-S, Macurak M, et al. Directional asymmetry of the zebrafish epithalamus guides dorsoventral innervation of the midbrain target. *Development*. 2005;132(21):4869–4881. Available from: https://doi.org/10.1242/dev.02046.
37. Dreosti E, Vendrell Llopis N, Carl M, Yaksi E, Wilson SW. Left-right asymmetry is required for the habenulae to respond to both visual and olfactory stimuli. *Curr Biol*. 2014;24(4):440–445. Available from: https://doi.org/10.1016/j.cub.2014.01.016.
38. Facchin L, Duboué ER, Halpern ME. Disruption of epithalamic left-right asymmetry increases anxiety in zebrafish. *J Neurosci*. 2015;35(48):15847–15859. Available from: https://doi.org/10.1523/JNEUROSCI.2593-15.2015.
39. Roussigné M, Blader P, Wilson SW. Breaking symmetry: the zebrafish as a model for understanding left-right asymmetry in the developing brain. *Dev Neurobiol*. 2012;72(3):269–281. Available from: https://doi.org/10.1002/dneu.20885.
40. Concha ML, Bianco IH, Wilson SW. Encoding asymmetry within neural circuits. *Nat Rev Neurosci*. 2012; 13(12):832–843. Available from: https://doi.org/10.1038/nrn3371.
41. Budaev S, Andrew RJ. Patterns of early embryonic light exposure determine behavioural asymmetries in zebrafish: a habenular hypothesis. *Behavioural Brain Res*. 2009;200(1):91–94. Available from: https://doi.org/10.1016/j.bbr.2008.12.030.
42. Dadda M, Domenichini A, Piffer L, Argenton F, Bisazza A. Early differences in epithalamic left-right asymmetry influence lateralization and personality of adult zebrafish. *Behavioural Brain Res*. 2010;206(2):208–215. Available from: https://doi.org/10.1016/j.bbr.2009.09.019.
43. Howe K, Clark MD, Torroja CF, et al. The zebrafish reference genome sequence and its relationship to the human genome. *Nature*. 2013;496(7446):498–503. Available from: https://doi.org/10.1038/nature12111.
44. Hüsken U, Carl M. The Wnt/beta-catenin signaling pathway establishes neuroanatomical asymmetries and their laterality. *Mech Dev*. 2013;130(6–8):330–335. Available from: https://doi.org/10.1016/j.mod.2012.09.002.
45. Neugebauer JM, Yost HJ. FGF signaling is required for brain left-right asymmetry and brain midline formation. *Dev Biol*. 2014;386(1):123–134. Available from: https://doi.org/10.1016/j.ydbio.2013.11.020.
46. Regan JC, Concha ML, Roussigné M, Russell C, Wilson SW. An Fgf8-dependent bistable cell migratory event establishes CNS asymmetry. *Neuron*. 2009;61(1):27–34. Available from: https://doi.org/10.1016/j.neuron.2008.11.030.
47. Concha ML, Burdine RD, Russell C, Schier AF, Wilson SW. A nodal signaling pathway regulates the laterality of neuroanatomical asymmetries in the zebrafish forebrain. *Neuron*. 2000;28(2):399–409. Available from: https://doi.org/10.1016/S0896-6273(00)00120-3.
48. Concha ML, Wilson SW. Asymmetry in the epithalamus of vertebrates. *J Anat*. 2001;199(Pt 1–2):63–84. Available from: https://doi.org/10.1046/j.1469-7580.2001.19910063.x.
49. Lopes SS, Lourenço R, Pacheco L, Moreno N, Kreiling J, Saúde L. Notch signalling regulates left-right asymmetry through ciliary length control. *Development*. 2010;137(21):3625–3632. Available from: https://doi.org/10.1242/dev.054452.
50. Raya A, Kawakami Y, Rodriguez-Esteban C, et al. Notch activity induces Nodal expression and mediates the establishment of left-right asymmetry in vertebrate embryos. *Genes Dev*. 2003;17(10):1213–1218. Available from: https://doi.org/10.1101/gad.1084403.

51. Takeuchi JK, Lickert H, Bisgrove BW, et al. Baf60c is a nuclear Notch signaling component required for the establishment of left-right asymmetry. *Proc Natl Acad Sci U S A*. 2007;104(3):846–851. Available from: https://doi.org/10.1073/pnas.0608118104.
52. Zhu P, Xu X, Lin X. Both ciliary and non-ciliary functions of Foxj1a confer Wnt/β-catenin signaling in zebrafish left-right patterning. *Biol Open*. 2015;4(11):1376–1386. Available from: https://doi.org/10.1242/bio.012088.
53. Aizawa H, Bianco IH, Hamaoka T, et al. Laterotopic representation of left-right information onto the dorso-ventral axis of a zebrafish midbrain target nucleus. *Curr Biol*. 2005;15(3):238–243. Available from: https://doi.org/10.1016/j.cub.2005.01.014.
54. Beretta CA, Dross N, Guiterrez-Triana JA, Ryu S, Carl M. Habenula circuit development: past, present, and future. *Front Neurosci*. 2012;6:51. Available from: https://doi.org/10.3389/fnins.2012.00051.
55. Chou M-Y, Amo R, Kinoshita M, et al. Social conflict resolution regulated by two dorsal habenular subregions in zebrafish. *Science*. 2016;352(6281):87–90. Available from: https://doi.org/10.1126/science.aac9508.
56. Kennedy DN, O'Craven KM, Ticho BS, Goldstein AM, Makris N, Henson JW. Structural and functional brain asymmetries in human situs inversus totalis. *Neurology*. 1999;53(6):1260–1265. Available from: https://doi.org/10.1212/WNL.53.6.1260.
57. Barth KA, Miklosi A, Watkins J, Bianco IH, Wilson SW, Andrew RJ. Fsi zebrafish show concordant reversal of laterality of viscera, neuroanatomy, and a subset of behavioral responses. *Curr Biol*. 2005;15(9):844–850. Available from: https://doi.org/10.1016/j.cub.2005.03.047.
58. Agetsuma M, Aizawa H, Aoki T, et al. The habenula is crucial for experience-dependent modification of fear responses in zebrafish. *Nat Neurosci*. 2010;13(11):1354–1356. Available from: https://doi.org/10.1038/nn.2654.
59. Siegel A, Bhatt S, Bhatt R, Zalcman SS. The neurobiological bases for development of pharmacological treatments of aggressive disorders. *Curr Neuropharmacol*. 2007;5(2):135–147. Available from: https://doi.org/10.2174/157015907780866929.
60. Nottebohm F. Ontogeny of bird song. *Science*. 1970;167(3920):950–956. Available from: https://doi.org/10.1126/science.167.3920.950.
61. Rogers LJ. Asymmetry of brain and behavior in animals: its development, function, and human relevance. *Genesis*. 2014;52(6):555–571. Available from: https://doi.org/10.1002/dvg.22741.
62. Valenti A, Sovrano VA, Zucca P, Vallortigara G. Visual lateralisation in quails (*Coturnix coturnix*). *Laterality*. 2003;8(1):67–78. Available from: https://doi.org/10.1080/713754470.
63. Güntürkün O. Evidence for a third primary visual area in the telencephalon of the pigeon. *Brain Res*. 1984; 294(2):247–254. Available from: https://doi.org/10.1016/0006-8993(84)91036-9.
64. von Fersen L, Güntürkün O. Visual memory lateralization in pigeons. *Neuropsychologia*. 1990;28(1):1–7. Available from: https://doi.org/10.1016/0028-3932(90)90081-x.
65. Yamazaki Y, Aust U, Huber L, Hausmann M, Güntürkün O. Lateralized cognition: asymmetrical and complementary strategies of pigeons during discrimination of the "human concept". *Cognition*. 2007;104(2):315–344. Available from: https://doi.org/10.1016/j.cognition.2006.07.004.
66. Rogers LJ. Development and function of lateralization in the avian brain. *Brain Res Bull*. 2008;76(3):235–244. Available from: https://doi.org/10.1016/j.brainresbull.2008.02.001.
67. Prior H, Wiltschko R, Stapput K, Güntürkün O, Wiltschko W. Visual lateralization and homing in pigeons. *Behavioural Brain Res*. 2004;154(2):301–310. Available from: https://doi.org/10.1016/j.bbr.2004.02.018.
68. Ulrich C. Left-hemispheric superiority for visuospatial orientation in homing pigeons. *Behavioural Brain Res*. 1999;104(1-2):169–178. Available from: https://doi.org/10.1016/s0166-4328(99)00062-5.
69. Wiltschko W, Traudt J, Güntürkün O, Prior H, Wiltschko R. Lateralization of magnetic compass orientation in a migratory bird. *Nature*. 2002;419(6906):467–470. Available from: https://doi.org/10.1038/nature00958.
70. Ventolini N, Ferrero EA, Sponza S, Della Chiesa A, Zucca P, Vallortigara G. Laterality in the wild: preferential hemifield use during predatory and sexual behaviour in the black-winged stilt. *Anim Behav*. 2005;69 (5):1077–1084. Available from: https://doi.org/10.1016/j.anbehav.2004.09.003.
71. Gülbetekin E, Güntürkün O, Dural S, Cetinkaya H. Visual asymmetries in Japanese quail (*Coturnix japonica*) retain a lifelong potential for plasticity. *Behav Neurosci*. 2009;123(4):815–821. Available from: https://doi.org/10.1037/a0016406.
72. Vallortigara G, Chiandetti C, Sovrano VA. Brain asymmetry (animal). *Wiley Interdiscip Rev Cogn Sci*. 2011; 2(2):146–157. Available from: https://doi.org/10.1002/wcs.100.

73. Diekamp B, Regolin L, Güntürkün O, Vallortigara G. A left-sided visuospatial bias in birds. *Curr Biol.* 2005; 15(10):R372–R373. Available from: https://doi.org/10.1016/j.cub.2005.05.017.
74. Vallortigara G, Andrew RJ. Differential involvement of right and left hemisphere in individual recognition in the domestic chick. *Behavioural Process.* 1994;33(1-2):41–57. Available from: https://doi.org/10.1016/0376-6357(94)90059-0.
75. Vallortigara G, Pagni P, Sovrano VA. Separate geometric and non-geometric modules for spatial reorientation: evidence from a lopsided animal brain. *J Cogn Neurosci.* 2004;16(3):390–400. Available from: https://doi.org/10.1162/089892904322926737.
76. Martinho A, Biro D, Guilford T, Gagliardo A, Kacelnik A. Asymmetric visual input and route recapitulation in homing pigeons. *Proc Biol Sci.* 2015;282(1816):20151957. Available from: https://doi.org/10.1098/rspb.2015.1957.
77. Dehaene S, Bossini S, Giraux P. The mental representation of parity and number magnitude. *J Exp Psychol Gen.* 1993;122(3):371–396. Available from: https://doi.org/10.1037/0096-3445.122.3.371.
78. Hevia MDD, Veggiotti L, Streri A, Bonn CD. At birth, humans associate "few" with left and "many" with right. *Curr Biol.* 2017;27(24):3879–3884.e2. Available from: https://doi.org/10.1016/j.cub.2017.11.024.
79. Rugani R, Vallortigara G, Priftis K, Regolin L. Animal cognition. Number-space mapping in the newborn chick resembles humans' mental number line. *Science.* 2015;347(6221):534–536. Available from: https://doi.org/10.1126/science.aaa1379.
80. Giurfa M, Marcout C, Hilpert P, Thevenot C, Rugani R. An insect brain organizes numbers on a left-to-right mental number line. *PNAS.* 2022;119(44). Available from: https://doi.org/10.1073/pnas.2203584119. e2203584119.
81. Rugani R, Vallortigara G, Regolin L. Mapping number to space in the two hemispheres of the avian brain. *Neurobiol Learn Mem.* 2016;133:13–18. Available from: https://doi.org/10.1016/j.nlm.2016.05.010.
82. Güntürkün O, Hahmann U. Visual acuity and hemispheric asymmetries in pigeons. *Behavioural Brain Res.* 1994;60(2):171–175. Available from: https://doi.org/10.1016/0166-4328(94)90144-9.
83. Blowers GH. To be a big shot or to be shot: Zing-Yang Kuo's other career. *Hist Psychol.* 2001;4(4):367–387. Available from: https://doi.org/10.1037/1093-4510.4.4.367.
84. Kuo ZY. Ontogeny of embryonic behavior in Aves. I. The chronology and general nature of the behavior of the chick embryo. *J Exp Zool.* 1932;61(3):395–430. Available from: https://doi.org/10.1002/jez.1400610304.
85. Buschmann J-UF, Manns M, Güntürkün O. "Let There be Light!" pigeon eggs are regularly exposed to light during breeding. *Behavioural Process.* 2006;73(1):62–67. Available from: https://doi.org/10.1016/j.beproc.2006.03.012.
86. Ramsdell AF, Yost H. Molecular mechanisms of vertebrate left–right development. *Trends Genet.* 1998;14(11):459–465. Available from: https://doi.org/10.1016/s0168-9525(98)01599-6.
87. Dunsirn S, Smyser C, Liao S, Inder T, Pineda R. Defining the nature and implications of head turn preference in the preterm infant. *Early Hum Dev.* 2016;96:53–60. Available from: https://doi.org/10.1016/j.earlhumdev.2016.02.002.
88. Deng C, Rogers LJ. Factors affecting the development of lateralisation in chicks. In: Rogers LJ, Andrew RJ, eds. *Comparative Vertebrate Lateralization.* Cambridge University Press; 2002:206–246.
89. Skiba M, Diekamp B, Güntürkün O. Embryonic light stimulation induces different asymmetries in visuoperceptual and visuomotor pathways of pigeons. *Behavioural Brain Res.* 2002;134(1-2):149–156. Available from: https://doi.org/10.1016/s0166-4328(01)00463-6.
90. Rogers LJ. Light input and the reversal of functional lateralization in the chicken brain. *Behavioural Brain Res.* 1990;38(3):211–221. Available from: https://doi.org/10.1016/0166-4328(90)90176-f.
91. Manns M, Güntürkün O. Monocular deprivation alters the direction of functional and morphological asymmetries in the pigeon's (*Columba livia*) visual system. *Behav Neurosci.* 1999;113(6):1257–1266. Available from: https://doi.org/10.1037/0735-7044.113.6.1257.
92. Lorenzi E, Tambalo S, Vallortigara G, Bifone A. Manganese-enhanced magnetic resonance imaging reveals light-induced brain asymmetry in embryo. *Elife.* 2023;12. Available from: https://doi.org/10.7554/eLife.86116.
93. Versace E, Sgadò P, George J, et al. Light-induced asymmetries in embryonic retinal gene expression are mediated by the vascular system and extracellular matrix. *Sci Rep.* 2022;12(1):12086. Available from: https://doi.org/10.1038/s41598-022-14963-8.
94. Manns M, Güntürkün O. Differential effects of ocular BDNF-injections onto the development of tectal cells characterized by calcium-binding proteins in pigeons. *Brain Res Bull.* 2005;66(4-6):475–478. Available from: https://doi.org/10.1016/j.brainresbull.2005.03.003.

95. Manns M, Güntürkün O, Heumann R, Blöchl A. Photic inhibition of TrkB/Ras activity in the pigeon's tectum during development: impact on brain asymmetry formation. *Eur J Neurosci*. 2005;22(9):2180–2186. Available from: https://doi.org/10.1111/j.1460-9568.2005.04410.x.
96. Chiandetti C, Regolin L, Rogers LJ, Vallortigara G. Effects of light stimulation of embryos on the use of position-specific and object-specific cues in binocular and monocular domestic chicks (*Gallus gallus*). *Behavioural Brain Res*. 2005;163(1):10–17. Available from: https://doi.org/10.1016/j.bbr.2005.03.024.
97. Andrew RJ, Johnston ANB, Robins A, Rogers LJ. Light experience and the development of behavioural lateralisation in chicks. II. Choice of familiar versus unfamiliar model social partner. *Behavioural Brain Res*. 2004;155(1):67–76. Available from: https://doi.org/10.1016/j.bbr.2004.04.016.
98. Vallortigara G, Andrew RJ. Lateralization of response by chicks to change in a model partner. *Anim Behav*. 1991;41(2):187–194. Available from: https://doi.org/10.1016/s0003-3472(05)80470-1.
99. Mascetti GG, Vallortigara G. Why do birds sleep with one eye open? Light exposure of the chick embryo as a determinant of monocular sleep. *Curr Biol*. 2001;11(12):971–974. Available from: https://doi.org/10.1016/s0960-9822(01)00265-2.
100. dos Reis M, Thawornwattana Y, Angelis K, Telford MJ, Donoghue PCJ, Yang Z. Uncertainty in the timing of origin of animals and the limits of precision in molecular timescales. *Curr Biol*. 2015;25(22):2939–2950. Available from: https://doi.org/10.1016/j.cub.2015.09.066.
101. Chiandetti C. Pseudoneglect and embryonic light stimulation in the avian brain. *Behav Neurosci*. 2011;125 (5):775–782. Available from: https://doi.org/10.1037/a0024721.
102. Letzner S, Güntürkün O, Lor S, Pawlik RJ, Manns M. Visuospatial attention in the lateralised brain of pigeons - a matter of ontogenetic light experiences. *Sci Rep*. 2017;7(1):15547. Available from: https://doi.org/10.1038/s41598-017-15796-6.
103. Nesbitt SJ. The early evolution of archosaurs: relationships and the origin of major clades. *Bull Am Mus Nat History*. 2011;352:1–292. Available from: https://doi.org/10.1206/352.1.
104. Xing L, Niu K, Ma W, Zelenitsky DK, Yang T-R, Brusatte SL. An exquisitely preserved in-ovo theropod dinosaur embryo sheds light on avian-like prehatching postures. *iScience*. 2022;25(1):103516. Available from: https://doi.org/10.1016/j.isci.2021.103516.
105. Chiandetti C, Galliussi J, Andrew RJ, Vallortigara G. Early-light embryonic stimulation suggests a second route, via gene activation, to cerebral lateralization in vertebrates. *Sci Rep*. 2013;3(1):2701. Available from: https://doi.org/10.1038/srep02701.
106. Tomonari S, Takagi A, Akamatsu S, Noji S, Ohuchi H. A non-canonical photopigment, melanopsin, is expressed in the differentiating ganglion, horizontal, and bipolar cells of the chicken retina. *Dev Dyn*. 2005;234(3):783–790. Available from: https://doi.org/10.1002/dvdy.20600.
107. Chiandetti C, Vallortigara G. Distinct effect of early and late embryonic light-stimulation on chicks' lateralization. *Neuroscience*. 2019;414:1–7. Available from: https://doi.org/10.1016/j.neuroscience.2019.06.036.
108. Zappia JV, Rogers LJ. Light experience during development affects asymmetry of forebrain function in chickens. *Dev Brain Res*. 1983;11(1):93–106. Available from: https://doi.org/10.1016/0165-3806(83)90204-3.
109. Nakane Y, Shimmura T, Abe H, Yoshimura T. Intrinsic photosensitivity of a deep brain photoreceptor. *Curr Biol*. 2014;24(13):R596–R597. Available from: https://doi.org/10.1016/j.cub.2014.05.038.
110. Rogers LJ, Zucca P, Vallortigara G. Advantages of having a lateralized brain. *Proc Biol Sci*. 2004;271(Suppl 6): S420–S422. Available from: https://doi.org/10.1098/rsbl.2004.0200.
111. Chiandetti C, Vallortigara G. Effects of embryonic light stimulation on the ability to discriminate left from right in the domestic chick. *Behavioural Brain Res*. 2009;198(1):240–246. Available from: https://doi.org/10.1016/j.bbr.2008.11.018.
112. Manns M, Römling J. The impact of asymmetrical light input on cerebral hemispheric specialization and interhemispheric cooperation. *Nat Commun*. 2012;3(1):696. Available from: https://doi.org/10.1038/ncomms1699.
113. Ströckens F, Freund N, Manns M, Ocklenburg S, Güntürkün O. Visual asymmetries and the ascending thalamofugal pathway in pigeons. *Brain Struct Funct*. 2013;218(5):1197–1209. Available from: https://doi.org/10.1007/s00429-012-0454-x.
114. Mouritsen H, Heyers D, Güntürkün O. The neural basis of long-distance navigation in birds. *Annu Rev Physiol*. 2016;78:133–154. Available from: https://doi.org/10.1146/annurev-physiol-021115-105054.

115. Güntürkün O, Hellmann B, Melsbach G, Prior H. Asymmetries of representation in the visual system of pigeons. *NeuroReport*. 1998;9(18):4127–4130. Available from: https://doi.org/10.1097/00001756-199812210-00023.
116. Letzner S, Patzke N, Verhaal J, Manns M. Shaping a lateralized brain: asymmetrical light experience modulates access to visual interhemispheric information in pigeons. *Sci Rep*. 2014;4(1):4253. Available from: https://doi.org/10.1038/srep04253.
117. Rogers LJ, Bolden SW. Light-dependent development and asymmetry of visual projections. *Neurosci Lett*. 1991;121(1-2):63–67. Available from: https://doi.org/10.1016/0304-3940(91)90650-i.
118. Rogers LJ, Sink HS. Transient asymmetry in the projections of the rostral thalamus to the visual hyperstriatum of the chicken, and reversal of its direction by light exposure. *Exp Brain Res*. 1988;70(2):378–384. Available from: https://doi.org/10.1007/bf00248362.
119. Vallortigara G, Rogers LJ. A function for the bicameral mind. *Cortex*. 2020;124:274–285. Available from: https://doi.org/10.1016/j.cortex.2019.11.018.
120. Costalunga G, Kobylkov D, Rosa-Salva O, Vallortigara G, Mayer U. Light-incubation effects on lateralisation of single unit responses in the visual Wulst of domestic chicks. *Brain Struct Funct*. 2022;227(2):497–513. Available from: https://doi.org/10.1007/s00429-021-02259-y.
121. Manns M. It is not just in the genes. *Symmetry*. 2021;13(10):1815. Available from: https://doi.org/10.3390/sym13101815.
122. Rogers LJ, Deng C. Light experience and lateralization of the two visual pathways in the chick. *Behavioural Brain Res*. 1999;98(2):277–287. Available from: https://doi.org/10.1016/s0166-4328(98)00094-1.
123. Morandi-Raikova A, Danieli K, Lorenzi E, Rosa-Salva O, Mayer U. Anatomical asymmetries in the tectofugal pathway of dark-incubated domestic chicks: rightwards lateralization of parvalbumin neurons in the entopallium. *Laterality*. 2021;26(1-2):163–185. Available from: https://doi.org/10.1080/1357650X.2021.1873357.
124. Güntürkün O. Avian visual lateralization: a review. *Neuroreport: An Int J Rapid Commun Res Neurosci*. 1997;8(6):iii–xi.
125. Güntürkün O, Ströckens F, Ocklenburg S. Brain lateralization: a comparative perspective. *Physiol Rev*. 2020;100(3):1019–1063. Available from: https://doi.org/10.1152/physrev.00006.2019.
126. Rogers LJ, Munro U, Freire R, Wiltschko R, Wiltschko W. Lateralized response of chicks to magnetic cues. *Behavioural Brain Res*. 2008;186(1):66–71. Available from: https://doi.org/10.1016/j.bbr.2007.07.029.
127. Xiao Q, Güntürkün O. Asymmetrical commissural control of the subdominant hemisphere in pigeons. *Cell Rep*. 2018;25(5):1171–1180, e3. Available from: https://doi.org/10.1016/j.celrep.2018.10.011.
128. Bloom JS, Hynd GW. The role of the corpus callosum in interhemispheric transfer of information: excitation or inhibition? *Neuropsychol Rev*. 2005;15(2):59–71. Available from: https://doi.org/10.1007/s11065-005-6252-y.
129. Letzner S, Simon A, Güntürkün O. Connectivity and neurochemistry of the commissura anterior of the pigeon (*Columba livia*). *J Comp Neurol*. 2016;524(2):343–361. Available from: https://doi.org/10.1002/cne.23858.
130. Weissman DH, Banich MT. The cerebral hemispheres cooperate to perform complex but not simple tasks. *Neuropsychology*. 2000;14(1):41–59. Available from: https://doi.org/10.1037//0894-4105.14.1.41.
131. Putnam MC, Wig GS, Grafton ST, Kelley WM, Gazzaniga MS. Structural organization of the corpus callosum predicts the extent and impact of cortical activity in the nondominant hemisphere. *J Neurosci*. 2008;28(11):2912–2918. Available from: https://doi.org/10.1523/JNEUROSCI.2295-07.2008.
132. Güntürkün O. Morphological asymmetries of the tectum opticum in the pigeon. *Exp Brain Res*. 1997;116(3):561–566. Available from: https://doi.org/10.1007/pl00005785.
133. Manns M, Güntürkün O. Natural' and artificial monocular deprivation effects on thalamic soma sizes in pigeons. *NeuroReport*. 1999;10(15):3223. Available from: https://doi.org/10.1097/00001756-199910190-00018.
134. Manns M, Güntürkün O. Light experience induces differential asymmetry pattern of GABA- and parvalbumin-positive cells in the pigeon's visual midbrain. *J Chem Neuroanat*. 2003;25(4):249–259. Available from: https://doi.org/10.1016/s0891-0618(03)00035-8.
135. Keysers C, Diekamp B, Güntürkün O. Evidence for physiological asymmetries in the intertectal connections of the pigeon (*Columba livia*) and their potential role in brain lateralisation. *Brain Res*. 2000;852(2):406–413. Available from: https://doi.org/10.1016/s0006-8993(99)02192-7.
136. Stacho M, Letzner S, Theiss C, Manns M, Güntürkün O. A GABAergic tecto-tegmento-tectal pathway in pigeons. *J Comp Neurol*. 2016;524(14):2886–2913. Available from: https://doi.org/10.1002/cne.23999.

137. Güntürkün O, Böhringer PG. Lateralization reversal after intertectal commissurotomy in the pigeon. *Brain Res*. 1987;408(1-2):1–5. Available from: https://doi.org/10.1016/0006-8993(87)90351-9.
138. Valencia-Alfonso C-E, Verhaal J, Güntürkün O. Ascending and descending mechanisms of visual lateralization in pigeons. *Philos Trans R Soc Lond B Biol Sci*. 2009;364(1519):955–963. Available from: https://doi.org/10.1098/rstb.2008.0240.
139. Verhaal J, Kirsch JA, Vlachos I, Manns M, Güntürkün O. Lateralized reward-related visual discrimination in the avian entopallium. *Eur J Neurosci*. 2012;35(8):1337–1343. Available from: https://doi.org/10.1111/j.1460-9568.2012.08049.x.
140. Güntürkün O, Hahmann U. Functional subdivisions of the ascending visual pathways in the pigeon. *Behavioural Brain Res*. 1999;98(2):193–201. Available from: https://doi.org/10.1016/s0166-4328(98)00084-9.
141. Folta K, Diekamp B, Güntürkün O. Asymmetrical modes of visual bottom-up and top-down integration in the thalamic nucleus rotundus of pigeons. *J Neurosci*. 2004;24(43):9475–9485. Available from: https://doi.org/10.1523/JNEUROSCI.3289-04.2004.
142. Freund N, Valencia-Alfonso CE, Kirsch J, Brodmann K, Manns M, Güntürkün O. Asymmetric top-down modulation of ascending visual pathways in pigeons. *Neuropsychologia*. 2016;83:37–47. Available from: https://doi.org/10.1016/j.neuropsychologia.2015.08.014.

CHAPTER

11

Hemispheric asymmetries over the lifespan: development and aging

OUTLINE

Introductory short story

Willow yawned. She was about to scan the brain of volunteer number 22 for the second project of her Ph.D. thesis using the 3-Tesla MRI scanner in the university hospital's basement. As always, she had first given the young woman named Abigail the participant information sheet to read through. After Willow had answered Abigail's question about the

The Lateralized Brain
DOI: https://doi.org/10.1016/B978-0-323-99737-9.00005-7

project, Abigail signed the informed consent form, filled out the Edinburgh Handedness Inventory and other laterality questionnaires, and then was ready to go into the scanner room. Willow asked her to empty her pockets and get rid of all metallic objects, as they may cause harm due to the strong magnetic field in the scanner room. After Abigail had removed her smartphone, keys, and all jewelry she had worn, Willow brought her into the scanner room, asked her to lie down on the scanner bed, and smoothly moved her into the humming machine. Next, Willow made herself comfortable in front of the MRI scanner's control computer and started the usual protocol. Absolutely nothing out of the ordinary had happened, and Willow was a bit bored. As an undergraduate student, she had always thought that neuroimaging had to be the most fascinating thing in the world. What could be more exciting than looking into people's brains and recording their thoughts? Now she was not so sure about that anymore. The day-to-day scanning routine was not all that exciting at all. Like always, her MRI scanning session started with a so-called localizer scan, a very short anatomical sequence that was important to localize the exact position of her volunteer's brain inside the MRI scanner. Willow had done this dozens of times before, but it looked different this time. Could this be? She grabbed the phone and called James, the scanner technician. "Hi James, can you come over to the scanner? I've got something here I need you to look at."

James was there about 2 min later. In disbelief, Willow and James looked at the computer screen with the localizer scan of Abigail's brain. Her right hemisphere looked perfectly normal, but where the left hemisphere should have been there was just an empty black space. Willow had never seen anything like it. She looked at James. "Could this be a technical mistake? Something wrong with the magnetic field of the scanner?" James shook his head. "This is remarkable, I also haven't seen anything like that, but I don't think this is a technical issue. I think she is just missing her complete left hemisphere." Willow looked at him. "But how can she speak, or even walk or use her right arm if she is missing her complete left hemisphere? I have seen stroke victims with small lesions in the left hemisphere who had severe language problems, and she seemed to speak perfectly normal." James suggested they call Inez, the neurologist who was on duty to review the scans and talk to volunteers about incidental findings that may be pathological. Inez came quickly, and Willow got Abigail out of the MRI scanner. "Is everything alright?" she asked, looking surprised. Inez looked at her and asked, "We noticed something unusual in your brain scans. Have you by any chance had any brain surgery before?" Abigail appeared a bit flustered. "Oh no! I completely forgot to tell you about it! I am so, so sorry! When I was a little child, I had these horrible seizures and none of the medications the doctors gave me helped at all. So, at some point, they decided to remove some part of my brain and the seizures stopped. In the beginning, I had some problems with walking and talking, but over the years it got much better and now I barely notice it, except sometimes I get these bad headaches." Inez smiled and looked at Willow. "The young lady is one of the few people who underwent a so-called hemispherectomy, a removal of one of the brain's hemispheres to stop treatment-resistant epilepsy. The fact that she has recovered so well that she sometimes forgets that she has lost half her brain shows you just how remarkable brain plasticity can be. The human brain truly is marvelous!"

Brain development and hemispheric asymmetries

Brain development in humans starts with the differentiation of neuronal progenitor cells 3 weeks after gestation.[1] The next 2 years are characterized by massive changes in neural structures until the human brain gains a shape and connectivity pattern that is largely like that of adults. However, this is not the end of brain development. The period around puberty and the time up to early adulthood are also characterized by important neural alterations. Moreover, our brain also transforms as we age. Our brain never completely stops changing so we can rightly say that brain development starts shortly after gestation and ends with death. Changes in neural structure and function happen throughout our entire lifetime. Many of these alterations affect hemispheric asymmetries. In the following chapter, we will highlight how developmental processes from pre- and postnatal early life events to aging affect hemispheric asymmetries.

Hemispherectomy: what happens if you lose half a brain?

In the preceding chapters, many forms of hemispheric asymmetries were described, in which one hemisphere is dominant for a specific cognitive function like language, fine motor behavior, face recognition, and so on. However, what happens if someone completely loses one hemisphere of their brain like the patient in the introductory short story? Are people always unable to speak without a left hemisphere, or always unable to recognize faces without a right hemisphere? The answers to these questions depend largely on the age of the individual at the time when they lost the hemisphere. The general rule is that the younger someone is, the more able they will be to compensate for some functions of the lost half of the brain. Fascinating insights into the effects of losing a hemisphere have been provided by studies in patients who underwent a so-called hemispherectomy.[2] Hemispherectomy is a neurosurgical technique in which one of the two cerebral hemispheres of the brain is removed (see Fig. 11.1).

Hemispherectomy was invented as a last-resort treatment for incurable and spreading gliomas, a type of brain cancer.[3] Later, it was adapted to treat incurable seizure diseases (similar to the callosotomy described in Chapter 3). Today this is its only use,[4] and it is not used in cancer therapy anymore. For those interested in a longer description of the development of a child who has undergone hemispherectomy, the book "Half a Brain is Enough – The Story of Nico" by Antonio M. Battro is an interesting read.[5] The book centers around Nico, a boy whose right hemisphere was removed at the age of 3 years to treat incurable epilepsy.

Most, but not all, patients who underwent hemispherectomy in childhood show some lifelong impairments in cognitive functions associated with the resected hemispheres, but some make an almost complete recovery like the patient in the introductory short story. Also, these functional impairments are typically less pronounced than if an older adult suffers extensive brain damage due to a stroke. Typically, damage to frontotemporal areas in the right hemisphere does not lead to language problems in adults, but damage to frontotemporal areas in

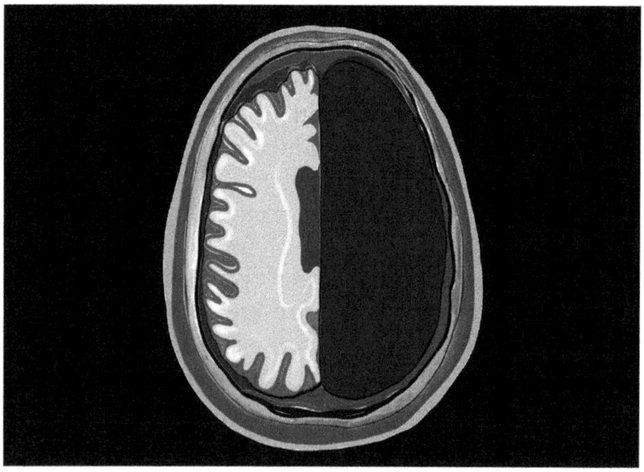

FIGURE 11.1 Schematic drawing of a typical MRI scan in a patient with hemispherectomy.

the left hemisphere often leads to aphasia, which patients rarely recover from.[6] Functional problems after hemispherectomy in early life also mirror functional lateralization patterns in the brain, but the severity of issues is typically much less pronounced than in adults who experienced brain damage. For example, it has been shown that right-sided hemispherectomy is associated with impairments of social functioning such as mentalizing the intent of others but not language problems.[7] In contrast, left-sided hemispherectomy is associated with language-related limitations but normal social functioning.[7–9] In general, the human brain is quite resilient to extensive unilateral lesions such as in hemispherectomy and shows an exceptional ability to rewire itself after injury if that injury happens early enough in development.[10,11] Typically, patients who underwent hemispherectomy early in life show extensive rewiring of sensory-motor, behavioral, and cognitive networks that helps in recovering function.[11] This exemplifies an important principle of how hemispheric asymmetries develop. They are not 100% genetically hardwired traits that are determined at conception. While genetic factors play a role, neuronal systems can dynamically readjust themselves in response to environmental events in early development which may lead to changes in hemispheric asymmetries over the course of a person's life.

Brain plasticity

The findings on hemispherectomy highlight a central principle in brain development and learning in all organisms that have a brain: brain plasticity. The very young human brain shows a sharp increase in white matter volume, while that of gray matter is already very high shortly after birth and then slowly starts to decline[12] (Fig. 11.2). Both trajectories taper off in adult years but never stop changing. Alterations in synapse numbers are more dramatic. After a rapid increase before birth and in the first months after delivery, massive activity-dependent synaptic pruning reduces synaptic density by

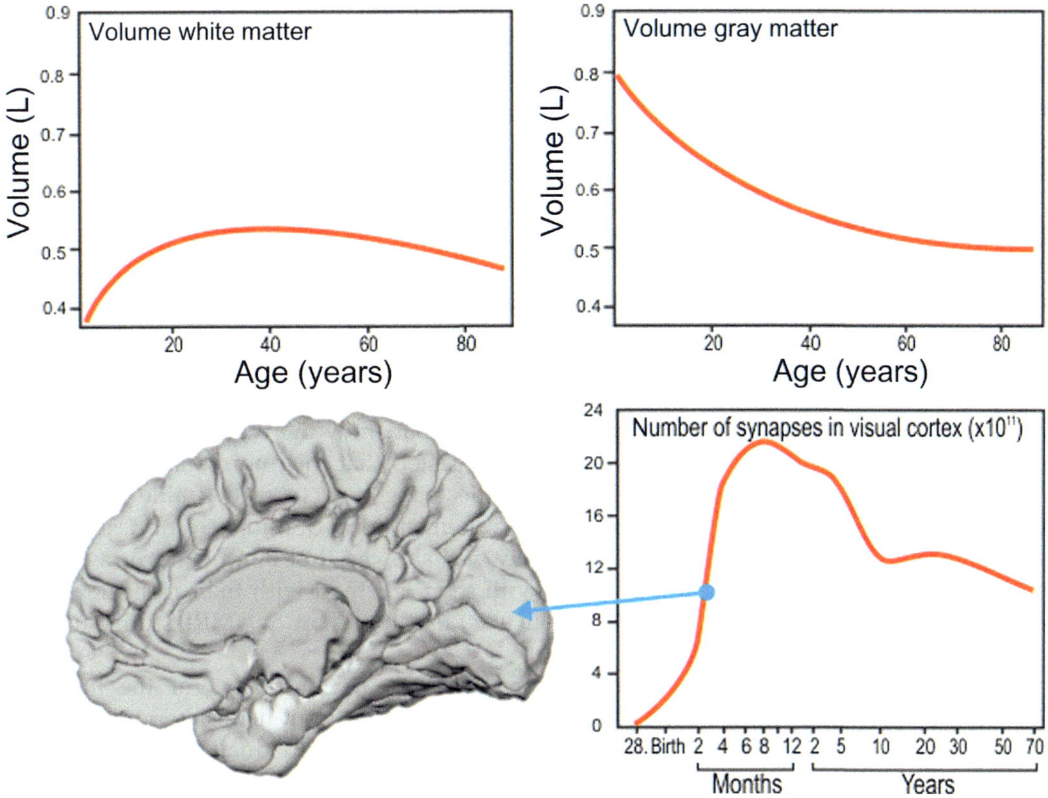

FIGURE 11.2 Developmental changes of overall cortical white matter and gray matter volume and synapse numbers in the primary visual cortex in humans. Note that the abscissa for the synapse number graph is nonlinear. The first data point is from the 28th week of gestation (28). The blue arrow illustrates the location of the sampling. Note that the changes are rapid during the early developmental period but never cease even in the old brain.[12,13] *Top, adapted from Lebel C, Gee M, Camicioli R, Wieler M, Martin W, Beaulieu C. Diffusion tensor imaging of white matter tract evolution over the lifespan.* Neuroimage. *2012;60:340–352 with permission from Elsevier; right lower graph adapted from Huttenlocher PR. Morphometric study of human cerebral cortex development.* Neuropsychologia. *1990;28:517–527 with permission of Elsevier.*

about 50%, thereby enabling the establishment of the typical cortical microarchitecture in humans.[13]

The mechanisms of synaptic growth and pruning during maturation overlap with those of learning in the mature brain.[14] Consequently, learning events are also associated with simultaneous growth and retraction of those dendrites and synapses that are part of the learning process.[15,16] Thus, the ability to produce and prune neural connections is first visible in development but then constitutes the neurobiological fundamentals of learning and cognition. Against this background, it is to be expected that hemispheric asymmetries cannot simply be largely fixed at birth but need to have the potential to continue changing. As will be shown below, at least some forms of hemispheric asymmetries can change during development, over the lifespan, and when aging.

Hemispheric asymmetries in individuals who are blind

One potential way to empirically assess the impact of plastic reorganization processes during development on hemispheric asymmetries is to investigate blind people. Congenital blindness leads to both structural and functional reorganization processes in the brain.[17] For example, the occipital cortex which is involved in the processing of visual input in sighted individuals is involved in auditory processing in blind individuals.[18] Interestingly, it has been shown that blindness affects several forms of functional hemispheric asymmetries. For example, a study of 1387 blind and 831 sighted children revealed significantly higher prevalences of left-handedness in blind children (male: 18.23%, female: 17.02%) than in sighted children (male: 11.02%, female 7.52%).[19] This suggests that visual perception plays a significant role in handedness development (see Chapter 6).

For language lateralization, a study in 66 children between 8 and 17 years of age[20] found that unlike sighted controls, blind children did not show the typical right ear advantage in the dichotic listening task. The authors suggested that this more bilateral language processing in blind individuals may be because learning how to read Braille is a highly spatial task, which may be affected by the right-hemispheric dominance for visuospatial processing. In line with this idea, an early study also showed that blind children were faster and more accurate when reading Braille with their left than with their right hand.[21] A reduction of leftward language lateralization was also found in an fMRI study in blind compared to sighted adults.[22]

For hemispheric asymmetries in spatial processing, a recent review of the literature showed that blind individuals, like sighted individuals, show right-hemispheric dominance for tasks such as haptic line bisection.[23] Interestingly, blind individuals showed stronger right-hemispheric dominance on a variety of spatial tasks compared to sighted individuals. The authors concluded that neuronal plasticity led to a stronger involvement of the right parietal and occipital cortices in blind individuals.

Hemispheric asymmetries in individuals who are deaf

Similar to blindness, congenital deafness is associated with cortical reorganization processes, with the recruitment of brain areas associated with hearing (e.g., in the temporal lobe) for intact sensory modalities such as vision or somatosensory processing.[24] Comparable to what has been found for blindness, deafness is associated with increased prevalence of left-handedness, but also non-right-handedness. A meta-analysis of 31 studies on handedness in deaf people found a prevalence of left-handedness of 14.7% and a prevalence of non-right-handedness of 17.7%.[25] Unlike for blindness, for which visuomotor learning of lateralized hand use may be an important influential factor on handedness during development, it is somewhat more unclear why there is a link between deafness and handedness. The authors of the meta-analysis suggested that the association between handedness and language lateralization may play a role in this association, with higher prevalences of left-handedness and non-right-handedness in deaf people potentially linked to a delay in language acquisition associated with deafness.

While typical paradigms used to assess language lateralization such as the dichotic listening task or fMRI acquisition during speech production or perception cannot be used in blind participants for obvious reasons, there is a substantial number of empirical neuroimaging studies on sign language. A recent activation likelihood estimation (ALE) meta-analysis of 23 neuroimaging studies on sign language found that similar to spoken and written language, sign language leads to activation of a predominantly left-hemispheric language network with Broca's area as the main hub.[26] This suggests that the left hemisphere is dominant for language processing independent of whether the language is spoken or signed. This also seems to be the case in individuals with unilateral deafness, as a study found that most patients with congenital right ear deafness still show the typical leftward language dominance.[27]

For hemispheric asymmetries in spatial processing, it has been found that unlike blind individuals and healthy controls, deaf individuals do not show the typical pseudoneglect (e.g., a tendency to bisect a visually or haptically presented line to the left of its veridical midpoint which reflects a right-hemispheric dominance for spatial processing).[28] The authors of the study concluded that hemispheric asymmetries in spatial perception may be altered in individuals who are deaf because of their stronger reliance on visual experience due to early auditory deprivation.

Early theories on the development of hemispheric asymmetries

Several theories have been proposed to answer the question to which extent hemispheric asymmetries are genetically hardwired at birth (e.g., there is a genetic predisposition for the left hemisphere to become language-dominant) or develop dynamically over the lifespan (e.g., the right hemisphere could become language-dominant). Before the widespread availability of modern neuroimaging methods, such theories often were informed by studies in patients with left- or right-hemispheric brain lesions. These include, for example, the abovementioned findings on neuronal rewiring after hemispherectomy, but also studies in other patient groups that suffered brain damage early in life. The most important early theories on the development of hemispheric asymmetries include:

1. The hemispheric equipotentiality theory: The idea of hemispheric equipotentiality posits that at birth, both hemispheres are equivalent in terms of basic language abilities and either could become the language-dominant hemisphere.[29,30] This line of thought has been shaped by the ideas of linguist Eric Heinz Lenneberg (1921–75). In his highly influential book "Biological Foundations of Language,"[31] Lenneberg suggested that language lateralization is not fully determined at birth but gradually develops until puberty based on a previous finding that left- and right-hemispheric brain lesions equally often lead to language issues if they occur before the age of 2 years.[32] Lenneberg concluded that language lateralization is undetermined before the age of 2, and that the two hemispheres are equipotential regarding language up to this point.
2. The differential cerebral maturation theory: The differential cerebral maturation theory assumes that the two hemispheres do not develop at the same pace, but that one hemisphere develops earlier and/or more rapidly than the other. Based on data from

temporal lobe epilepsy patients that showed that left-sided brain lesions were common in the first year and rare after 2 years, David C. Taylor suggested in a 1969 article in the medical journal *Lancet* that the left hemisphere was functionally less active in early life and the right hemisphere matured faster.[33] This idea was also supported by a later study showing a rightward asymmetry of cerebral blood flow between 1 and 3 years.[34] In contrast to these assumptions, Corballis and Beale later assumed that the left, not the right, hemisphere shows earlier or more rapid development.[35]

3. The crowding hypothesis: An influential hypothesis on the reorganization of language asymmetries after brain damage to the left hemisphere is the so-called "crowding hypothesis."[36–39] This hypothesis aims to explain why after epilepsy-related damage to the left hemisphere in early life, language ability is often largely unimpaired, but patients show impairments in typical nonverbal functions of the right hemisphere. The crowding hypothesis assumes that due to neural plasticity, the right hemisphere can take over language functions from the damaged left hemisphere if the damage occurs early enough in life. Since the amount of neural tissue is, however, limited, this reorganization comes at a price: Since a significant number of neural networks in the right hemisphere are needed to perform highly complex language functions, typically right-hemispheric functions are "crowded out" of the hemisphere, leading to functional impairments. This explains the paradoxical finding that early left-hemispheric damage does not necessarily result in language impairments but may also result in impairments of typically right-hemispheric functions.

In the following sections, we review the current knowledge on the development of structural and functional hemispheric asymmetries and will explore to what extent these early theories are reflected in empirical evidence. In general, modern laterality research has largely moved away from hypotheses that target the development of the left and the right hemisphere as a whole, as they are likely to underestimate the complexity of neural networks.

Development of structural hemispheric asymmetries

As described in Chapter 4, several different forms of structural hemispheric asymmetries can be differentiated, including molecular, microstructural, and macrostructural asymmetries,[40] but also asymmetries in structural connectivity.[41,42] There is convincing empirical evidence that structural hemispheric asymmetries emerge prenatally on several different of these levels.

On the molecular level, an RNA sequencing study in postmortem brain tissue samples of human embryos and fetuses showed that molecular hemispheric asymmetries emerge as early as 5 weeks after conception.[43] In the study, left-right differences in the transcriptomic profiles of brain structures in the left and the right hemispheres of six embryos (between 5 and 5.5 weeks after conception) and 21 fetuses (between 7.5 and 13 weeks after conception) were analyzed. Subtle left-right differences in expression profiles started to emerge already in the samples that had been collected 5 weeks after conception, with no clear overall left-right dominance, but distinct patterns for different brain regions. In the embryos, the hindbrain and midbrain showed faster right-hemispheric maturation, while the forebrain showed no asymmetry. In the fetuses, there was faster left-side maturation for the temporal lobe of the cortex, the diencephalon, the basal ganglia, and the choroid plexus. In contrast, structures of the cerebral cortex outside the temporal lobe showed faster transcriptional development in the right hemisphere. A previous RNA sequencing study in embryos between 4 and 8 weeks after conception had also shown that the left side of the spinal cord matured faster than the right in embryos.[44]

Further evidence for molecular asymmetries in the developing human brain came from a study on differential expression of microRNAs in the left and right hemispheres of fetuses 12 weeks after conception.[45] In this study, the authors identified 42 microRNAs that showed differential expression in the left and the right hemisphere at 12 weeks after conception, with the target genes of these asymmetrically expressed microRNAs being involved in different aspects of brain development such as the development of dopaminergic and GABAergic synapses and axon guidance. Moreover, a study investigating asymmetries of the proteome in brain tissue samples from fetuses in the late first trimester of pregnancy reported different protein expression patterns between the left and the right frontal lobe.[46]

Taken together, the empirical evidence suggests that several molecular asymmetries exist in the developing brain before birth. This finding is in contrast to the idea of differential maturation of the hemispheres described above,[34] since none of the two hemispheres developed faster than the other as a whole. It is also in contrast to the hemispheric equipotentiality theory, as it does suggest early hemispheric differences and not full equipotentiality. Instead, the findings suggest that structure-specific lateralized processes emerge very early during embryogenesis. This implies that hemispheric asymmetries in different cognitive domains (e.g., leftward asymmetry for language processing and rightward asymmetry for processing of faces) may already be reflected in these early molecular asymmetries and developed at domain-specific paces.

In addition to molecular asymmetries, macrostructural asymmetries have been shown to at least partly emerge before birth. In 1975, Juhn A. Wada (who invented the Wada test, see Chapter 5) and coworkers analyzed structural asymmetries in 100 adult and 100 infant brains and found that morphological asymmetries of the planum temporale and the frontal operculum could be measured 29 weeks after gestation.[47] Two years later, Je G. Chi from Harvard Medical School and coworkers conducted a larger postmortem study, investigating structural asymmetries in celloidin-embedded brains of 207 fetuses with a gestational age between 10 and 44 weeks.[48] Overall, the authors found a leftward asymmetry of the planum temporale and a rightward asymmetry of Heschl's gyrus in fetal brains, which became evident 31 weeks after gestation.

These early findings were largely confirmed by later studies. A recent scoping review of structural and functional hemispheric asymmetries in early development integrated data from 57 studies and reported that a leftward asymmetry of the planum temporale was the most commonly reported finding and that this asymmetry could be detected 29 weeks after gestation.[49] Taken together, these findings suggest that like molecular asymmetries, structural asymmetries in temporal speech regions emerge before birth. Specifically, they emerge in the third trimester of pregnancy, later than molecular asymmetries, which can already be detected in the first and second trimesters of pregnancy.

In addition to these studies on the development of structural hemispheric asymmetries before birth, large-scale neuroimaging studies on the development of structural asymmetries after birth have been conducted.[50–54] For example, cortical asymmetries in newborns have been investigated in 442 brain scans from healthy neonates from the Developing Human Connectome Project.[53] The authors determined hemispheric asymmetries in sulcal depth, surface area, and cortical thickness, and found the neonate cortex was markedly asymmetric, in line with the findings that molecular and macrostructural asymmetries already emerge before birth (see Fig. 11.3). For sulcal depth, the lateral fissure showed a leftward asymmetry, and the middle portion of the superior temporal sulcus showed a rightward asymmetry. The most pronounced asymmetries were found for the surface area, with both leftward and rightward asymmetries being observed in several areas across the whole cortex. A similar pattern but with a smaller magnitude was also observed for cortical thickness.

To investigate the development of structural asymmetries after birth, James M. Roe from the University of Oslo and coworkers conducted a large-scale analysis of 3937 longitudinal neuroimaging datasets from individuals between 4 and 89 years of age.[52] Areal asymmetries were largely stable across life in the investigated sample, suggesting that they are developmentally stable after the age of 4 years and that they develop early in life, in line with the previously discussed results. Thickness asymmetries grew in childhood and peaked in early adulthood, suggesting that they were influenced by the developmental growth of the brain to a larger extent than areal asymmetries.

In addition to structural asymmetries in gray matter, asymmetries in cortical connectivity have been investigated in the context of development. For example, in one structural connectivity study, neuroimaging data from 211 participants (between 6 and 21 years of age) were obtained and graph theory was used to quantify topological network metrics in the left and the right hemisphere.[41] The clustering of connections between regions showed a leftward asymmetry for the language network, but also the sensorimotor, default mode, and salience networks. The presence of hub nodes in the networks showed a rightward lateralization for the salience network and a leftward lateralization for the dorsal attention network. Importantly, these asymmetries were largely unaffected by age, suggesting that at the beginning of school age, white matter asymmetries had largely been established despite continuing changes in white matter properties throughout childhood and adolescence.[55]

Development of motor asymmetries

For functional hemispheric asymmetries, the empirical evidence suggests that the timepoint of their emergence strongly depends on the involved function. As described in Chapter 6, there is strong evidence that asymmetries in the motor domain first emerge before birth.[56] Specifically, human fetuses show a pronounced rightward movement bias even before birth, arguing for early determination of handedness driven by both genetic and epigenetic factors.[57–60] The fetus can independently move its arms at about 10 weeks of gestation, and this is also the age at which motor asymmetries emerge for the first time. A study in 72 fetuses at 10 weeks of gestation found a clear rightward asymmetry, with

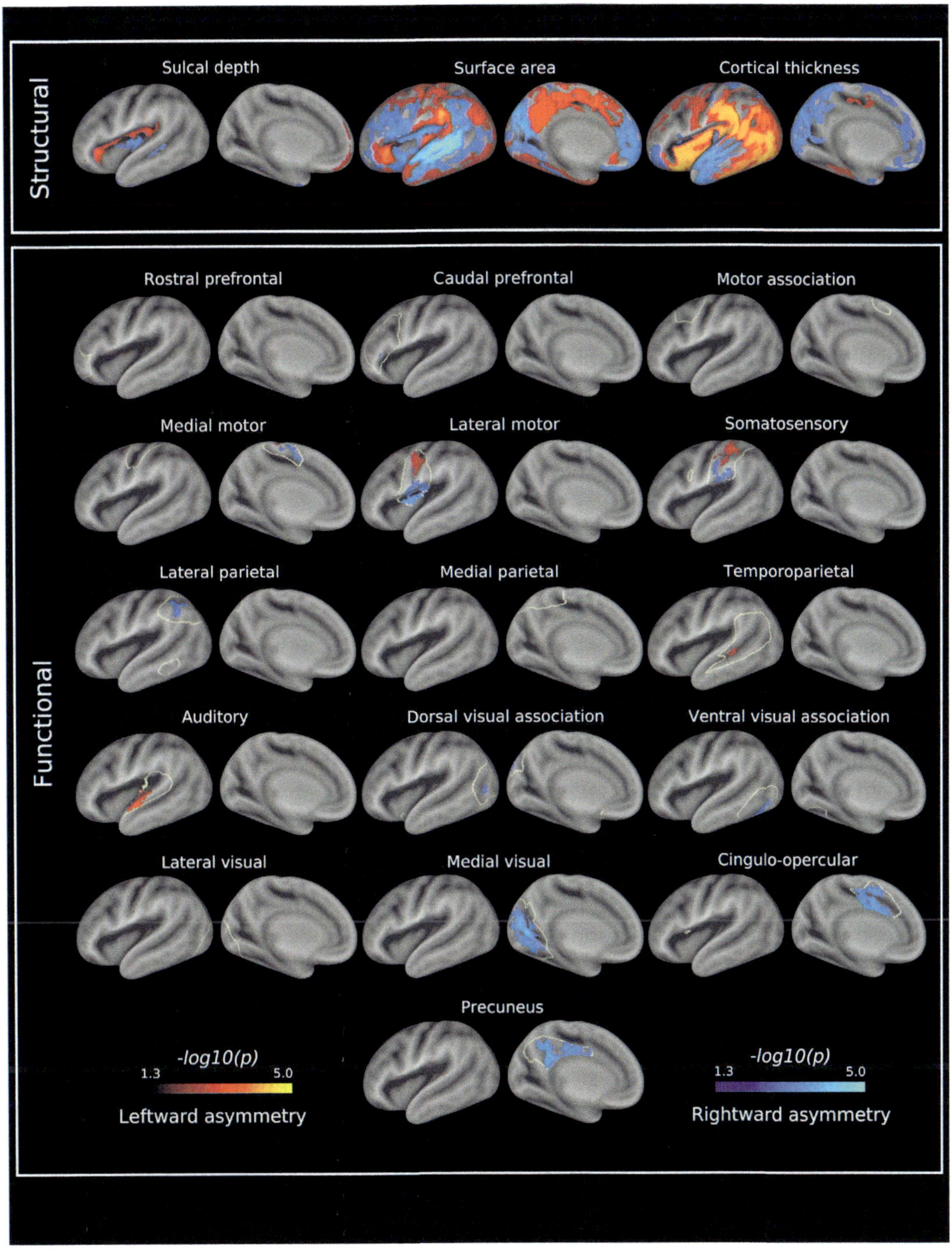

FIGURE 11.3 Structural asymmetries in the neonate's brain from the study by Williams et al. (2023).[53] Red to yellow colors indicate leftward asymmetry. Dark blue to light blue colors indicate rightward asymmetry. Values indicate $-\log_{10}(P)$ for all significant asymmetry effects. *Reused with permission by the publisher from Williams LZJ, Fitzgibbon SP, Bozek J, et al. Structural and functional asymmetry of the neonatal cerebral cortex.* Nat Hum Behav. *2023;7:942–955.*

85% of fetuses showing more movements with the right arm than with the left arm.[61] At 15 weeks of gestation, fetuses also show a clear preference to suck the right thumb.[62] A further study in fetuses between 12 and 27 weeks of gestation revealed that they showed more right-arm than left-arm movements at every gestational age.[63] In addition to arm and hand use asymmetries, a rightward bias in head-turning has been reported both prenatally and postnatally,[59,60,64,65] and it has been suggested that head-turning bias and handedness are associated (see Chapter 6 for details; also, two review articles have been published recently on developmental cascades relevant for handedness development[66,67]). Thus, the empirical evidence suggests that functional motor asymmetries, like molecular structural asymmetries, first emerge in the first trimester of pregnancy. In addition to handedness and head-turning preference, the development of motor asymmetries has also been investigated for footedness. It has been shown that an overall rightward preference for foot mobilization is largely stable across age in 5- to 20-year-old participants, suggesting that individual footedness is determined in early childhood and does not change much afterward.[68] A review article on foot laterality during childhood came to the conclusion that in children aged 3–11 years, the overall rightward preference was not as pronounced as for handedness and that about twice as many children show mixed-footedness than mixed-handedness.[69]

Development of language lateralization

For most forms of functional asymmetries outside the motor domain, development takes substantially longer, simply because the underlying cognitive systems need to develop first. One primary example is language lateralization,[51] since newborns do not speak and only gradually develop the ability to communicate using language over several years. It has been shown that language representation in the brain is far more diffuse in infants than in adults. Through language learning and synaptic pruning processes, this representation becomes more focal, until it resembles the adult pattern.[70]

One very interesting question is whether structural or functional asymmetries in the language domain occur first. Both scenarios are conceivable. In the first scenario, structural asymmetries, for example, in the planum temporale, would be the anatomical cause of functional language lateralization. In the second scenario, experience-dependent plasticity in the language system could affect its neuroanatomical correlates. Notably, structural asymmetries—for example, a larger left planum temporale and a deeper right temporal sulcus—can be observed even before birth.[71–73] In contrast, in newborns, functional lateralization of language processing is not yet established. In 2-day-old infants, language-related neural structures are active in both hemispheres and there is even enhanced activity in the right auditory cortex when exposed to normal speech. In contrast to adults, strong connectivity only exists between the two hemispheres, whereas intrahemispheric connections are weak.[74] The functional specialization of the left hemisphere develops during childhood, including functional reorganization of neuronal networks. Intrahemispheric circuits are strengthened especially within the left cortex.[74,75] While different language functions—for example, naming or word repetition—are only diffusely represented within the left and right cortex, language representation becomes more focal—mostly within the left hemisphere—as language skills develop.[70] Thus, acquiring language competence influences functional hemispheric asymmetries. Inherent predispositions to adopt specific functions may exist in the developing brain, but the determination of hemispheric specialization requires environmental input and is use-dependent. Importantly, it has been shown that structural asymmetries in language regions of the fetal brain have a high prognostic value for later language development. In a recent study of 38 children between ages 6 and 13 for whom fetal MRI diagnostics were available, asymmetries of the superior temporal sulcus were assessed in the fetal MRI scans.[76] The superior temporal sulcus showed a rightward asymmetry, and since it lies adjacent to the superior temporal gyrus that is highly relevant for language processing, the authors suggested that the superior temporal sulcus may be an anatomical marker for language abilities. Less rightward lateralization of the fetal superior temporal sulcus was linked to better verbal abilities in childhood, with a substantial amount of explained variance (40%). Moreover, children who showed less pronounced rightward asymmetries of the superior temporal sulcus before birth showed stronger leftward language localization in childhood. These findings suggest that even though children need to learn how to speak for several years before language abilities are fully developed, early structural asymmetries in language-relevant brain regions are highly relevant for language lateralization.

Development of spatial attention asymmetries

To assess spatial attention asymmetries in children, the so-called Teddy Bear Picnic task has been developed.[77] The Teddy Bear Picnic task is a child-friendly way to assess pseudoneglect. In the task, the participants see how the image of a knife is moved to different points of a food item such as a croissant or a cupcake. The participant is then asked to indicate when the knife was positioned in the middle of the food. The Teddy Bear Picnic task correlates highly with the results of the landmark task and participants on average show right-hemispheric asymmetry, as would be expected for a visuospatial task.

A meta-analysis of 31 line bisection studies in children aged 16 years or younger revealed somewhat inconsistent results.[78] In the overall sample including data from 2102 children, no pseudoneglect was observed. However, this effect was moderated by sex, with more leftward bias in studies that included more boys and hand use. Using the right hand resulted in rightward bisections and using the left hand resulted in leftward bisections. This suggests that task demands affect pseudoneglect in children to a larger extent than in adults.

Hemispheric functional segregation

One central question in research about the development of hemispheric asymmetries is to what extent different forms of asymmetries develop dependent on each other or independent of each other.[79,80] For example, does an individual who shows leftward language lateralization always show rightward lateralization for face processing? Intuitively, it makes sense to assume that it is unlikely that one hemisphere is dominant for all cognitive domains in a person. Just like there is typical and atypical dominance for specific cognitive domains (e.g., leftward language lateralization would be typical dominance and rightward language dominance would be atypical dominance), it has been proposed that there is a prototypical differentiation of several cognitive domains over the two hemispheres. Specifically, it has been suggested that prototypical functional segregation consists of leftward dominance for language generation and perception, praxis, and mathematical calculations, as well as rightward dominance for spatial attention, prosody, and face recognition.[81] Based on this assumption, Guy Vingerhoets from Ghent University in Belgium suggested in a recent review article[82] that there are three main phenotypes in hemispheric functional segregation in the human brain:

1. The typical functional segregation phenotype: Individuals with this phenotype show the prototypical functional segregation for all cognitive domains.
2. The reversed typical functional segregation phenotype: Individuals with this phenotype show a completely mirrored brain organization compared to prototypical functional segregation (e.g., rightward dominance for language generation and perception, praxis and mathematical calculations, and leftward dominance for spatial attention, prosody, and face recognition)
3. The atypical functional segregation phenotype: Individuals with this phenotype have at least one cognitive domain lateralized to the atypical hemisphere, while for the remaining cognitive domains, typical hemispheric segregation is maintained.

One commonly investigated association is the link between handedness and language lateralization.[83] For example, 4% of strong right-handers show atypical rightward language dominance, but 15% of ambidextrous individuals and 27% of strong left-handers do so.[84] Moreover, a recent large-scale study on dichotic listening performance in left-handers and right-handers found an increased right ear advantage in right-handers compared to left-handers.[85] These findings suggest a more pronounced leftward language lateralization in right-handers than in left-handers. These and similar findings led some authors to compare hemispheric functional segregation between left-handers and right-handers for other cognitive domains. For example, in a recent study, hemispheric asymmetries for visuospatial processing (using grayscales and colorscales tasks), face processing (using a chimeric face task), emotion processing (using an emotional dichotic listening task), and speech processing (using a consonant-vowel dichotic listening task) were compared between left-handers and right-handers.[86] Left-handers showed a reduction in typical asymmetries for face processing and in the consonant-vowel dichotic listening task.

Furthermore, it has been investigated to what extent left-handers with rightward language dominance differ from left-handers with leftward language dominance concerning other forms of hemispheric asymmetries to assess how common a complete mirror reversal of hemispheric asymmetries for different brain functions is.[87] In the study, 24 left-handers with atypical rightward language dominance were compared to 39 left-handers with typical leftward language dominance using fMRI paradigms for language lateralization, praxis, spatial attention, face recognition, and emotion processing. The results were quite striking: 50% of left-handers who showed rightward language dominance showed a complete mirror reversal of hemispheric asymmetries. A further 32% showed a reversal for all but one cognitive function. In contrast, 44% of left-handers who showed leftward language dominance showed a completely prototypical distribution of hemispheric asymmetries across cognitive domains. A further 35% deviated in only one function from a fully prototypical distribution of hemispheric asymmetries across cognitive domains. Interestingly, participants who showed atypical asymmetries in more than one cognitive domain showed impaired cognitive performance, suggesting an evolutionary advantage of typical hemispheric functional segregation.

What causes atypical hemispheric functional segregation? An fMRI study comparing functional segregation in four different localizer tasks (word generation, tool pantomiming, face recognition, landmark task) between individuals with situs inversus totalis and healthy controls found that situs inversus totalis was linked to higher prevalence of atypical hemispheric functional segregation.[81] Interestingly, stronger deviations from typical hemispheric functional segregation were also associated with lower cognitive performance. These findings suggest that the genetic factors involved in situs inversus may also affect the distribution of hemispheric asymmetries for different cognitive domains in the brain. Moreover, they further support the idea that typical hemispheric functional segregation seems to have evolutionary advantages, in that deviations from it are linked with lower cognitive performance.

To integrate the diverse findings on hemispheric segregation and develop a mechanistic model that could be used to generate hypotheses for future research, Robin Gerrits from Ghent University in Belgium developed the so-called Segregation Bias Model.[88] This model assumes that it is unlikely that different forms of hemispheric asymmetries develop completely independent of each other (in that case, there would not be a link between handedness and language lateralization), but it is also unlikely that they develop completely dependent on each other (in that case, there would be no individuals with the atypical functional segregation phenotype). To explain the hemispheric segregation patterns observed in empirical studies, the Segregation Bias Model assumes that two major factors influence the pattern of hemispheric asymmetries in the brain. In the first step, there is a domain-independent innate segregation bias that determines whether the individual has a bias toward the typical or the reversed-typical segregation phenotype. For individuals with the typical segregation phenotype, there is a strong probabilistic bias (but not a guarantee) that all cognitive domains show prototypical dominances. For individuals with the reversed-typical segregation phenotype, there is a strong probabilistic bias to develop a mirrored brain organization. In a second step, all individual dominances then develop in a probabilistic fashion (e.g., right-handers have a high chance, but are not guaranteed, to develop left-hemispheric language dominance). The model predicts that most people will develop a prototypical asymmetry pattern, but since both steps are probabilistic, occasionally individuals will show the atypical functional segregation phenotype. The model could in principle be used to assess any combinations of hemispheric asymmetries. In Fig. 11.4 it is shown for two forms of hemispheric asymmetries and in Fig. 11.5 for three forms of hemispheric asymmetries.

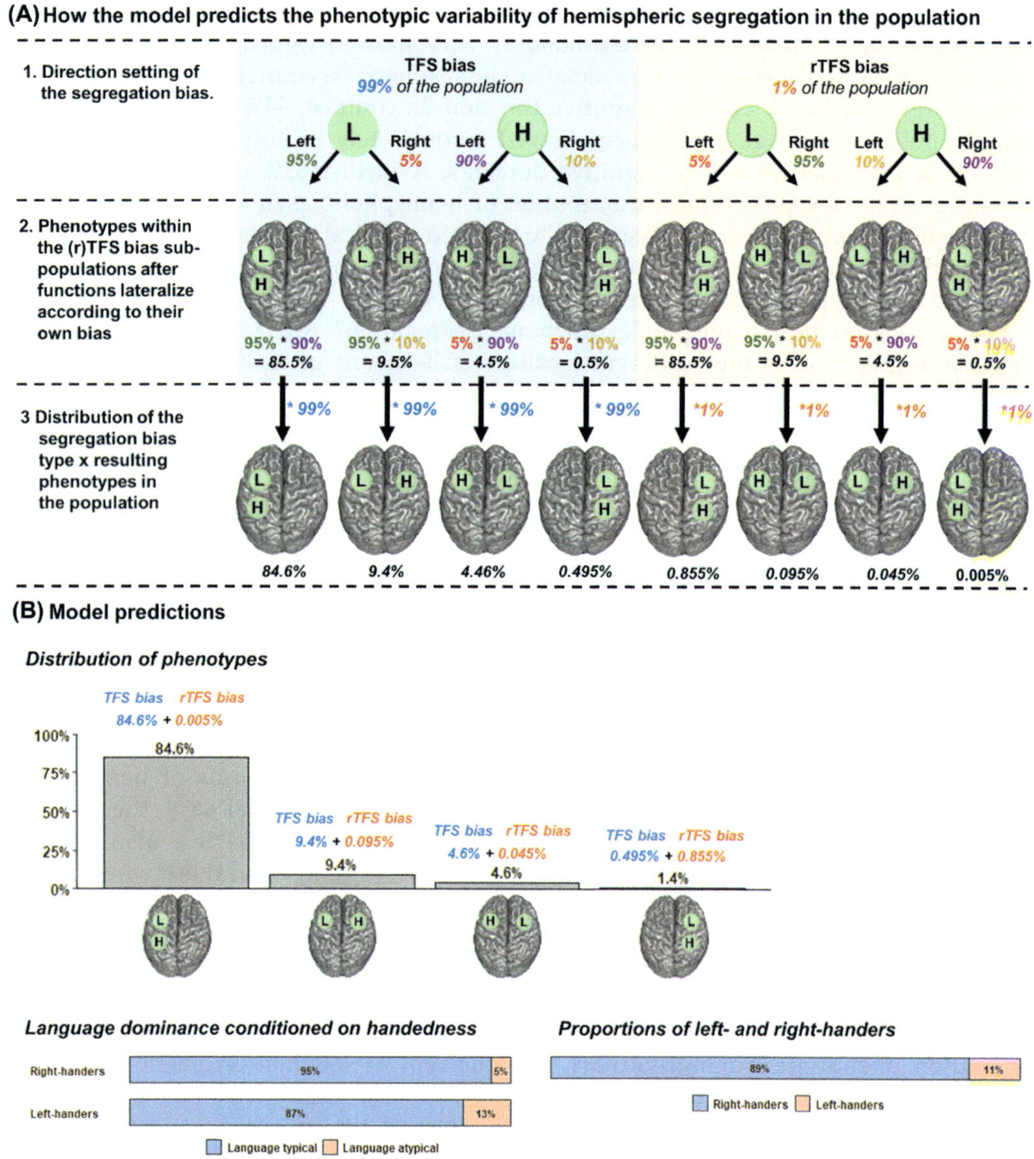

FIGURE 11.4 The Segregation Bias Model by Gerrits[88] for two asymmetry phenotypes (in this case handedness and language lateralization). (A) In the first step, segregation bias is set to either typical functional segregation (TFS) or reversed typical functional segregation (rTFS). In the second step, cognitive functions develop according to their own bias (e.g., a higher chance of left-hemispheric dominance for language, etc.). The third step shows the resulting distribution of phenotypes in the population. (B) Distribution of phenotypes for handedness and language lateralization according to the model. *Reused with permission by the publisher and the author from Gerrits R. Variability in hemispheric functional segregation phenotypes: a review and general mechanistic model.* Neuropsychol Rev. *2022; epub ahead of print.*

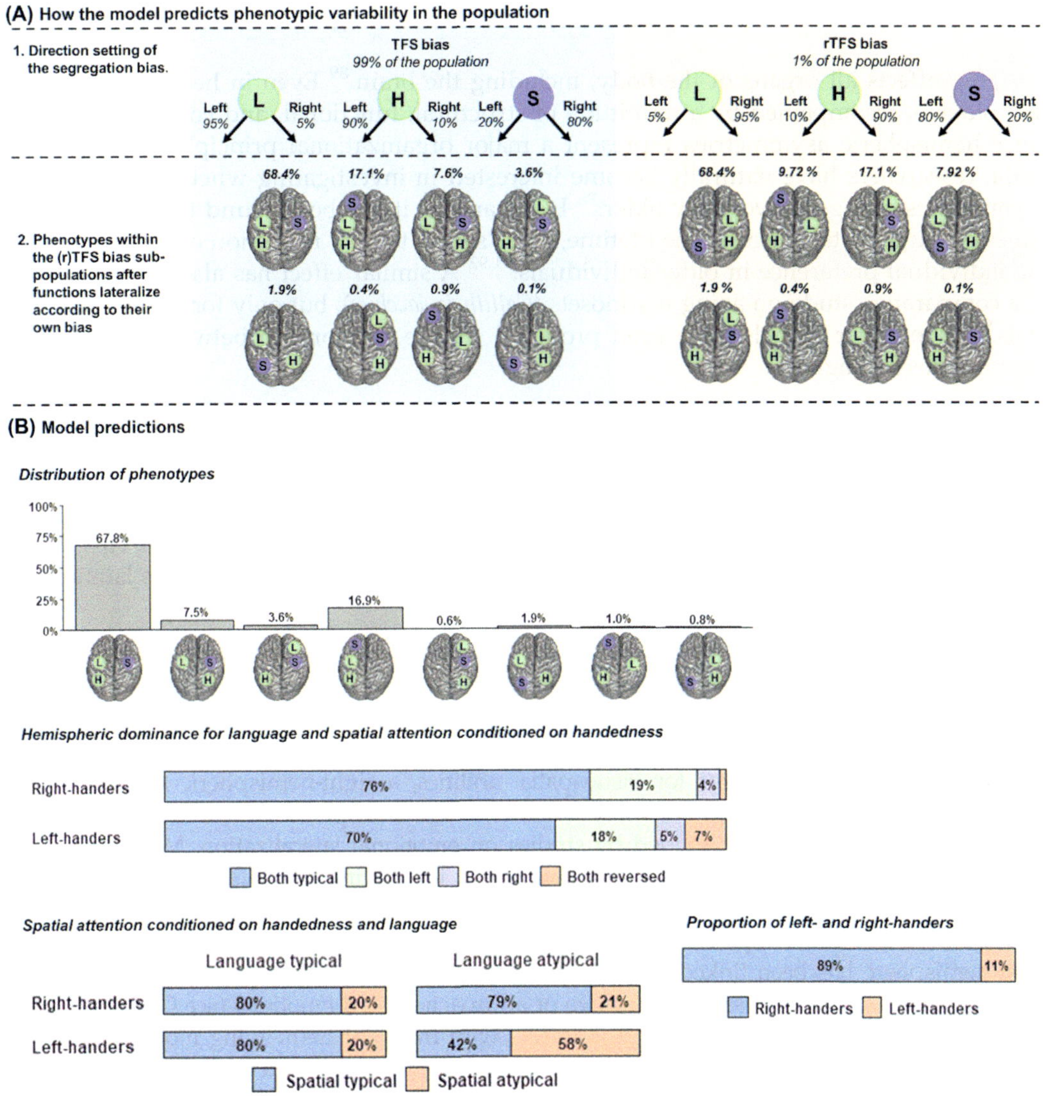

FIGURE 11.5 The Segregation Bias Model by Gerrits[88] for three asymmetry phenotypes (in this case handedness, language lateralization, and hemispheric asymmetry for spatial attention). (A) In the first step, segregation bias is set to either typical functional segregation (TFS) or reversed typical functional segregation (rTFS). In the second step, cognitive functions develop according to their own bias (e.g., a higher chance of left-hemispheric dominance for language, a higher chance of right-hemispheric dominance of spatial attention, etc.). The third step shows the resulting distribution of phenotypes in the population. (B) Distribution of phenotypes for handedness and language lateralization and spatial attention according to the model. *Reused with permission by the publisher and the author from Gerrits R. Variability in hemispheric functional segregation phenotypes: a review and general mechanistic model.* Neuropsychol Rev. *2022; epub ahead of print.*

Hemispheric asymmetries and aging

Aging affects all organs of the body, including the brain.[89] Even in healthy aging, getting older is accompanied by a multitude of structural, functional, and cognitive changes. Since hemispheric asymmetries represent a major organizational principle in the human brain, researchers have naturally become interested in investigating whether hemispheric asymmetries change as we grow older.[90] For example, it has been found that while handedness is largely stable across the lifetime, there seems to be a reduction of the strength of the individual preference in older individuals.[91,92] A similar effect has also been observed in a comparative study on aging marmosets (*Callithrix jacchus*), but only for left-handed animals.[93] Two major models have been proposed for the relationship between hemispheric asymmetries and aging[94]:

1. The right hemi-aging model: This model assumes that the right hemisphere is more vulnerable to the effects of aging than the left in the sense that both right and left hemisphere functions decline with age, but left hemisphere functions are less strongly impaired.[94–97]
2. The HAROLD model (Hemispheric Asymmetry Reduction in Older Adults): This model assumes that prefrontal activity during cognitive performances is less lateralized in older adults than in younger adults.[98]

Evidence for the right hemi-aging model was provided by studies on aging and intelligence. For example, in 1981, Gerald Goldstein and Carolyn Shelly from the University of Pittsburgh tested more than 1200 participants in six age groups from their 20s to their 70 s with a neuropsychological test battery specifically constructed to assess left- and right-hemispheric functions.[99] They found that scores for visuospatial abilities, a right-hemispheric function, were particularly impacted by aging. For verbal abilities, a left-hemispheric function, aging had less impact. More evidence was provided by studies on emotional lateralization. McDowell et al. reported that older participants were less accurate in identifying the valence of negative and neutral facial expressions than younger individuals.[100] In contrast, their performance for faces with positive emotional expressions was comparable to that of younger individuals. As the right hemisphere has been linked to the processing of negative or avoidant emotions and the left hemisphere to the processing of positive or approach-related emotions (see Chapter 9), this selective impairment for negative faces is in line with the right hemi-aging model. However, another study by Cherry et al. could not replicate this finding.[101] Interestingly, a large-scale investigation of structural brain asymmetries by the ENIGMA consortium recently reported findings that are partly in line with the right hemi-aging model.[102] While no age effects were observed in the overall sample, a reanalysis of datasets with an age range of at least 20 years found that increasing age was associated with a more pronounced leftward overall asymmetry of cortical thickness. This overall effect was partly driven by a more pronounced leftward cortical thickness asymmetry in the superior temporal gyrus. In line with the right hemi-aging model, this suggests that rightward cortical asymmetry reduces over time.

A large-scale ALE meta-analysis on neuroimaging studies on hemispheric asymmetries in gray matter loss during aging found results that partly supported and partly contradicted this finding.[103] In this analysis, more region-specific asymmetries changes were observed with healthy aging. Stronger gray matter loss in the right hemisphere was observed in the superior temporal gyrus. This finding matches the increase in leftward asymmetry in this brain area found in the ENIGMA study.[102] In addition, the ALE meta-analysis revealed a stronger leftward gray matter loss in the postcentral gyrus, which is at odds with the right hemi aging model.

The HAROLD model was developed by Roberto Cabeza from the Center for Cognitive Neuroscience at Duke University.[98] This model was developed in the context of hemispheric asymmetries in episodic memory. For younger adults, the so-called hemispheric encoding/retrieval asymmetry (HERA) model for hemispheric asymmetries associated with episodic memory has been suggested.[104] This model assumes that the left prefrontal cortex is dominant for encoding information into episodic memory, while the right prefrontal cortex is dominant for episodic memory retrieval. While supported by empirical evidence in younger adults, Cabeza noticed that the HERA model did not seem to reflect the data pattern in older adults very well. In a 1997 PET study, he and his coworkers observed that in young subjects, frontal brain activity was in line with the HERA model. It was left-lateralized during encoding and right-lateralized during recall. However, older subjects showed a different activation pattern. During encoding, they showed only little frontal activity, while during retrieval, a more bilateral pattern of frontal activation was observed.

The HAROLD model was supported by a 2013 study by Kevin Collins and Christine Mohr.[105] These authors tested 20 younger and 20 older adults with a lateralized lexical decision task to assess left-hemispheric function and a chimeric face sex decision task to assess right-hemispheric function. For both tasks, older subjects showed reduced hemispheric asymmetries, supporting the HAROLD model. Similar results were also obtained in a study by Hatta et al., who tested verbal memory and visuospatial task performance in 338 elderly participants between their 50 s and their 80 s.[96] They found a reduction of hemispheric asymmetries with increasing age, consistent with the HAROLD model.

For visuospatial asymmetries, a widely cited meta-analysis from 2000 by Jewell and McCourt also reported that the typical leftward bisection error in line bisection tasks (pseudoneglect) was shifted more rightward in adults over 50 years compared to adults aged under 40 years.[106] In line with the HAROLD model, this suggests a reduction of right-hemispheric dominance for spatial attention with increasing age. In contrast, a later meta-analysis that specifically assessed spatial attention bias in older adults revealed no asymmetry in the overall analysis, but a leftward bisection bias in an analysis of a subsample of studies, in which all individual participants were aged 50 years or older.[107] Thus, the findings of the overall analysis support the HAROLD model, as the typical leftward bias is reduced, but the findings of the second analysis do not support it. Aging has also been shown to cause reduced asymmetry in the motor domain—for example, in visuomotor adaptation[108] and intralimb coordination[109]—further supporting the HAROLD model.

Taken together, there is empirical evidence supporting both the HAROLD model and the right hemi-aging model, but also some findings that contradict these models. Florin Dolcos from Duke University and coworkers conducted a systematic review of all studies that had been published on the two models.[94] They concluded that the right hemi-aging model was supported by several behavioral studies on cognitive, affective, and sensorimotor processing. However, the overall evidence was rather mixed, as there were also studies that failed to show the effects predicted by the model. The HAROLD model was supported by evidence from functional neuroimaging studies, for example, on episodic memory, perception, and inhibitory control. Dolcos et al. concluded that the two models are not mutually exclusive, but might simply apply to different brain regions. While the HAROLD model is well-suited to predict the effect of aging on frontal asymmetries, the right hemi-aging model might apply to other brain regions.

Sample questions

- What is hemispherectomy and which effect does it have on brain function?
- Name and describe three early influential models for the development of hemispheric asymmetries.
- What are the main findings on early molecular hemispheric asymmetries in the embryonic and fetal brain?
- Please describe the main findings on the development of macrostructural asymmetries in the human brain before and after birth.
- What does the term hemispheric functional segregation mean? Please describe major empirical findings and one theoretical model.
- Name and describe the two major models for the impact of aging on hemispheric asymmetries.

References

1. Stiles J, Jernigan TL. The basics of brain development. *Neuropsychol Rev.* 2010;20(4):327–348. Available from: https://doi.org/10.1007/s11065-010-9148-4.
2. Pinto ALR, Lohani S, Bergin AMR, et al. Surgery for intractable epilepsy due to unilateral brain disease: a retrospective study comparing hemispherectomy techniques. *Pediatr Neurol.* 2014;51(3):336–343. Available from: https://doi.org/10.1016/j.pediatrneurol.2014.05.018.
3. Dandy WE. Removal of right cerebral hemisphere for certain tumors with hemiplegia. *JAMA.* 1928;90(11):823. Available from: https://doi.org/10.1001/jama.1928.02690380007003.
4. Lew SM. Hemispherectomy in the treatment of seizures: a review. *Transl Pediatr.* 2014;3(3):208–217. Available from: https://doi.org/10.3978/j.issn.2224-4336.2014.04.01.
5. Battro AM. *Half a Brain Is Enough: The Story of Nico.* University Press; 2001.
6. Holderbaum CS, Mansur LL, Salles JFde. Heterogeneity in semantic priming effect with a lexical decision task in patients after left hemisphere stroke. *Dement Neuropsychol.* 2016;10(2):91–97. Available from: https://doi.org/10.1590/S1980-5764-2016DN1002004.
7. Fournier NM, Calverley KL, Wagner JP, Poock JL, Crossley M. Impaired social cognition 30 years after hemispherectomy for intractable epilepsy: the importance of the right hemisphere in complex social functioning. *Epilepsy Behav.* 2008;12(3):460–471. Available from: https://doi.org/10.1016/j.yebeh.2007.12.009.
8. Stark RE, Bleile K, Brandt J, Freeman J, Vining EP. Speech-language outcomes of hemispherectomy in children and young adults. *Brain Lang.* 1995;51(3):406–421. Available from: https://doi.org/10.1006/brln.1995.1068.
9. Stark RE, McGregor KK. Follow-up study of a right- and a left-hemispherectomized child: implications for localization and impairment of language in children. *Brain Lang.* 1997;60(2):222–242. Available from: https://doi.org/10.1006/brln.1997.1800.
10. Sebastianelli L, Saltuari L, Nardone R. How the brain can rewire itself after an injury: the lesson from hemispherectomy. *Neural Regen Res.* 2017;12(9):1426–1427. Available from: https://doi.org/10.4103/1673-5374.215247.

11. Sebastianelli L, Versace V, Taylor A, et al. Functional reorganization after hemispherectomy in humans and animal models: what can we learn about the brain's resilience to extensive unilateral lesions? *Brain Res Bull.* 2017;131:156–167. Available from: https://doi.org/10.1016/j.brainresbull.2017.04.005.
12. Lebel C, Gee M, Camicioli R, Wieler M, Martin W, Beaulieu C. Diffusion tensor imaging of white matter tract evolution over the lifespan. *NeuroImage.* 2012;60(1):340–352. Available from: https://doi.org/10.1016/j.neuroimage.2011.11.094.
13. Huttenlocher PR. Morphometric study of human cerebral cortex development. *Neuropsychologia.* 1990;28 (6):517–527. Available from: https://doi.org/10.1016/0028-3932(90)90031-I.
14. Piochon C, Kano M, Hansel C. LTD-like molecular pathways in developmental synaptic pruning. *Nat Neurosci.* 2016;19(10):1299–1310. Available from: https://doi.org/10.1038/nn.4389.
15. Hawes SL, Evans RC, Unruh BA, et al. Multimodal plasticity in dorsal striatum while learning a lateralized navigation task. *J Neurosci.* 2015;35(29):10535–10549. Available from: https://doi.org/10.1523/JNEUROSCI.4415-14.2015.
16. Knafo S, Libersat F, Barkai E. Olfactory learning-induced morphological modifications in single dendritic spines of young rats. *Eur J Neurosci.* 2005;21(8):2217–2226. Available from: https://doi.org/10.1111/j.1460-9568.2005.04041.x.
17. Bridge H, Watkins KE. Structural and functional brain reorganisation due to blindness: the special case of bilateral congenital anophthalmia. *Neurosci Biobehav Rev.* 2019;107:765–774. Available from: https://doi.org/10.1016/j.neubiorev.2019.10.006.
18. Sabourin CJ, Merrikhi Y, Lomber SG. Do blind people hear better? *Trends Cogn Sci.* 2022;26(11):999–1012. Available from: https://doi.org/10.1016/j.tics.2022.08.016.
19. Caliskan E, Dane S. Left-handedness in blind and sighted children. *Laterality.* 2009;14(2):205–213. Available from: https://doi.org/10.1080/13576500802586251.
20. Larsen S, Håkonsen K. Absence of ear asymmetry in blind children on a dichotic listening task compared to sighted controls. *Brain Lang.* 1983;18(2):192–198. Available from: https://doi.org/10.1016/0093-934x(83)90014-7.
21. Hermelin B, O'Connor N. Functional asymmetry in the reading of Braille. *Neuropsychologia.* 1971;9(4):431–435. Available from: https://doi.org/10.1016/0028-3932(71)90007-8.
22. Lane C, Kanjlia S, Richardson H, Fulton A, Omaki A, Bedny M. Reduced left lateralization of language in congenitally blind individuals. *J Cogn Neurosci.* 2017;29(1):65–78. Available from: https://doi.org/10.1162/jocn_a_01045.
23. Rinaldi L, Ciricugno A, Merabet LB, Vecchi T, Cattaneo Z. The effect of blindness on spatial asymmetries. *Brain Sci.* 2020;10(10). Available from: https://doi.org/10.3390/brainsci10100662.
24. Glick H, Sharma A. Cross-modal plasticity in developmental and age-related hearing loss: clinical implications. *Hear Res.* 2017;343:191–201. Available from: https://doi.org/10.1016/j.heares.2016.08.012.
25. Papadatou-Pastou M, Sáfár A. Handedness prevalence in the deaf: meta-analyses. *Neurosci Biobehav Rev.* 2016;60:98–114. Available from: https://doi.org/10.1016/j.neubiorev.2015.11.013.
26. Trettenbrein PC, Papitto G, Friederici AD, Zaccarella E. Functional neuroanatomy of language without speech: an ALE meta-analysis of sign language. *Hum Brain Mapp.* 2021;42(3):699–712. Available from: https://doi.org/10.1002/hbm.25254.
27. van der Haegen L, Acke F, Vingerhoets G, et al. Laterality and unilateral deafness: patients with congenital right ear deafness do not develop atypical language dominance. *Neuropsychologia.* 2016;93(Pt B):482–492. Available from: https://doi.org/10.1016/j.neuropsychologia.2015.10.032.
28. Cattaneo Z, Rinaldi L, Geraci C, Cecchetto C, Papagno C. Spatial biases in deaf, blind, and deafblind individuals as revealed by a haptic line bisection task. *Q J Exp Psychol (Hove).* 2018;71(11):2325–2333. Available from: https://doi.org/10.1177/1747021817741288.
29. Beaton A. *Left Side, Right Side: A Review of Laterality Research.* Yale University Press; 1986.
30. Dennis M., Whitaker HA. 8 - Hemispheric equipotentiality and language acquisition. In: Segalowitz SJ, ed. Language Development and Neurological Theory [conference held in May, 1975 in St. Catharines, Ontario]. 4. [printing]. Academic Press; 1980:93-106.
31. Lenneberg EH. *Biological Foundations of Language.* Wiley; 1967.
32. Basser LS. Hemiplegia of early onset and the faculty of speech with special reference to the effects of hemispherectomy. *Brain.* 1962;85:427–460. Available from: https://doi.org/10.1093/brain/85.3.427.

33. Taylor DC. Differential rates of cerebral maturation between sexes and between hemispheres. *Evid Epilepsy Lancet*. 1969;2(7612):140–142. Available from: https://doi.org/10.1016/s0140-6736(69)92445-3.
34. Chiron C, Jambaque I, Nabbout R, Lounes R, Syrota A, Dulac O. The right brain hemisphere is dominant in human infants. *Brain*. 1997;120(Pt 6):1057–1065. Available from: https://doi.org/10.1093/brain/120.6.1057.
35. Corballis MC, Morgan MJ. On the biological basis of human laterality: I. Evidence for a maturational left–right gradient. *Behav Brain Sci*. 1978;1(2):261–269. Available from: https://doi.org/10.1017/S0140525X00074471.
36. Lansdell H. Verbal and nonverbal factors in right-hemisphere speech: relation to early neurological history. *J Comp Physiol Psychol*. 1969;69(4):734–738. Available from: https://doi.org/10.1037/h0028306.
37. Danguecan AN, Smith ML. Re-examining the crowding hypothesis in pediatric epilepsy. *Epilepsy Behav*. 2019;94:281–287. Available from: https://doi.org/10.1016/j.yebeh.2019.01.038.
38. Satz P, Strauss E, Hunter M, Wada J. Re-examination of the crowding hypothesis: effects of age of onset. *Neuropsychology*. 1994;8(2):255–262. Available from: https://doi.org/10.1037/0894-4105.8.2.255.
39. Strauss E, Satz P, Wada J. An examination of the crowding hypothesis in epileptic patients who have undergone the carotid amytal test. *Neuropsychologia*. 1990;28(11):1221–1227. Available from: https://doi.org/10.1016/0028-3932(90)90057-u.
40. Amunts K. Structural indices of asymmetry. In: Westerhausen R, Hugdahl K, eds. *The Two Halves of the Brain: Information Processing in the Cerebral Hemispheres*. MIT Press; 2010:145–176.
41. Craig BT, Geeraert B, Kinney-Lang E, et al. Structural brain network lateralization across childhood and adolescence. *Hum Brain Mapp*. 2023;44(4):1711–1724. Available from: https://doi.org/10.1002/hbm.26169.
42. Ocklenburg S, Friedrich P, Güntürkün O, Genç E. Intrahemispheric white matter asymmetries: the missing link between brain structure and functional lateralization? *Rev Neurosci*. 2016;27(5):465–480. Available from: https://doi.org/10.1515/revneuro-2015-0052.
43. Kovel CGF, de, Lisgo SN, Fisher SE, Francks C. Subtle left-right asymmetry of gene expression profiles in embryonic and foetal human brains. *Sci Rep*. 2018;8(1):12606. Available from: https://doi.org/10.1038/s41598-018-29496-2.
44. Kovel CGF, de, Lisgo S, Karlebach G, et al. Left-right asymmetry of maturation rates in human embryonic neural development. *Biol Psych*. 2017;82(3):204–212. Available from: https://doi.org/10.1016/j.biopsych.2017.01.016.
45. Miao N, Lai X, Zeng Z, Cai W, Chen W, Sun T. Differential expression of microRNAs in the human fetal left and right cerebral cortex. *Mol Biol Rep*. 2020;47(9):6573–6586. Available from: https://doi.org/10.1007/s11033-020-05708-9.
46. Zhao X, Liang W, Wang W, et al. Changes in and asymmetry of the proteome in the human fetal frontal lobe during early development. *Commun Biol*. 2022;5(1):1031. Available from: https://doi.org/10.1038/s42003-022-04003-6.
47. Wada JA, Clarke R, Hamm A. Cerebral hemispheric asymmetry in humans. Cortical speech zones 100 adults 100 infant brains. *Arch Neurol*. 1975;32(4):239–246. Available from: https://doi.org/10.1001/archneur.1975.00490460055007.
48. Chi JG, Dooling EC, Gilles FH. Left-right asymmetries of the temporal speech areas of the human fetus. *Arch Neurol*. 1977;34(6):346–348. Available from: https://doi.org/10.1001/archneur.1977.00500180040008.
49. Bisiacchi P, Cainelli E. Structural and functional brain asymmetries in the early phases of life: a scoping review. *Brain Struct Funct*. 2022;227(2):479–496. Available from: https://doi.org/10.1007/s00429-021-02256-1.
50. Dubois J, Benders M, Lazeyras F, et al. Structural asymmetries of perisylvian regions in the preterm newborn. *NeuroImage*. 2010;52(1):32–42. Available from: https://doi.org/10.1016/j.neuroimage.2010.03.054.
51. Reynolds JE, Long X, Grohs MN, Dewey D, Lebel C. Structural and functional asymmetry of the language network emerge in early childhood. *Dev Cogn Neurosci*. 2019;39:100682. Available from: https://doi.org/10.1016/j.dcn.2019.100682.
52. Roe JM, Vidal-Pineiro D, Amlien IK, et al. Tracing the development and lifespan change of population-level structural asymmetry in the cerebral cortex. *Elife*. 2023;12. Available from: https://doi.org/10.7554/eLife.84685.
53. Williams LZJ, Fitzgibbon SP, Bozek J, et al. Structural and functional asymmetry of the neonatal cerebral cortex. *Nat Hum Behav*. 2023;7(6):942–955. Available from: https://doi.org/10.1038/s41562-023-01542-8.
54. Zhong S, He Y, Shu H, Gong G. Developmental changes in topological asymmetry between hemispheric brain white matter networks from adolescence to young adulthood. *Cereb Cortex*. 2017;27(4):2560–2570. Available from: https://doi.org/10.1093/cercor/bhw109.

55. Goddings A-L, Roalf D, Lebel C, Tamnes CK. Development of white matter microstructure and executive functions during childhood and adolescence: a review of diffusion MRI studies. *Dev Cogn Neurosci.* 2021;51:101008. Available from: https://doi.org/10.1016/j.dcn.2021.101008.
56. Hepper PG. The developmental origins of laterality: fetal handedness. *Dev Psychobiol.* 2013;55(6):588–595. Available from: https://doi.org/10.1002/dev.21119.
57. Ocklenburg S, Beste C, Güntürkün O. Handedness: a neurogenetic shift of perspective. *Neurosci Biobehav Rev.* 2013;37(10 Pt 2):2788–2793. Available from: https://doi.org/10.1016/j.neubiorev.2013.09.014.
58. Ocklenburg S, Schmitz J, Moinfar Z, et al. Epigenetic regulation of lateralized fetal spinal gene expression underlies hemispheric asymmetries. *Elife.* 2017;6. Available from: https://doi.org/10.7554/eLife.22784.
59. Ververs IA, de Vries JI, van Geijn HP, Hopkins B. Prenatal head position from 12–38 weeks. I. Developmental aspects. *Early Hum Dev.* 1994;39(2):83–91. Available from: https://doi.org/10.1016/0378-3782(94)90157-0.
60. Ververs IA, de Vries JI, van Geijn HP, Hopkins B. Prenatal head position from 12–38 weeks. II. Eff fetal orientat placent localization. *Early Hum Dev.* 1994;39(2):93–100. Available from: https://doi.org/10.1016/0378-3782(94)90158-9.
61. Hepper PG, McCartney GR, Shannon EA. Lateralised behaviour in first trimester human foetuses. *Neuropsychologia.* 1998;36(6):531–534. Available from: https://doi.org/10.1016/s0028-3932(97)00156-5.
62. Hepper PG, Shahidullah S, White R. Origins of fetal handedness. *Nature.* 1990;347(6292):431. Available from: https://doi.org/10.1038/347431b0.
63. McCartney G, Hepper P. Development of lateralized behaviour in the human fetus from 12 to 27 weeks' gestation. *Dev Med Child Neurol.* 1999;41(2):83–86. Available from: https://doi.org/10.1017/s0012162299000183.
64. Michel GF. Right-handedness: a consequence of infant supine head-orientation preference? *Science.* 1981;212(4495):685–687. Available from: https://doi.org/10.1126/science.7221558.
65. Hopkins B, Lems W, Janssen B, Butterworth G. Postural and motor asymmetries in newlyborns. *Hum Neurobiol.* 1987;6(3):153–156.
66. Michel GF. Handedness development: a model for investigating the development of hemispheric specialization and interhemispheric coordination. *Symmetry.* 2021;13(6):992. Available from: https://doi.org/10.3390/sym13060992.
67. Nelson EL. Developmental cascades as a framework for primate handedness. *Front Behav Neurosci.* 2022;16:1063348. Available from: https://doi.org/10.3389/fnbeh.2022.1063348.
68. Bacelar AM, Teixeira LA. Footedness across ages: distinction between mobilization and stabilization tasks. *Laterality.* 2015;20(2):141–153. Available from: https://doi.org/10.1080/1357650X.2014.924959.
69. Gabbard C. Foot laterality during childhood: a review. *Int J Neurosci.* 1993;72(3-4):175–182. Available from: https://doi.org/10.3109/00207459309024106.
70. Bishop DVM. Cerebral asymmetry and language development: cause, correlate, or consequence? *Science.* 2013;340(6138):1230531. Available from: https://doi.org/10.1126/science.1230531.
71. van Essen DC. A population-average, landmark- and surface-based (PALS) atlas of human cerebral cortex. *NeuroImage.* 2005;28(3):635–662. Available from: https://doi.org/10.1016/j.neuroimage.2005.06.058.
72. Dubois J, Poupon C, Thirion B, et al. Exploring the early organization and maturation of linguistic pathways in the human infant brain. *Cereb Cortex.* 2016;26(5):2283–2298. Available from: https://doi.org/10.1093/cercor/bhv082.
73. Dehaene-Lambertz G, Spelke ES. The infancy of the human brain. *Neuron.* 2015;88(1):93–109. Available from: https://doi.org/10.1016/j.neuron.2015.09.026.
74. Perani D, Saccuman MC, Scifo P, et al. Neural language networks at birth. *PNAS.* 2011;108(38):16056–16061. Available from: https://doi.org/10.1073/pnas.1102991108.
75. Friederici AD. The brain basis of language processing: from structure to function. *Physiol Rev.* 2011;91(4):1357–1392. Available from: https://doi.org/10.1152/physrev.00006.2011.
76. Bartha-Doering L, Kollndorfer K, Schwartz E, et al. Fetal temporal sulcus depth asymmetry has prognostic value for language development. *Commun Biol.* 2023;6(1):109. Available from: https://doi.org/10.1038/s42003-023-04503-z.
77. Quin-Conroy JE, Chen Y, Bayliss DM, Badcock NA. Magic hats and teddy bear picnics: language and visuospatial lateralisation tasks for children. *Laterality.* 2022;27(2):232–256. Available from: https://doi.org/10.1080/1357650X.2021.2020808.

78. Kaul D, Papadatou-Pastou M, Learmonth G. A meta-analysis of the line bisection task in children. *Laterality*. 2023;28(1):48–71. Available from: https://doi.org/10.1080/1357650X.2022.2147941.
79. Badzakova-Trajkov G, Corballis MC, Häberling IS. Complementarity or independence of hemispheric specializations? A brief review. *Neuropsychologia*. 2016;93(Pt B):386–393. Available from: https://doi.org/10.1016/j.neuropsychologia.2015.12.018.
80. Badzakova-Trajkov G, Häberling IS, Roberts RP, Corballis MC. Cerebral asymmetries: complementary and independent processes. *PLoS One*. 2010;5(3):e9682. Available from: https://doi.org/10.1371/journal.pone.0009682.
81. Vingerhoets G, Gerrits R, Bogaert S. Atypical brain functional segregation is more frequent in situs inversus totalis. *Cortex*. 2018;106:12–25. Available from: https://doi.org/10.1016/j.cortex.2018.04.012.
82. Vingerhoets G. Phenotypes in hemispheric functional segregation? Perspectives and challenges. *Phys Life Rev*. 2019;30:1–18. Available from: https://doi.org/10.1016/j.plrev.2019.06.002.
83. Carey DP, Johnstone LT. Quantifying cerebral asymmetries for language in dextrals and adextrals with random-effects meta analysis. *Front Psychol*. 2014;5:1128. Available from: https://doi.org/10.3389/fpsyg.2014.01128.
84. Knecht S, Dräger B, Deppe M, et al. Handedness and hemispheric language dominance in healthy humans. *Brain*. 2000;123(Pt 12):2512–2518. Available from: https://doi.org/10.1093/brain/123.12.2512.
85. Karlsson EM, Hugdahl K, Hirnstein M, Carey DP. Analysis of distributions reveals real differences on dichotic listening scores between left- and right-handers. *Cereb Cortex Commun*. 2023;4(2):tgad009. Available from: https://doi.org/10.1093/texcom/tgad009.
86. Karlsson EM, Johnstone LT, Carey DP. The depth and breadth of multiple perceptual asymmetries in right handers and non-right handers. *Laterality*. 2019;24(6):707–739. Available from: https://doi.org/10.1080/1357650X.2019.1652308.
87. Gerrits R, Verhelst H, Vingerhoets G. Mirrored brain organization: statistical anomaly or reversal of hemispheric functional segregation bias? *PNAS*. 2020;117(25):14057–14065. Available from: https://doi.org/10.1073/pnas.2002981117.
88. Gerrits R. Variability in hemispheric functional segregation phenotypes: a review and general mechanistic model. *Neuropsychol Rev*. 2022. Available from: https://doi.org/10.1007/s11065-022-09575-y.
89. Raz N, Rodrigue KM. Differential aging of the brain: patterns, cognitive correlates and modifiers. *Neurosci Biobehav Rev*. 2006;30(6):730–748. Available from: https://doi.org/10.1016/j.neubiorev.2006.07.001.
90. Hill C, van Gemmert AWA, Fang Q, Hou L, Wang J, Pan Z. Asymmetry in the aging brain: a narrative review of cortical activation patterns and implications for motor function. *Laterality*. 2020;25(4):413–429. Available from: https://doi.org/10.1080/1357650X.2019.1707219.
91. Kalisch T, Wilimzig C, Kleibel N, Tegenthoff M, Dinse HR. Age-related attenuation of dominant hand superiority. *PLoS One*. 2006;1(1):e90. Available from: https://doi.org/10.1371/journal.pone.0000090.
92. Sebastjan A, Skrzek A, Ignasiak Z, Sławińska T. Age-related changes in hand dominance and functional asymmetry in older adults. *PLoS One*. 2017;12(5):e0177845. Available from: https://doi.org/10.1371/journal.pone.0177845.
93. Rogers LJ. Differential ageing of the brain hemispheres: evidence from a longitudinal study of hand preferences in common marmosets. *Symmetry*. 2021;13(12):2349. Available from: https://doi.org/10.3390/sym13122349.
94. Dolcos F, Rice HJ, Cabeza R. Hemispheric asymmetry and aging: right hemisphere decline or asymmetry reduction. *Neurosci Biobehav Rev*. 2002;26(7):819–825. Available from: https://doi.org/10.1016/s0149-7634(02)00068-4.
95. Salthouse TA. Differential age-related influences on memory for verbal-symbolic information and visual-spatial information? *J Gerontol B Psychol Sci Soc Sci*. 1995;50(4):P193–P201. Available from: https://doi.org/10.1093/geronb/50B.4.P193.
96. Hatta T, Iwahara A, Hatta T, et al. Developmental trajectories of verbal and visuospatial abilities in healthy older adults: comparison of the hemisphere asymmetry reduction in older adults model and the right hemi-ageing model. *Laterality*. 2015;20(1):69–81. Available from: https://doi.org/10.1080/1357650X.2014.917656.
97. Baltes PB, Lindenberger U. Emergence of a powerful connection between sensory and cognitive functions across the adult life span: a new window to the study of cognitive aging? *Psychol Aging*. 1997;12(1):12–21. Available from: https://doi.org/10.1037/0882-7974.12.1.12.
98. Cabeza R. Hemispheric asymmetry reduction in older adults: the HAROLD model. *Psychol Aging*. 2002;17(1):85–100. Available from: https://doi.org/10.1037/0882-7974.17.1.85.
99. Goldstein G, Shelly C. Does the right hemisphere age more rapidly than the left? *J Clin Neuropsychol*. 1981;3(1):65–78. Available from: https://doi.org/10.1080/01688638108403114.

100. McDowell CL, Harrison DW, Demaree HA. Is right hemisphere decline in the perception of emotion a function of aging? *Int J Neurosci*. 1994;79(1-2):1–11. Available from: https://doi.org/10.3109/00207459408986063.
101. Cherry BJ, Hellige JB, McDowd JM. Age differences and similarities in patterns of cerebral hemispheric asymmetry. *Psychol Aging*. 1995;10(2):191–203. Available from: https://doi.org/10.1037/0882-7974.10.2.191.
102. Kong X-Z, Mathias SR, Guadalupe T, et al. Mapping cortical brain asymmetry in 17,141 healthy individuals worldwide via the ENIGMA Consortium. *PNAS*. 2018;115(22):E5154–E5163. Available from: https://doi.org/10.1073/pnas.1718418115.
103. Minkova L, Habich A, Peter J, Kaller CP, Eickhoff SB, Klöppel S. Gray matter asymmetries in aging and neurodegeneration: a review and meta-analysis. *Hum Brain Mapp*. 2017;38(12):5890–5904. Available from: https://doi.org/10.1002/hbm.23772.
104. Nyberg L, Cabeza R, Tulving E. PET studies of encoding and retrieval: the HERA model. *Psychonomic Bull Rev*. 1996;3(2):135–148. Available from: https://doi.org/10.3758/BF03212412.
105. Collins K, Mohr C. Performance of younger and older adults in lateralised right and left hemisphere asymmetry tasks supports the HAROLD model. *Laterality*. 2013;18(4):491–512. Available from: https://doi.org/10.1080/1357650X.2012.724072.
106. Jewell G, McCourt ME. Pseudoneglect: a review and meta-analysis of performance factors in line bisection tasks. *Neuropsychologia*. 2000;38(1):93–110. Available from: https://doi.org/10.1016/s0028-3932(99)00045-7.
107. Learmonth G, Papadatou-Pastou M. A meta-analysis of line bisection and landmark task performance in older adults. *Neuropsychol Rev*. 2022;32(2):438–457. Available from: https://doi.org/10.1007/s11065-021-09505-4.
108. Wang J, Przybyla A, Wuebbenhorst K, Haaland KY, Sainburg RL. Aging reduces asymmetries in interlimb transfer of visuomotor adaptation. *Exp Brain Res*. 2011;210(2):283–290. Available from: https://doi.org/10.1007/s00221-011-2631-1.
109. Przybyla A, Haaland KY, Bagesteiro LB, Sainburg RL. Motor asymmetry reduction in older adults. *Neurosci Lett*. 2011;489(2):99–104. Available from: https://doi.org/10.1016/j.neulet.2010.11.074.

CHAPTER

12

Sex differences in hemispheric asymmetries

OUTLINE

Introductory short story

In the summer of 1992, Catrin Rode decided that she needed a change. Her initial approach to starting experiments for her Diploma thesis in Munich had ended in deep frustration. She was now keen to talk to several possible supervisors to select a topic that was both scientifically interesting and doable. At that time, she was enrolled as a student

DOI: https://doi.org/10.1016/B978-0-323-99737-9.00015-X

of psychology at the University of Konstanz in southern Germany. She had taken a seminar on brain asymmetries by Onur Güntürkün and had read articles that suggested that functional asymmetries could change during the menstrual cycle. Onur was fascinated by this possibility since it opened a way to study the functionality of the same person's brain in short succession with or without asymmetries of a certain cognitive system. Somehow, his fascination was contagious, and so Catrin visited him in his tiny office and discussed the option to study this topic. It was clear that they needed young female subjects with regular cycles who did not take oral contraceptives. In addition, they should test the same subjects once during menstruation and once during the luteal phase when progesterone and estradiol levels are high. Finally, hormone levels had to be controlled by taking repeated blood samples. It sounded fascinating, challenging, and doable. They both said, "Let's do it."

The next few months were filled with creating lateralized lexical and figural tasks, testing them in pilot studies, finding people who gave permission to take blood samples, locating a laboratory to run the immunoassays, and so forth. When everything was ready, the main obstacle of the study became visible: Most young women took contraceptives. Those who did not either had irregular cycles or were unwilling to participate and give blood samples. In despair, Catrin stocked up her sample size by approaching some young catholic nuns in Munich who worked as nurses. They accepted.

When all the subjects had been tested and the data were on a hard disk, Catrin was alone at home and conducted the statistical analysis. During menstruation, her subjects had evinced a clear right-hemispheric advantage in the figural recognition task. So, things worked as they should. Then she checked the data of the same subjects during the luteal phase. When the histograms popped up on the screen, she first thought that something was wrong. She checked all data points again and reran the calculation. It was the same result pattern. Her subjects had shifted from a right-hemispheric superiority during menstruation to a completely symmetrical result pattern when sex hormones were high. Within a few weeks, their functional asymmetry had vanished.

Introduction

Possibly, every person who is interested in the human brain has read some popular science literature on sex differences from time to time. This kind of writing often outlines that women and men think in quite different ways and that the female brain is more symmetrically organized than the male brain. Moreover, they often go on to say that a good part of sex differences in cognition and behavior results from this difference in brain asymmetry. Is this true? To find answers to these and related questions, we will proceed in four steps: First, we will look at the evidence for sex differences in hemispheric asymmetries. Second, we will outline that not only anatomical but also hormonal mechanisms can change asymmetries. In women, they do this during the menstrual cycle. Third, we will develop a mechanistic explanation of how variations in hormonal concentrations during the menstrual cycle could produce changes in functional left-right differences. Last but not least, we will combine all the evidence to see whether the combination of long-lasting neuroanatomical asymmetries and dynamic and fast-changing hormonal variations possibly constitute functional left-right differences between the sexes. Let us start by discussing the evidence for lateralization differences between women and men.

Do women and men differ in structural brain asymmetries?

Studies on sex differences in structural hemispheric asymmetries have a long history.[1,2] Wisniewski[3] provided an excellent overview of studies of the 20th century and concluded that the most replicable effects concerned sex differences in the size of several areas of the corpus callosum with higher relative values for females. A further meta-analysis also found that the corpus callosum is relatively larger in women than in men when taking brain size into account.[4] These findings could imply a stronger interhemispheric communication in females, but the author also highlights that the sex difference is small and that most men and women overlap in the relative size of the corpus callosum. Therefore the biological implications of this sex difference are largely unclear. Result patterns for other areas like the planum temporale (see Chapter 5) were less consistent and mostly could not be replicated in more recent meta-analyses.[5]

The largest studies on both cortical and subcortical structural hemispheric asymmetries have been conducted by the ENIGMA consortium (see Chapter 4), and have included sex as a potential influence factor in their analyses.[6] The main ENIGMA study on structural asymmetries in the human cortex included cortical thickness and surface area data obtained from MRI scans of 17,141 healthy participants from 99 datasets.[7] Together with age and intracranial volume, sex was one of the factors that was related to variability in cortical asymmetry. For cortical thickness, no sex difference in overall hemispheric asymmetry was observed, but there were significant sex differences in medial temporal regions. Specifically, men showed a more leftward/less rightward cortical thickness asymmetry in the entorhinal cortex and the parahippocampal gyrus. For the cortical surface area, the ENIGMA scientists observed a sex difference in global asymmetry. In general, there was a rightward asymmetry, with larger surface areas in the right hemisphere. Interestingly, men on average showed a more rightward asymmetry than women, suggesting stronger cortical surface asymmetries in men compared to women. Regional-specific surface area asymmetries underlying this overall effect were found in several areas of the frontal, temporal, and parietal lobes, as well as in the anterior cingulate cortex. A significantly stronger rightward asymmetry of gray matter surface in men compared to women has also been reported in an earlier study in 101 participants.[8] In line with these findings, a study of 457 healthy participants from 5 datasets reported that women's brains were globally more symmetric than men's brains.[9]

In addition to their study on cortical asymmetries, the ENIGMA consortium also conducted a similar study on structural asymmetries in subcortical structures, which also included analyses of sex differences. Guadalupe et al.[10] combined MRI data from 15,847 subjects and scrutinized this huge data set for structural asymmetries that did or did not differ according to the sex of the subjects. The most important asymmetry was discovered for the nucleus accumbens, which showed a rightward volume bias. However, this asymmetry did not differ between females and males. Only two areas of the basal ganglia, the globus pallidus and the putamen, evinced sex-variant asymmetries. While these two structures showed an overall bias to be larger on the left side, this asymmetry was less pronounced in males. Overall, sex differences in brain asymmetry exist, but their effect sizes are mostly small.

This conclusion is supported by recent analyses of the MRI data of more than 1,400 subjects.[11] This analysis revealed several diverse sex differences. However, as the authors correctly pointed out, one should not only look for the existence of singular findings but search for overall patterns of "maleness" or "femaleness" by combining observations. Then it becomes evident that the brains of most subjects were constituted by unique "mosaics" of structural features, of which a few are more prevalent in female subjects, while others occur more often in males. A more recent meta-synthesis of studies on sex differences in the human brain came to an even stronger conclusion.[12] Here, the authors reported that few sex differences survived correction for brain size and that overall sex accounted for only about 1% of the variance in laterality. This led them to the conclusion that neuroscientists should "dump the dimorphism," as thinking in terms of "male brains" and "female brains" does not properly reflect the biological reality. However, this extreme view has also been criticized, since even small differences in brain structure can have meaningful consequences for behavior.[13] Moreover, a study in which male and female brains were specifically matched for their size did find significantly stronger rightward asymmetry in men compared to women only in the cerebellum, but not in the cortex.[14] Thus sex differences in the structural hemispheric asymmetries exist, but effect sizes tend to be small and their biological significance is largely unclear. A simplified categorization into male- or female-type brains is, at least for the cortex, impossible.

What would be an alternative? Sophie Hodgetts and Markus Hausmann recently suggested that future research on sex or gender differences in hemispheric asymmetries focus on a biopsychosocial approach that includes biological factors like sex hormones, but also environmental factors, and socially derived gender roles and stereotypes.[2] While such a nonbinary approach is much more complex than a male/female dimorphism, it could hopefully resolve many inconsistencies and open questions in this research field.

Neural networks derive a good part of the functional specificity from their wiring patterns. Consequently, recent studies have turned their attention to analyses of sex differences in structural connectivity as revealed by diffusion tensor imaging.[15–17] In one of these studies,[15] the structural connectomes of 521 female and 428 male subjects were reconstructed and compared. The authors describe that telencephalic structures in males were characterized by a stronger shift toward an intrahemispheric connectivity pattern, while females' interhemispheric connections were more pronounced (Fig. 12.1).

This fits nicely to the results reported by Wisniewski.[3] That said, it is important to emphasize that the vast majority of connections are within the hemispheres and not between them, regardless of sex.[18] Nevertheless, data by Ingalhalikar et al.[15] imply that in females, there is a measurable shift toward increased interaction between the cortical hemispheres via the corpus callosum and the commissura anterior. The authors speculate that this connectional pattern could imply that males have a predominance of a connectivity type that links perception to action within a hemisphere, while females are more predisposed to integrate information from both hemispheres into their action planning. These and similar speculations[19] have to be considered with caution since they are presently based on extremely indirect measures of axonal pathways, often show substantial individual variations, and

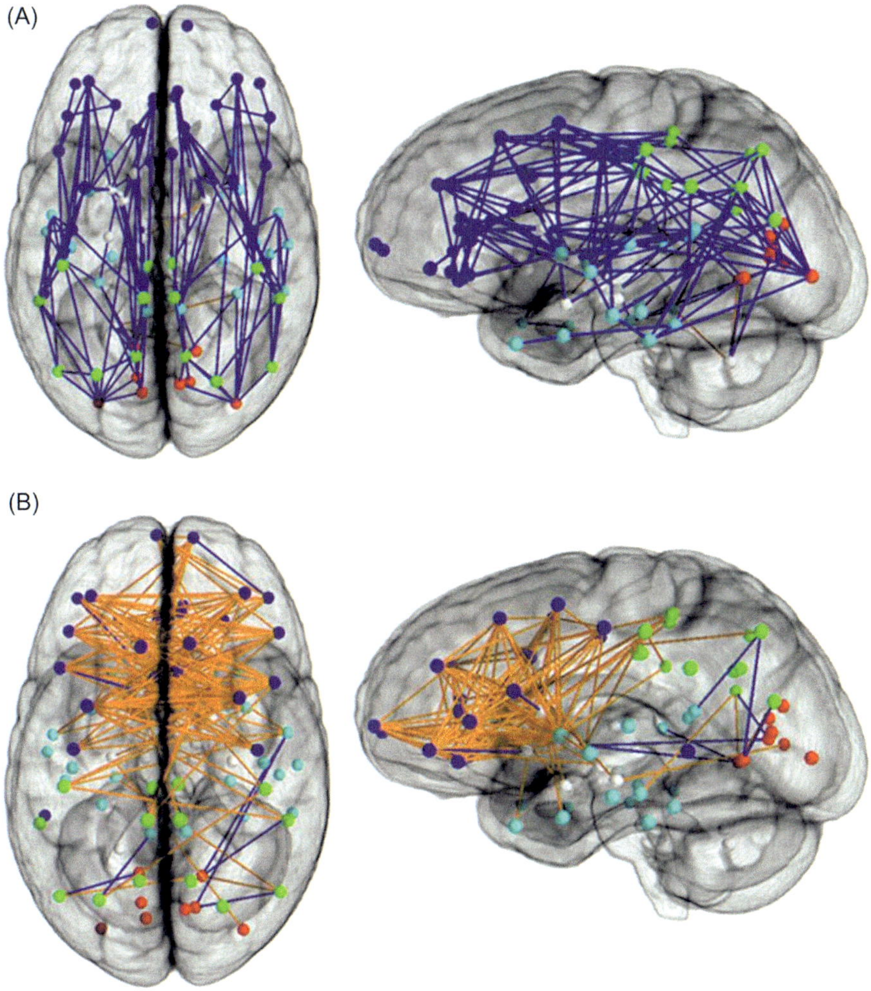

FIGURE 12.1 Connectivity patterns in the human brain show a higher intrahemispheric (*blue*) network structure in males (A) than in females (B). Interhemispheric connections are depicted in orange.[15] *Reproduced with permission from Ingalhalikar M, Smith A, Parker D, et al. Sex differences in the structural connectome of the human brain.* Proc Natl Acad Sci USA. *2014;111:823–828, with permission from the publisher.*

have small effect sizes. In addition, their most important problem is the difficulty of predicting the functional consequences of slight connectivity differences in such a bewilderingly complex system as the human brain. This does not devaluate the work and the hypotheses of these and further authors but should remind us that we are in a period in which we increasingly accumulate solid data on structural sex differences but mostly depend on mere speculations on their possible functional consequences.

There is one more point that is important to highlight about the results by Ingalhalikar et al.[15] Hänggi et al.[20] investigated axonal connectivity patterns in 138 large and small human brains and discovered a strong interaction between the size of a brain and the way it is wired. The larger the brain, the more it was connected *within* a single hemisphere instead of *between* hemispheres. This was sex-independent. Thus, men and women with large brains had relatively higher *intra*-hemispheric networks, while men and women with small brains had relatively stronger *inter*-hemispheric connections. However, since brain size is on average larger in men,[21] the results by Ingalhalikar et al.[15] are possibly not directly driven by the sex of the subjects but rather by their brain size. Still, the results by Ingalhalikar et al.[15] would be correct when applied to an average human sample in which we would expect female subjects to have smaller brains compared to males. However, based on Hänggi et al.,[20] we now would also expect females with large brains and males with small brains to have a male-like or female-like connectivity pattern, respectively. Thus the conclusions of Ingalhalikar et al.[15] are correct but for the wrong reasons.

Do women and men differ in functional brain asymmetries?

Handedness

At the beginning of the 20th century, a new youth culture spread over Germany: the cult of the novel human who can liberate each of the two brain hemispheres, increase hidden cognitive potential, and overcome old-fashioned cultural and useless biological boundaries. The followers of this culture were mostly young intellectuals who intensively trained their nondominant hands. They aimed to become ambidextrous and thereby unbind the hidden potential of the suppressed hemisphere that controlled their nondominant hand.[22] The once-famous physiologist Wilhelm Stekel is quoted as saying: "An immense perspective becomes visible. If humanity has indeed used only half its brain up to now, what humongous achievements can we expect from people who command twice the mental power! We so often lament the mental exhaustion of children; we complain that our brain, our nerves can't cope any longer with the onslaught of modern innovations; how would it be if there was a solution here that could be used to the benefit of mankind?" (according to Zauck,[22] pp. 264–265). Accordingly, whole schools in Germany taught their students to write with their left hand in order to double the cognitive potential of their brains. Never were any serious scientific analyses conducted to analyze whether these mass experiments were successful or had any impact on cognition. However, the culture of the novel human was paralleled by detailed analyses of the prevalence of handedness in different European nations.

In many cases, the prevalence of left-handers in Europe during this time hovered around 4%.[23] The scientists were not fully aware that this low estimate was due to massive social pressure on left-handers to switch to the other side, at least for writing (see Chapter 6). Scientists observed, however, that brain pathologies of all kinds were associated with a sharp increase in left-handedness. For example, Ganter[24] reported that about 20% of epileptic patients and intellectually challenged subjects were left-handed. These and similar reports strongly emphasized the view that left-handedness was associated with pathology or "cultural degeneration."[25] Consequently, it came as a bad surprise to the (mostly male) academic world when it became increasingly clear that males were significantly more often left-handed than females. For example, Smith[26] found that of 2055 schoolchildren, 2% of the girls and 5.5% of the boys were reported to be left-handed. Similarly, Gordon[27] reported that the prevalence of left-handedness in 1019 children was 2% for girls and 5% for boys.

Do these findings overlap with modern findings? Yes, they do, with the exception that today's best estimate for the prevalence of left-handedness is 10.6%.[28] In 2008, Papadatou-Pastou et al.[29] conducted a meta-analysis on sex differences in handedness that included data from 1,787,629 individuals extracted from 144 studies. They found a clear sex difference with higher prevalence of left-handedness in men. On average, the authors concluded that if the prevalence of female left-handedness was exactly 10%, we should expect an prevalence of male left-handedness of 12%. Sommer et al.[5] conducted a meta-analysis and practically came to the same conclusion. Interestingly, Grattan et al.[30] discovered a highly similar pattern when analyzing the preferred arm and leg movements of 36 newborns from several newborn nurseries in the United States. Thus we can safely conclude that handedness shows a clear sex difference with a higher prevalence of left-handedness in males.

Language

In 1977, Jeannette McGlone reported truly spectacular findings.[31] She reported that in 55 right-handed men and 47 right-handed women with unilateral brain damage, three times more men than women suffered from aphasia after left hemisphere lesions. This drastic sex difference in brain asymmetry was not related to age, education, and other controlled factors but could only be properly interpreted by assuming that females have a more bilateral organization of their language system and therefore had room for compensation after left-hemispheric damage. Understandably, such a bold finding stirred enormous interest. However, could McGlone's findings be replicated in further studies?

If the organization of the language system differs so substantially by sex, one should be able to demonstrate it by using fMRI. Frost et al.[32] tested 100 subjects (half of them female) in a language comprehension task. Both men and women showed strong left-hemispheric activation patterns, but no sex difference was found. Is this the end of the story? Certainly not. One main problem with the search for sex differences in language asymmetry is the vague usage of the term "language." Even handedness, arguably a much simpler function than language, is tested as a composite of different subfunctions in handedness questionnaires. In language, the subfunctions are so many and so differently organized that one should better test more than one such subfunction. Indeed, Harasty[33] argued that the specific task used by Frost et al.[32] would not be expected to reveal sex differences at all. Consequently, Knecht et al.[34] used a word generation paradigm that is usually associated with higher test scores in women. They tested 188 right-handed subjects. In addition, they employed a noninvasive quantitative technique to measure the relative proportion of left- and right-sided blood flow in the carotid arteries that supply oxygenated blood to the brain hemispheres. Again, no sex difference was visible.

However, scientists do not give up that easily. Now they tried to recruit a larger number of subjects for their experiments or to combine previous studies into meta-analyses to detect possible small effect sizes. Hiscock et al.[35] surveyed the results of 352 studies that tested the language domain of their subjects using dichotic and monaural listening experiments and found a small but significant sex difference with larger asymmetries for males. Boles[36] reported findings with 836 college students who were tested with various laterality procedures. In neither of the two language-related tasks could he detect a sex difference. Sommer et al.[5] identified a small sex difference with larger asymmetries for men in dichotic listening tasks that used the usual consonant-vowel stimuli. Hirnstein et al.[37] summarized their findings with exactly this technique by incorporating data from 1,782 participants. They could only discover a very small sex difference in brain asymmetry that also varied with age, with the largest effect (Cohen's $d = 0.31$) in adolescents. These observations accord with Bless et al.,[38] who assessed language lateralization in well over 4000 participants and found greater language lateralization in men than in women, albeit with a small effect of only Cohen's $d = 0.18$. Finally, Voyer[39] included the findings from 266 studies in his meta-analysis and concluded that verbal tasks in the auditory modality revealed a small but significant sex difference with men being more asymmetrically organized.

As repeatedly outlined in Chapter 5, language is a very complex ability that consists of a large number of diverse modules. These can be differently lateralized and some of these modules show sex differences, while others do not. This is nicely illustrated in the study by Meinschaefer et al.,[40] in which the authors tested the right-hemispheric superiority for processing of sentence prosody. Prosody processing goes beyond a simple analysis of the emotional undertone of speech, as it also involves aspects like syllable structure. That aspect is quite tricky in German where the syllable structure is predictable from the phonemic structure of a word. This is especially difficult with words that differ in the number of syllables but only minimally in phonemic structure—that is, those that differ in the presence versus absence of a neutral vowel such as the German words "breit" ("broad") and "bereit" ("ready"). Meinschaefer et al.[40] ran a dichotic listening study with several of these examples and found a clear right-hemispheric superiority. However, this lateralization was only present in males. Female subjects evinced no significant asymmetry.

Taken together, the very strong sex differences reported by McGlone[31] could not be replicated. Despite that, her conclusion that men are more lateralized in the verbal domain seems in principle to be correct. However, the overall effect size of this sex difference is so small that only studies with very large sample numbers, meta-analyses, or precise module-testing approaches can demonstrate the presence of this effect. Further below, we will discuss why these findings possibly overlooked an important variable that moderates some of the sex differences in brain asymmetry.

Figure recognition

For figure recognition, findings of studies are partly clearer. Voyer et al.[41] conducted a meta-analysis on sex differences in spatial abilities and concluded that the mental rotation test showed a strong and consistent difference favoring men. Some years later, Johnson et al.[42] showed that men displayed a higher degree of right-hemispheric activation using event-related potentials. A previous behavioral study had come to the same conclusion.[43]

Also, face recognition usually shows robust lateralization effects, with dominance of the right hemisphere.[44] This asymmetry is more pronounced in males.[45,46] For example,

Proverbio et al.[47] measured face-sensitive event-related potentials over the occipital and temporal cortices. They revealed a strong right-hemispheric dominance for men and a lack of such an asymmetry in women. The authors concluded that men display a salient asymmetric functioning of their associative visual cortex while decoding faces and expressions, while females show a more bilateral functioning. This finding was also confirmed by a more recent meta-analysis of 34 EEG and MEG studies ($n = 817$) on face processing and hemispheric asymmetries in the M/N170 response.[48] The results revealed a pronounced rightward asymmetry of bioelectrical activity in men. In contrast, women showed bilateral activity or leftward asymmetry.

Already in 1974, Paul Bakan[49] had reported that women tended to confuse the right and the left side of pictorially displayed body parts more often than males. In addition, the data hinted at stronger brain asymmetries in this task. Hirnstein et al.[50] could replicate the poorer performance of women compared to men in this task and demonstrated that this was related to the lateralized dichotic listening scores in women. In a subsequent study, Hjelmervik et al.[51] revealed that during a left-right discrimination task, brain activity patterns differed between sexes: both women and men activated the gyrus angularis in the inferior parietal cortex, but women, in addition, had stronger prefrontal activation patterns that are indicative of stronger top-down control during task execution. The results by Hjelmervik et al.[51] did not support a differential asymmetry pattern in women and men in this task, but the small number of subjects might not have been sufficient to detect such a difference.

Rasmjou et al.[52] studied visual illusions and reported an especially strong sex effect on figural asymmetry. Visual illusions are very interesting since they possibly represent the unplanned side effects of neural mechanisms that have evolved to increase the ability of animals to reveal the true three-dimensional structure of an object. Several such illusions appear to be lateralized and have been found to mainly deceive the right hemisphere.[53,54] Several clinical studies in patients with damage to only one hemisphere have provided further evidence for a dominance of the right hemisphere.[55] Based on these studies, Rasmjou et al.[52] used the herringbone illusion (Fig. 12.2), in which a vertical herringbone shading inside a square slants the sides, giving the impression of a trapeze. This illusion is perfect for laterality studies using visual half-field techniques (see Chapter 3 for description) since it does not depend on the perception of imaginary contours and thus is robust enough for lateral viewing. There were four different stimuli consisting of variations of two shapes: a square and a trapeze, each with two different shadings, either simple diagonal or herringbone. Eight pairs of different combinations of these four stimuli were presented, each with a fixation cross in between. The critical analysis of the results was only based on responses to the critical stimuli shown in Fig. 12.2.

The job of the subjects was basic: They had to fixate on the central cross, and if they spotted a trapezoid on one side, they had to press the switch on that side as quickly as possible. The results were clear-cut: There was no sex difference in the extent to which women or men were deceived by the illusion. However, there was an important difference concerning the deceived hemispheres. In women, both hemispheres contributed equally to the illusion. In men, however, only the right hemisphere was strongly deceived. The results of Rasmjou et al.[52] strongly support the existence of a sex difference of asymmetry in figure recognition with more pronounced left-right differences in males.

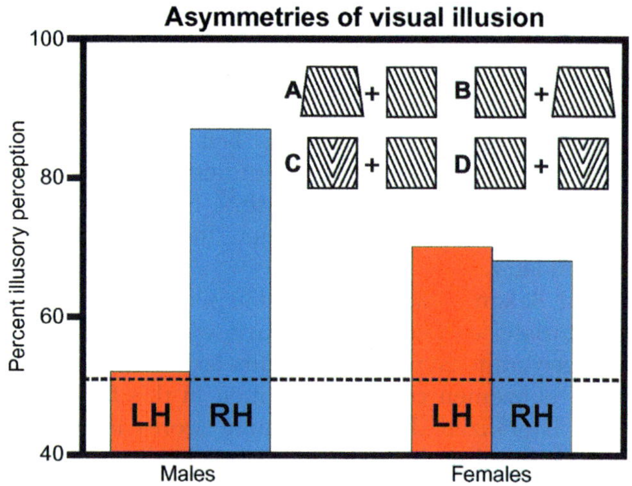

FIGURE 12.2 Result pattern by Rasmjou et al.[52] that reveals an asymmetry in the perception of the herringbone illusion in male subjects. Females show no hemispheric difference. The horizontal dotted line marks the point at which a hemisphere is not deceived. The inset shows some of the figure combinations used in the study. The plus sign shows the fixation point of the subjects. Two patterns are then flashed, one to the left (projecting to the right hemisphere) and one to the right of this point (represented in the left hemisphere). (A) A trapeze and a square with slanted lines inside; the subject should indicate that a trapeze is on the left. (B) The same in reversed order. (C) On both sides, a square is shown, but the herringbone pattern can deceive subjects to perceive a trapeze on the left. (D) The same in reversed order. LH: left hemisphere, RH: right hemisphere. *Reproduced with permission from Rasmjou S, Hausmann M, Güntürkün O. Hemispheric dominance and gender in the perception of an illusion.* Neuropsychologia. *1999;37:1041–1047.*

Taken together, there is some support for sex differences of functional brain asymmetries in figural tasks. This general conclusion is supported by meta-analyses. Hiscock et al.[56] analyzed articles from 368 issues of neuropsychology journals and discovered that male subjects showed stronger asymmetries in many tasks that employed visual stimuli such as faces. Voyer[39] also revealed that nonverbal tasks, both in the visual and the auditory modality, show sex differences in brain asymmetry.

Functional connectivity

Just as hemispheric asymmetries in structural connectivity between different brain areas can be investigated by comparing the volume or length of different white matter tracts, hemispheric asymmetries in functional connectivity can be assessed by comparing to what extent different brain areas are active together. A recent meta-analysis investigated sex differences in coactivation of homotopic brain areas in the left and right hemisphere.[57] The results of this meta-analysis on functional connectivity were similar to the previously described findings for structural connectivity.[15] Women showed stronger interhemispheric coactivation than men. The authors concluded that this finding suggested that women showed decreased hemispheric asymmetry and more integration of neural information between the hemispheres than men.

Conclusion

Overall, there is some support for sex differences in functional brain asymmetries. This effect is stronger for handedness than for other tasks. In handedness tasks, women showed more pronounced right-handedness. In other laterality tasks that tap language or visuospatial domains, female subjects showed a more bilateral pattern. However, in most of these experiments, a large sample size was needed to reveal this pattern. These findings are consistent, to some extent, with results by Joel et al.,[11] who conducted an imaging analysis of more than 1400 human brains and could not find an overall sex difference for gray and white matter as well as connectivity. Rather, their findings speak for a view according to which most brains are constituted of unique "mosaics" of features of which some are more common in females, while others are more often observed in males. The authors concluded that although sex differences in brain and function exist, human brains cannot easily be identified as belonging to one of two distinct categories: males and females.

This could have been the end of this chapter. However, results that slowly accumulated since the 1980s have forced us to take a second look at this matter.

Hormonal effects on functional brain asymmetries

Cerebral asymmetries in women not only seem to differ in scale but also tend to be more heterogeneous concerning their direction.[31,58] This is possibly due to another factor that interacts with the sex of the subject. What could this other factor be? Could hormones play a role? Very likely they do.[2,59] Before we start talking about sex hormones, we will discuss the two main ways in which hormones can alter brain function.

If we consult the textbooks to learn more about hormone effects on the brain, we often find a differentiation between the organizing and activating effects of gonadal hormones.[60] "Organizing" effects are defined as the developmental mechanisms with which hormones shape the anatomy of the developing brain and thus program the sex-specific behavior of individuals. The second mechanism is called "activation." It describes the acute effects of gonadal hormones on the fully developed nervous system. Activation effects are, for example, responsible for activating and maintaining sex-specific behavior in adulthood in a given situation.[61] Taken together, organizing effects set in early in ontogeny, shape the brain, and thus have long-lasting consequences. Activating effects set in during a certain situation and modulate behavior in this moment in a sex-specific way. However, this clear division of labor between "organizing" and "activating" effects meanwhile has started to erode. Many modern studies propose that sex hormones do both simultaneously. Hormones not only organize neural circuits but also activate them and go on to alter the organization of the brain.[62] In addition, the impact of the organizing effects of sex hormones does not only reach a very high level before birth but also more than a decade later—during puberty. As shown by Neufang et al.,[63] the surge of sex hormones during early adolescence initiates the emergence of important structural sex differences in the amygdala, the hippocampus, and many more areas of the brain. However, as shown by Barth et al.,[64] sex hormones go on changing the asymmetries of the brain in adulthood. These authors scanned the hippocampus of a single healthy female subject across two full menstrual cycles and analyzed fractional anisotropy (FA) values (a measure sensitive to changes in microstructural integrity, see Chapter 4) of left hippocampal fibers concerning fluctuations of estrogen during the cycle. They observed that left hippocampal FA values strongly, and positively, covary with the ups and downs of estrogen with a clear peak of FA during ovulation. If these FA values correlate with the functional properties of hippocampal fibers, this could indicate that fluctuations of estradiol during the cycle modify the functional properties of the left hippocampus in women.

Now that we have outlined what the organizing and activating effects of sex hormones are, let us go back to functional brain asymmetries. All of the sex differences in handedness, language, and visual figure recognition asymmetries could just be the results of organizing effects. Seen as that, we would expect them to be fully established at the end of puberty and not change thereafter. However, if organizing and activating effects are not separable, alterations of sex hormone concentrations during the menstrual cycle could also modify functional cerebral asymmetries during most of an adult's lifetime. If this was indeed the case, lateralization in women should be quite variable, since it constantly changes during the cycle. In addition, it would be likely that female asymmetry data are on average less pronounced than those of males since female subjects are possibly tested in different phases of their menstrual cycle. In these different phases, their asymmetries could sometimes be high, and sometimes low. This would further imply that cerebral asymmetries are not static but change dynamically over time. What a tantalizing possibility!

Before going into the scientific details, let us first recapitulate a few basic points about the menstrual cycle (Fig. 12.3).

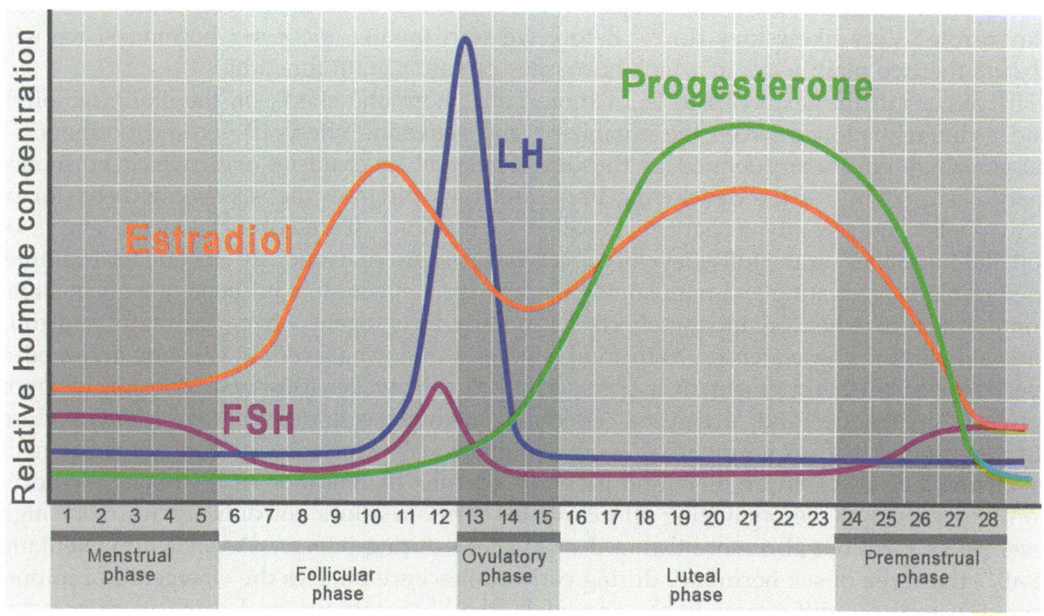

FIGURE 12.3 Hormonal fluctuations during a standard menstrual cycle. FSH: Follicle stimulating hormone. LH: Luteinizing hormone.

Each cycle can be roughly subdivided into five parts, and their succession is governed by hormonal changes. The first part is the menstrual phase. During this period, estradiol concentrations slowly start to rise, stop the bleeding, and start the follicular phase. Estradiol goes on to increase its concentration for a while but then decreases while the follicle-stimulating hormone (FSH) stimulates the gradual development of follicles in the ovary. Slightly later, the luteinizing hormone (LH) surges and the ovocyte is released. This event is called ovulation, and consequently, this part of the cycle is called ovulatory phase. The ovocyte survives only about 24 hours without fertilization. During this time, the concentrations of progesterone and estradiol increase to prepare the uterine lining for the potential embedding of an embryo. The onset of progesterone marks the beginning of the luteal phase. If no embryo develops, both progesterone and estradiol concentrations drop during the premenstrual phase, causing the onset of the next menstruation.

Toward the end of the 20th century, some publications reported that functional asymmetries seemed to change during the menstrual cycle. Heister et al.[65] found a linear decrease of the right-hemispheric advantage during the menstruation cycle in a face-perception task, while others[66,67] found pronounced asymmetries in the follicular phase. Overall, these studies suggested that high levels of female steroid hormones seemed to suppress or alter the generation of functional asymmetries. These initial reports were promising but mostly lacked proper validation with hormone measurements. Therefore Rode et al.[68] set out to test the idea that functional asymmetries change during the menstrual cycle.

As outlined above, estradiol increases its concentration up to its first peak before ovulation. During the luteal phase, estradiol reaches its second peak, this time accompanied by high levels of progesterone. Both estradiol and progesterone decrease 2–3 days before the onset of menses and remain at their lowest levels throughout menstruation. Obtaining maximal differences in hormonal status therefore requires testing once during the luteal (both estradiol and progesterone concentrations are high) and once during the menstrual cycle phase (both concentrations are low). While the beginning of the menstrual cycle is clearly visible due to the onset of bleeding, the luteal phase has to be verified by hormone concentration measurements. In addition, the order of testing (luteal phase, menstruation) has to be balanced among subjects, and these subjects should not take any hormone-based contraceptives. If all these requirements are met, testing can start.

Rode et al.[68] used both a lexical and a figure recognition test. While the lexical recognition test usually results in a left-hemispheric advantage in response speed and recognition, the figure task promotes a right-hemispheric advantage. Indeed, female subjects had a pronounced advantage in reaction time for the right visual field (left hemisphere) for the lexical task and left visual field (right hemisphere) superiority for the figure task. However, for the figure recognition task, this was only valid during menstruation. If the same subjects were tested during the luteal phase, no asymmetry was discernable (Fig. 12.4).

These data make it likely that functional asymmetries for figure recognition fluctuate during the menstrual cycle, concomitant with the ups and downs of sex hormones. This result could also provide a nice explanation for the higher variance of female asymmetry data and for the fact that many asymmetries in female subjects are less pronounced than those of males. Consider that you randomly ask 20 young women and 20 young men to participate in an asymmetry test. While the hormonal status of males will be stable, it will

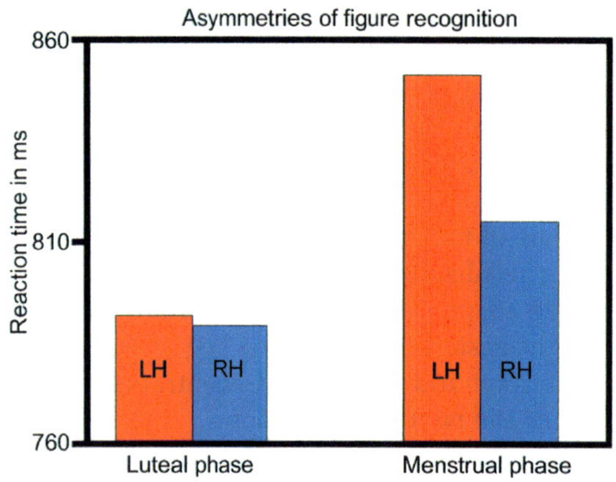

FIGURE 12.4 Schematic depiction of the experiment of Rode et al.[68] that revealed hemispheric asymmetries in figure recognition. The same subjects were tested twice: once during the luteal phase and once during menstruation. Performance was measured as reaction time in milliseconds. Faster reactions indicate superior performance and result in smaller numbers. As visible, left (LH) and right hemispheres (RH) are even during the luteal phase, while the right hemisphere evinces significantly faster reactions than the left hemisphere during menstruation *Reproduced with permission from Rode C, Wagner M, Güntürkün O. Menstrual cycle affects functional cerebral asymmetries. Neuropsychologia 1995;33:855–865.*

vary in females depending on their menstrual cycle phases. As a result, functional asymmetries in females will probably be more variable and less pronounced than in males. A higher variability of hemispheric asymmetries in women compared to men that was associated with the menstrual cycle was shown in a neuroimaging study on energy-metabolic asymmetries in frontal brain regions.[69] In this study, women were tested in the menstrual, follicular, and luteal phases of the menstrual cycle, and men at three similar time points. Creatine metabolism in frontal brain regions was assessed. While men showed stable creatine level asymmetries over all three time points, women showed an increase in left-hemispheric creatine in the follicular phase that was not observed in the luteal phase.

Around the time of the publication of the Rode et al.[68] paper, many comparable experiments were published, of which some discovered different result patterns (e.g., refs 67, 68, and 70–72) or found one or the other effect depending on task details.[73,74] Some of these conflicts possibly arose due to the problem that hormone fluctuations had not properly been tested. Such a lack of hormone measurement is a serious problem since only about 60% of younger women ovulate during their menstrual cycle.[75] In addition, fluctuations in sex hormone levels and the lengths of cycle phases can vary considerably among, and within, women. Studies that painstakingly measure hormone levels therefore usually have to exclude about half of their subjects since they are not in the expected cycle phase.[76,77] Therefore hormone measurements are essential to obtain valid experimental results.

Taken together, at the end of the 1990s, most scientists started to be convinced that functional asymmetries can change during the menstrual cycle. However, neither was a consensus reached on the direction of cycle-associated changes in asymmetry nor was it clear how sex hormones could alter brain asymmetries. Based on all of these considerations, Hausmann and Güntürkün[78] undertook a study, in which they recruited three groups of subjects: young women with a regular menstrual cycle, young men, and women after menopause. The last two groups were yoked with the first one. Thus, when a young woman had her menstruation, she was invited to the lab to participate in the experiment. At the same time, also the yoked young man and the postmenopausal woman were tested. Three visual half-field tasks were used that usually reveal a right-hemisphere (face discrimination and figural comparison) or a left-hemisphere advantage (lexical decision). All subjects were tested twice: younger women once during menses and once during the midluteal phase (about day 20–22); men and postmenopausal women in temporally corresponding sessions. This study revealed that young women show asymmetrical performance only during menses but not during the midluteal phase. Men and postmenopausal women evinced stable asymmetries for all three tasks. This overall pattern was true for all tests, regardless of being tested for left- or right-hemisphere functions. Thus the dynamic change of asymmetry during the cycle has to involve a mechanism that alters the overall balance between the hemispheres. Which neural mechanism could do that? One hint came from hormone measurements. At least for the figural comparison task, progesterone correlated with a decrease in functional asymmetry by enhancing the performance of the visuospatial nondominant left hemisphere.[78] Progesterone levels were also associated with decreased hemispheric asymmetries in a study on number comparison.[79] Could progesterone be the driving force behind these fluctuations of asymmetry?

Several subsequent studies could substantiate this. Hausmann et al.[80] analyzed functional cerebral asymmetries over a period of 6 weeks with 15 testing sessions by measuring in parallel progesterone, estradiol, testosterone, and the gonadotropins LH and FSH. The results of this study showed that especially progesterone was able to reduce asymmetries for the figural comparison task. One important prediction from these studies is that aging females should show changes in their cerebral asymmetries with different patterns before versus after menopause. No such alterations are to be expected in men. Indeed, as shown by Hausmann et al.[81] in young (mean age: 26.3 years) and elderly subjects (mean age: 63.8 years), a right-hemispheric advantage in a figural comparison task was marginally decreased with age in the men but significantly increased in the women. This is exactly what would be expected if especially left-hemispheric performance, as tested in the figural comparison task, is decreased due to a reduction of progesterone levels after menopause. In line with this finding, a study on functional hemispheric asymmetries and menopausal treatments found that in menopausal women, hormone therapy with a combination of estradiol hemihydrate and drospirenone (a so-called progestin, e.g., a drug that produces similar effects to progesterone) resulted in a reduction of hemispheric asymmetries in comparison to untreated controls of the same age.[82] Studies with different approaches to measure lateralization supported the conclusion that progesterone is a key hormone to induce changes in functional asymmetry[83–85] (see also ref. [86]). Support also came from studies that combined imaging and behavioral techniques.[87,88]

Unfortunately, all of these studies test subjects in different phases of the cycle without intervening in the system to understand the causality of the diverse effects. This was attempted by Hausmann et al.[89] by using transcranial magnetic stimulation (TMS). As a reminder, TMS can be used to stimulate small patches of the cortex. To deliver TMS pulses, a coil is placed near the scalp of the subject. For a short moment, current flows through the coil, producing a magnetic field perpendicular to the coil. Since the coil is only a few centimeters above the brain, a small electric current is induced in the target region of the brain via electromagnetic induction. When a TMS pulse is applied to the primary motor cortex around the representation of the hand, it evokes a response in the hand muscles contralateral to the stimulation. However, there is also a second effect: ipsilateral to the TMS pulse, a short period of suppression of ipsilateral hand muscle activity is generated.[90] This phenomenon is called the ipsilateral silent period (iSP). The reason for the iSP is fairly obvious: When the motor cortex of one hemisphere is excited by TMS, a volley of action potentials cross to the contralateral motor cortex via the corpus callosum and inhibit the contralateral pyramidal neurons. As a result, these neurons cannot activate their target muscles in the contralateral hand anymore. Therefore ipsilateral to the TMS pulse, the hand muscles slacken and an iSP can be measured. The length of the iSP reflects the functional integrity of transcallosal fibers that connect homotopic areas of the left and right motor cortices.[91] If progesterone or any other hormone could increase or decrease the efficacy of transmission via the corpus callosum, we should see that the length of iSP correlates with fluctuations of specific hormones during the menstrual cycle. Indeed, both increases in progesterone as well as in estradiol were able to predict a decrease in the iSP. However, while progesterone was effective during the luteal phase, estradiol exerted its effect only during the follicular phase. Remember that a short iSP is possibly related to less effective callosal transmission. Thus the data of Hausmann et al.[89] make it likely that both estradiol and progesterone reduce the efficacy of callosal transmission, albeit in different phases of the menstrual cycle. By reducing callosal transmission, they then can shift the female brain into a less asymmetric functional pattern (Fig. 12.5).

The study by Hausmann et al.[89] is one of several articles that describe that not only progesterone but also estradiol can alter callosal transmission and thus functional asymmetries.[87,92,93] So, both hormones seem to produce a similar pattern of action on functional asymmetries. At first glance, this is difficult to explain, since estradiol and progesterone have mostly opposite effects on glutamate and GABA receptors, the two main receptor systems in the cortex.[94] Let us therefore first revisit the old hypothesis on "progesterone-mediated hemispheric decoupling" by Hausmann and Güntürkün.[78] This hypothesis was formulated during a time when only progesterone was assumed able to change functional asymmetries. Subsequently, let us discuss the details of the actions of estradiol and progesterone.

Hormone-mediated decoupling of commissural interactions

Commissural interactions during menses

The corpus callosum is the largest commissural system of the human brain and is constituted of about 200 million axons[95] (see also Chapter 3). The majority of callosal fibers arise from excitatory and glutamatergic pyramidal neurons.[96,97] Despite the excitatory nature of corticocortical transmission, the main and longer-lasting effect of callosal activation is inhibitory.[98] The reason is simple: Most callosal axons synapse on dendrites of contralateral pyramidal neurons and excite these cells. These excited pyramidal cells subsequently activate GABAergic interneurons in their direct vicinity that are then able to induce a widespread inhibition around the activated pyramidal neuron.[96,99] This is nicely visible during movements of only one hand. During such action, the motor cortex ipsilateral to the moving hand is inhibited through the corpus callosum by the contralateral motor cortex. This transcallosal inhibition results in a short excitation, followed by a strong and long-lasting inhibition.[100,101] Thus the corpus callosum does not simply transmit excitatory or inhibitory action toward the contralateral hemisphere but rather induces a brief

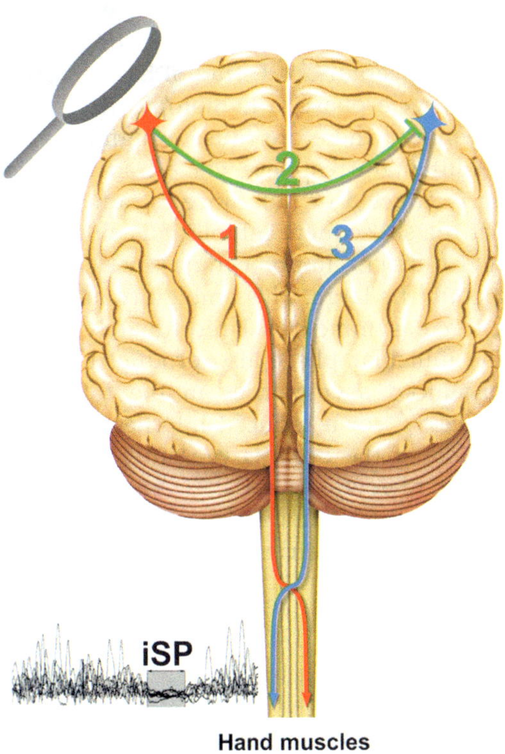

FIGURE 12.5 Schematic depiction of the experiment by Hausmann et al.[89] A TMS coil produces a magnetic field (shown as an hourglass-shaped transparent area) that induces an electric current in the hand region of the motor cortex. The descending corticospinal tract activates the contralateral hand muscles (1: Red pathway). In addition, an inhibitory signal is transferred to the motor cortex in the contralateral homotopic area (2: Green pathway). This results in a weakening of the descending motor signals, thereby producing an ipsilateral silent period (iSP) in the electromyogram of the hand muscles (3: Blue pathway). Since the green pathway (2) passes through the corpus callosum, the length of the iSP reflects the efficacy of callosal transmission. This efficacy varies with the concentration of progesterone and estradiol. *The iSP signal is reproduced with permission from Hausmann M, Tegenthoff M, Sänger J, Janssen F, Güntürkün O, Schwenkreis P. Transcallosal inhibition across the menstrual cycle: a TMS study. Clin Neurophysiol. 2006;117:26–32, with permission from the authors and publisher. Brain schema: ilusmedical/Shutterstock.com.*

and highly focal excitatory postsynaptic potential (EPSP), followed by prolonged and broad inhibition (IPSP).[97,102,103] The first excitatory response depends on the activation of glutamatergic AMPA and NMDA receptors, while the subsequent IPSP is due to the activation of $GABA_A$ and $GABA_B$ receptors.[102,104] Without the initial excitation, no subsequent inhibition is produced.[97]

As also outlined in Chapter 3, several neuropsychological models suggest that interhemispheric cross-talk is an essential mechanism of cerebral asymmetries.[105,106] The most widespread view in explaining cerebral asymmetries by callosal mechanisms is reciprocal inhibition, in which a stimulus-specific activation of one of the hemispheres inhibits the other one during task processing.[107] Now we will outline how sex hormones could alter transcallosal interaction and thus brain asymmetries.

Commissural interactions during the follicular phase

Sex hormones can substantially modulate the excitability of neurons. Already in 1885, the famous neurologist Sir William Richard Gowers observed that the seizure frequency of female epileptic patients was high during the follicular and low during the luteal phase.[108] Based on these observations, estradiol was discovered to promote epileptic seizures, while progesterone was later used as an anticonvulsant.[109]

As outlined before, the follicular phase is characterized by high levels of estradiol. Estradiol increases the response to glutamate by up to 86%.[110,111] This potentiating effect is thought to be mainly due to glutamatergic NMDA receptors.[112] Estradiol has no acute effects on GABA receptors.[113] During the follicular cycle phase, a homotopic activation of one cortical area through the corpus callosum should therefore cause a high activity peak due to the massive augmentation of the glutamate response via estradiol. It is unclear whether the subsequent broad inhibition is affected or not. Since estradiol does not modify GABA receptors, it is conceivable that the subsequent inhibition is similar to the situation during menses. If, however, the larger initial excitation can produce a stronger GABA response, the inhibitory surrounding could be large. This ambiguity makes it difficult to predict the effect of estradiol on functional asymmetries. Unfortunately, studies that tested functional asymmetries during the follicular phase also produced partly discrepant results. While some[66,67] found pronounced asymmetries in the follicular phase, others observed reduced left-right differences[87,92] or a reduced interhemispheric transfer efficacy during this period.[89,93] This period of the cycle is presently not well understood concerning functional asymmetries (for review, see ref. [114]). Thus we can safely conclude that estradiol can change some aspects of functional asymmetries in women, but we presently do not exactly know which asymmetries are altered by what kind of neurobiological mechanism.

Commissural interactions during the luteal phase

During the luteal phase, both estradiol and progesterone levels are very high. Progesterone receptors are abundant and widely distributed in the brain. Progesterone that circulates in the blood can easily enter all parts of the nervous system.[94] Progesterone acts very differently from estradiol by suppressing the glutamate-induced excitatory responses of neurons in a dose-dependent fashion.[115,116] Physiological doses of progesterone that are comparable to the midluteal phase can reduce the excitatory glutamate responses by 87%.[116] However, progesterone does not modulate all types of glutamate receptors equally. While it is well established that AMPA receptors are substantially modified by progesterone, there is no consensus on the effects on NMDA receptors. According to Kawaguchi,[102] progesterone does not modulate the response of NMDA receptors, while Kumar and Huguenard[97] observed at least a weak modulation.

However, progesterone does more than that. Its metabolite allopregnanolone also augments the inhibitory neuronal response to GABA, thereby increasing the effect of the long-lasting inhibition in the contralateral hemisphere.[117,118] Majewska et al.[119] could show that this is achieved both by increasing the effect size as well as the duration of the GABA response. Thus a callosal activation during the luteal phase would decrease the first excitatory EPSP. Since at least a good part of the subsequent broad and long-lasting GABA response depends on the presence of this EPSP, we could expect this inhibitory response to be diminished. Together, these processes may result in a relative functional decoupling of the hemispheres due to high progesterone concentrations during the midluteal phase. During the luteal phase, both estradiol and progesterone reach high concentrations. Since they have quite opposite effects on glutamate receptors, do these effects cancel each other

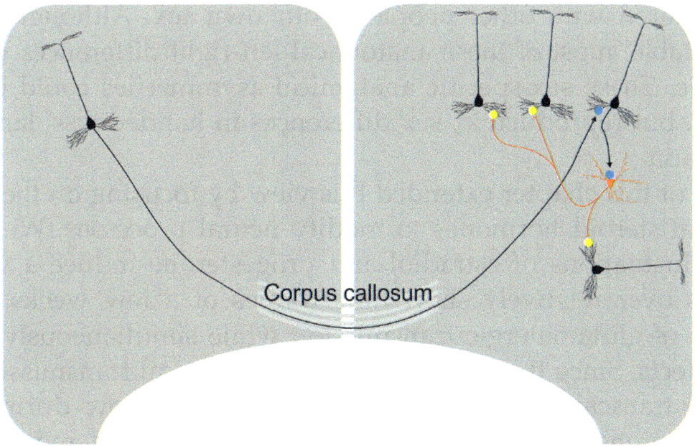

FIGURE 12.6 Schema of the "progesterone-mediated hemispheric decoupling" hypothesis by Hausmann and Güntürkün.[78] A pyramidal neuron in the left hemisphere activates via glutamate receptors (blue) a pyramidal neuron on the right which then excites an inhibitory interneuron (depicted in red). The inhibitory neuron inhibits a large number of pyramidal neurons in the surrounding via GABA receptors (yellow). High concentrations of progesterone during the luteal phase downregulate glutamate receptors, resulting in a diminished activation of the pyramidal neuron on the right side. As a result, the inhibitory interneuron is less activated, and the surrounding inhibition is diminished. *Reproduced with permission from Hausmann M, Güntürkün O. Steroid fluctuations modify functional cerebral asymmetries: the hypothesis of progesterone-mediated interhemispheric decoupling. Neuropsychologia. 2000;38:1362–1374.*

out? The answer is no. If estradiol and progesterone are applied in a combined fashion, progesterone seems to win the competition. Thus glutamate receptors are then mainly downregulated as with progesterone alone[116] (Fig. 12.6).

However, how would these progesterone effects ultimately alter functional asymmetries? Well, if progesterone dampens corticocortical transmission, then it is likely that it would reduce functional asymmetries. Remember that interhemispheric cross-talk is one of the most essential mechanisms in causing cerebral asymmetries (see Chapter 3). As also outlined by the "progesterone-mediated hemispheric decoupling" hypothesis by Hausmann and Güntürkün,[78] a downregulation of functional asymmetries during the peak time of progesterone would be expected. This is also what numerous studies demonstrated[68,80,81,83–85,87–89] (see also ref. [86]).

Summary: the dual coding of cerebral asymmetries

What are the neural fundamentals of our functional asymmetries that drive left-right differences in language, handedness, face recognition, spatial orientation, and so forth? In Chapter 4 we saw that the human brain is characterized by a multitude of structural asymmetries that very likely produce a lateralized process of information processing. In this chapter, we have shortly outlined that some of these left-right differences are different in women and men. These sex-specific anatomical asymmetries of our brain start to emerge before birth and then have a second peak of restructuration around puberty. When we enter our late adolescent period of life, we all have multiple brain asymmetries of which some are mainly shared with other people of our own sex. Although human brains are never completely stable, most of these anatomical left-right differences will possibly never change until we die. These sex-specific anatomical asymmetries could be key factors that produce the subtle, but reproducible, sex differences in handedness, language processing, and figure recognition.

The second part of this chapter extended this view by focusing on the importance of the potential of gonadal steroid hormones to modify neural processes. We have seen that in women, monthly fluctuations of estradiol and progesterone induce a variability of functional asymmetries over relatively short time frames of a few weeks. Progesterone can reduce the efficacy of glutamatergic transmission while simultaneously increasing inhibitory GABAergic effects. Since the initial step in corticocortical transmission relies on glutamatergic synapses, transcallosal interactions are likely at a low during the luteal cycle phase when progesterone concentrations are high. This should reduce functional brain asymmetries. Although some discrepant result patterns exist, this prediction seems to be correct overall. The situation for estradiol is presently more confusing. Estradiol augments glutamatergic effects, while not modifying inhibitory transmission. Thus estradiol should have mostly opposite effects to progesterone. The current state of findings tends to show the opposite, although the general picture is more variable for estradiol than for progesterone. However, it is very likely that not only progesterone but also estradiol can in principle modify functional asymmetries.

Do these sex differences in hemispheric asymmetries cause sex differences in cognitive abilities, for example, in mental rotation ability or other cognitive domains, as some popular science books suggest? Very likely not. In a comprehensive review of the last 40 years of research on sex differences in hemispheric asymmetries, Marco Hirnstein and coworkers concluded that sex differences in hemispheric asymmetries are small but robust.[1] Importantly, cognitive sex differences often emerge without sex differences in hemispheric asymmetries, and sex differences in hemispheric asymmetries are often observed without cognitive sex differences.

More research is needed on the effects of testosterone and other androgens on hemispheric asymmetries. Testosterone also has multiple effects on GABA receptors and is known to modify neural processes.[113] Testosterone shows fluctuations on a diurnal, seasonal, and age-related basis, although these are more modest and especially less abrupt than female gonadal hormones.[120] The empirical evidence for an effect of testosterone and other androgens on hemispheric asymmetries is inconclusive. For handedness, a few early studies suggest correlations between testosterone levels and handedness measures,[121,122] but a larger number of subsequent studies in sometimes very well-powered samples did not find convincing evidence for a link between testosterone or other androgens and handedness.[123–127] Thus testosterone is unlikely to be a major influence factor on handedness. In contrast to that, a few studies have suggested that testosterone levels affect different forms of functional hemispheric asymmetries such as asymmetries in global-local processing,[128] number processing,[79] mental rotation, and the chimeric faces task.[129] While men do not have a menstrual cycle, there are also variations in testosterone levels. For example, there is some evidence for cognitive changes that are driven by seasonal variations of testosterone. In the northern hemisphere, levels of testosterone in males are usually higher in fall than in spring. Possibly, this is an evolutionary adaptation to ensure that offspring is born in summer. Kimura and Hampson[130] indeed reported that fluctuating spatial abilities of young males correlated with the seasonal variations of testosterone. How such variations of testosterone could affect brain asymmetries is unknown.

Despite all these unanswered questions, we see that our view on brain asymmetries is up for an important change. This novel view is a much more dynamic, complex, and variable one. We suppose that both the anatomical and the hormone-dependent aspects together determine the functional architecture of asymmetries. Thus lateralized systems would be dually coded. One code is (relatively) static and would result from the continuing asymmetries resulting from anatomically lateralized streams of processing. The second code is dynamic and changes asymmetries by altering interhemispheric interactions. Functional cerebral asymmetries would then be constituted of static factors established in ontogeny and of multiple dynamic factors changing over short periods of time throughout the whole lifetime.

Sample questions

- Do males and females differ in their proportions of intra- and interhemispheric connections? Do you think that is a true sex difference, or could it be related to a further variable?
- Are functional asymmetries sex-dependent? Answer by giving examples from studies on handedness, language, and figure recognition tasks.
- What are the organizing and activating effects of hormones?
- How do functional asymmetries change over the menstrual cycle? Please give some examples from different perceptual or cognitive systems.
- What is the effect of progesterone and estradiol on neuronal functions?
- Please outline the hypothesis on progesterone-mediated hemispheric decoupling.

References

1. Hirnstein M, Hugdahl K, Hausmann M. Cognitive sex differences and hemispheric asymmetry: A critical review of 40 years of research. *Laterality*. 2019;24(2):204–252. Available from: https://doi.org/10.1080/1357650X.2018.1497044.
2. Hodgetts S, Hausmann M. Sex/Gender Differences in Brain Lateralisation and Connectivity. *Curr Top Behav Neurosci*. 2023;62:71–99. Available from: https://doi.org/10.1007/7854_2022_303.
3. Wisniewski A. Sexually-dimorphic patterns of cortical asymmetry, and the role for sex steroid hormones in determining cortical patterns of lateralization. *Psychoneuroendocrinology*. 1998;23(5):519–547. Available from: https://doi.org/10.1016/s0306-4530(98)00019-5.
4. Smith RJ. Relative Size versus Controlling for Size. *Curr Anthropology*. 2005;46(2):249–273. Available from: https://doi.org/10.1086/427117.
5. Sommer IE, Aleman A, Somers M, Boks MP, Kahn RS. Sex differences in handedness, asymmetry of the planum temporale and functional language lateralization. *Brain Res*. 2008;1206:76–88. Available from: https://doi.org/10.1016/j.brainres.2008.01.003.
6. Kong X-Z, Postema MC, Guadalupe T, et al. Mapping brain asymmetry in health and disease through the ENIGMA consortium. *Hum Brain Mapp*. 2022;43(1):167–181. Available from: https://doi.org/10.1002/hbm.25033.

7. Kong X-Z, Mathias SR, Guadalupe T, et al. Mapping cortical brain asymmetry in 17,141 healthy individuals worldwide via the ENIGMA Consortium. *PNAS*. 2018;115(22):E5154–E5163. Available from: https://doi.org/10.1073/pnas.1718418115.
8. Koelkebeck K, Miyata J, Kubota M, et al. The contribution of cortical thickness and surface area to gray matter asymmetries in the healthy human brain. *Hum Brain Mapp*. 2014;35(12):6011–6022. Available from: https://doi.org/10.1002/hbm.22601.
9. Núñez C, Theofanopoulou C, Senior C, et al. A large-scale study on the effects of sex on gray matter asymmetry. *Brain Struct Funct*. 2018;223(1):183–193. Available from: https://doi.org/10.1007/s00429-017-1481-4.
10. Guadalupe T, Mathias SR, vanErp TGM, et al. Human subcortical brain asymmetries in 15,847 people worldwide reveal effects of age and sex. *Brain Imaging Behav*. 2017;11(5):1497–1514. Available from: https://doi.org/10.1007/s11682-016-9629-z.
11. Joel D, Berman Z, Tavor I, et al. Sex beyond the genitalia: The human brain mosaic. *Proc Natl Acad Sci U S A*. 2015;112(50):15468–15473. Available from: https://doi.org/10.1073/pnas.1509654112.
12. Eliot L, Ahmed A, Khan H, Patel J. Dump the "dimorphism": Comprehensive synthesis of human brain studies reveals few male-female differences beyond size. *Neurosci Biobehav Rev*. 2021;125:667–697. Available from: https://doi.org/10.1016/j.neubiorev.2021.02.026.
13. Hirnstein M, Hausmann M. Sex/gender differences in the brain are not trivial-A commentary on Eliot et al. (2021). *Neurosci Biobehav Rev*. 2021;130:408–409. Available from: https://doi.org/10.1016/j.neubiorev.2021.09.012.
14. Kurth F, Thompson PM, Luders E. Investigating the differential contributions of sex and brain size to gray matter asymmetry. *Cortex*. 2018;99:235–242. Available from: https://doi.org/10.1016/j.cortex.2017.11.017.
15. Ingalhalikar M, Smith A, Parker D, et al. Sex differences in the structural connectome of the human brain. *PNAS*. 2014;111(2):823–828. Available from: https://doi.org/10.1073/pnas.1316909110.
16. Westerhausen R, Kompus K, Dramsdahl M, et al. A critical re-examination of sexual dimorphism in the corpus callosum microstructure. *NeuroImage*. 2011;56(3):874–880. Available from: https://doi.org/10.1016/j.neuroimage.2011.03.013.
17. Gong G, Rosa-Neto P, Carbonell F, Chen ZJ, He Y, Evans AC. Age- and gender-related differences in the cortical anatomical network. *J Neurosci*. 2009;29(50):15684–15693. Available from: https://doi.org/10.1523/JNEUROSCI.2308-09.2009.
18. Ocklenburg S, Friedrich P, Güntürkün O, Genç E. Intrahemispheric white matter asymmetries: the missing link between brain structure and functional lateralization. *Rev Neurosci*. 2016;27(5):465–480. Available from: https://doi.org/10.1515/revneuro-2015-0052.
19. Gong G, He Y, Evans AC. Brain connectivity: gender makes a difference. *Neuroscientist*. 2011;17(5):575–591. Available from: https://doi.org/10.1177/1073858410386492.
20. Hänggi J, Fövenyi L, Liem F, Meyer M, Jäncke L. The hypothesis of neuronal interconnectivity as a function of brain size-a general organization principle of the human connectome. *Front Hum Neurosci*. 2014;8:915. Available from: https://doi.org/10.3389/fnhum.2014.00915.
21. Ruigrok ANV, Salimi-Khorshidi G, Lai M-C, et al. A meta-analysis of sex differences in human brain structure. *Neurosci Biobehav Rev*. 2014;39:34–50. Available from: https://doi.org/10.1016/j.neubiorev.2013.12.004.
22. Zauck Linkskultur. Neue Bahnen. *Z der Reichsfachschaft IV*. 1922;33(8):262–265.
23. Ogle W. On Dextral Pre-eminence. *Med Chir Trans*. 1871;54:279–301. Available from: https://doi.org/10.1177/095952877105400114.
24. Ganter R. Ueber Linkshändigkeit bei Epileptischen, Schwachsinnigen und Normalen. *Allg Zsch f Psychiat*. 1919;75:689–700.
25. Mattauschek. Einiges über die Degeneration des bosnisch-herzegowinischen Volkes. *Jahrb f Psychiat*. 1908;29:113–115.
26. Smith LG. A Brief Survey of Right- and Left-Handedness. *Pedagogical Seminary*. 1917;24(1):19–35. Available from: https://doi.org/10.1080/08919402.1917.10534715.
27. Gordon K. Some notes on the mental status of the left-handed. *J Delinquency*. 1923;8:154–157.
28. Papadatou-Pastou M, Ntolka E, Schmitz J, et al. Human handedness: A meta-analysis. *Psychol Bull*. 2020;146(6):481–524. Available from: https://doi.org/10.1037/bul0000229.
29. Papadatou-Pastou M, Martin M, Munafò MR, Jones GV. Sex differences in left-handedness: a meta-analysis of 144 studies. *Psychol Bull*. 2008;134(5):677–699. Available from: https://doi.org/10.1037/a0012814.

30. Grattan MP, Vos E, Levy J, McClintock MK. Asymmetric Action in the Human Newborn: Sex Differences in Patterns of Organization. *Child Dev.* 1992;63(2):273–289. Available from: https://doi.org/10.1111/j.1467-8624.1992.tb01626.x.
31. McGlone J. Sex differences in the cerebral organization of verbal functions in patients with unilateral brain lesions. *Brain.* 1977;100(4):775–793. Available from: https://doi.org/10.1093/brain/100.4.775.
32. Frost JA, Binder JR, Springer JA, et al. Language processing is strongly left lateralized in both sexes. Evidence from functional MRI. *Brain.* 1999;122(Pt 2):199–208. Available from: https://doi.org/10.1093/brain/122.2.199.
33. Harasty J. Language processing in both sexes: evidence from brain studies. *Brain.* 2000;123(2):404–406. Available from: https://doi.org/10.1093/brain/123.2.404.
34. Knecht S, Deppe M, Dräger B, et al. Language lateralization in healthy right-handers. *Brain.* 2000;123(Pt 1):74–81. Available from: https://doi.org/10.1093/brain/123.1.74.
35. Hiscock M, Inch R, Jacek C, Hiscock-Kalil C, Kalil KM. Is there a sex difference in human laterality? I. An exhaustive survey of auditory laterality studies from six neuropsychology journals. *J Clin Exp Neuropsychol.* 1994;16(3):423–435. Available from: https://doi.org/10.1080/01688639408402653.
36. Boles DB. A large-sample study of sex differences in functional cerebral lateralization. *J Clin Exp Neuropsychol.* 2005;27(6):759–768. Available from: https://doi.org/10.1081/13803390590954263.
37. Hirnstein M, Westerhausen R, Korsnes MS, Hugdahl K. Sex differences in language asymmetry are age-dependent and small: a large-scale, consonant-vowel dichotic listening study with behavioral and fMRI data. *Cortex.* 2013;49(7):1910–1921. Available from: https://doi.org/10.1016/j.cortex.2012.08.002.
38. Bless JJ, Westerhausen R, Koss Torkildsen J, von, Gudmundsen M, Kompus K, Hugdahl K. Laterality across languages: Results from a global dichotic listening study using a smartphone application. *Laterality.* 2015;20 (4):434–452. Available from: https://doi.org/10.1080/1357650X.2014.997245.
39. Voyer D. On the magnitude of laterality effects and sex differences in functional lateralities. *Laterality.* 1996;1 (1):51–83. Available from: https://doi.org/10.1080/713754209.
40. Meinschaefer J, Hausmann M, Güntürkün O. Laterality effects in the processing of syllable structure. *Brain Lang.* 1999;70(2):287–293. Available from: https://doi.org/10.1006/brln.1999.2153.
41. Voyer D, Voyer S, Bryden MP. Magnitude of sex differences in spatial abilities: a meta-analysis and consideration of critical variables. *Psychol Bull.* 1995;117(2):250–270. Available from: https://doi.org/10.1037/0033-2909.117.2.250.
42. Johnson BW, McKenzie KJ, Hamm JP. Cerebral asymmetry for mental rotation: effects of response hand, handedness and gender. *NeuroReport.* 2002;13(15):1929. Available from: https://doi.org/10.1097/00001756-200210280-00020.
43. Voyer D, Bryden MP. Gender, level of spatial ability, and lateralization of mental rotation. *Brain Cognition.* 1990;13(1):18–29. Available from: https://doi.org/10.1016/0278-2626(90)90037-o.
44. Hausmann M, Güntürkün O. Sex differences in functional cerebral asymmetries in a repeated measures design. *Brain Cognition.* 1999;41(3):263–275. Available from: https://doi.org/10.1006/brcg.1999.1126.
45. Rizzolatti G, Buchtel HA. Hemispheric Superiority in Reaction Time to Faces: A Sex Difference. *Cortex.* 1977;13(3):300–305. Available from: https://doi.org/10.1016/s0010-9452(77)80039-7.
46. Ernest CH. Spatial ability and laterality effects on a face recognition task. *Personality Individ Differences.* 1997;23(5):839–848. Available from: https://doi.org/10.1016/s0191-8869(97)00076-7.
47. Proverbio AM, Brignone V, Matarazzo S, Del Zotto M, Zani A. Gender differences in hemispheric asymmetry for face processing. *BMC Neurosci.* 2006;7:44. Available from: https://doi.org/10.1186/1471-2202-7-44.
48. Proverbio AM. Sexual dimorphism in hemispheric processing of faces in humans: A meta-analysis of 817 cases. *Soc Cogn Affect Neurosci.* 2021;16(10):1023–1035. Available from: https://doi.org/10.1093/scan/nsab043.
49. Bakan P, Putnam W. Right-left discrimination and brain lateralization. *Sex differences Arch Neurol.* 1974;30 (4):334–335. Available from: https://doi.org/10.1001/archneur.1974.00490340062016.
50. Hirnstein M, Ocklenburg S, Schneider D, Hausmann M. Sex differences in left-right confusion depend on hemispheric asymmetry. *Cortex.* 2009;45(7):891–899. Available from: https://doi.org/10.1016/j.cortex.2008.11.009.
51. Hjelmervik H, Westerhausen R, Hirnstein M, Specht K, Hausmann M. The neural correlates of sex differences in left-right confusion. *NeuroImage.* 2015;113:196–206. Available from: https://doi.org/10.1016/j.neuroimage.2015.02.066.
52. Rasmjou S, Hausmann M, Güntürkün O. Hemispheric dominance and gender in the perception of an illusion. *Neuropsychologia.* 1999;37(9):1041–1047. Available from: https://doi.org/10.1016/s0028-3932(98)00154-7.

53. Clem RK, Pollack RH. Illusion magnitude as a function of visual field exposure. *Percept & Psychophys.* 1975;17 (5):450–454. Available from: https://doi.org/10.3758/BF03203292.
54. Rothwell B, Zaidel E. Visual field differences in the magnitude of the Oppel-Kundt illusion vary with processing time. *Percept & Psychophys.* 1990;47(2):180–190. Available from: https://doi.org/10.3758/bf03205982.
55. Houlard N, Fraisse P, Hécaen H. Effects of Unilateral Hemispheric Lesions on Two Types of Optico-Geometric Illusions. *Cortex.* 1976;12(3):232–240. Available from: https://doi.org/10.1016/s0010-9452(76)80004-4.
56. Hiscock M, Israelian M, Inch R, Jacek C, Hiscock-Kalil C. Is there a sex difference in human laterality? I I. An exhaustive survey of visual laterality studies from six neuropsychology journals. *J Clin Exp Neuropsychol.* 1995;17(4):590–610. Available from: https://doi.org/10.1080/01688639508405148.
57. Bonelli C, Mancuso L, Manuello J, Liloia D, Costa T, Cauda F. Sex differences in brain homotopic co-activations: a meta-analytic study. *Brain Struct Funct.* 2022;227(8):2839–2855. Available from: https://doi.org/10.1007/s00429-022-02572-0.
58. Hausmann M, Behrendt-Körbitz S, Kautz H, Lamm C, Radelt F, Güntürkün O. Sex differences in oral asymmetries during wordrepetition. *Neuropsychologia.* 1998;36(12):1397–1402. Available from: https://doi.org/10.1016/s0028-3932(98)00027-x.
59. Markus Hausmann DMichael Burt. Sex hormonal effects on brain lateralization. *Routledge International Handbook of Social Neuroendocrinology*. Routledge; 2018:357–370.
60. Cooke B, Hegstrom CD, Villeneuve LS, Breedlove SM. Sexual differentiation of the vertebrate brain: principles and mechanisms. *Front Neuroendocrinology.* 1998;19(4):323–362. Available from: https://doi.org/10.1006/frne.1998.0171.
61. Breedlove SM, Hampson E. Sexual differentiation of the brain and behavior. In: Becker JB, Breedlove SM, Crews D, McCarthy MM, eds. Behavioral endocrinology. 2. MIT Press; 2002:39–74.
62. Sisk CL, Foster DL. The neural basis of puberty and adolescence. *Nat Neurosci.* 2004;7(10):1040–1047. Available from: https://doi.org/10.1038/nn1326.
63. Neufang S, Specht K, Hausmann M, et al. Sex differences and the impact of steroid hormones on the developing human brain. *Cereb Cortex.* 2009;19(2):464–473. Available from: https://doi.org/10.1093/cercor/bhn100.
64. Barth C, Steele CJ, Mueller K, et al. In-vivo Dynamics of the Human Hippocampus across the Menstrual Cycle. *Sci Rep.* 2016;6(1):32833. Available from: https://doi.org/10.1038/srep32833.
65. Heister G, Landis T, Regard M, Schroeder-Heister P. Shift of functional cerebral asymmetry during the menstrual cycle. *Neuropsychologia.* 1989;27(6):871–880. Available from: https://doi.org/10.1016/0028-3932(89)90010-9.
66. Altemus M, Wexler BE, Boulis N. Changes in perceptual asymmetry with the menstrual cycle. *Neuropsychologia.* 1989;27(2):233–240. Available from: https://doi.org/10.1016/0028-3932(89)90174-7.
67. Hampson E. Estrogen-related variations in human spatial and articulatory-motor skills. *Psychoneuroendocrinology.* 1990;15(2):97–111. Available from: https://doi.org/10.1016/0306-4530(90)90018-5.
68. Rode C, Wagner M, Güntürkün O. Menstrual cycle affects functional cerebral asymmetries. *Neuropsychologia.* 1995;33(7):855–865. Available from: https://doi.org/10.1016/0028-3932(95)00015-u.
69. Hjelmervik H, Hausmann M, Craven AR, Hirnstein M, Hugdahl K, Specht K. Sex- and sex hormone-related variations in energy-metabolic frontal brain asymmetries: A magnetic resonance spectroscopy study. *NeuroImage.* 2018;172:817–825. Available from: https://doi.org/10.1016/j.neuroimage.2018.01.043.
70. McCourt ME W, Mark V, Radonovich KJ K, Willison S, Freeman P. The effects of gender, menstrual phase and practice on the perceived location of the midsagittal plane. *Neuropsychologia.* 1997;35(5):717–724. Available from: https://doi.org/10.1016/s0028-3932(96)00115-7.
71. Weekes NY, Zaidel E. The effects of procedural variations on lateralized Stroop effects. *Brain Cognition.* 1996;31(3):308–330. Available from: https://doi.org/10.1006/brcg.1996.0049.
72. Bibawi D, Cherry B, Hellige JB. Fluctuations of perceptual asymmetry across time in women and men: Effects related to the menstrual cycle. *Neuropsychologia.* 1995;33(1):131–138. Available from: https://doi.org/10.1016/0028-3932(94)00103-v.
73. Mead LA, Hampson E. Asymmetric effects of ovarian hormones on hemispheric activity: Evidence from dichotic and tachistoscopic tests. *Neuropsychology.* 1996;10(4):578–587. Available from: https://doi.org/10.1037/0894-4105.10.4.578.
74. Sanders G, Wenmoth D. Verbal and music dichotic listening tasks reveal variations in functional cerebral asymmetry across the menstrual cycle that are phase and task dependent. *Neuropsychologia.* 1998;36 (9):869–874. Available from: https://doi.org/10.1016/s0028-3932(98)00022-0.

75. Metcalf MG, Mackenzie JA. Incidence of ovulation in young women. *J Biosoc Sci.* 1980;12(3):345–352. Available from: https://doi.org/10.1017/s002193200001289x.
76. Gordon HW, Corbin ED, Lee PA. Changes in Specialized Cognitive Function Following Changes in Hormone Levels. *Cortex.* 1986;22(3):399–415. Available from: https://doi.org/10.1016/s0010-9452(86)80004-1.
77. Hodgetts S, Weis S, Hausmann M. Sex hormones affect language lateralisation but not cognitive control in normally cycling women. *Hormones Behav.* 2015;74:194–200. Available from: https://doi.org/10.1016/j.yhbeh.2015.06.019.
78. Hausmann M, Güntürkün O. Steroid fluctuations modify functional cerebral asymmetries: the hypothesis of progesterone-mediated interhemispheric decoupling. *Neuropsychologia.* 2000;38(10):1362–1374. Available from: https://doi.org/10.1016/s0028-3932(00)00045-2.
79. Pletzer B, Jäger S, Hawelka S. Sex hormones and number processing. *Progesterone testosterone relate hemispheric asymmetries number comparison Hormones Behav.* 2019;115:104553. Available from: https://doi.org/10.1016/j.yhbeh.2019.07.001.
80. Hausmann M, Becker C, Gather U, Güntürkün O. Functional cerebral asymmetries during the menstrual cycle: a cross-sectional and longitudinal analysis. *Neuropsychologia.* 2002;40(7):808–816. Available from: https://doi.org/10.1016/s0028-3932(01)00179-8.
81. Hausmann M, Güntürkün O, Corballis M. Age-related changes in hemispheric asymmetry depend on sex. *Laterality.* 2003;8(3):277–290. Available from: https://doi.org/10.1080/13576500244000201.
82. Digesu GA, Riemma G, Torella M, et al. Functional Brain Asymmetry and Menopausal Treatments: Is There a Link? *Medicina (Kaunas).* 2022;58(5). Available from: https://doi.org/10.3390/medicina58050616.
83. Alexander GM, Altemus M, Peterson BS, Wexler BE. Replication of a premenstrual decrease in right-ear advantage on language-related dichotic listening tests of cerebral laterality. *Neuropsychologia.* 2002;40(8):1293–1299. Available from: https://doi.org/10.1016/s0028-3932(01)00220-2.
84. Wadnerkar MB, Whiteside SP, Cowell PE. Dichotic listening asymmetry: sex differences and menstrual cycle effects. *Laterality.* 2008;13(4):297–309. Available from: https://doi.org/10.1080/13576500701821106.
85. Bayer U, Hausmann M. Menstrual cycle-related changes of functional cerebral asymmetries in fine motor coordination. *Brain Cognition.* 2012;79(1):34–38. Available from: https://doi.org/10.1016/j.bandc.2012.02.003.
86. Compton RJ, Costello C, Diepold J. Interhemispheric integration during the menstrual cycle: failure to confirm progesterone-mediated interhemispheric decoupling. *Neuropsychologia.* 2004;42(11):1496–1503. Available from: https://doi.org/10.1016/j.neuropsychologia.2004.03.006.
87. Fernández G, Weis S, Stoffel-Wagner B, et al. Menstrual Cycle-Dependent Neural Plasticity in the Adult Human Brain Is Hormone, Task, and Region Specific. *J Neurosci.* 2003;23(9):3790–3795. Available from: https://doi.org/10.1523/JNEUROSCI.23-09-03790.2003.
88. Weis S, Hausmann M, Stoffers B, Sturm W. Dynamic changes in functional cerebral connectivity of spatial cognition during the menstrual cycle. *Hum Brain Mapp.* 2011;32(10):1544–1556. Available from: https://doi.org/10.1002/hbm.21126.
89. Hausmann M, Tegenthoff M, Sänger J, Janssen F, Güntürkün O, Schwenkreis P. Transcallosal inhibition across the menstrual cycle: a TMS study. *Clin Neurophysiol.* 2006;117(1):26–32. Available from: https://doi.org/10.1016/j.clinph.2005.08.022.
90. Ferbert A, Priori A, Rothwell JC, Day BL, Colebatch JG, Marsden CD. Interhemispheric inhibition of the human motor cortex. *J Physiol.* 1992;453(1):525–546. Available from: https://doi.org/10.1113/jphysiol.1992.sp019243.
91. Buchmann J, Wolters A, Haessler F, Bohne S, Nordbeck R, Kunesch E. Disturbed transcallosally mediated motor inhibition in children with attention deficit hyperactivity disorder (ADHD. *Clin Neurophysiol.* 2003;114(11):2036–2042. Available from: https://doi.org/10.1016/s1388-2457(03)00208-6.
92. Weis S, Hausmann M, Stoffers B, Vohn R, Kellermann T, Sturm W. Estradiol modulates functional brain organization during the menstrual cycle: an analysis of interhemispheric inhibition. *J Neurosci.* 2008;28(50):13401–13410. Available from: https://doi.org/10.1523/JNEUROSCI.4392-08.2008.
93. Bayer U, Hausmann M. Effects of sex hormone therapy on interhemispheric crosstalk in postmenopausal women. *Neuropsychology.* 2009;23(4):509–518. Available from: https://doi.org/10.1037/a0015436.
94. Reddy DS. Neurosteroids and their role in sex-specific epilepsies. *Neurobiol Dis.* 2014;72(Pt B):198–209. Available from: https://doi.org/10.1016/j.nbd.2014.06.010.
95. Aboitiz F, Scheibel AB, Fisher RS, Zaidel E. Fiber composition of the human corpus callosum. *Brain Res.* 1992;598(1-2):143–153. Available from: https://doi.org/10.1016/0006-8993(92)90178-c.

96. Conti F, Manzoni T. The neurotransmitters and postsynaptic actions of callosally projecting neurons. *Behavioural Brain Res.* 1994;64(1-2):37–53. Available from: https://doi.org/10.1016/0166-4328(94)90117-1.
97. Kumar SS, Huguenard JR. Properties of excitatory synaptic connections mediated by the corpus callosum in the developing rat neocortex. *J Neurophysiol.* 2001;86(6):2973–2985. Available from: https://doi.org/10.1152/jn.2001.86.6.2973.
98. van der Knaap LJ, van der Ham IJM. How does the corpus callosum mediate interhemispheric transfer? A review. *Behavioural Brain Res.* 2011;223(1):211–221. Available from: https://doi.org/10.1016/j.bbr.2011.04.018.
99. White EL, Czeiger D. Synapses made by axons of callosal projection neurons in mouse somatosensory cortex: emphasis on intrinsic connections. *J Comp Neurol.* 1991;303(2):233–244. Available from: https://doi.org/10.1002/cne.903030206.
100. Wassermann EM, Pascual-Leone A, Hallett M. Cortical motor representation of the ipsilateral hand and arm. *Exp Brain Res.* 1994;100(1):121–132. Available from: https://doi.org/10.1007/BF00227284.
101. Ziemann U, Hallett M. Hemispheric asymmetry of ipsilateral motor cortex activation during unimanual motor tasks: further evidence for motor dominance. *Clin Neurophysiol.* 2001;112(1):107–113. Available from: https://doi.org/10.1016/S1388-2457(00)00502-2.
102. Kawaguchi Y. Receptor subtypes involved in callosally-induced postsynaptic potentials in rat frontal agranular cortex in vitro. *Exp Brain Res.* 1992;88(1):33–40. Available from: https://doi.org/10.1007/bf02259126.
103. Isaacson JS, Scanziani M. How inhibition shapes cortical activity. *Neuron.* 2011;72(2):231–243. Available from: https://doi.org/10.1016/j.neuron.2011.09.027.
104. McCarthy MM, Frank A. Functional significance of steroid modulation of GABAergic neurotransmission: analysis at the behavioral, cellular, and molecular levels, Beach Award*Hormones Behav.* 1995;29(2):131–140. Available from: https://doi.org/10.1006/hbeh.1995.1010.
105. Chiarello C, Maxfield L. Varieties of interhemispheric inhibition, or how to keep a good hemisphere down. *Brain Cognition.* 1996;30(1):81–108. Available from: https://doi.org/10.1006/brcg.1996.0006.
106. Ocklenburg S, Ball A, Wolf CC, Genç E, Güntürkün O. Functional cerebral lateralization and interhemispheric interaction in patients with callosal agenesis. *Neuropsychology.* 2015;29(5):806–815. Available from: https://doi.org/10.1037/neu0000193.
107. Genç E, Ocklenburg S, Singer W, Güntürkün O. Abnormal interhemispheric motor interactions in patients with callosal agenesis. *Behavioural brain Res.* 2015;293:1–9. Available from: https://doi.org/10.1016/j.bbr.2015.07.016.
108. Gowers WR. *Epilepsy and Other Chronic Convulsive.* Diseases:Their Causes, Symptoms & Treatment. *N.Y.* William Wood; 1885.
109. Bäckström T, Zetterlund B, Blom S, Romano M. Effects of intravenous progesterone infusions on the epileptic discharge frequency in women with partial epilepsy. *Acta Neurologica Scandinavica.* 1984;69(4):240–248. Available from: https://doi.org/10.1111/j.1600-0404.1984.tb07807.x.
110. Smith SS, Waterhouse BD, Woodward DJ. Sex steroid effects on extrahypothalamic CNS. I. Estrogen augments neuronal responsiveness to iontophoretically applied glutamate in the cerebellum. *Brain Res.* 1987;422(1):40–51. Available from: https://doi.org/10.1016/0006-8993(87)90538-5.
111. Smith SS, Waterhouse BD, Woodward DJ. Locally applied estrogens potentiate glutamate-evoked excitation of cerebellar Purkinje cells. *Brain Res.* 1988;475(2):272–282. Available from: https://doi.org/10.1016/0006-8993(88)90615-4.
112. Smith SS, Woolley CS. Cellular and molecular effects of steroid hormones on CNS excitability. *Cleve Clin J Med.* 2004;71(Suppl 2):S4–S10. Available from: https://doi.org/10.3949/ccjm.71.suppl_2.s4.
113. Taubøll E, Sveberg L, Svalheim S. Interactions between hormones and epilepsy. *Seizure.* 2015;28:3–11. Available from: https://doi.org/10.1016/j.seizure.2015.02.012.
114. Hausmann M. Why sex hormones matter for neuroscience: A very short review on sex, sex hormones, and functional brain asymmetries. *J Neurosci Res.* 2017;95(1-2):40–49. Available from: https://doi.org/10.1002/jnr.23857.
115. Smith SS, Waterhouse BD, Chapin JK, Woodward DJ. Progesterone alters GABA and glutamate responsiveness: a possible mechanism for its anxiolytic action. *Brain Res.* 1987;400(2):353–359. Available from: https://doi.org/10.1016/0006-8993(87)90634-2.
116. Smith SS, Waterhouse BD, Woodward DJ. Locally applied progesterone metabolites alter neuronal responsiveness in the cerebellum. *Brain Res Bull.* 1987;18(6):739–747. Available from: https://doi.org/10.1016/0361-9230(87)90209-7.

117. Smith SS. Progesterone administration attenuates excitatory amino acid responses of cerebellar Purkinje cells. *Neuroscience*. 1991;42(2):309–320. Available from: https://doi.org/10.1016/0306-4522(91)90377-Z.
118. Peters JA, Kirkness EF, Callachan H, Lambert JJ, Turner AJ. Modulation of the GABAA receptor by depressant barbiturates and pregnane steroids. *Br J Pharmacology*. 1988;94(4):1257–1269. Available from: https://doi.org/10.1111/j.1476-5381.1988.tb11646.x.
119. Majewska MD, Harrison NL, Schwartz RD, Barker JL, Paul SM. Steroid hormone metabolites are barbiturate-like modulators of the GABA receptor. *Science*. 1986;232(4753):1004–1007. Available from: https://doi.org/10.1126/science.2422758.
120. Smith RP, Coward RM, Kovac JR, Lipshultz LI. The evidence for seasonal variations of testosterone in men. *Maturitas*. 2013;74(3):208–212. Available from: https://doi.org/10.1016/j.maturitas.2012.12.003.
121. Gouchie C, Kimura D. The relationship between testosterone levels and cognitive ability patterns. *Psychoneuroendocrinology*. 1991;16(4):323–334. Available from: https://doi.org/10.1016/0306-4530(91)90018-o.
122. Moffat SD, Hampson E. Salivary testosterone concentrations in left-handers: An association with cerebral language lateralization. *Neuropsychology*. 2000;14(1):71–81. Available from: https://doi.org/10.1037/0894-4105.14.1.71.
123. Beaton AA, Rudling N, Kissling C, Taurines R, Thome J. Digit ratio (2D:4D), salivary testosterone, and handedness. *Laterality*. 2011;16(2):136–155. Available from: https://doi.org/10.1080/13576500903410369.
124. Richards G, Beking T, Kreukels BPC, Geuze RH, Beaton AA, Groothuis T. An examination of the influence of prenatal sex hormones on handedness: Literature review and amniotic fluid data. *Horm Behav*. 2021;129:104929. Available from: https://doi.org/10.1016/j.yhbeh.2021.104929.
125. Richards G, Medland SE, Beaton AA. Digit ratio (2D:4D) and handedness: A meta-analysis of the available literature. *Laterality*. 2021;26(4):421–484. Available from: https://doi.org/10.1080/1357650X.2020.1862141.
126. Richards G, Tan DW, Whitehouse AJO, et al. A longitudinal examination of perinatal testosterone, estradiol and vitamin D as predictors of handedness outcomes in childhood and adolescence. *Laterality*. 2022;27(6):547–580. Available from: https://doi.org/10.1080/1357650X.2022.2109656.
127. Richardson T. No association between adult sex steroids and hand preference in humans. *Am J Hum Biol*. 2022;34(2):e23605. Available from: https://doi.org/10.1002/ajhb.23605.
128. Hausinger T, Pletzer B. Sex hormones modulate sex differences and relate to hemispheric asymmetries in a divided visual field Navon task. *Behavioural brain Res*. 2021;408:113281. Available from: https://doi.org/10.1016/j.bbr.2021.113281.
129. Beking T, Geuze RH, van Faassen M, Kema IP, Kreukels BPC, Groothuis TGG. Prenatal and pubertal testosterone affect brain lateralization. *Psychoneuroendocrinology*. 2018;88:78–91. Available from: https://doi.org/10.1016/j.psyneuen.2017.10.027.
130. Kimura D, Hampson E. Cognitive Pattern in Men and Women Is Influenced by Fluctuations in Sex Hormones. *Curr Dir Psychol Sci*. 1994;3(2):57–61. Available from: https://doi.org/10.1111/1467-8721.ep10769964.

CHAPTER

13

Altered hemispheric asymmetries in neurodevelopmental, mental, and neurological disorders

OUTLINE

Introductory short story

With a stern face, Dr. Samuel Torrey Orton looked at the boy. "Please read this sentence, Robby." Hunched over the book, the pupil struggled, almost visibly. "Th...th...the dab bog ra...ra...ran a...a...away..." Orton looked at the worn and battered book he had placed on the desk in front of Robby. The page he had opened showed a scruffy dog that was chasing

The Lateralized Brain
DOI: https://doi.org/10.1016/B978-0-323-99737-9.00008-2

after a ball. Underneath the picture, it read "The bad dog ran away." "This is remarkable," Orton thought. He had spent the last 2 weeks evaluating the cognitive abilities of several pupils from the Greene County area, whom their teachers had referred to his clinic since they had failed at school. What was remarkable about these cases was that despite normal intelligence, the pupils all showed similar problems when reading. In particular, they made a lot of left-right confusions and reversals such as mistaking the letter "b" for the letter "d." Their ability to distinguish left from right seemed to be impaired. Another observation Orton considered striking was that many of these kids were left- or mixed-handed.

Born on October 15, 1879, in Columbus, Ohio, Orton was the son of Edward Orton Sr., a famous oil geology professor who was the first president of Ohio State University and had also been president of the Geological Society of America and of the American Association for the Advancement of Science. Having grown up in an intellectual environment that encouraged progressive thinking, Orton developed a radical new theory for the development of the reading problems he observed in his clinic in Greene County, Iowa, in 1925. Based on these and other observations, Orton concluded that reading problems were caused by a failure to establish the left-right sense, which in turn was caused by a failure to establish correct cerebral dominance. The first widely recognized theory that linked atypical lateralization and a developmental disorder was born! While scientists have developed new theories on how asymmetries and dyslexia are linked, Orton's work has inspired a whole research field: clinical laterality research.

Inspired by [1–3].

The clinical perspective on laterality

Laterality represents a basic organizational principle in both vertebrate and invertebrate nervous systems.[4] As such, it has been the focus of many basic research studies that have been outlined in the preceding chapters. In addition, one of the main factors generating interest in laterality research is the fact that several neurodevelopmental, mental, and neurological disorders have been associated with altered hemispheric asymmetries.[5] In this chapter, we will therefore highlight research on altered hemispheric asymmetries in different neurodevelopmental, mental, and neurological disorders. The association between hemispheric asymmetries and these disorders can be conceptualized in different ways,[6] which include the following:

1. Nonspecific association: Altered asymmetries are associated with a generally increased risk to develop neurodevelopmental, mental, and neurological disorders, independent of a specific diagnosis. There is no mechanistic coupling between altered asymmetries and specific diagnoses.
2. Diagnosis-specific association: Altered asymmetries are associated with a specific diagnosis, but not with others (e.g., more mixed-handedness in schizophrenia, but not in depression).
3. Symptom-specific association: Altered asymmetries are associated with a specific symptom group (e.g., hallucinations, affective symptoms, or cognitive symptoms), but independent of diagnosis.

Neurodevelopmental disorders

Dyslexia

As described in the introductory short story, the first researcher to make a widely recognized link between atypical lateralization and neurodevelopmental disorders was Samuel Torrey Orton (1879–1948). Orton was an American physician who investigated the development of dyslexia, a disorder characterized by difficulties in reading, but normal intelligence, in the 1920s and 1930s. While Orton's ideas somewhat fell into obscurity after World War II, decades later another prominent theory linking atypical lateralization and dyslexia emerged. In the mid-1980s, Norman Geschwind and Al Galaburda from the Department of Neurology at Harvard Medical School, USA, published a three-paper series in the scientific journal *Archives of Neurology*,[7–9] preceded by a short article by Geschwind and Peter Behan in *PNAS*.[10] In these articles, the three authors described what would later be known as the Geschwind–Behan–Galaburda (GBG) model. Combined together, the four papers have been cited more than 2,700 times, probably rendering the GBG model the most widely cited theory on the association between atypical lateralization and pathology. As indicated by the close to 90 pages the four articles are composed of, it is not a simple model. It was conceived to explain the following observations and assumptions:

- Left-handedness is more common in males than females (see Chapter 12 for a detailed description of sex differences in hemispheric asymmetries).
- Dyslexia and other neurodevelopmental disorders—such as autism—are more common in males than females.
- Females often show better verbal abilities (a left-hemispheric function) than males, but males show better visuospatial abilities (a right-hemispheric function) than females.
- Left-handers and individuals with dyslexia often show superior right-hemispheric functions.
- Left-handedness is more common in individuals with dyslexia than in the general population.
- Immune diseases are more common in left-handers than in right-handers.

The GBG model assumes that testosterone is the factor that links these observations. Specifically, it assumes that intrauterine testosterone levels and sensitivity to testosterone differently affect the developmental trajectories of the left and right hemisphere. High testosterone levels selectively delay the maturation of the left hemisphere and lead to a stronger focus on right-hemispheric function and thus atypical functional lateralization. This assumption could explain why there are more male than female left-handers. Moreover, high intrauterine testosterone levels are also thought to cause abnormalities in the development of left-hemispheric speech regions in the temporal lobe, resulting in atypical language lateralization and higher prevalence of dyslexia. In addition to these direct effects on handedness and language capabilities, the GBG model also suggests a major impact of intrauterine testosterone levels on immune function and immune disorders. The GBG model made a lot of predictions that could be tested empirically and that were sometimes, but not always, supported by the data. Thus, the GBG model has been discussed controversially ever since its publication. For example, in 1994, Phil Bryden, Chris McManus, and Barbara Bulman-Fleming published an article of more than 60 pages in the scientific journal *Brain and Cognition* in which they systematically evaluated the empirical support for the GBG model.[11] While they found the empirical evidence for a link between atypical lateralization and dyslexia to be fairly strong, their analysis revealed that there was almost no compelling evidence supporting the idea that specifically fetal testosterone was the factor linking dyslexia and atypical lateralization. Ultimately, they concluded that the GBG model in its entirety was not supported by empirical evidence, a view shared by other authors.[12] Others, however, have highlighted that despite not having withstood the rigorous empirical testing by Bryden, McManus, and Bulman-Fleming, the GBG model, or at least some of its parts, still made an important and useful contribution to the understanding of how dyslexia and atypical lateralization are linked, mainly by conceptualizing their association in a developmental framework.[13,14]

If testosterone is not the developmental factor that links atypical lateralization and dyslexia, what is? One idea that had initially been brought forward within the larger framework of Annett's monogenic right-shift model for the development of handedness (see Chapter 6) is the assumption that there might be a genetic link between non-right-handedness and dyslexia. Specifically, the model assumed that a single, hypothetical gene determines right-handedness and left-hemispheric language lateralization.[15] If due to a specific allelic configuration in this gene, an individual lacked the typical left-hemispheric language dominance, it was assumed that this individual would have a disadvantage when it comes to language learning and a higher chance of developing dyslexia. While the idea that handedness could be determined by a single gene has been disproven by molecular genetic evidence,[16] the assumption that there is at least some genetic link between handedness and dyslexia is supported by molecular studies. When looking at the genes that have been related to dyslexia (e.g., *DCDC2*, *ROBO1*, *DYX1C1*, and *KIAA0319*),[17] it becomes evident that there is at least some overlap with genes that have been related to hemispheric asymmetries. For example, *KIAA0319*, a gene that encodes a transmembrane protein that likely plays a role in cortical development, neuronal migration, and cell adhesion, has been associated with both dyslexia[17] and language lateralization.[18] Moreover, it has been suggested by William M. Brandler and Silvia Paracchini that genes involved in the biological pathways establishing bodily left-right asymmetries in the embryo might affect both handedness and dyslexia.[19,20] For example, genes involved in cilia formation are thought to also affect dyslexia. As outlined in Chapter 10, rotation of primary cilia in the node plays a critical role in nodal-induced symmetry breaking in the vertebrate embryo.[21] Interestingly, length and signaling of primary cilia in neurons are affected by overexpression of *DCDC2*, a gene that has been implicated in neuronal migration and dyslexia.[22] Additionally, the zebrafish orthologue of the dyslexia candidate gene *DYX1C1* has been shown to play an essential role for cilia growth and function, as knockout mutants lost both the outer and the inner dynein arms, two structures required for cilia motility.[23] Moreover, *PCSK6*, a gene encoding a protein convertase that cleaves nodal into its active form, has been related to handedness specifically in a dyslexia cohort.[24]

While the idea that genes involved in ciliogenesis might represent the link between dyslexia and atypical lateralization appears to be a very promising working hypothesis,[19,20] more research is needed to unravel the exact molecular processes involved in the generation of both phenotypes.

So, what does the data actually suggest with regard to the relation of atypical lateralization and dyslexia? For handedness, the original paper by Geschwind and Behan in 1982 suggested that strong left-handers had a substantially higher prevalence (9.5%) of learning disorders (dyslexia and stuttering) than strong right-handers (0.8%).[10] Several other studies have investigated this relationship, but results have been far less conclusive. Two efforts for systematic integration of results across studies have been conducted. The first was published in a 1990 book by Dorothy Bishop and contained an analysis of 25 studies,[25] which was not significant. Only when the largest study was removed from the meta-analysis, a significant effect with 11.2% of left-handers in dyslexia patients compared to 5.8% of left-handers in healthy controls was observed.[26] In 1994, the 25 studies included in Bishop's analysis were reanalyzed by Eglinton and Annett with different meta-analysis statistics.[27] These authors found that the number of non-right-handers was significantly larger for individuals with dyslexia than for healthy controls, thus indicating an increase of left-handedness in dyslexia. More recently, a large-scale study including data from 2503 cases with language and/or reading impairments and 4316 controls obtained from ten different cohort studies also showed an elevated prevalence of non-right-handedness in individuals with language and/or reading impairment.[28] Since not all individuals included in this study had a dyslexia diagnosis, the findings tentatively suggest that there is an association between altered hemispheric asymmetries and language/reading problems that is not necessarily limited to the specific diagnosis of dyslexia.

Research on atypical lateralization in dyslexia is not limited to handedness. As dyslexia is a developmental disorder of the language system, several authors also have investigated its relation to language lateralization. On the behavioral level, this was mainly done using the dichotic listening task (see Chapter 5 for a detailed description). Here, results have been somewhat mixed, with several—but not all—studies showing a reduced right-ear advantage and hence reduced left-hemispheric language dominance in patients compared to controls.[29,30] Interestingly, symptom severity seems to be a crucial factor in the relation of language lateralization and dyslexia. Turid Helland from the University of Bergen in Norway reported that a cohort of dyslexic individuals who experienced few symptoms and responded well to educational support performed similarly to healthy controls in the dichotic listening task.[31] In contrast, a second cohort of dyslexic individuals who showed stronger symptoms and did not respond to educational support showed reduced language lateralization (see Fig. 13.1). These findings support the idea of a symptom-specific association between altered hemispheric asymmetries and neurodevelopmental disorders.

Atypical language lateralization in dyslexia has also been shown on a neurophysiological level. For example, it has been reported that, on average, children with developmental dyslexia do not show the typical left lateralization in core language regions (Broca's area, supramarginal gyrus, and superior temporal gyrus) during lexico-semantic and phonological language fMRI tasks.[32,33] Importantly, no lateralization differences between patients and controls were observed for all other brain regions. Dyslexia has also been related to atypical structural asymmetries in language-related brain areas, particularly in the planum temporale. Here, individuals with dyslexia show a reduction of the typical leftward asymmetry and an increase of atypical rightward asymmetry.[34] Taken together, dyslexia has been related to atypical asymmetries on the behavioral, neurophysiological, and structural level. While the GBG model assumed that atypical lateralization clearly is the cause for dyslexia and this view is widely cited in the literature, this idea has recently been called into question (see Box 13.1).

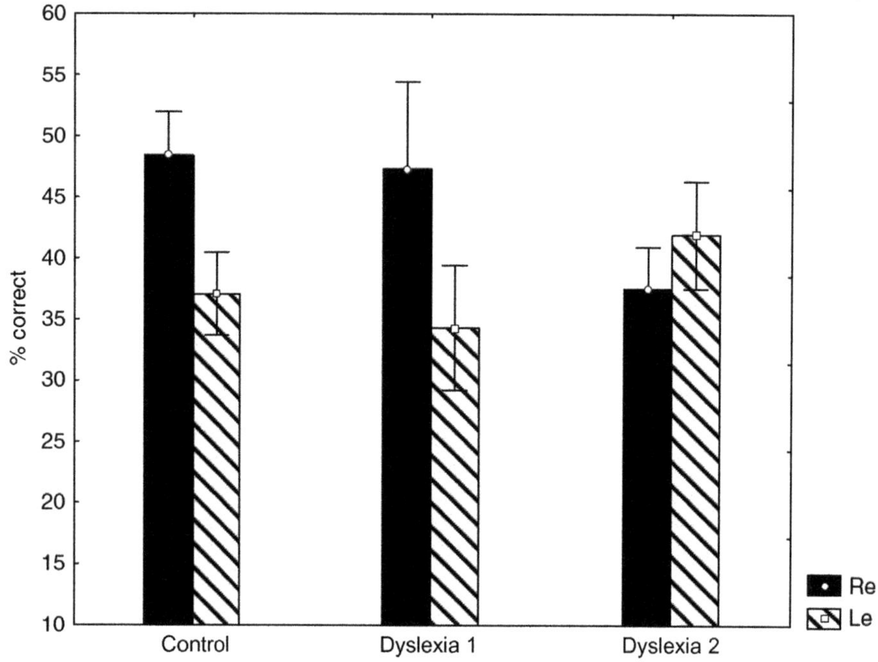

FIGURE 13.1 Results of the study by Helland and coworkers.[31] Percentage of right-ear (Re) and left-ear (Le) answers in the dichotic listening task for healthy controls, dyslexia patients with low symptom severity (Dyslexia 1), and dyslexia patients with high symptom severity (Dyslexia 2). Dyslexia patients with high symptom severity show a reduction of the typical right-ear advantage compared to healthy controls and dyslexia patients with low symptom severity. *Source: Reproduced with permission of the authors and the publisher from Helland T, Asbjørnsen AE, Hushovd AE, Hugdahl K. Dichotic listening and school performance in dyslexia.* Dyslexia. *2008;14 (1):42–53. https://doi.org/10.1002/dys.343.*

Autism spectrum disorders

In addition to dyslexia, another group of developmental disorders that has been associated with atypical lateralization is autism spectrum disorders (ASD), for example, autism and Asperger syndrome. Patients typically show deficits in social and communicative skills as well as stereotyped behavioral patterns. ASD has been related to non-right-handedness by several authors. A meta-analysis of 12 studies—with an overall sample size of 497 individuals with ASD and healthy controls—revealed a significant difference between the two groups.[37] In the healthy control group, 79% of individuals were right-handed, 16% were mixed-handed, and 5% were left-handed. ASD patients showed a marked reduction of right-handedness and an increase in mixed- and left-handedness. In this group, only 40% were right-handed, but 44% were mixed-handed, and 16% were left-handed.

As deficits in communicative abilities are one of the core characteristics of ASD, language lateralization is likely to be affected. Indeed, language lateralization seems to be the

BOX 13.1

Critical thinking: atypical lateralization—cause or consequence of neurodevelopmental disorders?

In 2013, Dorothy Bishop from the Department of Experimental Psychology at the University of Oxford, UK, published an article in the prestigious scientific journal *Science* that questioned the widely cited idea that atypical lateralization might be one of the causative factors for neurodevelopmental disorders such as dyslexia.[35] In this paper, she argued that atypical language lateralization in dyslexia is not the cause of the disorder but—at least partly—a consequence of poor language development. How did she come to this conclusion? First, she noted that if atypical lateralization were indeed the cause of dyslexia, it should at least be as heritable as the disorder. Moreover, both traits should be affected by largely the same genes. However, this is not the case. While dyslexia shows a heritability of about 70%, heritability indices for most forms of hemispheric asymmetries are below 30%. Moreover, while there is some overlap in the genes involved in determining the two phenotypes (see section on the Brandler–Paracchini model), the majority of genes implicated in dyslexia have not been related to language lateralization and vice versa. Bishop also noted that atypical lateralization was not necessarily linked to cognitive impairments. Many atypically lateralized individuals in the healthy population showed normal cognitive capabilities, rendering a strong causal link between atypical lateralization and cognitive impairments unlikely. Moreover, it was argued that hemispheric asymmetries were not fixed from birth but changed over the course of cortical development (see Chapter 11). Based on these findings, atypical lateralization is more likely a consequence rather than a cause of dyslexia.

An interesting contribution to the cause or consequence debate came from a functional transcranial Doppler sonography study, in which participants had to perform three different language tasks which relied on different language subprocesses.[36] In this study, inconsistent language lateralization across the three tasks (e.g., leftward lateralization in one task and rightward lateralization in the two other tasks) was shown to be much more common in individuals with a developmental disorder affecting language than in healthy controls. Over the whole sample, 84% of those individuals showing inconsistent language lateralization were in the disorder group, leading the author to conclude that inconsistent lateralization may be a cause for language disorder, but may also be caused by compensatory reorganization and thus an indirect consequence of the disorder. While more research, particularly on genetic and epigenetic determinants of both lateralization and language disorders, is needed before any final conclusions can be drawn, these findings certainly provide food for thought for laterality researchers interested in pathology.

primarily altered form of hemispheric asymmetries in ASD,[38–40] while right-hemispheric functions such as visuospatial skills seem to be less affected.[41] Several studies have shown that in ASD, the typical leftward dominance in the language system is frequently reduced or even reversed such that patients show no or rightward asymmetries.[42–44] On the behavioral level, ASD patients have a higher prevalence of left-ear advantage in dichotic listening tasks, indicating rightward language dominance.[45] Moreover, neuroimaging studies have revealed that ASD patients show significantly reduced leftward activation dominance for both perception and production of speech.[46,47]

On the structural level, an early study indicated rightward volume asymmetry in Broca's area of children with ASD, while healthy controls show leftward asymmetry.[48] Moreover, a significantly reduced leftward asymmetry of the planum temporale has been reported in ASD.[49,50] The largest study on altered structural asymmetries in ASD was conducted by the ENIGMA consortium.[51] In this study, the authors analyzed data from 1774 individuals with ASD and 1809 controls, obtained from 54 independent cohorts. The results were quite striking (see Fig. 13.2). Individuals with ASD show significantly different asymmetries in cortical thickness compared to controls in a widespread network of brain areas involving medial frontal, orbitofrontal, cingulate, and inferior temporal areas. Moreover, a significant asymmetry difference was observed for the surface area of the orbitofrontal cortex. Importantly, all these cortical asymmetries were reduced in individuals with ASD compared to controls and their effect sizes were small (the largest Cohen's d was −0.13). This shows that altered structural asymmetries are widespread in ASD, but small in size, which makes large-scale datasets necessary to detect them. On the subcortical level, an increased asymmetry of the putamen was observed. Since the orbitofrontal cortex was the only brain area that showed changes in asymmetry for both cortical thickness and surface area, the authors of the study suggested that lateralization of this brain area may show the most pronounced disruption of laterality in ASD. The orbitofrontal cortex plays a role for executive functions and thus may be relevant for ASD symptoms related to repetitive and stereotyped behaviors.

The ENIGMA consortium has also published a second study on asymmetries in ASD.[52] In this study, asymmetries in structural networks were assessed in data obtained from 43 independent cohorts. Individuals with ASD showed altered asymmetries of brain networks involving the medial orbitofrontal cortex, rostral middle frontal cortex, and fusiform cortex compared to controls. By comparing these networks with the results of previously published functional neuroimaging meta-analyses, the authors concluded that ASD is associated with altered asymmetries in networks related to executive functions, language, and sensorimotor processing.

Interestingly, it has been reported that in ASD patients, altered structural asymmetries in the thalamus correlate significantly with Autism Behavior Checklist language scores, supporting the idea that symptom severity is a relevant factor for the association between ASD and hemispheric asymmetries.[53]

In addition to altered gray matter asymmetries, ASD has also been shown to be accompanied by structural alterations in language-relevant white matter tracts. For example, the typical leftward volume asymmetry of the arcuate fasciculus seems to be reversed to a rightward asymmetry in ASD patients.[54] Moreover, ASD patients exhibited significantly reduced fractional anisotropy in the left but not in the right uncinate fasciculus.[55] Interestingly, differences in neuronal connectivity between ASD patients and controls have not only been observed for structural but also for functional networks. A recent resting state fMRI study revealed significant asymmetry differences between ASD patients and controls for 10 out of 17 functional networks—for example, not only for language networks but also for visual, auditory, motor, executive, and attentional networks.[56] All of

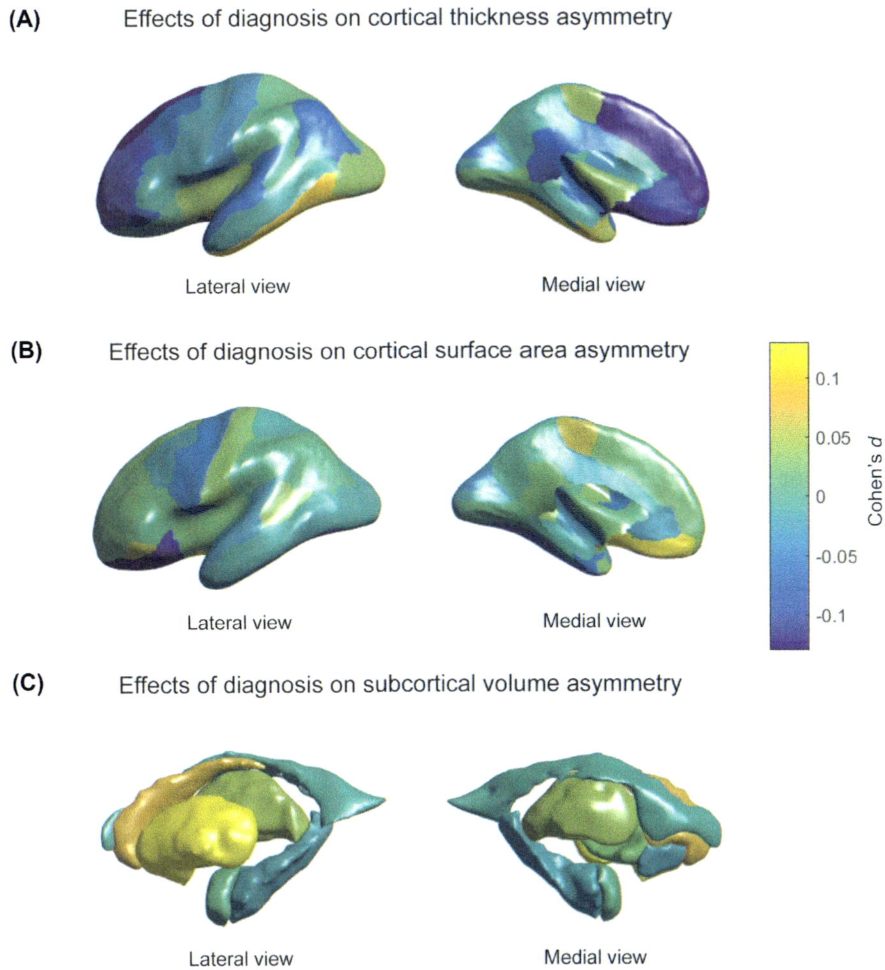

FIGURE 13.2 Alterations of structural asymmetries in ASD for cortical thickness (panel A), cortical surface area (panel B), and subcortical volume (panel C). Color indicates Cohen's d as the effect size of the comparison between individuals with ASD and controls. Positive values of Cohen's d (yellow color) indicate a shift toward greater leftward/reduced rightward asymmetry in individuals with ASD compared to controls. In contrast, negative Cohen's d values (blue color) indicate a shift toward reduced leftward/greater rightward asymmetry in individuals with ASD compared to controls. *ASD*, Autism spectrum disorders. *Licensed under a Creative Commons Attribution 4.0 International License and reproduced with permission of the authors from Postema MC, van Rooij D, Anagnostou E, et al. Altered structural brain asymmetry in autism spectrum disorder in a study of 54 datasets.* Nat Commun. *2019;10(1):4958. https://doi.org/10.1038/s41467-019-13005-8.*

these networks showed a rightward shift in the patient group. This finding is particularly interesting because atypical lateralization in frontotemporal brain networks has been related to clinical severity in toddlers with ASD, with individuals with reversed asymmetries showing the most severe symptoms.[57]

Based on these and similar findings, Annukka K. Lindell and Kristelle Hudry from the La Trobe University in Melbourne, Australia, proposed an integrative model for the relationship between reduced functional and structural lateralization and language impairment in ASD[38] (see Fig. 13.3).

This model can be outlined as follows:

- ASD-specific genetic or environmental factors lead to reduced or reversed asymmetries in the language system.
- These factors lead to higher prevalence of non-right-handedness.
- They also lead to abnormal structural lateralization.
- For gray matter, this affects the inferior frontal region (Broca's area), the posterior superior temporal region, and the planum temporale.
- For white matter, this affects the arcuate fasciculus, uncinate fasciculus, and the cingulum.
- Atypical gray matter asymmetries lead to attenuated or reversed left-hemispheric dominance for semantic processing and perception of speech sounds.

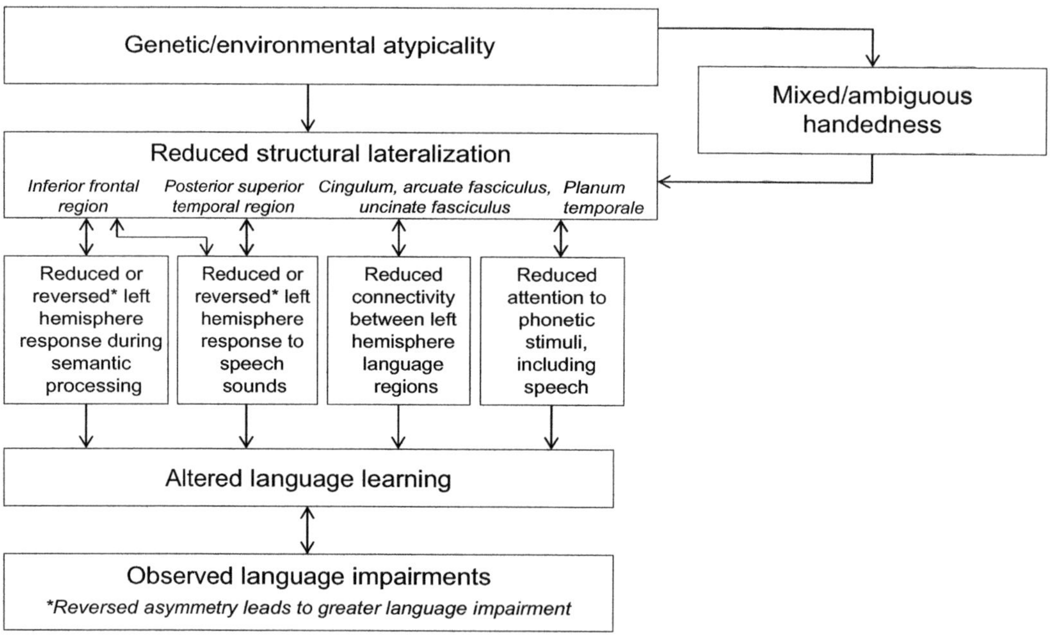

FIGURE 13.3 The Lindell and Hudry model[38] for the relationship between reduced structural lateralization, reduced functional lateralization, and language impairment in ASD. *ASD*, Autism spectrum disorder. *Source: Reproduced with permission of the authors and the publisher from Lindell AK, Hudry K. Atypicalities in cortical structure, handedness, and functional lateralization for language in autism spectrum disorders.* Neuropsychol Rev. *2013;23(3):257-270. https://doi.org/10.1007/s11065-013-9234-5.*

- Atypical white matter asymmetries lead to weaker connectivity within the left-hemispheric language network.
- Together, these substantial changes in the neuroanatomy of the language system lead to alterations in language learning.
- These lead to the typical communicative impairments in ASD, with reversed lateralization leading to greater language impairment than reduced lateralization.

Stuttering

Another developmental disorder that affects the language system and has been associated with atypical asymmetries is stuttering. Stuttering is a speech disorder that usually begins in early childhood and has a lifetime prevalence of about 5%. Affected individuals show a disrupted flow of speech, characterized by repetitions of sounds as well as pauses and/or temporary inability to produce specific sounds. One of the first reports for atypical hemispheric asymmetries in stuttering individuals was provided by Kypros Chrysanthis, a doctor from Cyprus who investigated handedness in stuttering school children in the 1940s.[58] In his sample, only 4.2% of the nonstuttering children were left-handed, whereas about 23.8% of the stuttering children were left-handed—a more than fourfold increase! However, later studies in much larger samples of stutterers failed to find any significant association between stuttering and handedness, reporting comparable prevalences for left-handedness in stutterers as in the general population.[59,60]

Historically, it has also been assumed that stuttering is not associated with increased left-handedness per se but that forcing left-handers to write with their right hand could cause stuttering. This forced switching was a common practice in schools in many Western countries including the United States and Germany in the early to mid-20th century. In principle, it was assumed that a child's wish to write with the left instead of the right hand did not reflect an inherited motor preference but a defiant personality.[61] Therefore, it needed to be corrected with sometimes brutal methods such as tying up the left hand or beating up children when caught writing with the left hand. However, as described by Howard I. Kushner from Emory University in Atlanta, USA, in a historical article, not everyone agreed with the gruesome practice.[61] Between the 1920s and the 1950s, the aforementioned physician Samuel Torrey Orton and psychologist Lee Edward Travis put forward the idea that forcing natural left-handers to write with their right hands could explain the etiology of stuttering in some cases. This idea was based on the observation that about 50% of the observed stutterers were natural left-handers who had been forced to write with their right hands. Orton concluded that forced relearning led to confusion of cerebral dominance that then led to stuttering. These observations were complemented by the finding that at least in some naturally left-handed stutterers who had been forced to write with their right hands, a reversal of relearning (e.g., writing with the left hand again) led to substantial improvement of stuttering symptoms.

Since most countries abandoned the practice of forced handedness relearning in schools before the 1970s, this line of research has been largely forgotten in stuttering research today. What, however, has repeatedly been confirmed by modern research is that individuals who stutter show atypical hemispheric asymmetries particularly in the language domain. In 1980, it was shown that individuals who stutter showed decreased EEG alpha-band activity in the right hemisphere during verbal and nonverbal tasks.[62] Since EEG alpha-band activity is thought to reflect an absence of focused mental activity, these findings were interpreted to indicate overactivation of the right hemisphere during these tasks in stutterers.

The notion of a hyperactive right hemisphere during language processing in stutterers is also supported by modern neuroimaging studies.[63] An early PET study by Peter T. Fox from the University of Texas, USA, indicated that stutterers showed right-hemispheric overactivation in the motor system but a lack of left-hemispheric activation in the auditory system during speech production.[64] A subsequent PET analysis added to these findings by showing that in individuals who stutter, left-hemispheric activity during language production was associated with stuttered speech, while right-hemispheric activity was related to fluent speech, suggesting a compensatory role for right-hemispheric activation.[65] Following these studies, several PET and fMRI studies comparing stutterers to healthy controls have been published. These works were statistically integrated by Steven Brown and coworkers from the University of Texas Health Science Center at San Antonio, USA,[66] who used a technique called ALE to meta-analyze these empirical imaging studies. ALE is a statistical technique that allows for the integration of several neuroimaging studies essentially by determining how often a specific locus in the brain is active across different studies.[67,68] Brown and his colleagues conducted two parallel ALE analyses, one that integrated findings on stuttered production in stutterers and another one that integrated findings on fluent production of speech in healthy subjects. While essentially the same brain areas were involved in stuttered and fluent speech, there was overactivation of motor areas in stuttered speech. Moreover, atypical rightward activation asymmetries in the frontal operculum, Rolandic operculum, and the anterior insula emerged, indicating a rightward shift of brain activity during stuttering. In addition, while controls showed activation in the auditory association cortex, this was not the case for stutterers.

Interestingly, atypical language lateralization in stuttering is also reflected on the structural level. Individuals who stutter show reduced leftward structural asymmetry in the planum temporale[69,70] and atypical leftward structural asymmetry in the caudate,[71] a subcortical structure linked to motor defects in speech planning in stuttering. Moreover, in stutterers, increased white matter volume has been reported for the right planum temporale and the right inferior frontal gyrus—included are the homotopic areas to pars triangularis of Broca's area in the right hemisphere—but also in motor areas related to face and mouth and in the anterior frontal gyrus.[72] In addition, healthy controls were observed to have larger left than right volume of the occipital lobe and a reversed pattern in the frontal lobe. These asymmetries were not observed in stutterers.[73]

Taken together, stuttering has been related to atypical handedness and language lateralization on multiple levels. Whether these atypical asymmetries are indeed a causative factor in the ontogeny of stuttering as had been suggested in the early 20th century or whether they are a consequence of brain reorganization processes related to stuttering is an interesting question for future research.

Attention deficit/hyperactivity disorder

Atypical lateralization has also been linked to developmental disorders outside the domain of speech-related pathology, for example, to attention deficit/hyperactivity disorder (ADHD). ADHD is characterized by inattention, hyperactivity, and impulsivity. It usually starts between 3 and 6 years of age and can continue into adulthood. With a prevalence between 5% and 7.1%, it is one of the most common neurodevelopmental disorders in children and adolescents.[74]

Findings from studies investigating the prevalence of left-handedness in ADHD patients have yielded somewhat inconsistent results. While an increase of left-handedness in ADHD patients has been reported,[75] other authors have failed to find such an association.[76] A recent meta-analysis found a significantly higher prevalence of non-right-handedness in individuals with ADHD compared to controls, with a non-right-handedness prevalence of 27.32% compared to 18.1% in controls.[77] However, handedness is not related to inattention and hyperactivity symptoms in ADHD.[78]

As a deficit of attention is one of the core symptoms of ADHD, and since the right hemisphere is dominant for visuospatial attention,[79] ADHD has been suggested to link to a deficit of the right hemisphere.[80–82] The first evidence for this notion was provided by lesion studies showing that lesions of the right hemisphere could lead to attention problems similar to those experienced by patients with ADHD.[83] Subsequently, Kenneth M. Heilman from the Department of Neurology at the University of Florida, USA, suggested that ADHD patients exhibit a dysfunction of right-sided frontostriatal systems.[84] This idea has been supported by various empirical findings. For example, children with ADHD were reported to show a disruption of right-hemispheric attentional processes.[85] Moreover, fMRI studies have revealed that response inhibition, a cognitive function mediated by frontostriatal brain circuits, is impaired in ADHD.[86]

The largest study on structural hemispheric asymmetries in ADHD has been conducted by the ENIGMA consortium.[87] In this study, cortical and subcortical asymmetries were assessed in neuroimaging data from 1933 individuals with ADHD and 1829 controls. This dataset was obtained from 39 independent cohorts. In line with the idea of a right-hemispheric deficit in ADHD, children with ADHD showed less rightward asymmetry of the total surface area of the hemispheres. Moreover, individuals with ADHD (children, adolescents, and adults combined) showed a reduced asymmetry of the medical orbitofrontal area, similar to what has been reported in ASD.[51] Also similar to ASD, effect sizes were small (Cohen's d between −0.18 and 0.18). This led the authors of the study to conclude that altered structural asymmetries in ADHD provide interesting insights into the neurobiology of the disorder, but are unlikely to be a useful biomarker, since their predictive value is low.

Interestingly, symptom severity is a relevant factor for the association between ADHD and hemispheric asymmetries. For example, caudate volume asymmetry has been shown to predict symptom severity in ADHD.[88] Moreover, a recent meta-analysis of seven ADHD neuroimaging datasets suggested a link between altered asymmetries in ADHD and symptom severity.[89]

While some of these findings support the idea of right-hemispheric dysfunction in ADHD, ADHD does not affect the right hemisphere alone, given that neuroscientific research has also indicated ADHD-related changes in the left hemisphere. Instead, especially interhemispheric networks governed by the right hemisphere have been related to ADHD outcome measures.[90]

Mental disorders

Schizophrenia

The most widely investigated mental disorder in laterality research is certainly schizophrenia, a severe mental disorder with a lifetime prevalence ranging between 0.3% and 0.7%. Schizophrenia is characterized by delusions and hallucinations as well as cognitive and motivational problems.[91] The notion that atypical lateralization might be a causative factor for psychiatric disorders such as schizophrenia first emerged in the 19th century. As outlined in Chapter 1, Arthur Wigan, a British M.D. who had practices in London and Brighton, published a book called *The duality of the Mind* in 1844, proposing that the two hemispheres are capable of separate cognition and volition.[92,93] Interestingly, his ideas transcended the notion of different cognitive capabilities for the two hemispheres. The full title of the book is *A New View of Insanity: The Duality of the Mind Proved by the Structure, Functions and Diseases of the Brain and by the Phenomena of Mental Derangement and Shown to be Essential to Moral Responsibility* and already hinted at Wigan's radical new theory on how mental disorders developed. In his book, he proposed that in the healthy brain, one hemisphere is always dominant and controls behavior. When this interhemispheric control is lacking, two diverging trains of thought develop and compete in attempting to control behavior. This leads to a lack of control over—and coherence of—behavior which manifests as mental illness.[93] With his theory, Wigan essentially was the first widely read author proposing that a reduction of hemispheric asymmetries could lead to mental illness. Wigan's general idea of the relationship between hemispheric asymmetries and mental illness was taken up by other authors in the decades following the publication of his book. In 1969, Pierre Flor-Henry, a French psychiatrist who at that time worked at Bethlem Royal and Maudsley Hospital in London, UK, published an article in the scientific journal *Epilepsia*, in which he investigated a link between psychosis and temporal lobe epilepsy.[94] He compared the laterality of epileptic foci in epilepsia patients with and without schizophrenia. In epileptic patients with schizophrenia, he found an overrepresentation of left-sided foci—right hemisphere: 9.5%; left hemisphere 43%; bilateral 47.5%—compared to nonpsychotic epilepsy patients—right hemisphere: 44%; left hemisphere: 22%; bilateral: 33%. He concluded that schizophrenia was related to disturbance of the dominant left hemisphere, while in nonpsychotic controls, the distribution of foci was essentially random. This finding sparked the interest of the scientific community in the relationship between hemispheric asymmetries and schizophrenia, and hundreds of studies investigating this relationship have since been conducted. Findings have indicated that in schizophrenia, atypical lateralization can be observed on the behavioral level, in brain activation patterns, and on the structural level.

On the behavioral level, one of the most robust findings is that schizophrenia patients are more often non-right-handed than controls, a result that to date has been confirmed by a vast number of empirical studies and three independent meta-analyses.[95–97] How big is this difference between schizophrenia patients and healthy controls? Meta-analyses often indicate differences between patient cohorts and healthy controls using so-called odds ratios. This is an association measure that basically compares the odds of being non-right-handed in the control groups with the odds of being non-right-handed in the patient groups. Meta-analyses have consistently yielded odds ratios between 1.5 and 1.8 for the relation of schizophrenia and handedness.[95–97] Thus, if we assume that about 10% of healthy controls are non-right-handed, 15%–18% of schizophrenia patients are as well. Interestingly, this does not only apply to schizophrenia but also to schizotypy—for example, nonclinical schizotypal personality tendencies.[98] Non-right-handed individuals showed significantly higher schizotypy than right-handed individuals.

In addition to an increase in non-right-handedness, schizophrenia patients show a reduction of the typical left-hemispheric dominance in language tasks. For example, patients show a decreased right-ear advantage in the dichotic listening task as confirmed by a meta-analysis[99] (see Fig. 13.4, but see Box 13.2 for an alternative explanation for this finding).

Behavioral findings have been complemented with several different neuroscientific techniques. For example, EEG[103,104] as well as PET[105] and fMRI[106–114] studies assessing language perception or production in schizophrenia patients and healthy controls all found reduced leftward language lateralization. Interestingly, aspects of language that are lateralized to the right in the majority of the population—for example, prosody, intonation, and rhythm—also show atypical lateralization in schizophrenia patients, for example, reduced rightward or inverted leftward asymmetries.[115–118] This indicates that atypical language lateralization in schizophrenia is not limited to left-hemispheric language aspects.

Aside from changes in functional language lateralization, schizophrenia has also been linked to changes in structural lateralization of speech-related areas. While the planum temporale shows strong leftward volume asymmetry in healthy individuals, this asymmetry is either absent or reversed in schizophrenia patients.[97,119] Moreover, the Sylvian fissure, which typically is longer in the left than in the right hemisphere, shows reduced asymmetry in schizophrenia patients.[97] Schizophrenia-related changes in structural asymmetries also extend to white matter structures, often showing a reduction of leftward structural asymmetries relative to healthy controls.[120] This has, for example, been reported for the arcuate fasciculus,[121,122] the uncinate fasciculus,[123] and the superior occipitofrontal fasciculus.[124] Moreover, the corpus callosum, the large white matter tract connecting the left and the right hemisphere, exhibits structural changes in schizophrenia. For example, patients with schizophrenia show significant decreases in white matter density in both the genu and the truncus of the corpus callosum.[125]

The largest study on alterations of structural asymmetries in schizophrenia has been conducted by the ENIGMA consortium.[126] In this study, macrostructural asymmetries in cortical thickness and surface area, as well as subcortical volume asymmetries were assessed in MRI data obtained from 46 different datasets. Overall, neuroimaging data from 5080 patients with schizophrenia and 6015 controls were included in the analysis. The results are shown in Fig. 13.6. On the level of individual brain areas, significant differences between patients with schizophrenia and controls were observed in the rostral

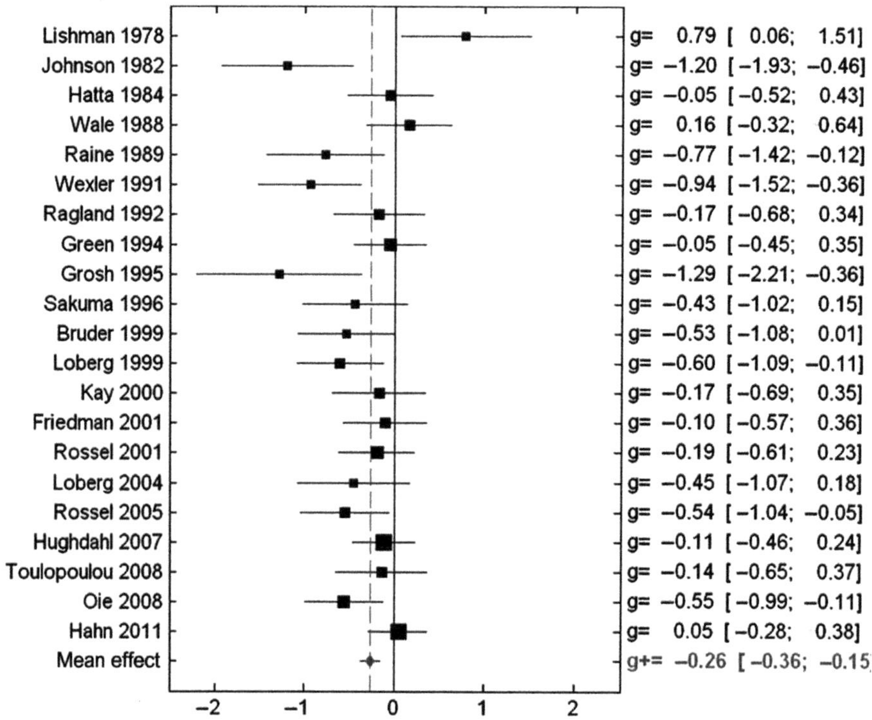

FIGURE 13.4 Results of a meta-analysis comparing dichotic listening performance in patients suffering from schizophrenia and healthy controls.[99] Squares indicate study effect sizes, and horizontal lines indicate 95% confidence intervals for the difference in language lateralization between schizophrenia patients and healthy controls. Negative values indicate that patients had a lower laterality index than controls, and positive values indicate that controls had a lower laterality index than patients. The mean effect over all studies is indicated by the gray diamond in the bottom row. The sizes of the squares representing the individual studies indicate the weight of the study in the calculation of the mean effect. It is based on the number of participants in each study. The numbers at the right margin show the individual study effect sizes with the numbers in parentheses indicating the lower and upper boundaries of the 95%-confidence interval. Overall, the plot shows that patients suffering from schizophrenia show reduced language lateralization as determined with the dichotic listening task. *Source: Reproduced with permission of the authors and the publisher from Ocklenburg S, Westerhausen R, Hirnstein M, Hugdahl K. Auditory hallucinations and reduced language lateralization in schizophrenia: a meta-analysis of dichotic listening studies. J Int Neuropsychol Soc.* 2013;*19(4):410–418. https://doi.org/10.1017/s1355617712001476.*

anterior cingulate cortex and the middle temporal gyrus. As the middle temporal gyrus is involved in the language network, this finding mirrors earlier studies that found changes in hemispheric asymmetries in language-relevant regions in schizophrenia. Even more interesting than this analysis on the regional level was a second multivariate analysis that was conducted in the study. In this analysis, 7% of the variance across all structural asymmetries was explained by whether an individual was a patient or a control. This suggests that changes of asymmetries in schizophrenia may not be limited to a few language-related regions but that there may be subtle changes in asymmetries across the whole brain.

BOX 13.2

Critical thinking: reduced right-ear advantage in schizophrenia—the role of nicotine

Confounding variables—that is, factors that can explain the relation between two variables such as diagnosis of schizophrenia and a reduced right-ear advantage in the dichotic listening task—are a big problem in neuropsychiatric research. Since mental illness is typically a very complex process that can affect virtually all levels of behavior in patients, it is easy to omit potentially important variables when planning an experimental design. Use or abuse of legal or illicit drugs can be such a variable. For example, about two to four times as many schizophrenia patients than individuals from the general population are smokers, presumably because smoking may be a form of self-medication.[100] Smoking has been shown to reduce the right-ear advantage in men but not in women,[101] presumably through sex-specific disruption of neuronal function. Constanze Hahn and coworkers from Ruhr University Bochum in Germany investigated the role of smoking for right-ear advantage reduction during dichotic listening in male and female schizophrenia patients and healthy controls.[102] They found that non-smoking men had a stronger right-ear advantage than smoking men, regardless of whether or not they suffered from schizophrenia (see Fig. 13.5). Nicotine consumption thus seems to be an important factor that should be taken into account when assessing the relationship between mental illness and atypical lateralization.

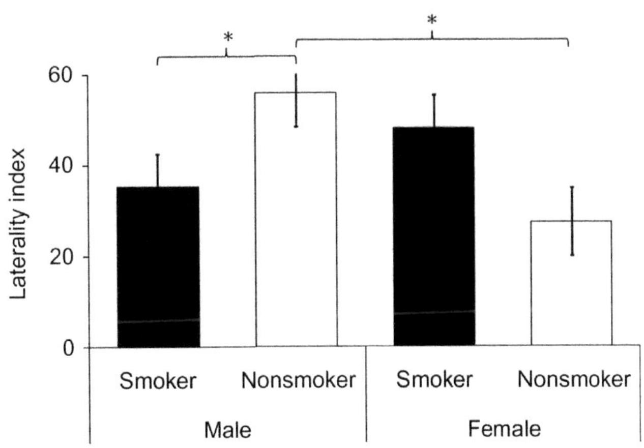

FIGURE 13.5 The effect of smoking on dichotic listening performance in the study by Hahn and coworkers.[102] Dichotic listening laterality index is shown for male (left) and female (right) participants who smoke (black) or do not smoke. Compared to nonsmoking men, smoking men showed a reduced right-ear advantage. There was no significant difference in women. Note that schizophrenia patients and healthy controls have been merged in this figure. *Source: Reproduced with permission of the authors and the publisher from Hahn C, Neuhaus AH, Pogun S, et al. Smoking reduces language lateralization: a dichotic listening study with control participants and schizophrenia patients.* Brain Cognition. *2011;76(2):300–309. https://doi.org/10.1016/j.bandc.2011.03.015.*

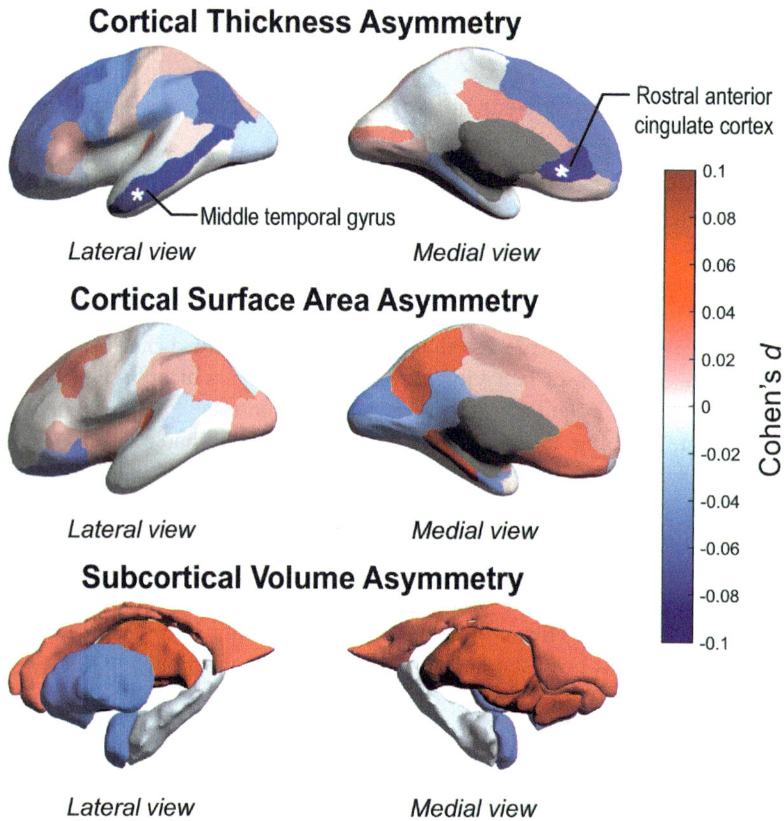

FIGURE 13.6 Cortical and subcortical asymmetry differences between patients with schizophrenia and controls from the study by the ENIGMA consortium.[126] Colors indicate Cohen's d effect size differences. Positive effects are shown in red (larger leftward or smaller rightward asymmetries in patients). Negative effects are shown in blue (smaller leftward or larger rightward asymmetries in patients). Asterisks indicate regions that are significant after correction for multiple comparisons. *Source: Licensed under a Creative Commons Attribution 4.0 International License and reproduced with permission from Schijven D, Postema MC, Fukunaga M, et al. Large-scale analysis of structural brain asymmetries in schizophrenia via the ENIGMA consortium.* PNAS. *2023;120(14):e2213880120. https://doi.org/10.1073/pnas.2213880120.*

A second analysis of structural hemispheric asymmetries in schizophrenia by the ENIGMA consortium focused on the shape of seven deep brain structures: thalamus, caudate, putamen, pallidum, accumbens, hippocampus, and amygdala.[127] In this study, patients with schizophrenia showed stronger asymmetries than controls in hippocampus, amygdala, and thalamus, but weaker asymmetries than controls in the ventral striatum and ventral and dorsal thalamus.

Based on the strong empirical evidence for the association of atypical asymmetries and schizophrenia, several authors have put forward models assuming some causal role of altered laterality for the emergence of schizophrenia. For instance, the hemispheric imbalance model by Gruzelier[128] assumes a three-syndrome model of symptom expression in schizophrenia, with these three syndromes being differentially related to so-called "hemispheric imbalances":

1. The "active syndrome": This syndrome comprises positive symptoms of schizophrenia such as affective delusions or maniac ideas and is thought to be caused by overly strong left-hemispheric activation.
2. The "withdrawn syndrome": This syndrome comprises negative symptoms such as catatonia and social withdrawal and is thought to be caused by overly strong right-hemispheric activation.
3. The "unreality syndrome": This syndrome includes unreality symptoms such as auditory hallucinations and is thought to be unrelated to hemispheric asymmetries. Gruzelier assumes that genetic and hormonal influences interact with early experiences—particularly regarding stress—in forming the so-called lateral imbalances in thalamocortical and callosal arousal systems that cause the aforementioned hemispheric imbalances.[129] However, the model does not elaborate on specific molecular processes.

More recently, Tim Crow proposed the so-called "Big Bang" theory of the origin of schizophrenia and its relation to altered language asymmetries.[130] This theory assumes that:

1. About 160,000 years ago, a genetic speciation event occurred that allowed for the development of language and thus gave rise to the modern human (the so-called "Big Bang").
2. This speciation event led to the development of hemispheric asymmetries which enabled the development of language. Thus, both lateralization and language are seen as uniquely human.
3. The gene pair Protocadherin11X and Y is thought to play an essential role in this process, as it was subject to a change within the hominid line at that time point.
4. Schizophrenia is thought to be "the price that Homo Sapiens paid for language,"[130] as it is supposed to reflect the disadvantageous component of the speciation event. If the lateralized organization of the brain breaks down, psychosis can develop.

While this model has been widely cited in the literature, recent studies are not in line with it. Specifically, there are two main issues with the model:

1. The idea that lateralization reflects a speciation event in humans is in conflict with a large number of recent studies showing that hemispheric asymmetries exist in all vertebrate classes[131,132] (see Chapter 2).
2. In a large-scale genome-wide association study on schizophrenia with more than 36,000 patients and more than 113,000 controls, 128 independent genetic associations were significant at the genome-wide significance threshold.[133] While three single nucleotide polymorphisms, that is, frequent genetic variants in the general population, located on the X-chromosome reached genome-wide significance, there was no association with the loci suggested by Crow, rendering a big role in the ontogenesis of schizophrenia unlikely.

Nevertheless, the basic ideas of the "Big Bang" theory that lateralization may be crucial for the development of speech and that atypical lateralization might increase the risk of developing mental illness still provide interesting food for thought for researchers in the field.

In addition to models suggesting a link between the diagnosis of schizophrenia and atypical lateralization, it has been proposed that a symptom-based approach might be a more meaningful way of understanding this relationship (see Fig. 13.7).[134]

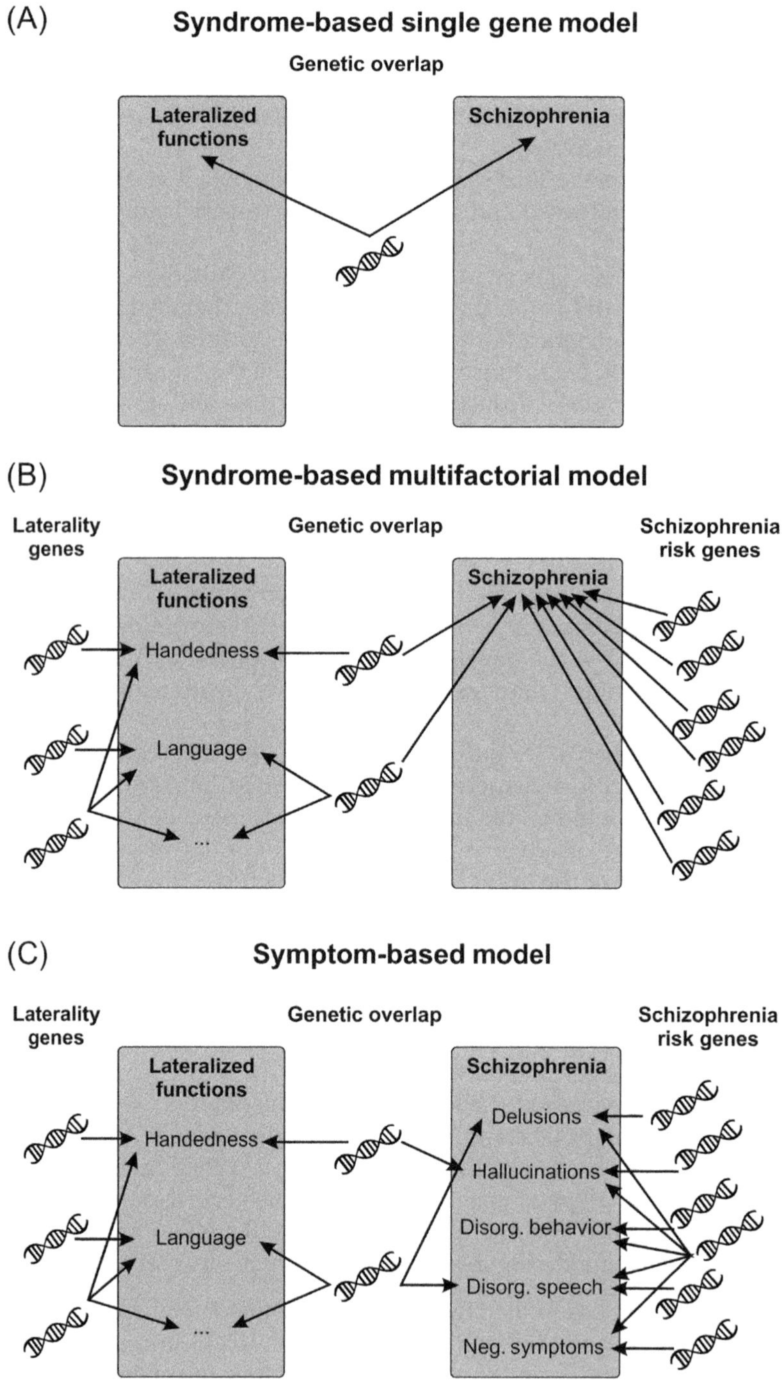

FIGURE 13.7 Different models for the association of schizophrenia with atypical lateralization.[134] (A) A syndrome-based single gene model assumes that one gene determines both laterality and schizophrenia. (B) A syndrome-based multifactorial model assumes that both traits are determined by a number of different genetic and nongenetic influences. (C) A symptom-based approach assumes that different schizophrenia symptoms are determined by distinct genetic networks which are differentially linked to lateralization. *Source: Reproduced with permission of the authors and the publisher from Ocklenburg S, Güntürkün O, Hugdahl K, Hirnstein M. Laterality and mental disorders in the postgenomic age—A closer look at schizophrenia and language lateralization.* Neurosci Biobehav Rev. *2015;59:100–110. https://doi.org/10.1016/j.neubiorev.2015.08.019.*

Within this approach, it is assumed that it is easier to understand the relationship of atypical laterality and schizophrenia on a symptom level—for example, the relationship between auditory hallucinations and language lateralization.

This idea is based on the following thoughts:

1. Schizophrenia is a very heterogeneous disorder and two patients with the same diagnosis can exhibit very different symptoms[135]—for example, one patient can have mostly negative symptoms such as catatonia, while another patient displays mostly positive symptoms such as auditory hallucinations.
2. Genome-wide association studies in schizophrenia indicate that different clusters of symptoms are related to distinct genetic networks.[136]
3. Both schizophrenia and lateralization are complex traits determined by a multitude of genetic and nongenetic factors.[133,137] While some genetic influence factors are shared between the two traits, their genetic determinants are largely independent.
4. Symptom severity is a critical factor for the extent of atypical lateralization in schizophrenia. For example, it has been shown that schizophrenia patients with low overall symptom severity and no auditory hallucinations do not show a reduced right-ear advantage in the dichotic listening task.[138] However, patients who show auditory hallucinations have a stronger reduction of the right-ear advantage than the general population of schizophrenia patients[99] (see Fig. 13.8).

Taken together, the relationship between atypical lateralization and schizophrenia has been investigated for more than 150 years and still is not well understood. Whether or not atypical lateralization is a cause or—as suggested by Dorothy Bishop for dyslexia[35]—a consequence of schizophrenia remains a matter of debate. A combination of modern neurogenetic research techniques and symptom-based theoretical approaches might help to resolve this issue.

Mood disorders

Hemispheric asymmetries have also been investigated in major depressive disorder and other forms of mood disorders. The link between handedness and depression is much weaker than for schizophrenia. Two large studies have used depression questionnaires in nonclinical samples and found that left-handedness was related to depressive symptoms.[139,140] One study found the effect for both men and women, the other only for men. Findings from clinical samples are also inconsistent, with some reporting a relation of non-right-handedness and major depression[141] and others not.[142] A recent meta-analysis of 87 studies did not report any statistically significant differences between patients with depression and controls regarding left-handedness, mixed-handedness, and non-right-handedness.[143] This finding strongly suggests that unlike what has been found for schizophrenia, there is no increase of mixed- or left-handedness in depression.

This null finding on the behavioral level was also mirrored in a large-scale study on macrostructural asymmetries in depression by the ENIGMA consortium.[144] In this study, the authors investigated asymmetries in cortical thickness and surface area in cortical 34 brain regions in data pooled from 31 datasets containing 2256 patients with major

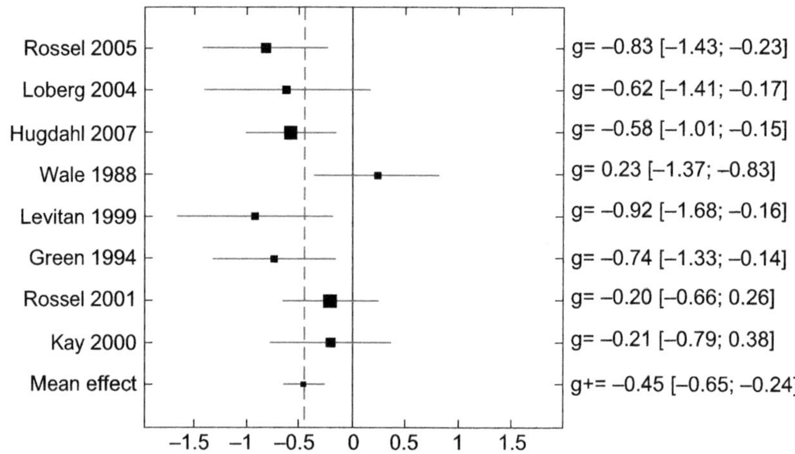

FIGURE 13.8 Results of a meta-analysis comparing dichotic listening performance between schizophrenia patients who experience auditory hallucinations and nonhallucinating controls.[99] Squares indicate study effect sizes, and horizontal lines indicate 95% confidence intervals for the difference in language lateralization between schizophrenia patients who experience auditory hallucinations and nonhallucinating controls. Negative values indicate that schizophrenia patients who experience auditory hallucinations had a lower laterality index compared to nonhallucinating controls, and positive values indicate that nonhallucinating controls had a lower laterality index than hallucinating patients. The mean effect over all studies is indicated by the black square in the bottom row. The sizes of the squares representing the individual studies indicate the weight of the study in the calculation of the mean effect. It is based on the number of participants in each study. The numbers at the right margin show the individual study effect sizes with the numbers in parentheses indicating the lower and upper boundaries of the 95% confidence interval. Overall, the plot shows that schizophrenia patients who experience auditory hallucinations show reduced language lateralization compared to nonhallucinating controls as determined with the dichotic listening task. *Source: Reproduced with permission of the authors and the publisher from Ocklenburg S, Westerhausen R, Hirnstein M, Hugdahl K. Auditory hallucinations and reduced language lateralization in schizophrenia: a meta-analysis of dichotic listening studies.* J Int Neuropsychol Soc. *2013;19(4):410–418. https://doi.org/10.1017/s1355617712001476.*

depression and 3504 controls. No statistically significant differences between patients and controls were observed, suggesting that macrostructural brain asymmetries are not altered in major depression.

One of the most widely investigated forms of hemispheric asymmetries in patients with depression are asymmetries in frontal resting state EEG alpha, an EEG parameter that is thought to be an inverse measure of cognitive activity.[145–149] Typically, healthy subjects show rightward asymmetry in alpha activity in frontal regions, indicating greater activation of the left hemisphere. In depressed patients, it has been suggested that this asymmetry is reduced or even reversed, indicating stronger activation of right frontal areas during resting state. This is in line with the idea that the right hemisphere is dominant for the processing of negative emotions. However, two recent meta-analyses have shown that there is no statistically significant difference between patients with depression and controls in frontal alpha asymmetry.[150,151]

A hyperactivation of the right and hypoactivation of the left frontal lobe in depression has been investigated in studies in which depressive patients and controls had to perform cognitive tasks. In these studies, evidence supporting the idea was found. For example, in one EEG study, depressed adolescents and healthy controls performed two complementing emotional Go-NoGo tasks, one in which they had to respond to negative faces and to withhold responses to positive faces and one in which this pattern was reversed.[152] The N2, a frontocentral event-related potential component linked to response inhibition, was left-lateralized in healthy controls but showed atypical right lateralization in depressed adolescents, indicating hypoactivation of the left frontal lobe. Similar findings have also been reported in fMRI studies. For example, in one study, depressed and nondepressed individuals had to identify the color in which emotional words were printed.[153] Depressed subjects showed more right-hemispheric activation in the dorsolateral prefrontal cortex, which further confirmed a more right-hemispheric processing style in depressed patients.

David Hecht from University College London, UK, suggested that functional hemispheric asymmetries for different cognitive processes relevant for depression could explain this pattern.[154] For example, the right hemisphere has been suggested to be dominant for the processing of negative emotions and pessimistic thoughts,[155] two activities depressive patients typically overengage in. In contrast, the left hemisphere has been suggested to be dominant for the processing of positive emotions and pleasurable experiences, two activities depressed patients typically engage in less than healthy controls. According to Hecht, this differential involvement of the hemispheres in positively and negatively valenced cognition could explain the more right-hemispheric activation pattern in depressive patients.

In general, several meta-analyses in the past 5 years indicated that the link between depression and hemispheric asymmetries is weaker than previously thought. One key issue in this context seems to be that fact that depression has a high prevalence and patients that are included in different studies may differ significantly in terms of symptom severity and symptom profiles. Thus, it has been suggested that future laterality studies in patients with depression should focus much more on a detailed assessment of the severity of individual symptoms and symptom profiles.[156] Comparable to what has been found in dyslexia and schizophrenia, only patients with severe depression symptoms may show pronounced alterations of hemispheric asymmetries.

Addiction and substance abuse

Addiction and abuse of legal and illicit drugs have also been investigated in relation to atypical lateralization. In principle, all of the studies followed the basic premise that atypical lateralization increases the risk for psychopathology. The first researcher to report an association between alcoholism and handedness was Paul Bakan, who reported that in a small sample of 47 people with alcohol addiction, 12 were non-right-handed (left-handed or ambidextrous). Thus, about 25.5% of these patients showed atypical handedness, a ratio that is substantially higher than that in the general population. While this finding was rather anecdotal since no systematic comparison with healthy controls was performed, subsequent studies have confirmed the assumption that left-handers have a higher risk of developing an alcohol addiction.[157–159] The largest study on handedness and drinking behavior has been conducted by Kevin Denny from University College Dublin, Ireland.[160] The author used a dataset with more than 27,000 participants over the age of 50 from 12 different European countries. Self-reports of alcohol consumption habits were linked to handedness in showing that left-handers drank more frequently than right-handers. This effect was mainly driven by the fact that left-handers were less likely to drink rarely—or not at all—than right-handers. In addition to drinking behavior, left-handedness has also been linked to abuse of illicit drugs. One Italian study with over 1000 participants reported that after controlling for several confounding variables such as sex and age, left-handers were significantly more likely to have had experiences with the use of heroin, amphetamines, and hallucinogens.[161] Moreover, left-handers were at a higher risk to develop social network disorders—for example, pathological use of social network websites and apps—resembling an addiction.[162]

Alcoholism has also been related to changes in other forms of hemispheric asymmetries than handedness—for example, language lateralization,[163] lateralization of face perception,[164] or visuospatial lateralization.[165] Atypical hemispheric asymmetries in people with alcohol addiction have also been reported on the structural level, for example, reflected in reduced left-right asymmetry of the insula[166] or increased rightward asymmetry in the hippocampus.[167] The largest study on cortical and subcortical asymmetries in individuals with substance dependences was conducted by the ENIGMA consortium.[168] In this study, neuroimaging data from 22 independent cohorts were analyzed. The dataset included data from 1,796 individuals who had a dependence on either alcohol, nicotine, cocaine, methamphetamine, or cannabis, and 996 controls. The results indicated that individuals with a dependence showed a significantly less rightward asymmetry of the nucleus accumbens compared to controls. The nucleus accumbens is a key structure for goal-based and reward-based behavior and has been shown to be involved in addiction and substance misuse.[169] This suggests that altered structural asymmetries in substance dependence are focused on a brain structure that is highly relevant for the symptoms experienced in substance dependence such as altered processing in the reward system.

Neurological disorders

Parkinson's disease

Research on the relationship between hemispheric asymmetries and neurological disorders has focused less on whether hemispheric asymmetries are reduced in patients suffering from the disorders. Instead, the core question in this field is to what extent left- or right-hemispheric lesions affect patient outcome. A prime example is, of course, visuospatial neglect, in which the vast majority of patients present with right-hemispheric lesions (see Chapter 7). Another well-known example is Parkinson's disease. Parkinson's disease is a neurodegenerative disorder that leads to cell death in the substantia nigra of the midbrain, resulting in a breakdown of dopaminergic transmission in the motor system. This leads to severe behavioral motor symptoms, including tremor (shaking), bradykinesia (slowness of movements), and rigidity. Interestingly, the motor symptoms in Parkinson's disease typically show a unilateral onset, and while both sides of the body are eventually affected when the disorder progresses, the side initially afflicted shows stronger symptoms.[170] This asymmetry in motor symptoms is caused by an underlying asymmetry in dopamine depletion.[171] Importantly, the side of onset influences the severity of both motor and nonmotor symptoms related to Parkinson's disease. Therefore, a distinction between RPD—that is, patients with right body onset of motor symptoms—and LPD—that is, patients with left body onset of motor symptoms—has been proposed for the diagnosis of Parkinson's disease.[170] With regard to motor symptoms, patients with right-sided onset and tremor as the main symptom have a better prognosis than patients with left-sided symptom onset with bradykinesia or rigidity as the main symptom. Moreover, left-sided onset of tremor has been associated with neuropsychological deficits, while patients with right-sided onset of tremor typically show less cognitive decline.[172] RPD and LPD Parkinson patients have been shown to differ in cognitive performance in a number of tasks.[173] RPD patients often have problems with verbal memory and language, while LPD patients have problems with visuospatial attention and mental imagery.[173] One of the most striking findings regarding asymmetries in Parkinson's disease is a relationship between the side of motor symptom onset and handedness. In a meta-analysis of 10 Parkinson's disease studies assessing handedness, Anouk van der Hoorn from the University of Groningen in the Netherlands found a substantial association between handedness and the side of motor symptom onset.[174] For left-handed patients, 40.8% showed right-dominant and 59.2% showed left-dominant Parkinson symptoms. This relation was reversed in right-handed patients, with 59.5% showing right-dominant and only 40.5% showing left-dominant Parkinson's symptoms. Thus, Parkinson motor symptoms emerge significantly more often on the side of the dominant hand. Van der Hoorn suggested that handedness leads to increased metabolic demand in the contralateral hemisphere that causes oxidative stress which could result in stronger dopaminergic defects in this hemisphere. These defects, in turn, could affect the side of motor symptom onset.

Beyond the diagnosis: transdiagnostic approaches in clinical laterality research

Besides clinical studies that focus on hemispheric asymmetries in patients with a specific diagnosis compared to controls, there also have been studies that investigated hemispheric asymmetries across different diagnoses. For example, one article suggested a perspective motivated by the NIMH Research Domain Criteria (RDoC) to investigate frontal EEG asymmetries across traditional disorder diagnoses.[175] RDoC is a research framework for investigating mental disorders that focuses on six functional symptom domains (social processes, negative valence, positive valence, arousal, sensorimotor, and cognitive) rather than on diagnoses.[176] In the RDoC asymmetry article, the authors suggested that it would be beneficial for clinical laterality research to move beyond the classifications of mood and anxiety disorders and instead focus on the symptom clusters. While not many clinical laterality studies have used an RDoC approach, it certainly provides an interesting alternative.

Other studies have investigated hemispheric asymmetries in different disorder groups to identify overlaps and discrepancies between patient groups. In one recent ALE meta-analysis, data from 169 neuroimaging studies including 3517 patients with schizophrenia, 1575 patients with bipolar disorder, 3280 patients with major depressive disorder, and 9733 controls were analyzed.[177] The results indicated a leftward shift of asymmetries in schizophrenia and bipolar disorders, but at different locations in the brain. For major depressive disorder, a rightward shift of asymmetries was observed instead. These data clearly suggest that the disorders show distinct patterns of altered asymmetries, as opposed to a common pattern across disorder, which clearly argues against a nonspecific association between laterality and disorders.

The role of stress

The largest GWAS on handedness identified some genetic overlaps between handedness and schizophrenia and bipolar disorder.[178] Nevertheless, it is unlikely that the association between hemispheric asymmetries and mental and neurodevelopmental disorders is largely driven by genetic factors, since most genes identified by whole-genome association studies do not overlap much between different disorders nor with the genes associated with handedness and language lateralization. Therefore, it is likely that nongenetic factors are relevant for altered asymmetries in mental and neurodevelopmental disorders. Stress is a factor that has been shown to affect molecular epigenetic processes during brain development, and this is also related to the ontogenesis of both lateralization and several of the disorders described in this chapter.[179] A recent meta-analysis of handedness in PTSD patients revealed a higher prevalence of non-right-handedness and mixed-handedness in PTSD, suggesting altered asymmetries in this stress-related disorder.[180] Interestingly, a comparative study on turning behavior in rats showed strong effects of severe early life stress on motor asymmetries.[181] In this study, control animals and animals that were stressed for shorter periods of time during development showed no significant turning asymmetries. In contrast, an experimental group that was stressed for a prolonged period during development showed a significant leftward turning bias, indicating greater right-hemispheric involvement. Thus, chronic stress exposure during early life alters motor asymmetries in later life in rats. In humans, a large-scale study investigating early life factors affecting handedness also found significant effects of two potentially stress-related factors, birthweight and being part of a multiple birth.[182] It is possible that lower birthweight and being part of a birth with one or more other children may reflect increased physiological stress. In contrast to these findings about early life stress and motor asymmetries, evidence for an effect of acute psychosocial stress on hemispheric asymmetries is less clear-cut.[183–185] This suggests that particularly chronic early life stress is a relevant factor for the association between hemispheric asymmetries and mental and neurodevelopmental disorders, but more research is needed.

Conclusion

One of the biggest unsolved scientific questions in laterality research is why so many seemingly divergent disorders are associated with altered hemispheric asymmetries. In the last decade, tremendous advances were made to answer this question, but clearly, we are still far away from having any form of mechanistic understanding. Identifying how genetic and epigenetic mechanisms interact during the ontogenesis of both lateralization and the various disorders mentioned in this chapter probably represents one of the biggest challenges in laterality research that will need to be tackled over the next decades. Animal models of psychiatric disorder may play a key role in this undertaking,[186] as they are crucial to study the effects of early life stress on hemispheric asymmetries in an experimentally controlled way. Moreover, several studies reviewed in this chapter have indicated that symptom severity may be a highly relevant factor in clinical laterality research, with higher symptom burden linked to stronger alterations of asymmetries. Thus, moving beyond comparing patients to controls and including multidimensional symptom data will be crucial for the advancement of clinical laterality research in the next decade.

Sample questions

- Please describe the GBG model for the relation of atypical hemispheric asymmetries and dyslexia.
- Is there evidence for a genetic link between dyslexia and handedness?
- Please describe the Lindell and Hudry model for the relationship between reduced structural lateralization, reduced functional lateralization, and language impairment in ASD.
- What models have been suggested for the association of atypical lateralization and schizophrenia? Name and describe at least three.
- What is currently known about the association of atypical lateralization and mood disorders?
- What is currently known about the association of atypical lateralization and addiction?
- How does handedness influence the symptoms of Parkinson's disease?
- What have studies by the ENIGMA consortium contributed to our understanding of the association between mental and neurodevelopmental disorders and hemispheric asymmetries?
- What is known about the role of early life stress for handedness?

References

1. Corballis MC. Left brain, right brain: facts and fantasies. *PLOS Biol.* 2014;12(1):e1001767. Available from: https://doi.org/10.1371/journal.pbio.1001767.
2. Willows DM, Kruk R, Corcos E. *Visual Processes in Reading and Reading Disabilities.* Taylor and Francis; 2012.
3. Reading Orton ST. *Writing and Speech Problems in Children. & Co.* W. W. Norton; 1937.
4. Vallortigara G, Rogers LJ. A function for the bicameral mind. *Cortex.* 2020;124:274–285. Available from: https://doi.org/10.1016/j.cortex.2019.11.018.
5. Mundorf A, Ocklenburg S. *The Clinical Neuroscience of Lateralization.* Routledge; 2021.
6. Mundorf A, Peterburs J, Ocklenburg S. Asymmetry in the central nervous system: a clinical neuroscience perspective. *Front Syst Neurosci.* 2021;15:733898. Available from: https://doi.org/10.3389/fnsys.2021.733898.
7. Geschwind N, Galaburda AM. Cerebral lateralization. Biological mechanisms, associations, and pathology: I. A hypothesis and a program for research. *Arch Neurol.* 1985;42(5):428–459. Available from: https://doi.org/10.1001/archneur.1985.04060050026008.
8. Geschwind N, Galaburda AM. Cerebral lateralization. Biological mechanisms, associations, and pathology: II. A hypothesis and a program for research. *Arch Neurol.* 1985;42(6):521–552. Available from: https://doi.org/10.1001/archneur.1985.04060060019009.

9. Geschwind N, Galaburda AM. Cerebral lateralization. Biological mechanisms, associations, and pathology: III. A hypothesis and a program for research. *Arch Neurol.* 1985;42(7):634–654. Available from: https://doi.org/10.1001/archneur.1985.04060070024012.
10. Geschwind N, Behan P. Left-handedness: association with immune disease, migraine, and developmental learning disorder. *Proc Natl Acad Sci U S A.* 1982;79(16):5097–5100. Available from: https://doi.org/10.1073/pnas.79.16.5097.
11. Bryden MP, McManus IC, Bulman-Fleming MB. Evaluating the empirical support for the Geschwind-Behan-Galaburda model of cerebral lateralization. *Brain Cognition.* 1994;26(2):103–167. Available from: https://doi.org/10.1006/brcg.1994.1045.
12. Berenbaum SA, Denburg SD. Evaluating the empirical support for the role of testosterone in the Geschwind-Behan-Galaburda model of cerebral lateralization: commentary on Bryden, McManus, and Bulman-Fleming. *Brain Cognition.* 1995;27(1):79–83, discussion 94–7. Available from: https://doi.org/10.1006/brcg.1995.1005.
13. Previc FH. Assessing the legacy of the GBG model. *Brain Cognition.* 1994;26(2):174–180. Available from: https://doi.org/10.1006/brcg.1994.1047.
14. Hellige JB. Babies, bath water, and the chicken's way out. *Brain Cognition.* 1994;26(2):228–235. Available from: https://doi.org/10.1006/brcg.1994.1056.
15. Handedness Annett M. *and Brain Asymmetry: The Right Shift Theory*. Psychology Press; 2002.
16. Armour JAL, Davison A, McManus IC. Genome-wide association study of handedness excludes simple genetic models. *Heredity.* 2014;112(3):221–225. Available from: https://doi.org/10.1038/hdy.2013.93.
17. Paracchini S, Scerri T, Monaco AP. The genetic lexicon of dyslexia. *Annu Rev Genomics Hum Genet.* 2007;8:57–79. Available from: https://doi.org/10.1146/annurev.genom.8.080706.092312.
18. Pinel P, Fauchereau F, Moreno A, et al. Genetic variants of FOXP2 and KIAA0319/TTRAP/THEM2 locus are associated with altered brain activation in distinct language-related regions. *J Neurosci.* 2012;32(3):817–825. Available from: https://doi.org/10.1523/JNEUROSCI.5996-10.2012.
19. Brandler WM, Paracchini S. The genetic relationship between handedness and neurodevelopmental disorders. *Trends Mol Med.* 2014;20(2):83–90. Available from: https://doi.org/10.1016/j.molmed.2013.10.008.
20. Paracchini S, Diaz R, Stein J. Advances in dyslexia genetics—new insights into the role of brain asymmetries. In: Friedmann T, Dunlap JC, Goodwin SF, eds. *Advances in Genetics.* Elsevier Science; 2016:53–97.
21. Blum M, Feistel K, Thumberger T, Schweickert A. The evolution and conservation of left-right patterning mechanisms. *Development.* 2014;141(8):1603–1613. Available from: https://doi.org/10.1242/dev.100560.
22. Massinen S, Hokkanen M-E, Matsson H, et al. Increased expression of the dyslexia candidate gene DCDC2 affects length and signaling of primary cilia in neurons. *PLOS ONE.* 2011;6(6):e20580. Available from: https://doi.org/10.1371/journal.pone.0020580.
23. Chandrasekar G, Vesterlund L, Hultenby K, Tapia-Páez I, Kere J. The zebrafish orthologue of the dyslexia candidate gene DYX1C1 is essential for cilia growth and function. *PLOS ONE.* 2013;8(5):e63123. Available from: https://doi.org/10.1371/journal.pone.0063123.
24. Scerri TS, Brandler WM, Paracchini S, et al. PCSK6 is associated with handedness in individuals with dyslexia. *Hum Mol Genet.* 2011;20(3):608–614. Available from: https://doi.org/10.1093/hmg/ddq475.
25. Bishop DVM. *Handedness and Developmental Disorder.* Mac Keith Press; 1990.
26. Bishop DVM. Using non-preferred hand skill to investigate pathological left-handedness in an unselected population. *Dev Med Child Neurol.* 1984;26(2):214–226. Available from: https://doi.org/10.1111/j.1469-8749.1984.tb04434.x.
27. Eglinton E, Annett M. Handedness and dyslexia: a meta-analysis. *Percept Mot Skills.* 1994;79(3 Pt 2):1611–1616. Available from: https://doi.org/10.2466/pms.1994.79.3f.1611.
28. Abbondanza F, Dale PS, Wang CA, et al. Language and reading impairments are associated with increased prevalence of non-right-handedness. *Child Dev.* 2023. Available from: https://doi.org/10.1111/cdev.13914.
29. Hakvoort B, van der Leij A, van Setten E, Maurits N, Maassen B, van Zuijen T. Dichotic listening as an index of lateralization of speech perception in familial risk children with and without dyslexia. *Brain Cognition.* 2016;109:75–83. Available from: https://doi.org/10.1016/j.bandc.2016.09.004.
30. Obrzut JE, Boliek CA. Lateralization characteristics in learning disabled children. *J Learn Disabilities.* 1986;19(5):308–314. Available from: https://doi.org/10.1177/002221948601900511.
31. Helland T, Asbjørnsen AE, Hushovd AE, Hugdahl K. Dichotic listening and school performance in dyslexia. *Dyslexia.* 2008;14(1):42–53. Available from: https://doi.org/10.1002/dys.343.

32. de Guibert C, Maumet C, Jannin P, et al. Abnormal functional lateralization and activity of language brain areas in typical specific language impairment (developmental dysphasia. *Brain*. 2011;134(Pt 10):3044–3058. Available from: https://doi.org/10.1093/brain/awr141.
33. Sun Y-F, Lee J-S, Kirby R. Brain imaging findings in dyslexia. *Pediatrics & Neonatology*. 2010;51(2):89–96. Available from: https://doi.org/10.1016/s1875-9572(10)60017-4.
34. Beaton AA. The relation of planum temporale asymmetry and morphology of the corpus callosum to handedness, gender, and dyslexia: a review of the evidence. *Brain Lang*. 1997;60(2):255–322. Available from: https://doi.org/10.1006/brln.1997.1825.
35. Bishop DVM. Cerebral asymmetry and language development: cause, correlate, or consequence. *Science*. 2013;340(6138):1230531. Available from: https://doi.org/10.1126/science.1230531.
36. Bradshaw AR, Woodhead ZVJ, Thompson PA, Bishop DVM. Investigation into inconsistent lateralisation of language functions as a potential risk factor for language impairment. *Eur J Neurosci*. 2020;51(4):1106–1121. Available from: https://doi.org/10.1111/ejn.14623.
37. Rysstad AL, Pedersen AV. Brief report: non-right-handedness within the autism spectrum disorder. *J Autism Dev Disord*. 2016;46(3):1110–1117. Available from: https://doi.org/10.1007/s10803-015-2631-2.
38. Lindell AK, Hudry K. Atypicalities in cortical structure, handedness, and functional lateralization for language in autism spectrum disorders. *Neuropsychol Rev*. 2013;23(3):257–270. Available from: https://doi.org/10.1007/s11065-013-9234-5.
39. Stefanatos GA, Baron IS. The ontogenesis of language impairment in autism: a neuropsychological perspective. *Neuropsychol Rev*. 2011;21(3):252–270. Available from: https://doi.org/10.1007/s11065-011-9178-6.
40. Corballis MC. Of mice and men - and lopsided birds. *Cortex*. 2008;44(1):3–7. Available from: https://doi.org/10.1016/j.cortex.2007.10.001.
41. Floris DL, Lai M-C, Auer T, et al. Atypically rightward cerebral asymmetry in male adults with autism stratifies individuals with and without language delay. *Hum Brain Mapp*. 2016;37(1):230–253. Available from: https://doi.org/10.1002/hbm.23023.
42. Preslar J, Kushner HI, Marino L, Pearce B. Autism, lateralisation, and handedness: a review of the literature and meta-analysis. *Laterality*. 2014;19(1):64–95. Available from: https://doi.org/10.1080/1357650X.2013.772621.
43. Cauda F, Costa T, Palermo S, et al. Concordance of white matter and gray matter abnormalities in autism spectrum disorders: a voxel-based meta-analysis study. *Hum Brain Mapp*. 2014;35(5):2073–2098. Available from: https://doi.org/10.1002/hbm.22313.
44. Joseph RM, Fricker Z, Fenoglio A, Lindgren KA, Knaus TA, Tager-Flusberg H. Structural asymmetries of language-related gray and white matter and their relationship to language function in young children with ASD. *Brain Imaging Behav*. 2014;8(1):60–72. Available from: https://doi.org/10.1007/s11682-013-9245-0.
45. Hayashi M, Takamura I, Kohara H, Yamazaki K. A neurolinguistic study of autistic children employing dichotic listening. *Tokai J Exp Clin Med*. 1989;14(4):339–345.
46. Kleinhans NM, Müller R-A, Cohen DN, Courchesne E. Atypical functional lateralization of language in autism spectrum disorders. *Brain Res*. 2008;1221:115–125. Available from: https://doi.org/10.1016/j.brainres.2008.04.080.
47. Minagawa-Kawai Y, Naoi N, Kikuchi N, Yamamoto J, Nakamura K, Kojima S. Cerebral laterality for phonemic and prosodic cue decoding in children with autism. *NeuroReport*. 2009;20(13):1219–1224. Available from: https://doi.org/10.1097/WNR.0b013e32832fa65f.
48. de Fossé L, Hodge SM, Makris N, et al. Language-association cortex asymmetry in autism and specific language impairment. *Ann Neurol*. 2004;56(6):757–766. Available from: https://doi.org/10.1002/ana.20275.
49. Rojas DC, Bawn SD, Benkers TL, Reite ML, Rogers SJ. Smaller left hemisphere planum temporale in adults with autistic disorder. *Neurosci Lett*. 2002;328(3):237–240. Available from: https://doi.org/10.1016/s0304-3940(02)00521-9.
50. Rojas DC, Camou SL, Reite ML, Rogers SJ. Planum temporale volume in children and adolescents with autism. *J Autism Dev Disord*. 2005;35(4):479–486. Available from: https://doi.org/10.1007/s10803-005-5038-7.
51. Postema MC, van Rooij D, Anagnostou E, et al. Altered structural brain asymmetry in autism spectrum disorder in a study of 54 datasets. *Nat Commun*. 2019;10(1):4958. Available from: https://doi.org/10.1038/s41467-019-13005-8.
52. Sha Z, van Rooij D, Anagnostou E, et al. Subtly altered topological asymmetry of brain structural covariance networks in autism spectrum disorder across 43 datasets from the ENIGMA consortium. *Mol Psychiatry*. 2022;27(4):2114–2125. Available from: https://doi.org/10.1038/s41380-022-01452-7.

53. Say GN, Sahin B, Aslan K, Akbaş S, Lncesu L, Ceyhan M. increased laterality of the thalamus in children and adolescents with Asperger's disorder: an MRI and proton spectroscopy study. *Psychiatry Investig*. 2014;11 (3):237–242. Available from: https://doi.org/10.4306/pi.2014.11.3.237.
54. Wan CY, Marchina S, Norton A, Schlaug G. Atypical hemispheric asymmetry in the arcuate fasciculus of completely nonverbal children with autism. *Ann N Y Acad Sci*. 2012;1252:332–337. Available from: https://doi.org/10.1111/j.1749-6632.2012.06446.x.
55. Aoki Y, Abe O, Nippashi Y, Yamasue H. Comparison of white matter integrity between autism spectrum disorder subjects and typically developing individuals: a meta-analysis of diffusion tensor imaging tractography studies. *Mol Autism*. 2013;4(1):25. Available from: https://doi.org/10.1186/2040-2392-4-25.
56. Cardinale RC, Shih P, Fishman I, Ford LM, Müller R-A. Pervasive rightward asymmetry shifts of functional networks in autism spectrum disorder. *JAMA Psychiatry*. 2013;70(9):975–982. Available from: https://doi.org/10.1001/jamapsychiatry.2013.382.
57. Conti E, Calderoni S, Gaglianese A, et al. Lateralization of brain networks and clinical severity in toddlers with autism spectrum disorder: a HARDI Diffusion MRI Study. *Autism Res*. 2016;9(3):382–392. Available from: https://doi.org/10.1002/aur.1533.
58. Chrysanthis K. Stammering and handedness. *Lancet*. 1947;249(6442):270–271. Available from: https://doi.org/10.1016/s0140-6736(47)92286-1.
59. Webster WG, Poulos M. Handedness distributions among adults who stutter. *Cortex*. 1987;23(4):705–708. Available from: https://doi.org/10.1016/s0010-9452(87)80062-x.
60. Salihović N, Sinanović O. Mucanje i ljevorukost. *Med Arh*. 2000;54(3):173–175.
61. Kushner HI. Retraining left-handers and the aetiology of stuttering: the rise and fall of an intriguing theory. *Laterality*. 2012;17(6):673–693. Available from: https://doi.org/10.1080/1357650X.2011.615127.
62. Moore WH, Haynes WO. Alpha hemispheric asymmetry and stuttering: some support for a segmentation dysfunction hypothesis. *J Speech Hear Res*. 1980;23(2):229–247. Available from: https://doi.org/10.1044/jshr.2302.229.
63. Büchel C, Sommer M. What causes stuttering. *PLOS Biol*. 2004;2(2):E46. Available from: https://doi.org/10.1371/journal.pbio.0020046.
64. Fox PT, Ingham RJ, Ingham JC, et al. A PET study of the neural systems of stuttering. *Nature*. 1996;382 (6587):158–161. Available from: https://doi.org/10.1038/382158a0.
65. Braun AR, Varga M, Stager S, et al. Altered patterns of cerebral activity during speech and language production in developmental stuttering. An H2(15)O positron emission tomography study. *Brain*. 1997;120(Pt 5):761–784. Available from: https://doi.org/10.1093/brain/120.5.761.
66. Brown S, Ingham RJ, Ingham JC, Laird AR, Fox PT. Stuttered and fluent speech production: an ALE meta-analysis of functional neuroimaging studies. *Hum Brain Mapp*. 2005;25(1):105–117. Available from: https://doi.org/10.1002/hbm.20140.
67. Turkeltaub PE, Eden GF, Jones KM, Zeffiro TA. Meta-analysis of the functional neuroanatomy of single-word reading: method and validation. *NeuroImage*. 2002;16(3 Pt 1):765–780. Available from: https://doi.org/10.1006/nimg.2002.1131.
68. Eickhoff SB, Bzdok D, Laird AR, Kurth F, Fox PT. Activation likelihood estimation meta-analysis revisited. *NeuroImage*. 2012;59(3):2349–2361. Available from: https://doi.org/10.1016/j.neuroimage.2011.09.017.
69. Foundas AL, Bollich AM, Corey DM, Hurley M, Heilman KM. Anomalous anatomy of speech-language areas in adults with persistent developmental stuttering. *Neurology*. 2001;57(2):207–215. Available from: https://doi.org/10.1212/WNL.57.2.207.
70. Foundas AL, Bollich AM, Feldman J, et al. Aberrant auditory processing and atypical planum temporale in developmental stuttering. *Neurology*. 2004;63(9):1640–1646. Available from: https://doi.org/10.1212/01.WNL.0000142993.33158.2A.
71. Foundas AL, Mock JR, Cindass R, Corey DM. Atypical caudate anatomy in children who stutter. *Percept Mot Skills*. 2013;116(2):528–543. Available from: https://doi.org/10.2466/15.10.PMS.116.2.528-543.
72. Jäncke L, Hänggi J, Steinmetz H. Morphological brain differences between adult stutterers and non-stutterers. *BMC Neurol*. 2004;4(1):23. Available from: https://doi.org/10.1186/1471-2377-4-23.
73. Foundas AL, Corey DM, Angeles V, Bollich AM, Crabtree-Hartman E, Heilman KM. Atypical cerebral laterality in adults with persistent developmental stuttering. *Neurology*. 2003;61(10):1378–1385. Available from: https://doi.org/10.1212/01.WNL.0000094320.44334.86.
74. Willcutt EG. The prevalence of DSM-IV attention-deficit/hyperactivity disorder: a meta-analytic review. *Neurotherapeutics*. 2012;9(3):490–499. Available from: https://doi.org/10.1007/s13311-012-0135-8.

75. Niederhofer H. Hand preference in attention deficit hyperactivity disorder. *Percept Mot Skills*. 2005;101 (3):808–810. Available from: https://doi.org/10.2466/pms.101.3.808-810.
76. Biederman J, Milberger S, Faraone SV, Lapey KA, Reed ED, Seidman LJ. No confirmation of Geschwind's hypothesis of associations between reading disability, immune disorders, and motor preference in ADHD. *J Abnorm Child Psychol*. 1995;23(5):545–552. Available from: https://doi.org/10.1007/BF01447660.
77. Nastou E, Ocklenburg S, Hoogman M, Papadatou-Pastou M. Handedness in ADHD: Meta-Analyses. *Neuropsychol Rev*. 2022;32(4):877–892. Available from: https://doi.org/10.1007/s11065-021-09530-3.
78. Ghanizadeh A. Lack of association of handedness with inattention and hyperactivity symptoms in ADHD. *J Atten Disord*. 2013;17(4):302–307. Available from: https://doi.org/10.1177/1087054711429789.
79. Heilman KM, van den Abell T. Right hemisphere dominance for attention: the mechanism underlying hemispheric asymmetries of inattention (neglect. *Neurology*. 1980;30(3):327–330. Available from: https://doi.org/10.1212/WNL.30.3.327.
80. Stefanatos GA, Wasserstein J. Attention deficit/hyperactivity disorder as a right hemisphere syndrome. Selective literature review and detailed neuropsychological case studies. *Ann N Y Acad Sci*. 2001;931: 172–195.
81. Rentería ME. Cerebral asymmetry: a quantitative, multifactorial, and plastic brain phenotype. *Twin Res Hum Genet*. 2012;15(3):401–413. Available from: https://doi.org/10.1017/thg.2012.13.
82. Klimkeit EI, Bradshaw JL. Anomalous lateralisation in neurodevelopmental disorders. *Cortex*. 2006;42 (1):113–116. Available from: https://doi.org/10.1016/s0010-9452(08)70334-4.
83. Heilman KM, Bowers D, Valenstein E, Watson RT. The right hemisphere: neuropsychological functions. *J Neurosurg*. 1986;64(5):693–704. Available from: https://doi.org/10.3171/jns.1986.64.5.0693.
84. Heilman KM, Voeller KK, Nadeau SE. A possible pathophysiologic substrate of attention deficit hyperactivity disorder. *J Child Neurol*. 1991;6(Suppl:):S76–S81. Available from: https://doi.org/10.1177/0883073891006001S09.
85. Chan E, Mattingley JB, Huang-Pollock C, et al. Abnormal spatial asymmetry of selective attention in ADHD. *J Child Psychol Psychiatry*. 2009;50(9):1064–1072. Available from: https://doi.org/10.1111/j.1469-7610.2009.02096.x.
86. Casey BJ, Castellanos FX, Giedd JN, et al. Implication of right frontostriatal circuitry in response inhibition and attention-deficit/hyperactivity disorder. *J Am Acad Child & Adolesc Psychiatry*. 1997;36(3):374–383. Available from: https://doi.org/10.1097/00004583-199703000-00016.
87. Postema MC, Hoogman M, Ambrosino S, et al. Analysis of structural brain asymmetries in attention-deficit/hyperactivity disorder in 39 datasets. *J Child Psychol Psychiatry*. 2021;62(10):1202–1219. Available from: https://doi.org/10.1111/jcpp.13396.
88. Schrimsher GW, Billingsley RL, Jackson EF, Moore BD. Caudate nucleus volume asymmetry predicts attention-deficit hyperactivity disorder (ADHD) symptomatology in children. *J Child Neurol*. 2002;17 (12):877–884. Available from: https://doi.org/10.1177/08830738020170122001.
89. He N, Palaniyappan L, Linli Z, Guo S. Abnormal hemispheric asymmetry of both brain function and structure in attention deficit/hyperactivity disorder: a meta-analysis of individual participant data. *Brain Imaging Behav*. 2022;16(1):54–68. Available from: https://doi.org/10.1007/s11682-021-00476-x.
90. Hale TS, Loo SK, Zaidel E, Hanada G, Macion J, Smalley SL. Rethinking a right hemisphere deficit in ADHD. *J Atten Disord*. 2009;13(1):3–17. Available from: https://doi.org/10.1177/1087054708323005.
91. van Os J, Kapur S. Schizophrenia. *Lancet*. 2009;374(9690):635–645. Available from: https://doi.org/10.1016/S0140-6736(09)60995-8.
92. Clarke B. Arthur Wigan and the duality of the mind. *Psychol Med Monogr Suppl*. 1987;11:1–52. Available from: https://doi.org/10.1017/s0264180100000102.
93. Wigan AL. A new view of insanity: the duality of the mind proved by the structure, functions, and diseases of the brain, and by the phenomena of mental derangement, and shewn to be essential to moral responsibility. with an appendix. Longman, Brown, Green, and Longmans; 1844.
94. Flor-Henry P. Psychosis and temporal lobe epilepsy. A controlled investigation. *Epilepsia*. 1969;10(3):363–395. Available from: https://doi.org/10.1111/j.1528-1157.1969.tb03853.x.
95. Dragovic M, Hammond G. Handedness in schizophrenia: a quantitative review of evidence. *Acta Psychiatrica Scandinavica*. 2005;111(6):410–419. Available from: https://doi.org/10.1111/j.1600-0447.2005.00519.x.
96. Hirnstein M, Hugdahl K. Excess of non-right-handedness in schizophrenia: meta-analysis of gender effects and potential biases in handedness assessment. *Br J Psychiatry*. 2014;205(4):260–267. Available from: https://doi.org/10.1192/bjp.bp.113.137349.

97. Sommer I, Ramsey N, Kahn RS, Aleman A, Bouma A. Handedness, language lateralisation and anatomical asymmetry in schizophrenia: meta-analysis. *Br J Psychiatry*. 2001;178:344–351. Available from: https://doi.org/10.1192/bjp.178.4.344.
98. Somers M, Sommer I, Boks MP, Kahn RS. Hand-preference and population schizotypy: a meta-analysis. *Schizophrenia Res*. 2009;108(1-3):25–32. Available from: https://doi.org/10.1016/j.schres.2008.11.010.
99. Ocklenburg S, Westerhausen R, Hirnstein M, Hugdahl K. Auditory hallucinations and reduced language lateralization in schizophrenia: a meta-analysis of dichotic listening studies. *J Int Neuropsychol Soc*. 2013;19(4):410–418. Available from: https://doi.org/10.1017/s1355617712001476.
100. Kumari V, Postma P. Nicotine use in schizophrenia: the self medication hypotheses. *Neurosci & Biobehav Rev*. 2005;29(6):1021–1034. Available from: https://doi.org/10.1016/j.neubiorev.2005.02.006.
101. Hahn C, Pogun S, Güntürkün O. Smoking modulates language lateralization in a sex-specific way. *Neuropsychologia*. 2010;48(14):3993–4002. Available from: https://doi.org/10.1016/j.neuropsychologia.2010.10.014.
102. Hahn C, Neuhaus AH, Pogun S, et al. Smoking reduces language lateralization: a dichotic listening study with control participants and schizophrenia patients. *Brain Cognition*. 2011;76(2):300–309. Available from: https://doi.org/10.1016/j.bandc.2011.03.015.
103. Angrilli A, Spironelli C, Elbert T, Crow TJ, Marano G, Stegagno L. Schizophrenia as failure of left hemispheric dominance for the phonological component of language. *PLOS ONE*. 2009;4(2):e4507. Available from: https://doi.org/10.1371/journal.pone.0004507.
104. Jalili M, Meuli R, Do KQ, Hasler M, Crow TJ, Knyazeva MG. Attenuated asymmetry of functional connectivity in schizophrenia: a high-resolution EEG study. *Psychophysiology*. 2010;47(4):706–716. Available from: https://doi.org/10.1111/j.1469-8986.2009.00971.x.
105. Artiges E, Martinot JL, Verdys M, et al. Altered hemispheric functional dominance during word generation in negative schizophrenia. *Schizophr Bull*. 2000;26(3):709–721. Available from: https://doi.org/10.1093/oxfordjournals.schbul.a033488.
106. Alary M, Delcroix N, Leroux E, et al. Functional hemispheric lateralization for language in patients with schizophrenia. *Schizophrenia Res*. 2013;149(1-3):42–47. Available from: https://doi.org/10.1016/j.schres.2013.06.003.
107. Bleich-Cohen M, Sharon H, Weizman R, Poyurovsky M, Faragian S, Hendler T. Diminished language lateralization in schizophrenia corresponds to impaired inter-hemispheric functional connectivity. *Schizophrenia Res*. 2012;134(2-3):131–136. Available from: https://doi.org/10.1016/j.schres.2011.10.011.
108. Dollfus S, Razafimandimby A, Delamillieure P, et al. Atypical hemispheric specialization for language in right-handed schizophrenia patients. *Biol Psychiatry*. 2005;57(9):1020–1028. Available from: https://doi.org/10.1016/j.biopsych.2005.01.009.
109. Oertel V, Knöchel C, Rotarska-Jagiela A, et al. Reduced laterality as a trait marker of schizophrenia—evidence from structural and functional neuroimaging. *J Neurosci*. 2010;30(6):2289–2299. Available from: https://doi.org/10.1523/JNEUROSCI.4575-09.2010.
110. Razafimandimby A, Maïza O, Hervé P-Y, et al. Stability of functional language lateralization over time in schizophrenia patients. *Schizophrenia Res*. 2007;94(1-3):197–206. Available from: https://doi.org/10.1016/j.schres.2007.04.011.
111. Weiss EM, Hofer A, Golaszewski S, et al. Brain activation patterns during a verbal fluency test-a functional MRI study in healthy volunteers and patients with schizophrenia. *Schizophrenia Res*. 2004;70(2-3):287–291. Available from: https://doi.org/10.1016/j.schres.2004.01.010.
112. Weiss EM, Hofer A, Golaszewski S, Siedentopf C, Felber S, Fleischhacker WW. Language lateralization in unmedicated patients during an acute episode of schizophrenia: a functional MRI study. *Psychiatry Res*. 2006;146(2):185–190. Available from: https://doi.org/10.1016/j.pscychresns.2005.11.003.
113. Sommer I, Ramsey NF, Kahn RS. Language lateralization in schizophrenia, an fMRI study. *Schizophrenia Res*. 2001;52(1-2):57–67. Available from: https://doi.org/10.1016/s0920-9964(00)00180-8.
114. Sommer I, Ramsey NF, Mandl R, Kahn RS. Language lateralization in female patients with schizophrenia: an fMRI study. *Schizophrenia Res*. 2003;60(2-3):183–190. Available from: https://doi.org/10.1016/s0920-9964(02)00300-6.
115. Mitchell RLC, Ross ED. fMRI evidence for the effect of verbal complexity on lateralisation of the neural response associated with decoding prosodic emotion. *Neuropsychologia*. 2008;46(12):2880–2887. Available from: https://doi.org/10.1016/j.neuropsychologia.2008.05.024.

116. Alba-Ferrara L, de Erausquin GA, Hirnstein M, Weis S, Hausmann M. Emotional prosody modulates attention in schizophrenia patients with hallucinations. *Front Hum Neurosci.* 2013;7:59. Available from: https://doi.org/10.3389/fnhum.2013.00059.
117. Mitchell RLC, Elliott R, Barry M, Cruttenden A, Woodruff PWR. Neural response to emotional prosody in schizophrenia and in bipolar affective disorder. *Br J Psychiatry.* 2004;184(3):223–230. Available from: https://doi.org/10.1192/bjp.184.3.223.
118. Mitchell RLC, Crow TJ. Right hemisphere language functions and schizophrenia: the forgotten hemisphere? *Brain.* 2005;128(Pt 5):963–978. Available from: https://doi.org/10.1093/brain/awh466.
119. Shapleske J, Rossell S, Woodruff P, David A. The planum temporale: a systematic, quantitative review of its structural, functional and clinical significance. *Brain Res Rev.* 1999;29(1):26–49. Available from: https://doi.org/10.1016/s0165-0173(98)00047-2.
120. Ribolsi M, Daskalakis ZJ, Siracusano A, Koch G. Abnormal asymmetry of brain connectivity in schizophrenia. *Front Hum Neurosci.* 2014;8:1010. Available from: https://doi.org/10.3389/fnhum.2014.01010.
121. Abdul-Rahman MF, Qiu A, Woon PS, Kuswanto C, Collinson SL, Sim K. Arcuate fasciculus abnormalities and their relationship with psychotic symptoms in schizophrenia. *PLOS ONE.* 2012;7(1):e29315. Available from: https://doi.org/10.1371/journal.pone.0029315.
122. Geoffroy PA, Houenou J, Duhamel A, et al. The Arcuate Fasciculus in auditory-verbal hallucinations: a meta-analysis of diffusion-tensor-imaging studies. *Schizophrenia Res.* 2014;159(1):234–237. Available from: https://doi.org/10.1016/j.schres.2014.07.014.
123. Kubicki M, Westin C-F, Maier SE, et al. Uncinate fasciculus findings in schizophrenia: a magnetic resonance diffusion tensor imaging study. *Am J Psychiatry.* 2002;159(5):813–820. Available from: https://doi.org/10.1176/appi.ajp.159.5.813.
124. Kunimatsu N, Aoki S, Kunimatsu A, et al. Tract-specific analysis of the superior occipitofrontal fasciculus in schizophrenia. *Psychiatry Res.* 2008;164(3):198–205. Available from: https://doi.org/10.1016/j.pscychresns.2008.03.004.
125. Hulshoff Pol HE, Schnack HG, Mandl RCW, et al. Focal white matter density changes in schizophrenia: reduced inter-hemispheric connectivity. *NeuroImage.* 2004;21(1):27–35. Available from: https://doi.org/10.1016/j.neuroimage.2003.09.026.
126. Schijven D, Postema MC, Fukunaga M, et al. Large-scale analysis of structural brain asymmetries in schizophrenia via the ENIGMA consortium. *PNAS.* 2023;120(14). Available from: https://doi.org/10.1073/pnas.2213880120. e2213880120.
127. Gutman BA, van Erp TGM, Alpert K, et al. A meta-analysis of deep brain structural shape and asymmetry abnormalities in 2,833 individuals with schizophrenia compared with 3,929 healthy volunteers via the ENIGMA Consortium. *Hum Brain Mapp.* 2022;43(1):352–372. Available from: https://doi.org/10.1002/hbm.25625.
128. Gruzelier JH. Syndromes of schizophrenia and schizotypy, hemispheric imbalance and sex differences: implications for developmental psychopathology. *Int J Psychophysiol.* 1994;18(3):167–178. Available from: https://doi.org/10.1016/0167-8760(94)90002-7.
129. Gruzelier JH, Wilson L, Richardson A. Cognitive asymmetry patterns in schizophrenia: retest reliability and modification with recovery. *Int J Psychophysiol.* 1999;34(3):323–331. Available from: https://doi.org/10.1016/s0167-8760(99)00089-6.
130. Crow TJ. The 'big bang' theory of the origin of psychosis and the faculty of language. *Schizophrenia Res.* 2008;102(1-3):31–52. Available from: https://doi.org/10.1016/j.schres.2008.03.010.
131. Ocklenburg S, Ströckens F, Güntürkün O. Lateralisation of conspecific vocalisation in non-human vertebrates. *Laterality.* 2013;18(1):1–31. Available from: https://doi.org/10.1080/1357650X.2011.626561.
132. Ströckens F, Güntürkün O, Ocklenburg S. Limb preferences in non-human vertebrates. *Laterality.* 2013;18(5):536–575. Available from: https://doi.org/10.1080/1357650X.2012.723008.
133. Schizophrenia Working Group of the Psychiatric Genomics Consortium. Biological insights from 108 schizophrenia-associated genetic loci. *Nature.* 2014;511(7510):421–427. Available from: https://doi.org/10.1038/nature13595.
134. Ocklenburg S, Güntürkün O, Hugdahl K, Hirnstein M. Laterality and mental disorders in the postgenomic age–A closer look at schizophrenia and language lateralization. *Neurosci Biobehav Rev.* 2015;59:100–110. Available from: https://doi.org/10.1016/j.neubiorev.2015.08.019.
135. Picardi A, Viroli C, Tarsitani L, et al. Heterogeneity and symptom structure of schizophrenia. *Psychiatry Res.* 2012;198(3):386–394. Available from: https://doi.org/10.1016/j.psychres.2011.12.051.

136. Arnedo J, Svrakic DM, Del Val C, et al. Uncovering the hidden risk architecture of the schizophrenias: confirmation in three independent genome-wide association studies. *Am J Psychiatry*. 2015;172(2):139–153. Available from: https://doi.org/10.1176/appi.ajp.2014.14040435.
137. Ocklenburg S, Beste C, Güntürkün O. Handedness: a neurogenetic shift of perspective. *Neurosci Biobehav Rev*. 2013;37(10 Pt 2):2788–2793. Available from: https://doi.org/10.1016/j.neubiorev.2013.09.014.
138. Løberg E-M, Jørgensen HA, Hugdahl K. Functional brain asymmetry and attentional modulation in young and stabilised schizophrenic patients: a dichotic listening study. *Psychiatry Res*. 2002;109(3):281–287. Available from: https://doi.org/10.1016/s0165-1781(02)00019-7.
139. Denny K. Handedness and depression: evidence from a large population survey. *Laterality*. 2009;14 (3):246–255. Available from: https://doi.org/10.1080/13576500802362869.
140. Elias LJ, Saucier DM, Guylee MJ. Handedness and Depression in University Students: A sex by handedness interaction. *Brain Cognition*. 2001;46(1-2):125–129. Available from: https://doi.org/10.1016/s0278-2626(01)80048-8.
141. Biederman J, Lapey KA, Milberger S, Faraone SV, Reed ED, Seidman LJ. Motor preference, major depression and psychosocial dysfunction among children with attention deficit hyperactivity disorder. *J Psychiatr Res*. 1994;28(2):171–184. Available from: https://doi.org/10.1016/0022-3956(94)90028-0.
142. Moscovitch M, Strauss E, Olds J. Handedness and dichotic listening performance in patients with unipolar endogenous depression who received ECT. *Am J Psychiatry*. 1981;138(7):988–990. Available from: https://doi.org/10.1176/ajp.138.7.988.
143. Packheiser J, Schmitz J, Stein CC, et al. Handedness and depression: A meta-analysis across 87 studies. *J Affect Disord*. 2021;294:200–209. Available from: https://doi.org/10.1016/j.jad.2021.07.052.
144. Kovel CGF, de, Aftanas L, Aleman A, et al. No alterations of brain structural asymmetry in major depressive disorder: an ENIGMA Consortium Analysis. *Am J Psychiatry*. 2019;176(12):1039–1049. Available from: https://doi.org/10.1176/appi.ajp.2019.18101144.
145. Debener S, Beauducel A, Nessler D, Brocke B, Heilemann H, Kayser J. Is resting anterior EEG alpha asymmetry a trait marker for depression? Findings for healthy adults and clinically depressed patients. *NPS*. 2000;41(1):31–37. Available from: https://doi.org/10.1159/000026630.
146. Henriques JB, Davidson RJ. Left frontal hypoactivation in depression. *J Abnorm Psychol*. 1991;100(4):535–545. Available from: https://doi.org/10.1037/0021-843x.100.4.535.
147. Kano K, Nakamura M, Matsuoka T, Iida H, Nakajima T. The topographical features of EEGs in patients with affective disorders. *Electroencephalography Clin Neurophysiol*. 1992;83(2):124–129. Available from: https://doi.org/10.1016/0013-4694(92)90025-d.
148. Roemer RA, Shagass C, Dubin W, Jaffe R, Siegal L. Quantitative EEG in elderly depressives. *Brain Topogr*. 1992;4(4):285–290. Available from: https://doi.org/10.1007/BF01135566.
149. Saletu B, Brandstätter N, Metka M, et al. Hormonal, syndromal and EEG mapping studies in menopausal syndrome patients with and without depression as compared with controls. *Maturitas*. 1996;23(1):91–105. Available from: https://doi.org/10.1016/0378-5122(95)00946-9.
150. Horato N, Quagliato LA, Nardi AE. The relationship between emotional regulation and hemispheric lateralization in depression: a systematic review and a meta-analysis. *Transl Psychiatry*. 2022;12(1):162. Available from: https://doi.org/10.1038/s41398-022-01927-9.
151. Van der Vinne N, Vollebregt MA, Van Putten MJAM, Arns M. Frontal alpha asymmetry as a diagnostic marker in depression: fact or fiction? A meta-analysis. *Neuroimage Clin*. 2017;16:79–87. Available from: https://doi.org/10.1016/j.nicl.2017.07.006.
152. Trinkl M, Greimel E, Bartling J, Grünewald B, Schulte-Körne G, Grossheinrich N. Right-lateralization of N2-amplitudes in depressive adolescents: an emotional go/no-go study. *J Child Psychol Psychiatry*. 2015;56 (1):76–86. Available from: https://doi.org/10.1111/jcpp.12282.
153. Herrington JD, Heller W, Mohanty A, et al. Localization of asymmetric brain function in emotion and depression. *Psychophysiology*. 2010;47(3):442–454. Available from: https://doi.org/10.1111/j.1469-8986.2009.00958.x.
154. Hecht D. Depression and the hyperactive right-hemisphere. *Neurosci Res*. 2010;68(2):77–87. Available from: https://doi.org/10.1016/j.neures.2010.06.013.
155. Silberman EK, Weingartner H. Hemispheric lateralization of functions related to emotion. *Brain Cognition*. 1986;5(3):322–353. Available from: https://doi.org/10.1016/0278-2626(86)90035-7.

156. Jesulola E, Sharpley CF, Bitsika V, Agnew LL, Wilson P. Frontal alpha asymmetry as a pathway to behavioural withdrawal in depression: research findings and issues. *Behav Brain Res*. 2015;292:56–67. Available from: https://doi.org/10.1016/j.bbr.2015.05.058.
157. McNamara P, Blum D, O'Quin K, Schachter S. Markers of cerebral lateralization and alcoholism. *Percept Mot Skills*. 1994;79(3 Pt 2):1435–1440. Available from: https://doi.org/10.2466/pms.1994.79.3f.1435.
158. Sperling W, Frank H, Martus P, et al. The concept of abnormal hemispheric organization in addiction research. *Alcohol Alcohol*. 2000;35(4):394–399. Available from: https://doi.org/10.1093/alcalc/35.4.394.
159. Mandal MK, Bhushan B, Kumar A, Gupta P. Side-bias in alcohol and heroin addicts. *Alcohol Alcohol*. 2000;35 (4):381–383. Available from: https://doi.org/10.1093/alcalc/35.4.381.
160. Denny K. Handedness and drinking behaviour. *Br J Health Psychol*. 2011;16(Pt 2):386–395. Available from: https://doi.org/10.1348/135910710X515705.
161. Preti A, Usai I, Pintus E, Sardu C, Petretto DR, Masala C. Left-handedness is statistically linked to lifetime experimentation with illicit drugs. *Laterality*. 2012;17(3):318–339. Available from: https://doi.org/10.1080/1357650X.2011.575786.
162. Bouna-Pyrrou P, Mühle C, Kornhuber J, Lenz B. Internet gaming disorder, social network disorder and laterality: handedness relates to pathological use of social networks. *J Neural Transm*. 2015;122(8):1187–1196. Available from: https://doi.org/10.1007/s00702-014-1361-5.
163. Drake AI, Hannay HJ, Gam J. Effects of chronic alcoholism on hemispheric functioning: an examination of gender differences for cognitive and dichotic listening tasks. *J Clin Exp Neuropsychology*. 1990;12(5):781–797. Available from: https://doi.org/10.1080/01688639008401019.
164. Yoon HW, Chung J-Y, Oh J-H, et al. Differential activation of face memory encoding tasks in alcohol-dependent patients compared to healthy subjects: an fMRI study. *Neurosci Lett*. 2009;450(3):311–316. Available from: https://doi.org/10.1016/j.neulet.2008.12.011.
165. Fabian MS, Parsons OA, Sheldon MD. Effects of gender and alcoholism on verbal and visual-spatial learning. *J Nerv Ment Dis*. 1984;172(1):16–20. Available from: https://doi.org/10.1097/00005053-198401000-00005.
166. Jung YC, Jang D-P, Namkoong K, et al. Shape deformation of the insula in alcoholics: reduction of left-right asymmetry. *NeuroReport*. 2007;18(17):1787–1791. Available from: https://doi.org/10.1097/WNR.0b013e3282f193b4.
167. Medina KL, Schweinsburg AD, Cohen-Zion M, Nagel BJ, Tapert SF. Effects of alcohol and combined marijuana and alcohol use during adolescence on hippocampal volume and asymmetry. *Neurotoxicology Teratology*. 2007;29(1):141–152. Available from: https://doi.org/10.1016/j.ntt.2006.10.010.
168. Cao Z, Ottino-Gonzalez J, Cupertino RB, et al. Mapping cortical and subcortical asymmetries in substance dependence: findings from the ENIGMA Addiction Working Group. *Addict Biol*. 2021;26(5):e13010. Available from: https://doi.org/10.1111/adb.13010.
169. Yan H, Shlobin NA, Jung Y, et al. Nucleus accumbens: a systematic review of neural circuitry and clinical studies in healthy and pathological states. *J Neurosurg*. 2023;138(2):337–346. Available from: https://doi.org/10.3171/2022.5.JNS212548.
170. Cronin-Golomb A. Parkinson's disease as a disconnection syndrome. *Neuropsychol Rev*. 2010;20(2):191–208. Available from: https://doi.org/10.1007/s11065-010-9128-8.
171. Kempster PA, Gibb WR, Stern GM, Lees AJ. Asymmetry of substantia nigra neuronal loss in Parkinson's disease and its relevance to the mechanism of levodopa related motor fluctuations. *J Neurol Neurosurg Psychiatry*. 1989;52(1):72–76. Available from: https://doi.org/10.1136/jnnp.52.1.72.
172. Riederer P, Sian-Hülsmann J. The significance of neuronal lateralisation in Parkinson's disease. *J Neural Transm*. 2012;119(8):953–962. Available from: https://doi.org/10.1007/s00702-012-0775-1.
173. Verreyt N, Nys GMS, Santens P, Vingerhoets G. Cognitive differences between patients with left-sided and right-sided Parkinson's disease. A review. *Neuropsychol Rev*. 2011;21(4):405–424. Available from: https://doi.org/10.1007/s11065-011-9182-x.
174. van der Hoorn A, Burger H, Leenders KL, de Jong BM. Handedness correlates with the dominant Parkinson side: a systematic review and meta-analysis. *Mov Disord*. 2012;27(2):206–210. Available from: https://doi.org/10.1002/mds.24007.
175. Nusslock R, Walden K, Harmon-Jones E. Asymmetrical frontal cortical activity associated with differential risk for mood and anxiety disorder symptoms: An RDoC perspective. *Int J Psychophysiol*. 2015;98(2 Pt 2):249–261. Available from: https://doi.org/10.1016/j.ijpsycho.2015.06.004.

176. Insel T, Cuthbert B, Garvey M, et al. Research domain criteria (RDoC): toward a new classification framework for research on mental disorders. *Am J Psychiatry*. 2010;167(7):748–751. Available from: https://doi.org/10.1176/appi.ajp.2010.09091379.
177. Huang K, Kang Y, Wu Z, Wang Y, Cai S, Huang L. Asymmetrical alterations of grey matter among psychiatric disorders: a systematic analysis by voxel-based activation likelihood estimation. *Prog Neuropsychopharmacol Biol Psychiatry*. 2021;110:110322. Available from: https://doi.org/10.1016/j.pnpbp.2021.110322.
178. Cuellar-Partida G, Tung JY, Eriksson N, et al. Genome-wide association study identifies 48 common genetic variants associated with handedness. *Nat Hum Behav*. 2021;5(1):59–70. Available from: https://doi.org/10.1038/s41562-020-00956-y.
179. Ocklenburg S, Korte SM, Peterburs J, Wolf OT, Güntürkün O. Stress and laterality - The comparative perspective. *Physiol & Behav*. 2016;164(Pt A):321–329. Available from: https://doi.org/10.1016/j.physbeh.2016.06.020.
180. Borawski J, Papadatou-Pastou M, Packheiser J, Ocklenburg S. Handedness in post-traumatic stress disorder: a meta-analysis. *Neurosci Biobehav Rev*. 2023;145:105009. Available from: https://doi.org/10.1016/j.neubiorev.2022.105009.
181. Mundorf A, Matsui H, Ocklenburg S, Freund N. Asymmetry of turning behavior in rats is modulated by early life stress. *Behav Brain Res*. 2020;393:112807. Available from: https://doi.org/10.1016/j.bbr.2020.112807.
182. Kovel CGF, de, Carrión-Castillo A, Francks C. A large-scale population study of early life factors influencing left-handedness. *Sci Rep*. 2019;9(1):584. Available from: https://doi.org/10.1038/s41598-018-37423-8.
183. Berretz G, Packheiser J, Höffken O, Wolf OT, Ocklenburg S. Dichotic listening performance and interhemispheric integration after administration of hydrocortisone. *Sci Rep*. 2021;11(1):21581. Available from: https://doi.org/10.1038/s41598-021-00896-1.
184. Berretz G, Packheiser J, Wolf OT, Ocklenburg S. Acute stress increases left hemispheric activity measured via changes in frontal alpha asymmetries. *iScience*. 2022;25(2):103841. Available from: https://doi.org/10.1016/j.isci.2022.103841.
185. Berretz G, Packheiser J, Wolf OT, Ocklenburg S. Improved interhemispheric connectivity after stress during lexical decision making. *Behav Brain Res*. 2022;418:113648. Available from: https://doi.org/10.1016/j.bbr.2021.113648.
186. Mundorf A, Ocklenburg S. Hemispheric asymmetries in mental disorders: evidence from rodent studies. *J Neural Transm*. 2023;. Available from: https://doi.org/10.1007/s00702-023-02610-z.

Index

Note: Page numbers followed by "*f*," "*t*," and "*b*" refer to figures, tables, and boxes, respectively.

B

I

M

O

P

T

Y

Z

9780323997379